THE
AMERICAN
PRESIDENCY

★ ★ ★

The
AMERICAN
PRESIDENCY

★ ★ ★ ★ ★ ★ ★ ★ ★ ★ ★ ★ ★

Edited by

Alan Brinkley and Davis Dyer

Houghton Mifflin Company

BOSTON • NEW YORK

2004

For information about permission to reproduce selections from
this book, write to Permissions, Houghton Mifflin Company,
215 Park Avenue South, New York, New York 10003.
Visit our Web site: www.houghtonmifflinbooks.com.

Library of Congress Cataloging-in-Publication Data

The American presidency / edited by
Alan Brinkley and Davis Dyer.
p.   cm.
Includes bibliographical references and index.
ISBN 0-618-38273-9
1. Presidents — United States — History. 2. United States —
Politics and government. 3. Political leadership — United
States — History. 4. Executive power — United States — His-
tory. 5. Presidents — United States — Biography. I. Brinkley,
Alan. II. Dyer, Davis.
E176.1.A653 2004
973'.09'9 — dc22   2003062513

PRINTED IN THE UNITED STATES OF AMERICA

Book design by Robert Overholtzer

MP 10 9 8 7 6 5 4 3 2 1

Contents

Contributors

Joyce Appleby University of California at Los Angeles
Jean H. Baker Goucher College
Paula Baker University of Pittsburgh
Alan Brinkley Columbia University
Joel Brinkley *New York Times*
Catherine Clinton Baruch College
Robert Dallek Boston University
Matthew Dickinson Middlebury College
Davis Dyer The Winthrop Group
David Farber University of New Mexico
Paula S. Fass University of California at Berkeley
Eric Foner Columbia University
Ronald P. Formisano University of Kentucky
William E. Gienapp Harvard University
Steven M. Gillon University of Oklahoma
Lewis L. Gould University of Texas at Austin
James A. Henretta, University of Maryland
Michael F. Holt University of Virginia
Richard R. John University of Illinois at Chicago
Michael Kazin Georgetown University
Thomas J. Knock Southern Methodist University
Marc W. Kruman Wayne State University
Walter LaFeber Cornell University
Mark H. Leff University of Illinois at Urbana-Champaign
Drew R. McCoy Clark University
William McFeely
Michael McGerr Indiana University at Bloomington
Yanek Mieczkowski Dowling College
Roger Morris
Karen Orren University of California at Los Angeles

David M. Oshinsky University of Texas at Austin
Michael S. Sherry Northwestern University
Joel H. Silbey Cornell University
Herbert Sloan Barnard College
David L. Stebenne Ohio State University
Michael B. Stoff University of Texas at Austin
Alan Taylor University of California at Davis
Gil Troy McGill University
Michael Vorenberg Brown University
Harry L. Watson University of North Carolina at Chapel Hill

Introduction

Article II of the United States Constitution provides a spare, even skeletal description of the role of the president of the United States. The president, it says, will be vested with "executive power," will be commander in chief of the nation's military forces, and will have the power to make treaties and appoint judges and executive officers with the advice and consent of the Senate. "He shall from time to time give to the Congress information on the State of the Union" and recommend measures for the legislature's consideration. The president will receive ambassadors and will "take care that the Laws be faithfully executed." Otherwise, the Framers had almost nothing to say about what the president would do or what kind of person the president would be.

Through most of American history, however, the presidency has been much more than a simple instrument of executive power. Presidents, far from merely executing laws conceived and passed by others, have been the source of some of the most important shifts in the nation's public policy and political ideology. They have played not only political, but social and cultural roles in American life. They have experienced tremendous variations in their power and prestige. The presidency has hidden its occupants behind a vast screen of delegated powers and deliberate image-making. And the office has been critically shaped not just by individuals but by powerful social, economic, and cultural forces over which leaders have little or no control. Characterizing the American presidency — the task that this book has set for itself — is, as a result, very challenging.

We start by distinguishing the presidency from the presidents, the office from those who held it. This book is not, then, a collection of presidential biographies, although it provides much biographical information about each of the forty-two men who have served as president. Rather, its focus is how these individuals have perceived and used the office, and how the office has changed as a result.

Since George Washington's Inauguration in 1789, there have been periods of greater and lesser change, of turbulence and calm, of advance and retreat in the American presidency. Across these many years, however, four broad themes stand out: the symbolic importance of the presidency, which transcends its formal constitutional powers; the wide swings in its fortunes; the influence wielded not only by the president but also by his advisers; and the role of contingency and context in shaping the office and particular presidencies.

Among the salient characteristics of the American presidency is that it has usually played a role in American life that extends well beyond the formal responsibilities of the office. Almost all presidents—whatever they have or have not achieved — have occupied positions of enormous symbolic and cultural importance in American life. They have become the secular icons of the republic — emblems of nationhood and embodiments of the values that Americans have claimed to cherish.

Exaggerated images of the virtues (and occasionally the sins) of American presidents have helped shape the nation's picture of itself. Stories of presidential childhoods and youths have become staples of popular culture and instructional literature. Parson Weems's early-nineteenth-century life of Washington, with its invented stories of chopping down a cherry tree and throwing a silver dollar across the river, contributed greatly to the early self-image of the American nation. The popular Boys' Life of Theodore Roosevelt influenced generations of young Americans and helped form twentieth-century images of the presidency and of the nature of leadership.

Just as Americans have often exaggerated the virtues of their presidents, so they have often exaggerated their flaws. Charges of presidential misconduct and moral turpitude have repeatedly mesmerized the nation for two centuries. The scandals that plagued Ulysses S. Grant, Warren G. Harding, Richard M. Nixon, and Bill Clinton have molded both scholarly and popular views of those presidencies. But these most famously bedeviled presidencies are hardly alone. Thomas Jefferson, one of the most revered of all Americans, was savagely attacked in his lifetime as a revolutionary, a tyrant, and a miscegenist. John Adams and John Quincy Adams, pillars of personal rectitude, were harried throughout their presidencies by accusations of corruption, fraud, and abuses of power. Rutherford B. Hayes, a paragon of propriety (and sobriety), was known during his unhappy administration as "His Fraudulency" for having allegedly stolen the 1876 election from Samuel Tilden. Harry Truman, a folk hero today for what Americans like to remember as his plainspeaking honesty, was buffeted for years by charges of "cronyism" and "corruption"—for creating what Richard Nixon and many others in 1952 liked to call the "mess in Washington."

Almost everything a president does, in the end, seems to much of the nation to be larger than life, even that least dignified of political activities: running for office. During the first century of the American republic, most Americans considered the presidency so august a position that candidates for the office were expected not only to refrain from campaigning, but to display no desire at all for the office. In reality, of course, most of those seeking the presidency did a great deal to advance their own candidacies. In public, however, they accepted their nominations and, if successful, their elections as if they were gifts from the people. In the twentieth century, campaigning for president became almost a full-time job, both before and after election, and no one could hope to be elected in our time by pretending to have no interest in the White House. But the perpetual campaigning has given the presidency a different kind of symbolic importance; for presidents, and presidential candidates, are now ubiquitous figures in our media culture, their presumed personalities and their projected, carefully crafted images a focus of almost obsessive attention and fascination.

The gap between the image and reality of the president and his office has been enormous at some moments, relatively narrow at others, but always there. The reality of George Washington's life was for many years almost completely replaced by the hagiographic myths created by Weems. Abraham Lincoln, widely and justly regarded as America's greatest president, became soon after his death a kind of national saint, his actual character as an intensely and brilliantly political man obscured behind generations of paeans to his humility and strength—by his law partner William Herndon, by his White House aides John G. Nicolay and John Hay (whose account occupied ten volumes), by the poet Carl Sandburg (who limited his to three). Franklin Roosevelt, a wily figure whose evasiveness and inconsistency infuriated even his closest allies, became and for generations remained "Our Friend," the heroic battler against depression and tyranny. But even less exalted figures have inspired their myths—the aristocratic William Henry Harrison portrayed as the simple product of a log cabin who liked hard cider from a jug; the gruff, stubborn, and ultimately rather ineffectual Ulysses S. Grant considered, for a time, a great and noble conciliator; the cool, detached, intensely pragmatic John Kennedy, who became a symbol of passionate idealism and commitment.

*   *   *

Another distinctive characteristic of the American presidency is the tremendous variation in the fortunes of the office. At times it has been a position of great power and enormous prestige — an almost majestic office whose occupant dominated and helped shape the public life of the age. At other times it has been weak and relatively ineffectual. The number of issues with which presidents must deal has, of course, continuously expanded as the nation has grown larger, wealthier, more powerful, and more interconnected with other nations and international institutions. But the ability of presidents to deal effectively with those issues has ebbed and flowed. In the early years of the republic, presidents had great influence over the behavior of the relatively small federal government and were usually able to win support for their goals from Congress. The three presidents who served during the turbulent 1850s found themselves — because of their own limitations and the character of their time — almost powerless in the face of an increasingly assertive and fractious Congress. Lincoln seized wartime authority that gave him unprecedented, many believed almost dictatorial, power. Only a few years later, Andrew Johnson and Ulysses Grant found themselves almost entirely subordinated to the will of Congress — a situation that continued through most of the rest of the nineteenth century.

The twentieth century witnessed a dramatic expansion of the presidency's importance and power, and also a significant growth in the constraints facing individual presidents. In their relations with Congress, twentieth-century presidents — like their nineteenth-century counterparts — included men who have been both commanding and disastrously weak. Social crises strengthened some presidents and weakened or destroyed others. War and cold war exalted some presidents and brought others to ruin. Some modern presidents — Woodrow Wilson, Lyndon Johnson, Richard Nixon — experienced moments of towering achievement, only to suffer painful defeats and humiliations soon after. Others, most notably Harry Truman, rose from what seemed the ruins of their presidencies to political success and historical regard. A few, like Franklin Roosevelt and John F. Kennedy, enjoyed cordial relations with the growing press corps in Washington, but most of their fellow presidents found those relations contentious and difficult to manage.

The president, the political scientist Clinton Rossiter wrote exuberantly in 1956, is "a kind of magnificent lion, who can roam freely and do great deeds so long as he does not try to break loose from his broad reservation. . . . There is virtually no limit to what the president can do if he does it for democratic ends and through democratic means." Rossiter's view fit comfortably into the exalted image of the presidency at the height of the cold war. But a generation later, in 1980, Godfrey Hodgson, an astute British observer of American life, offered a starkly different evaluation in a book whose subtitle, *The False Promise of the American Presidency,* summarized the prevailing assumptions of his time. "Never has one office had so much power as the president of the United States possesses," he wrote. "Never has so powerful a leader been so impotent to do what he wants to do, what he is pledged to do, what he is expected to do, and what he knows he must do." And yet, a year after Hodgson wrote, Ronald Reagan entered the White House, reasserted the centrality and authority of the office, and changed the image of the presidency once again.

It is not just the gap between image and reality that makes American presidents elusive and intriguing figures. It is also the problem that both contemporaries and historians experience in trying to separate the president as a person from the things done in his name. Even in the nineteenth century, when presidents worked with tiny staffs, wrote their own speeches, and lived relatively openly, the president was never wholly master of his own fate. His cabinet, his party, the party's newspapers, and many others acted at times under the authority of the presidency and created a haze over the president's own intentions and motives. Historians have often struggled to separate George Washington's own actions, desires, and achievements from those of his powerful ally Alexander Hamilton. Andrew Jackson's towering image, during his lifetime and since, is the product not only of his own imposing personality but of the actions of such powerful associates as John Eaton, Roger Taney, Amos Kendall, and Martin Van Buren.

In the twentieth century, the president came to be served by a vast and sprawling staff, presided over an even vaster and more sprawling bureaucracy, and became subject to increasingly sophisticated methods for shaping his image. And the biographers of modern presidents often find it difficult to chronicle a subject's life in the White House because the man himself seems often to disappear into the sheer enormity of his office. Changes in the once simple process of housing a president's papers is one indication of how the office has grown. The papers of every president until Herbert Hoover are housed in relatively confined spaces in the Library of Congress. The papers of every president beginning with Hoover are housed in large presidential libraries, which are not just testimonies to the eagerness of presidents to memorialize themselves, but also to the enormity of the records of modern administrations. The paper trail of modern presidents is not a result of presidents themselves writing more; in fact, most of them probably write less than their nine-teenth-century predecessors did. It is a result of the steady increases in presidential staffs and the dispersion of presidential power into many corners of government.

How can we separate those parts of the New Deal or wartime leadership for which Franklin Roosevelt is himself responsible from those that are the work of the large assemblage of talented, energetic, and ambitious men and women who acted in his name? Did John Foster Dulles shape the foreign policy of the Eisenhower years, as many people believed in the 1950s, or was Eisenhower himself at the center of the process, as some historians have argued since? Was Ronald Reagan the bold and decisive leader his admirers describe, or the passive, uninformed, detached president that his critics — and some of his closest associates — have portrayed? There are no simple answers to any of these questions because the modern presidency is the creation of a single man and of many people whose actions the president may never see.

Assessing the presidency is, finally, complicated by the difficulty of evaluating the importance of political leadership in relation to other forces in shaping historical events. This is a long-standing subject of debate among historians and other scholars, but it is also a concrete historical problem. Most historians would agree that events are seldom inevitable, that the specific actions of individuals can affect, sometimes even fundamentally shape, the course of history. But most historians today would also agree that the decisions of individual leaders are not the only, and often far from the most important, factors in explaining the past. Sometimes tiny contingencies exert enormous effects, as did the butterfly ballots and hanging chads in the 2000 election. But time and again, large social, economic, cultural, and demographic changes limit the options and overwhelm the assumptions of leaders. American presidents are not only figures of power, therefore, but also products — sometimes as beneficiaries, sometimes as victims — of the character of their times, as they themselves have often noted. "If during the lifetime of a generation," Theodore Roosevelt once observed ruefully, "no crisis occurs sufficient to call out in marked manner the energies of the strongest leader, then of course the world does not and cannot know of the existence of such a leader; and in consequence there are long periods in the history of every nation during which no man appears who leaves an indelible mark in history. . . . If there is not the war, you don't get the great general; if there is not the great occasion, you don't get the great statesman; if Lincoln had lived in times of peace, no one would know his name now."

Might we remember the talented and intelligent James Garfield as among our greatest presidents if he, rather than Lincoln, had presided over the Civil War? Would Franklin Roosevelt be the enormous historical figure he has become if he had not had a depression and a war to fight? Would Richard Nixon have fallen so ignominiously if he had not been president during a time of great turbulence and social division, a time he described in his memoirs, self-servingly and extravagantly but not entirely falsely, as a "season of mindless terror," an "epidemic of unprecedented domestic terrorism" driven by "highly organized and highly skilled revolutionaries dedicated to the violent destruction of our democratic system"? Would Ronald Reagan have enjoyed such extraordinary political success if he had not entered office on the heels of an era of intense economic anxiety and international humiliation? Would George W. Bush have become a towering international leader, at least for a while, without the events of September 11, 2001? Understanding the presidency requires, of course, taking seriously the role that individuals play in history, and there are many occasions in which one could imagine a very different history if a different leader had been in place. At the same time, however, the fates of the various presidencies are inexplicable without attention to the character of their times.

The essays collected in this book, the work of a distinguished group of scholars, present concise and thorough accounts of the important events of each of the forty-three presidencies. But they also attempt to do more than that. They are deliberately and frankly interpretive, offering assessments of individual men and of both great and small events. They are also, in varying degrees, contextual, situating presidents in their time and suggesting how the character of American society and culture shaped the character of presidential performance. Each of the scholars contributing to this volume has had to contend with a large, at times overwhelmingly vast, literature on almost every American president. But we believe that these essays have something fresh and new to say about all of them and about that most awesome and frustrating of offices: the presidency of the United States.

THE EDITORS

THE
AMERICAN
PRESIDENCY

★　★　★

George Washington

1789–1797

b. February 22, 1732
d. December 14, 1799

GEORGE WASHINGTON, the first president of the United States, spent two-thirds of his life as a British subject. Forty-four by the time the colonies declared their independence, he was fifty-five when the Constitutional Convention created the presidency, an office designed by men who intended that he would be the first to fill it. No other president so embodies the epochal events out of which the United States arose; none played a more important part in them. In many ways, Washington would invent the presidency, just as he helped to create the United States.

Washington was a Virginian long before he became an American, and the marks of a Virginia upbringing remained with him throughout his life. He was born on February 22, 1732, at a plantation on the banks of the Potomac in Westmoreland County, the first child of Augustine Washington (1694–1743) and his second wife, Mary Ball Washington (circa 1708–1789). By Augustine's day, the Washingtons were solid members of the Virginia gentry, the interlocking group of families that had emerged from the turbulent conditions of seventeenth-century Virginia as a coherent ruling group, dominating politics and owning the lion's share of resources, whether in land or in slaves and indentured servants.

Despite the stories famously recorded by Washington's early biographer Parson Weems (the demise of the cherry tree, the silver dollar hurled across the Rappahannock), little is known about Washington's childhood. What must have been important was his position within the family: he was a younger son, and in the Virginia of that day younger sons lacked the life chances enjoyed by firstborn males. Washington's limited prospects were confirmed by his father's will. Augustine Washington died in 1743, when Washington was eleven, leaving the largest share of his estate (*in toto* it comprised about ten thousand acres of land and fifty slaves) to Lawrence, the eldest son. Washington himself received a 260-acre farm, other tracts of land, and ten slaves. What mattered now, in shaping Washington's future, was a combination of his own intense ambition and his ability to attract patrons willing to instruct him in the ways of the world and ease his hoped-for ascent.

Washington was fortunate in attracting the support of his older half brother Lawrence, the first of the essential sponsors who furthered his career. Lawrence took his fatherless half brother into his family at Mount Vernon. There Washington completed what little formal education he received. It was practical rather than classical, and it stressed mathematics, surveying, and accounting — in short, things useful to someone expected to make his own way in the world of mid-eighteenth-century Virginia.

The adolescent Washington struggled to acquire the accomplishments that would make up for his lack of fortune. The story of Washington's continuous self-fashioning is one of the great tales of early American history, best illustrated, perhaps, by his 1747 manuscript, which he called "Rules of Civility & Decent Behaviour In Company and Conversation."

The crucial part of his early training came when he began to learn surveying. Virginia in the middle of the eighteenth century was still largely wild, and surveying was a significant calling. It gave those who went out into the woods firsthand acquaintance with the quality of the land and a considerable advantage when it came to locating new settlements. Over the course of his life, these were skills that would serve Washington well. In 1748, at sixteen, he began working as a surveyor. Using the insider's information he had been accumulating, Washington made his first purchase of land in 1750, of 1,459 acres in the Shenandoah Valley; in the years to come, he would buy many more, and at his death he owned some 56,000 acres.

Washington's career was interrupted when his half brother developed a serious lung ailment. In the fall of 1751, Washington went with Lawrence to Barbados in search of a cure. While there, Washington caught smallpox, acquiring immunity to a disease often fatal in the eighteenth century. Lawrence's health did not improve, however, and in 1752 he died, leaving most of his property to his wife and daughter, who died in 1761 and 1754, respectively. His will provided that Washington would inherit Mount Vernon if he survived the other heirs.

By itself, surveying could not guarantee the rise in status Washington desired. Politics might, but here he was at a disadvantage. Unlike many in the House of Burgesses, he was not a lawyer. Nor did he yet have the substantial landholdings that would legitimate a bid for office. But another route was opening up as Washington came of age: the military. Virginia's territorial goals in the Ohio Valley were threatened in the 1740s and 1750s by the French and their Indian allies, and Washington was to be involved in the French and Indian War (1754 to 1763) from the outset.

Washington began his military career in the fall of 1753, traveling to the Forks of the Ohio (present-day Pittsburgh) and then to Fort Le Boeuf (near Lake Erie) to assess the situation and deliver Virginia's demand that the French abandon their positions in the Ohio country. The journey brought Washington a measure of fame when his account of the mission, published in Williamsburg in 1754 as *The Journal of Major George Washington,* was reprinted in Massachusetts and Maryland and then as a pamphlet in London.

Hostilities broke out in earnest in the spring of 1754. The Virginia militia was the front line of imperial defense, and Washington seized his chance. Having received a commission as a lieutenant colonel, he went off at twenty-two to defend Virginia's claims to the Ohio region. Defeating a small enemy force in western Pennsylvania on May 28, he built Fort Necessity but was compelled by the French and Indians to surrender it on July 3. Worse, he had to sign a capitulation (written in French) acknowledging that the English had trespassed on French territory and that he had "assassinated" the French commander in the encounter on May 28. Washington knew no French and denied having made the admissions, later claiming that he had been misled by his translator.

Again Washington's activities attracted notice beyond Virginia. His letter describing the May 28 affray appeared in London, and Washington enjoyed a brief transatlantic celebrity. "I heard the Bulletts whistle and believe me there was something charming in the sound," he had written to his brother John Augustine, words that supposedly led the king, George II, to remark that "he would not say so, if he had been used to hear many." Washington's first command had ended in failure, but his military career went on.

He received his commission as a colonel of the Virginia forces on August 14, 1755, and for the next three and a half years he struggled to defend the frontier with an inadequate and undependable militia and without the supplies his troops needed. Endlessly frustrating, the experiences of these years broadened Washington's horizons. Conferences with the British commanders in North America took him to Boston in 1756 and to Philadelphia in 1757. In 1758 he led the Virginia contingent in the successful campaign that finally took Fort Duquesne from the French. Only twenty-six in 1758, Washington had already become a person of note in the colonial world.

Nonetheless, once it was clear that his goal of a commission in the British army was unattainable, Washington turned his attention to his long-term aim of acquiring a substantial position in the ranks of the gentry. In 1758, the voters of Frederick County elected him to the legislature. He held the seat until 1765, then won election in his home county of Fairfax and served for that constituency until the outbreak of the Revolution.

Another prize came his way in 1758, perhaps the most important of all: the widow Martha Dandridge Custis, whom he married in January 1759, shortly after resigning his commission in December 1758. He married up. Though herself of undistinguished family, Martha Dandridge Custis had bettered her position appreciably when she married Daniel Parke Custis, who came into a large inheritance soon after the wedding in 1749. When Daniel died in 1757, Martha found herself in the marriage market, and it did not take long for Washington to win her.

By marrying, Washington obtained the financial base that would support his ambitions, and while it would be wrong to say that he married for money alone, the bride's wealth was clearly a consideration. Now that he had a family and the means, he moved to establish a home of his own, and he chose as his residence a property with strong family connections. He had already leased Mount Vernon from his half brother's widow in 1754; following her death in 1761, he inherited it outright under the terms of Lawrence's will. Washington now began to modify the existing house on the property, though Mount Vernon as we know it, with the cupola and the pillared piazza, would not be finished until the 1780s.

Washington's political activity intensified in the second half of the 1760s. An orthodox Whig in his prerevolutionary politics, espousing the core ideas of the republican philosophy that guided him throughout his life, he worked with his neighbor George Mason to mobilize Fairfax County's response to British measures that, as they and others saw it, infringed on colonial liberties. Throughout this period, Washington's correspondence reveals his unquestioning acceptance of the basic notions that inspired colonial resistance to British measures and, ultimately, the Revolution. Suspicious of power exercised from London by those over whom the Virginia gentry had no control, he shared the widespread view that concessions on matters of principle would be fatal.

The early 1770s brought a relaxation in tensions between the colonies and the mother country, but the underlying problems remained, and Parliament's imposition of the Tea Act in 1773 reopened the conflict. Once the news of London's response to the Boston Tea Party — the Intolerable Acts of 1774 — arrived in Virginia, Washington was again active in urging resistance. His standing among his fellow gentry was confirmed by his election as one of the Virginia delegates to the First Continental Congress, held in Philadelphia in the fall of 1774. Chosen again to represent his colony at the Second Continental Congress in the spring of 1775, and then chosen to command the Virginia forces, Washington left for Philadelphia on May 4, beginning a six-year absence from Mount Vernon that would last until the summer of 1781. Attending Congress in his military uniform, Washington consciously or unconsciously advertised his availability for higher command. On June 14, 1775, John Adams of Massachusetts nominated him as commander of the Continental forces. Unanimously confirmed the following day, Washington accepted the position on June 16.

The mature Washington, who was forty-three when Congress offered him command of the army, cut an impressive figure. At least six feet tall, weighing between 175 and 200 pounds, and possessed of legendary physical strength, Washington literally towered over most of his contemporaries. A superb horseman, fond of hunting (fox hunts were one of his principal recreations), he was happiest when out-of-doors. His health was good, apart from his teeth, which already troubled him, and he would survive the Revolution without serious illness.

Congress's choice of Washington to lead its armies was the turning point in his life. In conventional terms, he was not a success. Between the conclusion of the siege of Boston in the spring of 1776 and the effective end of the war in America with the siege of Yorktown in the fall of 1781, there were few victories to report. The British had no difficulty in the summer of 1776 ousting him from New York. Nor was he successful in defending Philadelphia in 1777; here, too, the British overcame American resistance, though the battle at Germantown on October 4, 1777, was hard fought and at first Washington appeared to be winning. There was another major encounter in 1778 at the controversial battle of Monmouth Courthouse in New Jersey, but once again the British prevailed, as they usually did when confronting Washington's forces. For the rest of the war, until the Yorktown campaign in 1781, the troops under Washington's direct command played little part in shaping the military outcome.

The odds against Washington were enormous. With inadequate supplies, with men whose conditions led them to mutiny, especially in 1780 and 1781, with officers increasingly alienated from their ineffectual civilian superiors in Congress, it was all Washington could do to keep the army together. In the end, that turned out to be what mattered.

Yet the effort Washington and his troops expended would have been in vain without the French. Eager to avenge its defeat in the Seven Years War, France formally joined the war in 1778, after news of the American victory at Saratoga persuaded Louis XVI and his advisers that the American revolt was worth backing. French support meant two things, each critical in shaping the war's outcome: more supplies for American armies, and the diversion of enemy resources as the French threatened the British West Indies and turned the contest into a global conflict.

Washington was finally able to achieve victory, thanks in large part to the French navy. Defeating a British fleet on its way to relieve Lord Cornwallis in a battle off the Chesapeake capes on September 5, 1781, French ships under Admiral de Grasse prepared the way for the successful siege at Yorktown, the last military engagement of the Revolutionary War over which Washington presided. There, on October 19, in a moment long remembered in American legend, the British trooped out to surrender as the band allegedly played "The World Turned Upside Down."

Washington retired from the army in 1783, but he could not retire from public life. By the end of his service as commander in chief, no American was better known or more widely revered. For the rest of his life, Washington was lionized wherever he went. In 1787, his trips to and from the Constitutional Convention in Philadelphia took on the character of public occasions. The same was true, on an even larger scale, in the spring of 1789 when he traveled north from Mount Vernon to the capital in New York City to be sworn in as president; on this journey, Washington participated in elaborate "entries," formal receptions as he arrived in towns and cities which are best understood as republicanized versions of the ceremonial practices of European royalty.

For those who wanted souvenirs of Washington, there were many options. The hero's image (often in military uniform) appeared on paper in various media, and on textiles and ceramics as well. Washington repeatedly sat for the leading American painters, among them Charles Willson Peale and Gilbert Stuart. His portraits had wide circulation as prints and engravings. Washington was also one of the first Americans to be sculpted. For his American contemporaries, the living Washington was like no one else.

Despite his great popular stature in the nation, Washington shifted his focus for a time after the Revolution to more local and personal concerns. He took an interest in regional economic development and joined in projects to improve the Potomac region and Virginia more generally. He also attended to his own affairs. Mount Vernon — like his other property — had suffered from his long absence, and the Revolution had had a serious impact on his fortune. Convinced that no sacrifice for the glorious cause was too great, Washington resented the way those sacrifices fell more heavily on some (himself, for example), for during the war he had incurred major losses when his debtors used the tender law to pay him off in worthless paper money.

The Washington who returned to Mount Vernon in 1784 was no longer simply a local notable and the master of a Virginia plantation. His home was fast becoming an obligatory stop for traveling Americans and foreigners seeking a glimpse of the hero in repose. He was already, as he himself understood as well as anyone, a historical figure, and he engaged a secretary to copy and organize his wartime papers. But in the background, never far from the center of his concern, was the state of the nation.

The union of the new American states, weakly joined by the Articles of Confederation (ratified in 1781), seemed on the verge of collapsing before the experiment had received a fair trial. The states did not comply with Congress's directives; the British still retained several key posts on the western frontier — an especially sensitive point for Washington, with his dreams of westward expansion. In the 1780s, a decade of economic difficulties, the state legislatures created paper money and otherwise interfered in what Washington understood to be the proper relation between creditor and debtor. The instability of these years seemed to him to jeopardize the country's continued existence. "Anarchy" became Washington's standard description of what menaced the nation in the 1780s.

In 1786 and 1787, with the outbreak of Shays's Rebellion, an uprising by disgruntled Massachusetts farmers against their state government, these sentiments crystallized and moved Washington, and others, to action. As news of the Shaysites reached him, Washington's alarm grew. The uprising had to be suppressed, of course, but the very fact that it had broken out was reason enough to demand new measures. And so, after the failure of earlier attempts to reform the Confederation, Washington endorsed the call for a convention to meet in Philadelphia in the summer of 1787 to propose amendments to the Articles, amendments that would strengthen the Confederation and provide Congress with essential powers it lacked, above all the power to tax.

Washington would have preferred not to attend the convention, but it was clear to him and others that his presence would confer a legitimacy far greater than any resolution of Congress. Chosen president of the convention on May 25, 1787, he presided for the rest of the summer, taking no part in the floor debates. Once the delegates decided to craft a new document, rather than simply repair the old one, and once they moved to create a strong one-man executive, it was inevitable that Washington would be the first president. No other national figure enjoyed so much prestige, at home or abroad. Powers could safely be entrusted to a single leader, the delegates thought, because it was obvious that the president would be Washington. On the much debated question of whether the president should be eligible for reelection, Washington was again a key factor. No one feared Washington's becoming a de facto monarch, for his conduct during the Revolution had shown that he had no dangerous ambitions. And his lack of offspring and thus of a direct heir provided additional assurances: Washington could be the father of his country in part because he was, literally, the father of no one.

Hammered out over the summer, the Constitution was ready for signature early in September. Washington happily signed it. It was not perfect, he argued, but it was the best that could be had under the circumstances. Washington's endorsement of the document was critical to its success; its proponents took full advantage of his prestige to insist that the nation had nothing to fear from something the immortal Washington accepted.

Among the Founders, Washington often appears the least intellectually interesting. He served the new nation less as a fount of ideas than as the living exemplar of its political and moral values. To his contemporaries, he seemed a reincarnation of the ancient Roman hero Cincinnatus. By the end of the Revolution, most of the American people regarded him as indispensable to their political existence, and he knew it, although he was never entirely comfortable with the implications.

Washington remained at Mount Vernon during the ratification process, receiving reports from his correspondents as the Constitution made its way through the state conventions. Each new ratification was reason for rejoicing, each setback cause for dismay. By the end of June 1788, ten states had ratified, including Virginia, ensuring that the new charter would go into effect.

The elections of 1788 and 1789 produced few surprises. Washington's selection as president was a foregone conclusion; interest in the doings of the electoral college centered on the choice of vice president. More important as evidence of the nation's mood were the elections for the House of Representatives and the Senate, the latter in the hands of the state legislatures. Washington was relieved when the House elections, including those in Virginia, returned a majority of friends to the new Constitution, although his own state's choice of senators (both of them Anti-Federalists) was disappointing.

Well before he was formally notified on April 14, 1789, of his election as president, Washington began to consider an inaugural address, perhaps the first innovation in national politics under the Constitution. For while Article II stipulated that the president furnish an annual message on the state of the union, it made no mention of an inaugural address. Yet Washington understood the importance of setting the tone for the renewed experiment in republican government by giving the people an expression of the principles underlying it. Patriotically garbed in a suit of American cloth, Washington read his address — largely the work of James Madison — from the balcony of New York's City Hall (Congress's temporary home) on April 30.

The weather was good, the crowd was large, and by all accounts the first Inauguration went off without incident. Although Washington had confessed before leaving Mount Vernon that his feelings were those of a "culprit . . . going to his place of execution," once in New York he rose to his task. "The preservation of the sacred fire of liberty, and the destiny of the Republican model of Government, are justly considered as deeply, perhaps as finally staked, on the experiment entrusted to the hands of the American people," he told his listeners. And he warned Congress against those "local prejudices, or attachments," "separate views," and "party animosities" that would "misdirect the comprehensive and equal eye which ought to watch over this great Assemblage of communities and interests."

Politics, for Washington and many others in 1789, was the province of a virtuous and disinterested elite of white men whose talents and stations in life uniquely qualified them to rule. As he explained it in 1796, "the People" (by which he meant white males qualified to vote) had "the power and right . . . to establish Government," but this "presuppos[ed] the duty of every Individual to obey the established Government." Liberty did not imply license; the United States was to be an ordered republic, not an unruly democracy. The national good, discerned by the people's representatives, rather than the people themselves, was unitary. "Faction" and party were anathema in this mental universe.

Still, in 1789 no such set of assumptions, no matter how firmly held, could determine what it meant to be president. Washington was conscious that a major part of his duty as chief executive was to define and invent that role. From the outset, he realized that everything he did as president would be scrutinized by the public, and on matters large and small he sought the advice of those he trusted. Establishing the ceremonial side of the presidency — the Constitution said nothing about this — proved difficult, and not all of Washington's initial decisions met with approval. First there was the controversy over titles. How was the president to be addressed? The Senate proposed the unfortunate title of "His Highness the President of the United States of America, and the Protector of their Liberties." Wiser heads prevailed, but the damage had been done.

Controversy continued when Washington attempted to establish formal rules of etiquette for the presidential dinners and receptions. At a distance, not least from the perspective of some of his neighbors in Virginia, it looked as though the president was intent on adopting the trappings of royalty, or at least a ceremonial role not in keeping with their notions of republican simplicity. Washington never quite lived down this stir; in later years, with the growth of an opposition press, such features of social life in the capital as the annual celebration of the president's birthday (continued by the Federalists even after Washington retired from office in 1797) were constantly cited as proof of his monarchical inclinations and, even more, of those of some of his supporters.

On other fronts as well, Washington was beginning to investigate the limitations of the Constitution and discover what would and what would not work in his new role. In the summer of 1789, he met with the Senate to seek its constitutionally required advice and consent on negotiations then under way with a delegation from the Creek Nation. The effect of Washington's presence on the upper house was to stop discussion rather than stimulate it, and Washington was not at all pleased when one of the senators moved to refer the president's documents to a committee. "This defeats every purpose of my coming here," he is recorded as having said. Henceforth, the executive would conduct diplomatic negotiations on its own; advice and consent effectively became consent only.

Likewise, the presidential veto was part of the original constitutional scheme; and here too there was the problem of knowing what it was to mean in practice. Washington followed a conservative course, and the model of restraint he established would determine the behavior of his successors down to the time of Andrew Jackson. Early on, he concluded that he would not veto bills simply because he disapproved of them; only if his duty to uphold the Constitution required it would he use the veto. He exercised his right only once, vetoing the congressional reapportionment bill of 1792.

Washington's decision to use the senior executive officers as a cabinet may seem natural, even inevitable, in retrospect, and indeed it was common in those days at the state level for the governor to be advised by a council. Yet the Constitutional Convention had rejected proposals to create such a body. Washington's practice, clearly influenced by his reliance on councils of war on the battlefield, was a significant addition to the constitutional structure. The small size of the cabinet in his day meant that it could function as a genuinely deliberative body; similarly, the fact that it was composed mostly of men of national stature with whom Washington had long-standing relationships meant that the president could place a good deal of confidence in the advice he received. And while the intense rivalry between Hamilton and Jefferson, each of them bidding feverishly for the president's support, at times made the cabinet less harmonious than Washington wished, on the whole it was a productive body.

Selection of the secretaries, in fact, was high on Washington's list of priorities in 1789. Once Congress authorized the creation of executive departments (Treasury, State, and War; there was also an attorney general, although the post was not at first annexed to a specific department), Washington could proceed with his choices. He seems to have had few doubts about the cabinet-level officers: Alexander Hamilton for Treasury, Thomas Jefferson for State, and Henry Knox (a holdover from the Confederation) for War. Edmund Randolph, once an opponent of the Constitution but now reconciled to it, was to be attorney general. Candidates for the Supreme Court, led by John Jay, secretary for foreign affairs under the Confederation, were identified, nominated, and confirmed, as was a host of minor officeholders. Washington's appointment policy in the opening stages of his first term was straightforward: he wished to secure the services of those he called the "best characters," above all men who had proved themselves by their services during the Revolution and by their support of the new Constitution.

Lack of money had hobbled government under the Articles; the Constitution, with its grant of taxing power to Congress, was intended to remedy that. In 1789, the substantial debt from the Revolution, not to mention additional loans from Dutch bankers during the Confederation, posed a major problem. Reaching agreement on what to do about the revolutionary debt thus became the first substantive test of the new system's viability. Hamilton presented a controversial plan to Congress in January 1790 that rewarded speculators who had bought up the debt at deep discounts and seemed to treat some states more favorably than others. Washington firmly supported it, despite the outrage it caused in Virginia, because he believed it offered the best promise of restoring public credit. In the complex negotiations for the compromise known as the "Dinner Table Bargain," which led to the acceptance of a modified version of Hamilton's original plan, Washington's role, though shadowy, was surely significant. The horse-trading that produced the result involved an issue as near and dear to his heart as establishing public credit: the placement of the future permanent national capital on the banks of the Potomac.

Throughout his terms and afterward, Washington devoted considerable attention to the creation of the new Federal City. He selected its site, and in his Masonic regalia he presided over the laying of the Capitol's cornerstone. Despite his efforts, the project was plagued with difficulties. Neither Virginia nor Maryland (both of which had contributed the territory that made up the original ten-mile square of the District of Columbia) was willing to subsidize development, and federal appropriations were so modest that some feared the government would not relocate in 1800 as planned from the temporary capital in Philadelphia, then the country's largest city.

Just how important it had been to put finances on a sound footing became clear as Washington confronted problems on the frontier that lasted well into his second term. In 1783, the peace treaty with Britain had brought the Americans the northwest, the area beyond the Appalachians and north of the Ohio, but until the United States subdued the region's Native Americans there would be little chance of fully opening the Ohio country to white settlement. During the later 1780s, the Confederation's inability to support a military force on the frontier had been painfully apparent, and even after the revival of national authority under the Constitution, the situation continued to deteriorate. Matters did not improve until Anthony Wayne's victory on August 20, 1794, at the battle of Fallen Timbers (near the site of present-day Toledo, Ohio) broke the logjam in the northwest and led to the 1795 Treaty of Greenville, which quieted that frontier for the next decade and more. None of this could have happened without the tax dollars Alexander Hamilton was collecting at the Treasury.

The events of 1791 and 1792 made it clear how many illusions the Founders had entertained about the conduct of politics under the new regime. No one had imagined that political parties — universally condemned as "factions" — would have a role to play in the new republic, but that is precisely what now happened, much to Washington's chagrin. Arising from sectional interests, differences over foreign policy, and alternative visions for the new nation, groupings began to form, at first in Congress, then at the state and local levels. If not quite a replay of the contest between the Federalists and Anti-Federalists of 1787 and 1788 (Madison, for one, found himself switching sides), there was nevertheless considerable continuity in the basic positions. Supporters of Hamilton, taking the name of Federalists, argued for an energetic policy that would reduce the role of the states and use the resources of government for national economic development; their opponents, who called themselves Republicans, were suspicious of power, favored the states over the national government, and in general worried that the Federalists were far too enamored of the British model of government, even of the monarchical system.

Ideology and interest both played roles in shaping the first party system in the early 1790s, and the result was a bitter politics that left Washington at a loss. Never a party leader in the modern sense, he knew himself to be a good republican and a staunch patriot, and it was difficult for him to come to terms with the idea that he, too, was a partisan figure. In fact he never did, and his 1794 denunciation of the "self-created societies" that dared to organize opposition to the constituted and constitutional authorities came directly from the heart.

By the summer of 1792, the divisions in his cabinet were common knowledge. Insinuating articles appeared in the press, as Hamilton and Jefferson (or his surrogates) each denounced the other. Washington was appalled; convinced that the split did no good either to the country or to those personally involved, he entreated his warring secretaries to make peace. Carefully considering and replying to both Jefferson's charges against Hamilton (the secretary of state claimed, among other things, that Hamilton was intent on introducing the British system of political management by "corruption") and the latter's point-by-point rebuttal, Washington hoped that he had patched up matters by the fall of 1792. At best, however, it was a truce with no prospect of lasting; things had gone too far, and the divisions were too deep, for there to be any chance of a permanent resolution. Partisanship was now a fact, and both sides — the Federalists, as Hamilton's supporters were known, and the Republicans, as the Jeffersonians called themselves — knew that Washington alone supplied the façade of national unity. Recognizing the gravity of the situation, Washington consented to stand for office again. As in the first election, he had no rivals in 1792, winning all of the 132 votes cast in the electoral college.

The second term was full of incident, but like so many second terms in the future, it also seemed to mark a falling off from the accomplishments of the first. The fatigue of office took its toll on Washington. The original cabinet departed for private life, leaving the aging president to soldier on without them. None of the replacements who eventually served Washington in the second term had the stature of their first-term predecessors.

Foreign relations dominated the second term. Britain's entry into the war against revolutionary France early in 1793 presented the United States with an endless series of international problems. Washington had watched the progress of the Revolution in France from the outset, often suspending judgment. The cabinet divided bitterly: Jefferson hailed the progress of liberty in France; Hamilton made no attempt to hide his hatred of the Revolution.

In the spring of 1793, with France and Britain now at war, the president and his cabinet were forced to consider what course to take. The United States was France's ally, for the treaty of alliance of 1778 was still in effect, and under its terms the United States was bound to come to France's aid if summoned. Too active a policy in favor of France, however, threatened to bring on war with Britain, and that was exactly what the president and cabinet did not want. The Neutrality Proclamation of April 22, 1793, the result of contentious cabinet negotiations, satisfied Washington but raised a storm of controversy.

The arrival that spring of the new minister from republican France, Edmond Charles Genêt, did nothing to reduce tensions. "Citizen Genêt," an ardent advocate of spreading revolution, at first found himself cheered by the crowds that turned out to celebrate the progress of the French Revolution. Expecting an equally warm welcome from Washington, Genêt found instead a president who appeared to keep up the rituals of the ancien régime, a dubious republican hostile to revolutionary France and anything but a friend of the people. Genêt attacked the Neutrality Proclamation as treason to the cause of liberty, and over the weeks that followed he proceeded to commit blunder after blunder, alienating the president and exhausting even Jefferson's reservoir of friendship. Matters came to a head when Genêt had boasted that he would appeal to the people to reverse the executive's decisions. The president exploded and demanded Genêt's recall.

The difficulty over, at least for the moment, Washington faced the year 1794 with the hope that matters might take a smoother course. But he was to be disappointed. The French had been the source of trouble in 1793; now it was the turn of the British, whose seizures of American ships in the West Indies brought cries of protest and unleashed two years of continuous crisis that would last through the spring of 1796. Sensing their opportunity, the Republicans attacked what they saw as the pro-British course of foreign policy. (Behind it, they discerned what they considered the malign influence of Alexander Hamilton.)

Pressure in Congress mounted during the early months of 1794. The House adopted a motion for an embargo. War with Britain seemed imminent. Leading Federalists urged the president to send a mission to London. They hoped to buy time, allow passions to cool, and perhaps in the interim the British would adopt more sensible courses. On April 16, 1794, Washington nominated Chief Justice Jay as his special envoy, and Jay sailed for London, where he concluded an agreement with the British in November. For months, Washington and his cabinet and Congress waited for news, but they heard nothing of substance until a completed treaty arrived nearly a year later.

In the meantime, long-simmering unrest in western Pennsylvania over the tax on distilled spirits came to a boil, and Washington received reports of armed resistance to the excise law. It was Shays's Rebellion all over again, the president and his advisers believed. Predictably, Hamilton urged strong measures, and Washington concurred. The president issued proclamations against the insurgents in August and September, and the militia in the middle states was called out to put down what was termed the Whiskey Rebellion. Washington took command of the twelve-thousand-man force himself, accompanying it in October on the first stages of the journey west. The show of strength — combined with negotiations and some needed reforms — defused the situation. Washington's considered view of the matter was presented in his annual message to Congress that fall; its denunciation of those who fomented resistance to the laws was widely — and correctly — understood as a condemnation of the Republicans.

When it finally arrived, in March 1795, Jay's treaty with the British turned out to be explosive. At first glance, the chief justice appeared to have surrendered everything an American envoy ought to have insisted on. But Washington and his advisers realized that Jay had gained as much as could be expected, given the disparities of power between the United States and Great Britain. Accordingly, the treaty was submitted to the Senate, which approved it in special session in June 1795 by a vote of 20 to 10, and then only after stipulating that an article limiting the size of ships in the West India trade be dropped. Senator Stevens T. Mason, a Virginia Republican, leaked the text, which was supposed to remain confidential until the exchange of ratifications. Published on July

1, 1795, by a Republican newspaper, the treaty's contents created a sensation. Jay rather than Washington was the immediate target of public anger, but the treaty's effects on the president and on politics were far from over.

Several of the treaty's provisions required appropriations, and under the Constitution the House of Representatives had to initiate all money bills. The Republicans therefore insisted that the Senate's advice and consent was not enough in this case. The House, they held, had a constitutional right to judge the treaty — and reject it, if need be. After all, they said, if the president and the Senate could jointly approve treaties that required appropriations, then the constitutional design of lodging the power of the purse in the popular branch would be fatally wounded. This case again showed the Framers' failure to anticipate; for Washington, it would be the supreme constitutional crisis of his two terms in office.

Confident that their superior numbers in the House assured victory, the Republicans launched an assault on the treaty once Congress reassembled in December 1795. On March 25, 1796, the House called on the president to submit the papers relating to Jay's negotiation. Washington refused, noting the need for secrecy in diplomatic matters and pointedly mentioning the Senate's role in treaty-making. In April, after a prolonged struggle in which the Republicans' strength melted away, the House finally approved the appropriations to support the treaty.

Washington had had enough. Exhausted after years of public service, disgusted by the rise of partisanship, and longing to return permanently to Mount Vernon, he announced in the fall of 1796 that he would not run for a third term. A farewell address drafted in 1792 by Madison was dusted off, and Washington, now estranged from his fellow Virginian, chose Hamilton to help with revisions. Delivered on September 17, 1796, the farewell address was a lecture on the dangers of party and factionalism.

Later generations have seen it as a warning against entangling alliances and European involvements. Washington did insist that connections with foreign powers were dangerous, but most of all because "a passionate attachment of one Nation for another produces a variety of evils," including domestic divisions. Domestic divisions — the growth of faction and party — were what Washington, never more a traditionalist than in this message, feared, especially when those divisions threatened to organize themselves "on Geographical discriminations." The outgoing president gave "the baneful effects of the Spirit of Party" the full treatment: "in [governments] of the popular form it is seen in its greatest rankness and is truly their worst enemy," he thundered.

The address, then, emphasized core republican themes as Washington understood them, themes he hoped would provide a basis for unity. If Federalists found those arguments self-evident, the Republicans in 1796 believed that the people were wholeheartedly behind them and that only the dark forces of "corruption" kept the popular will from prevailing. The address did nothing to diminish partisanship, and the election results that fall showed that the voters had failed to heed Washington's warnings.

Washington was still the nation's hero when he left office in March 1797, but there were signs that his reputation was beginning to fray. If many of his contemporaries had virtually deified first the general and then the president, that view had never been uncontested. On the contrary, almost from the outset of his Continental career in 1775, there were always some unfavorable opinions. Apart from the Loyalists, there had been military critics during the war; civilians like John Adams and Benjamin Rush, who thought they knew who had really done the work of the Revolution, privately deplored the praise heaped on Washington. After 1789 critics began to complain, first of the president's apparently monarchical pomp and circumstance and later of his politics, especially when those politics seemed to favor Britain rather than France. Yet the critics never prevailed. Even after Washington had become identified with the interests of the Federalist party, it was still difficult to attack him directly and escape unscathed.

Washington's retirement was brief. In July 1798, as the undeclared naval war with France began to heat up, he reluctantly accepted President Adams's plea that he command the forces being raised to defend the country against the possibility of invasion. Of Washington's many public services, this one was the least happy. Controversies over the appointment of officers were bitter; Secretary of War James McHenry's incompetence led to complications. Apart from a trip to Philadelphia in November 1798 to consult with Adams and others, Washington remained at Mount Vernon, commanding from his study.

Washington knew he was aging. And he was not satisfied with the way things were going at Mount Vernon. Repeated absences over the previous two decades had not helped, and he gradually came to conclude that, whatever its other evils, slavery no longer worked as a labor system. He had too many slaves; his profits would be greater, he thought, with only half the slaves now at Mount Vernon. Like a number of other Virginians in the 1790s, he expressed himself in favor of gradual emancipation, though only in private and with trusted correspondents.

Washington understood that it was time to put his affairs in order and in July 1799 rewrote his will, a long and meticulous document that provided for his wife and relatives and disposed of his real estate and other property. He offered one of his slaves, William Lee, who had accompanied him throughout the Revolution, the option of immediate freedom; the other 124 slaves at Mount Vernon who were his to dispose of were to be freed after Martha Washington's death. But the remaining human chattels at Mount Vernon, the 153 "dower" slaves, were the property of the Custis estate, and neither Washington nor his wife could legally emancipate them.

Washington spent December 12, 1799, inspecting his plantation in miserable weather and caught a serious cold. His condition degenerated rapidly, and he soon realized that he was dying. The doctors bled him profusely, and on the night of December 14, he expired at age sixty-seven. He was interred in the family tomb at Mount Vernon on December 18. News of Washington's death spread rapidly, and public mourning was intense.

Virginia Federalist Representative Henry Lee, whose district included Mount Vernon, established the pattern of subsequent commentary in his oration for Congress on December 26. Washington, "Light Horse Harry" declared, was "first in war, first in peace and first in the hearts of his countrymen." Washington still stands near the head of the list when historians and political scientists rank the presidents; first in chronological order, he remains near the top in order of importance as well. Washington may not have been everything his mythologists claim, but no one else could have performed so well in the unique role that was his.

— HERBERT SLOAN
Barnard College

Washington's Leadership Style

We picture him in uniform, and indeed, George Washington's political instincts were distinctly military in origin. Unlike every president who followed, he never had to play the role of national candidate. Instead, Washington forged his style of leadership on the battlefield.

He made an early, conscious decision to elevate his social standing through military achievement. At first, the plan worked admirably: in 1755, by the age of twenty-three, he had risen to the rank of colonel in the Virginia militia and was acting as commander in chief of the colony's troops in the frontier war against France. But he was denied recognition in the form of a commission in the regular British army. The rejection stung deeply and caused Washington to reassess his career choice. Still, when he attended the Second Continental Congress in 1775, he came in uniform and readily assumed command of the army.

Washington recognized that defeating the British would require fashioning ragtag American troops into a regular army. Holding that "an Army without Order, Regularity and Discipline, is no better than a Commission'd Mob," he implemented strict regulations and stiff punishments. He drilled and paraded his troops in textbook European style. And he lobbied hard against relying on militia units.

Given the advantages of the British army, this strategy was ambitious and risky. In fact, Washington struggled mightily to keep his army in the field, executing a campaign that amounted for the most part to a string of tactical retreats and counterattacks, while scrapping constantly for enough supplies, provisions, and money to keep going. But he believed that if he maintained an army in the field long enough, the enemy would eventually wear down. More fundamentally, he was convinced that only a regular war, waged by a regular army, could achieve and sustain American independence. He would preside over the new nation with the same convictions about organization, discipline, and the value of enduring institutions.

John Adams

1797–1801

b. October 30, 1735
d. July 4, 1826

BORN IN 1735, John Adams grew up in the country town of Braintree (in the part that subsequently became Quincy), an agricultural suburb of Boston in the colony of Massachusetts. The son of a prosperous and locally respected farmer, young John longed to become a great and famous man and saw his opportunity in the study of law. After graduating from Harvard in 1755 Adams taught school for three years before winning admission to the bar in 1758. Thereafter, he worked tirelessly to master both the law and political philosophy and to "spread an opinion of myself as a lawyer of distinguished genius, learning, and virtue." In 1764 he married Abigail Smith, who was his equal in intellect if not education, and who, throughout life, fiercely defended her beloved partner.

Determined to secure public acclaim, Adams saw his opportunity in the swelling colonial resistance to British imperial rule. During the 1760s and early 1770s he won renown throughout Massachusetts by drafting spirited resolutions and essays critical of British taxation and regulation. In 1774 Adams became a delegate to the Continental Congress, which first met in Philadelphia that fall. By the time Congress reconvened in the spring of 1775, war had erupted between the thirteen colonies and the British Empire. The immense difficulties and innumerable details of military and political administration required interminable hours of committee work that exhausted the congressmen. No one worked harder or more effectively than Adams, who rendered himself indispensable to the war effort. Earlier than most of his colleagues, Adams recognized that reconciliation with Great Britain was futile. In early 1776 he worked tirelessly to promote a complete break with Britain by a formal declaration of independence. On July 2 Congress so voted and two days later adopted the Declaration of Independence based upon a draft submitted by a committee chaired by Adams and including Thomas Jefferson, who was the principal author.

John Adams also took a keen interest in the construction of republican constitutions for the various states, especially Massachusetts, which adopted a government largely based on his design. In two long and complex works, *Thoughts on Government* (1776) and *Defence of the Constitutions of Government of the United States* (1787), he urged his countrymen to frame complicated institutional arrangements meant to provide checks and balances on the exercise of the popular will. During the later years of the war, he also took a leading part in American foreign relations. In 1778 he went to Paris as one of the commissioners to negotiate an alliance with France and he remained in Europe to secure financial assistance from Dutch bankers and to help negotiate the peace treaty ending the war with Great Britain in 1783. In 1785 he became the first American ambassador to Great Britain, where he spent three frustrating years dealing with a resentful king and a contemptuous Parliament.

Adams returned home in 1788 and won election as the first vice president of the United States under the new Constitution — whose bicameral legislature and powerful executive showed the influence of Adams's political thinking. He was the first to discover the political futility and empty honor of the vice-presidency, serving two inconsequential terms while President George Washington and his cabinet conducted policy without him. The ambitious and domineering secretary of the treasury, Alexander Hamilton, guided the administration, to the horror of the opposition party, the Republicans, led by Thomas Jefferson, James Madison, and Aaron Burr. Adams's most conspicuous moment came in 1789 when he argued that the new officials of the nation required titles of dignity to command public esteem. In particular, he proposed that Congress decree that Washington be called "His Majesty the President." This proposal offended staunch Republicans, who charged that Adams was a closet monarchist. They also gibed that the stout vice president should be hailed by the title "His Rotundity." Thereafter, Adams remained largely silent in public affairs, but associated himself vaguely with the Federalist party, which supported Washington's administration.

In 1796, when Washington declined reelection to a third term, the Federalists rallied to Adams as their presidential candidate to stave off Jefferson's bid. The most partisan Federalists would have preferred Hamilton (who had retired from the cabinet), but his foreign birth meant that he was constitutionally ineligible for the presidency. With grudging, bad grace, Hamilton and his friends supported Adams as the lesser of two evils. On the one hand, they regretted that he lacked the gravity and popularity that had rendered Washington such a formidable front man for Federalist policies. They doubted Adams's effectiveness against their formidable foes, the shrewd Jefferson and his cunning running mate, Aaron Burr. On the other hand, the Federalist elite feared that Adams was too vain, opinionated, unpredictable, and stubborn to follow their directions. In private, Hamilton disparaged Adams's independence as "disgusting egotism," and Oliver Wolcott insisted that Adams had "far less real abilities than he believes he possesses."

Honoring the late-eighteenth-century republican orthodoxy that candidates should not overtly seek or campaign for office, Adams quietly spent the presidential campaign at home in Quincy. In private letters he made much of his diffidence, announcing, "I am determined to be a Silent Spectator of the Silly and Wicked game." Indeed, he did not even contribute to the selection of his running mate, Thomas Pinckney of South Carolina, a favorite of Alexander Hamilton. Meanwhile the local partisans of both parties worked in their home states to secure electors who would cast presidential ballots for their preferred candidates. At the end of December, Adams learned that he had won the presidency, but by the narrowest of margins: 71 electoral votes for Adams to 68 for Jefferson (who became the vice president). As expected, Jefferson prevailed in the southern states, while Adams dominated New England and edged Jefferson in the Middle Atlantic states. In a letter to his wife, an unusually ebullient Adams celebrated victory, writing that "John Adams never felt more serene in his life." The serenity would not last, however, because the close election revealed the considerable power of the Republican opposition; because the sniping by Hamilton and his friends suggested that Adams would face fights on both political flanks; because his self-doubting but prickly personality was incapable of prolonged equanimity; and because Adams did not understand that he had to play the popular politician to be an effective president. Indeed, Adams insisted that he would "quarrel with both parties and every Individual in each, before I would subjugate my understanding, or prostitute my tongue or pen to either."

Naively hoping to disarm partisanship, Adams reached out to his Republican opponents. During his first month in office, he proposed that Jefferson or Madison head an American diplomatic delegation bound to France for critical negotiations. But Madison balked, Jefferson demurred, and Adams's Federalist cabinet ministers threatened to resign. The president backed away, committing himself to a Federalist administration that retained Washington's cabinet, including Secretary of State Timothy Pickering, Secretary at War James McHenry, and Hamilton's successor as Secretary of the Treasury, Oliver Wolcott. Admirers of Hamilton, they repeatedly did their best to frustrate Adams's search for the political middle ground. Jefferson slyly observed, "The Hamiltonians who surround him are only a little less hostile to him than to me."

As president, Adams had to operate in the intense, superheated partisan politics of the 1790s, when the fate of the fragile new federal republic seemed to teeter on the brink of disaster. Unlike twentieth-century Americans, the leaders of the early republic could not comfort themselves with a long history of enduring union and republican government. Politicians of every stripe expected the worst of one another and of events. Such beliefs tended to become self-fulfilling prophecies as both Federalists and Republicans magnified every sign of opposition into a crisis that imperiled the nation. Lacking any tradition of a limited but accepted political opposition, both the Federalists and the Republicans believed that they alone represented the public good of the republican nation. Consequently, their opponents could only be insidious conspirators determined to destroy both freedom and union. Federalists saw the Republicans as libertines and anarchists eager to replicate the terror of the French Revolution. In turn, the Republicans depicted the Federalists as cryptomonarchists preparing to destroy the republic and to restore British domination. In the early republic, only George Washington (and only for his first term) could hope to float above the bitter partisanship.

But Adams was no George Washington. Short, stout, balding, and unprepossessing, Adams could not command attention and respect as had the tall, strong, and dignified Washington. And, unlike the first president, Adams did not know when to shut up. Voluble, opinionated, censorious, impulsive, vain, and touchy, Adams possessed the gift for political disaster in a republican government. Relentlessly hard on himself, he readily applied the same impossible standards to others. And unforgiving of his own foibles, he could not abide criticism by others. When faulted, he became thoroughly self-righteous and self-pitying. In 1783 he remarked, "It is a sad Thing that Simple Integrity should have so many Enemies in this world, without deserving one."

Congenitally incapable of dissembling, he made a better political theorist than a politician. He has endeared himself to historians, as he offended contemporaries, with his blunt, relentless, and tactless honesty. In a 1789 letter he savaged his countrymen for "bawling about a Republicanism which they understand not." For his harsh candor, Adams reaped political vilification. In 1787 he reflected, "Popularity was never my Mistress, nor was I ever, or shall I ever be a popular Man." In 1812 he sadly observed to a friend, "From the year 1761, now more than Fifty years, I have constantly lived in an enemies Country."

Adams abundantly recognized his own shortcomings, often writing variations on the theme: "Oh! that I could wear out of my mind every mean and base affectation, conquer my natural Pride and Self Conceit, expect no more deference from my fellows than I deserve, acquire that meekness and humility, which are the sure marks and Characters of a great and generous Soul." Ruing his tactless egotism, he vowed "never to shew my own Importance or Superiority, by remarking the Foibles, Vices, or Inferiority of others." Of course, he never mastered himself, which led to further self-recrimination: "I am constantly forming, but never executing good resolutions."

As the nation's second president, Adams recognized that his principal challenge was to preserve Washington's policy of American neutrality in the struggle between the great warring powers of France and Great Britain. Peace was essential to preserving prosperity and concord within the fragile but promising United States. Given a prolonged peace and continuing growth in population and wealth, in another generation, the United States would become too powerful for either France or Britain to intimidate. In the meantime, preserving neutrality would not be easy because, as Adams well understood, neither France nor Great Britain would relinquish any effort or intrigue that could compel the United States to take its side.

The French enjoyed greater popularity in the United States during the early 1790s because of gratitude for their help in securing American independence; because of widespread sympathy for the recent French Revolution as a common cause with American republicanism; and because most Americans remained distrustful of British designs. Consequently, the treaty of rapprochement with Great Britain negotiated by John Jay in 1794 ignited widespread public protests that it betrayed America's formal alliance with France. Only Washington's immense prestige secured ratification of the treaty in the United States Senate in 1795. Early in 1797 the infuriated French government began to seize and confiscate American merchant ships trading with the British. The French Directory hoped that the crisis would induce a pro-French American public to topple their Federalist rulers.

Determined to avoid war, Adams proposed sending a commission to negotiate a better understanding with France. Still seeking some bipartisanship in foreign policy, he chose as his commissioners the Republican Elbridge Gerry (an old friend of Adams's), as well as the Federalists Charles Cotesworth Pinckney and John Marshall. To prepare for the possibility of diplomatic failure, Adams also urged Congress to expand the navy, arm merchant ships, and augment the army.

On March 4, 1798, news reached Philadelphia that the French foreign minister, Talleyrand, had refused to receive the American commissioners unless they offered him a bribe and a loan to the French government. Pickering urged an American declaration of war, but Adams balked, hoping to buy time. Skeptical Republicans in Congress forced his hand by demanding the release of the envoys' dispatches. To the Republicans' chagrin, publication proved that Talleyrand had indeed made the insulting demands. Consequently, publication aroused public outrage, which stiffened the resolve of staunch Federalists to prepare the nation for war (but not yet to declare it). Adams bent under the pressure from his own party and, for a time, embraced the martial enthusiasm, issuing truculent addresses to the American people. Indeed, although he had no military experience, he began to appear in public wearing a full military uniform, including a sword strapped to his side. A majority in Congress swung into line, authorizing construction of twelve new frigates for the navy and bolstering the army to at least ten thousand men, while authorizing a further buildup to fifty thousand if the president deemed that necessary.

Command of the army proved troublesome. Politically, if not militarily, the obvious choice for the overall command was the elderly George Washington. The problem came when Washington insisted upon having Hamilton as his second-in-command. Over the years, Washington had grown dependent upon Hamilton as his administrator. The former president's demand was doubly difficult because more experienced generals would balk at serving under Hamilton and because Adams had learned to despise him. Not without reason, Adams deemed Hamilton "the most restless, impatient, artful, indefatigable and unprincipled Intriguer in the United States." But unable politically to resist Washington's ultimatum, the president capitulated. Because Washington intended to remain at home until the unlikely event of a French invasion, Hamilton exercised active command of the expanding army. Consequently, Adams fretted about the malign purposes for which he feared Hamilton might employ the army. The president shared the Republicans' dread that, in the event of a political crisis, Hamilton would not resist the temptation to seize power. After all, Hamilton once assured Jefferson that the Roman military despot Julius Caesar was the greatest man in history.

Adams derived far greater satisfaction from the new American navy, in which he took a special pride and interest. Although no war had been formally declared, on the high seas American and French warships engaged in deadly combat during the three years 1798 to 1800. Although small and new, the American navy proved surprisingly effective at countering French privateers and warships along the Atlantic seaboard and in the West Indies. In the most significant action, on February 9, 1799, in the Caribbean, the United States frigate *Constellation,* under the command of Thomas Truxton, defeated and captured a larger French warship, *L'Insurgente.* The American naval victories strengthened Adams's hand in pursuing a diplomatic resolution with France.

At home the Republicans refused to support the war effort, insisting that the Federalists were overreacting in a cunning bid to enhance their power. The Republican press savaged the Federalists as corrupt monarchists out to dominate and exploit common American taxpayers. In turn, the Federalists truly overreacted by denouncing their Republican critics as French partisans bent on subverting American defense and American independence. Federalists charged that the Republicans meant to seize power by provoking mob violence. Alarmed, Adams stockpiled arms to help defend the presidential mansion.

In June 1798, the Federalist majority in Congress passed four acts meant to suppress Republican dissent. Because most immigrants (especially Irishmen) voted Republican, the Federalist Congress, with the Naturalization Act, doubled to fourteen years the period required to naturalize the foreign-born to American citizenship. The Alien Friends Act and the Alien Enemies Act permitted the chief executive to deport any foreigner that he deemed dangerous to the United States. Finally, the Federalists enacted the Sedition Act, which criminalized public criticism of the federal government, imposing punishments of two to five years in prison and fines of $2,000 to $5,000. Although the president had neither designed nor promoted the measures, he signed all four into law. The thin-skinned Adams had no objection to legislation meant to punish those who abused his government. Indeed, he sometimes urged on the prosecution of his critics. The administration secured at least fourteen indictments under the Sedition Act. The indicted and convicted included the leading Republican journalists James T. Callender and William Duane, as well as Matthew Lyon, a Republican congressman from Vermont, who was convicted, fined $1,000, and sentenced to four months in jail.

By overplaying their hand, the Federalists undermined the public's war fever. The Alien and Sedition Acts proved almost as unpopular as the increased taxes needed to fund the expanding and expensive army and navy. By 1799 the national government under Adams was spending twice as much as had the Washington administration in its last year. Republicans were outraged by the Alien and Sedition Acts and terrified that a Federalist program of heavy taxes and a large professional army would combine to suppress the common people and, eventually, replace the republic with a monarchy. This rhetoric played increasingly well with a public that reflexively hated taxes.

In late 1798 the southern Republicans began to look for relief from federal power by insisting, in resolutions adopted by Kentucky and Virginia, that state governments had the right and power to judge the constitutionality of federal laws. Written by Jefferson (Kentucky) and Madison (Virginia), the resolutions denounced the Alien and Sedition Acts as unconstitutional and hinted that the states could and would nullify their enforcement. Preparing for a potential military showdown with Hamilton and his federal army, Virginia's Republican-dominated state government began to stockpile munitions and to reorganize the state militia. Learning of these preparations, Hamilton proposed sending troops into that state "and then let measures be taken to act upon the laws and put Virginia to the Test of resistance." Leading Virginians, including Jefferson, privately anticipated secession and civil war.

Popular resistance to Federalist measures was not confined to the south. In Republican-dominated locales throughout the Middle Atlantic states and even in pockets of New England, the disaffected erected liberty poles — a signal that they were again prepared for a revolution against authority. In early 1799 attempts to collect a new federal land tax provoked mob resistance in parts of two Pennsylvania counties. This violence was hyperbolically called Fries Rebellion after its supposed leader, Captain John Fries, who was arrested, tried, convicted of treason, and sentenced to die.

Although ready to suppress dissent in America, neither Adams nor the Federalist majority in Congress was yet willing to declare war against France. Consequently, the nation remained in limbo through the summer and fall of 1798, preparing for a war that its leaders could neither embrace nor relinquish. Adams despaired at the prolonged impasse and at life in the national capital of Philadelphia, which was rancorous, hot, humid, and disease-infested, and far from his beloved home and wife, who for health considerations had returned to Quincy. "I am old — very Old and never shall be very well — certainly while in this office,"

Adams brooded in December. He felt politically trapped between, on the one hand, the seditious and perhaps secessionist Republicans and, on the other, Hamilton and the Federalist hard-liners. Civil war loomed if he could not find an escape.

Adams was intrigued, therefore, by hints from Paris in early 1799 that the French government would reconsider its policy toward the United States. Without consulting either his cabinet or the Federalists in Congress, Adams resolved to send another delegation to Paris to seek peace. Because the most partisan Federalists regarded the war fever as their greatest political asset, they were furious at the president's surprise announcement, but they dared not openly oppose the request for fear of confirming the Republican charge that they were reckless warmongers. In late February the Senate reluctantly approved the mission, entrusted to William Vans Murray, William Davie, and Oliver Ellsworth — all Federalists. In a concession to his cabinet and Congress, Adams delayed dispatching the mission to France, pending a clearer signal from the French.

In March, rather than waiting in Philadelphia for that signal, Adams bolted from the capital to return to Abigail in Quincy. Concerned for her tenuous health and dreading a return to the capital and its political dilemmas, he remained away for eight months, more than twice as long as Washington's longest absence. Because it was a period of suspense and crisis in American domestic politics and foreign relations, Adams's prolonged absence aroused considerable comment and criticism. And it enabled Hamilton's allies in the cabinet to govern with virtual independence. As a consequence, Pickering, McHenry, and Wolcott continued to delay the envoys' departure for France.

Despite the swelling complaints, Adams remained in defiant but impotent isolation until September, when his loyal secretary of the navy, Benjamin Stoddert, wrote to warn that, in his prolonged absence, "artful, designing men" were subverting the overture to France. Recognizing that Stoddert referred to Hamilton's friends in the cabinet, Adams hastened to the capital. Only the specter of Hamiltonian intrigue could dislodge Adams from home and send him to detested Philadelphia. There in October, and over the protests of Hamilton and his three protégés in the cabinet, Adams at last dispatched the American commissioners to Europe. Despite his private fury, Adams temporized about removing the wayward cabinet secretaries until May 1800, when he at last discharged two of them. He replaced Pickering and McHenry with John Marshall as secretary of state and Samuel Dexter as secretary of war. That spring, Adams also ordered the demobilization of Hamilton's army and the pardon of John Fries, sparing him from execution for treason and Virginia from invasion. Outraged by all three moves, Hamilton fumed that Adams was "as wicked as he [was] mad."

In fact, Adams was shrewder than Hamilton, who was behaving with increasing irrationality in anger over his eroding political power. The president belatedly recognized that his own political fortunes in the approaching fall election depended upon a swing to the political middle. And his moves also defused the crisis with Virginia that had threatened the union. Northern Republicans helped as well by rejecting the Kentucky and Virginia resolutions. Instead, northern state legislators, Republican as well as Federalist, reaffirmed their commitment to union. Preparing a renewed bid for the presidency, Jefferson recognized that he needed northern Republican support, especially in New York. Consequently, he quietly dropped his more extreme states' rights rhetoric of 1798. Finally, in the summer of 1800 white Virginians narrowly escaped a formidable slave uprising planned by a charismatic slave named Gabriel. Frightened leading southern Republicans cooled their divisive rhetoric, lest they provoke a civil war that would provide a renewed opportunity for slaves to liberate themselves.

In the fall presidential race, John Adams and his running mate, Charles Cotesworth Pinckney, confronted the Republican duo of Jefferson and Burr. Behind the scenes Hamilton intrigued to sabotage Adams in hopes of boosting the more pliant Pinckney to the presidency. In October, Hamilton recklessly published a vitriolic pamphlet denouncing Adams as mad with egotism and rage. The publication divided and demoralized the Federalists at a critical juncture in the campaign. To make matters worse, in November, Abigail and John Adams moved from Philadelphia into the still unfinished presidential mansion in the newly established capital district on the banks of the Potomac. They had thought that nothing could be worse than life in Philadelphia — until they moved to raw Washington City, then a miasmic, muddy, and shabby village with absurd pretensions to grandeur.

In December, Adams learned that he had lost narrowly to Jefferson by 65 to 73 electoral votes. As in 1796, Adams prevailed in New England, and Jefferson carried the south, but this time Jefferson captured the critical Middle Atlantic state of New York, thanks largely to the astute management of Aaron Burr. However, the cunning Burr detected an opportunity to steal the presidency from Jefferson at the last moment by exploiting a flaw in the Constitution. Until subsequently amended, the Constitution did not permit the electors to distinguish between president and vice president in casting their two ballots. Instead, the Constitution simply stipulated that the top vote-getter would become president and the second would settle for the vice presidency. As a consequence of this awkward arrangement, the Republican Jefferson had served as vice president in the presidential administration of his rival Adams. And as a further consequence, when both Burr and Jefferson finished with 73 electoral votes, the presidential election was thrown into the lame-duck Federalist House of Representatives. Despite the fact that the Republican electors had intended that their ballots would elect Jefferson as president and Burr as vice president, many Federalist congressmen could not resist the temptation to make trouble for their rivals by supplanting Jefferson with Burr. After a prolonged deadlock, the House narrowly honored the will of the electorate and made Jefferson the president and Burr the vice president. That decision appeased anxious Republican leaders who had spoken of resorting to armed force if the Federalist House stole the election from Jefferson. The election of 1800 proved fateful, for the Federalists lost control of the incoming Congress as well as the presidency and would never regain either. Instead, the Federalists declined steadily until 1808, when they staged a partial recovery only to collapse utterly after 1814.

Humiliated and embittered by the electoral rejection, Adams rarely left the White House and declined to receive visitors during the last three months of his presidency. But, ever dutiful, he worked late into the night at his official responsibilities. In particular, he assiduously filled the vacancies in the federal judiciary with Federalists, including John Marshall as the new chief justice of the Supreme Court. These lame-duck appointments, especially the selection of Marshall, infuriated Jefferson. Although Marshall and Jefferson were cousins, they had become bitter political enemies. At noon on March 4, 1801, Adams's presidency ended and Jefferson's began; rather than remain for the inauguration, a sullen Adams quietly slipped out of the capital before dawn.

Meanwhile, war with France had been averted by prolonged negotiations that culminated in an accord signed on October 3, 1800. The agreement was a compromise that satisfied the American demands for French recognition of their rights as a neutral to trade in nonmilitary goods with Great Britain, but provided no indemnification for the American vessels seized by the French during the crisis. Unfortunately, formal news of the accord reached the United States too late in the fall to avert Adams's presidential defeat. In early February 1801 the Senate ratified the treaty (with reservations).

Although Adams had lost his election, he had belatedly secured peace. Characteristically, even his triumph redounded to Adams's detriment rather than to his credit. The peace restored American prosperity and helped ensure the success of Jefferson's first term. The new president was able to cut military expenditures, reduce taxes, consign the Alien and Sedition Acts to oblivion, and negotiate the purchase of the immense Louisiana Territory from France. All of these proved immensely popular and, as Jefferson's popularity waxed, the reputation of the Adams administration waned. In 1805 Adams grumbled that his presidency seemed "condemned to everlasting . . . infamy." Just as Adams suffered in a comparison with his extraordinarily popular predecessor, George Washington, he also suffered from the dramatic success of his popular successor, Thomas Jefferson.

Characteristically, Adams perversely cherished his defeat and unpopularity as the ultimate proofs of his virtue and independence. He insisted that he had sacrificed himself for his country's good; that by reaching for peace in 1799 he had divided his party, which destroyed his bid for reelection but rescued the nation from collapse. Most historians have taken Adams at his word. In fact, his bid for peace in 1799 and 1800 was less self-sacrificing and more politically astute than Adams liked to recall. By rejecting Hamilton's increasingly unpopular policies of confrontation at home and abroad, Adams revived his otherwise moribund prospects for reelection in the fall of 1800. Adams ultimately lost because, far from acting boldly, he had waited too long to distance himself from Hamilton and the extreme war measures of the Federalist Congress. Adams paid an appropriate political price for temporizing so long at such great risk to the peace, prosperity, and union of the nation.

Adams retired to Quincy, dabbled in agriculture, and found comfort in books, in the company of Abigail, and in the ascending political fortunes of their son John Quincy Adams. Still furious at Hamilton and the Hamiltonians, Adams and his son abandoned the Federalist party to become conservative Republicans at the end of the decade. In 1812 this shift helped Adams and Jefferson to renew their old friendship by commencing a free-ranging and voluminous correspondence during the remaining fourteen years of their lives. The exchange had a healing effect on Adams's resentments and reputation, as the public came to think of him as Jefferson's friend and revolutionary colleague rather than as his Federalist foe. His last years brought both profound grief and vicarious triumph: Abigail died in 1818, but John Quincy Adams won the presidency in 1824.

Living a long time proved to be the best political revenge. Adams survived almost all of his Federalist enemies (most notably Hamilton, who, to Adams's grim satisfaction, had died in a duel with Aaron Burr in

1804). Consequently, Adams was alive and available in the 1820s when a new generation of Americans began to mythologize the Founding Fathers of their now (it seemed) secure and thriving republic. And Adams's beneficial association with Jefferson became complete with the exquisite mutual timing of their deaths on July 4, 1826, the fiftieth anniversary of the Declaration of Independence. That sensational coincidence confirmed many Americans in the conviction that their republic was providentially ordained and that Jefferson and Adams were, along with Washington, its greatest founders.

A hard man for his contemporaries to love, John Adams has enjoyed a far better press among twentieth-century historians. When surveyed, they indulgently place the Adams administration in the "near great" category, primarily because they credit his claim that he gave of himself to secure the nation's peace. Moreover, the same qualities of biting honesty, prolix writing, and determined independence that so offended colleagues have endeared Adams to scholars. They delight in his vivid quotations, exhaustive documentation, and utter inability to hide his feelings or cover his tracks. He is such a remarkably instructive and cooperative historical source precisely because he was so difficult for most of his contemporaries to work with.

— ALAN TAYLOR
University of California at Davis

Thomas Jefferson

1801–1809

b. April 13, 1743
d. July 4, 1826

T HE LIVES OF THOMAS JEFFERSON and his presidential predecessor, John Adams, became entwined in fascinating ways. Both served on the committee to draft a declaration of independence in early 1776. It was Adams who urged that Jefferson be made the committee's penman because of the "felicity" of his style. The Revolutionary War brought new responsibilities to each, as Jefferson became governor of Virginia and Adams joined Benjamin Franklin in Paris to help negotiate the treaty that brought the Revolution to a successful conclusion. Then with Jefferson's being named Franklin's successor as minister to France and Adams's appointment as America's first envoy to Great Britain, they were again in partnership as their nation's principal diplomats from 1785 to 1788. For both men, the extraordinary political events of their young adulthood pushed them onto a world stage when their birth, as in Jefferson's case, prepared him only to be a Virginia planter, lawyer, local leader, and slaveholder.

Neither Jefferson nor Adams attended the Constitutional Convention in Philadelphia — the most prominent revolutionary leaders to be absent from it. Earlier, they had both gained a continental reputation through their pamphleteering in the years leading up to the Revolution. The organization of Washington's first administration, after the ratification of the Constitution, brought them together again when Adams served as vice president and Jefferson as secretary of state. What neither could have anticipated was that within three years, Jefferson would begin to mobilize an opposition to the Federalists, who sought stability for the new central government in the support of a vigorous, national elite composed of the rich, educated, and wellborn, a view that thoroughly alarmed Jefferson. Having created an energetic national government with powers to clip the wings of the old sovereign states, the Federalists set about reclaiming from ordinary voters the deference that they had enjoyed before the wartime turmoil.

Jefferson had been an eyewitness to the revolutionary actions of 1789 when France's ancient Estates General had transformed itself into a modern legislative body. Memories of the Tennis Court Oath, the storming of the Bastille, the abolition of feudal privileges, and the adoption of the Declaration of the Rights of Man and the Citizen were fresh in his mind when he arrived in New York City in 1790 to take up his duties as secretary of state. Coming from the most radical milieus in the Old World he landed in one of the most conservative in the New, for while the French were busy destroying their old authoritarian institutions, the Federalists were busy creating new ones. At least that is the way it appeared to Jefferson as he became acquainted with his fellow cabinet members.

An elegant social life had sprung into being in Philadelphia during the sessions of Congress. For many Federalists the round of presidential levees, balls, and dinner parties signaled the consolidation of a national elite ready to exercise as much authority over matters of taste as they would in policies of state and firmly attached to the interests of the new central government. Jefferson, however, was dismayed by the lavish entertaining that Federalists flaunted before the good people of Philadelphia. In his eyes, they were deliberately creating a veritable aristocracy to place ordinary Americans in the same kind of thrall to their leaders that he had seen in the impoverished and ignorant masses of Europe.

Through seven years, the new nation witnessed a succession of rancorous disputes as the old revolutionary leaders and new citizens coming of age struggled to define what the American Revolution meant. Jefferson's challenge to Adams's succession in 1796 won him the vice-presidency. But the awkwardness of this provision of the Constitution, that the second vote-getting candidate for president earned the number-two spot, became manifest during Adams's presidency when Jefferson found himself an unhappy presider over a Federalist Senate bent on silencing its opponents. When Congress passed the Alien and Sedition Acts, which made it harder for immigrants to become citizens and beefed up the machinery for prosecuting sedition, Jefferson rushed to Monticello and secretly composed resolutions calling the states to disallow "unconstitutional" reaches of congressional power. Only two states responded, but the Virginia and Kentucky resolutions of 1798 laid the groundwork for Jefferson's later reputation as a states' rights leader. When Jefferson triumphed in 1800, denying Adams his anticipated second term, the cordiality they had once enjoyed vanished. Adams decamped for Massachusetts before the Inauguration ceremony took place, leaving behind a curt message about the state of the White House stables.

In 1812 a mutual friend got the two old heroes, now retired elder statesmen, to resume their once friendly correspondence. This fascinating, thirteen-year exchange of letters fills an entire published volume, their commentaries roaming widely over the colonial resistance movement, their opinions of Napoleon Bonaparte, and their enduring political commitments. Adams was characteristically crusty, using his words to jab and thrust at the stupidities of the world, while Jefferson calmly steered his philosophical ship into untroubled waters.

Like his native New England, Adams held a view of human nature that owed more to John Calvin than John Locke, while Jefferson retained a measured optimism befitting his role as an Enlightenment thinker. One wonders if Jefferson caught the irony when Adams commented that his taste was "judicious in liking better the dreams of the Future, than the History of the Past." Destined never to see each other again, the two men ended their correspondence on July 4, 1826, when they both died, amazingly, on the fiftieth anniversary of the Declaration of Independence. Needless to say, the country's great orators then outdid themselves in paying tribute to these giants of the nation's founding. At the end, Adams's thoughts had turned to his old friend and political rival. His last words were: "Thomas Jefferson still survives." So too do Thomas Jefferson's ideals and the tensions they generated.

The Adams-Jefferson friendship offers an illuminating frame for the presidency of Thomas Jefferson, not just because of its unusual combination of affection and antagonism, but also because Adams's great-grandson, Henry Adams, decisively shaped how historians have viewed Jefferson's presidency. Writing in 1883, Henry Adams put forth the interpretation that, despite Jefferson's vigorous opposition to the principles and programs of Washington and Adams, in office he had "outfederalized" the Federalists. What his great-grandfather had lost at the polls, Henry helped recover in the classroom when his history became a classic. Twenty years after its publication, Henry's brother, Brooks, equally critical of Jefferson, was writing a biography of their grandfather, another Adams president denied a second term. After looking at the manuscript, Henry wrote Brooks with some exasperation, "For God Almighty's sake, leave Jefferson alone!"

The Adamses could not, any more than the country, leave Jefferson alone, for he was the principal spokesman for its twin ideals of liberty and equality. But those very ideals produced a wedge that divided the United States from its inception. At the very time that the new government under the Constitution was being formed, individual Northern states abolished slavery leaving the South to defend its peculiar institution alone. With the split between North and South deepening in the 1840s and 1850s, contemporaries began viewing Jefferson as a spokesman for states' rights. Thus implicated in the Civil War, his reputation remained in the shadows until Franklin Delano Roosevelt restored its luster with his linkage of Jefferson to the founding of the Democratic party and with the erection of the Jefferson Memorial alongside those of Lincoln and Washington in the nation's capital.

Jefferson had no idea of founding a party in the 1790s. He intended rather to return the country to its true purposes. It had not been wholly clear what the implications were of declaring independence by saying that "we hold these truths to be self-evident that all men are created equal and endowed by their creator with certain inalienable rights among which are life, liberty and the pursuit of happiness." For most of the revolutionary leaders there were few radical implications to that affirmation of natural rights. They expected continued deference, and they supported the authority of patriarchal families, established churches, and active government. The novelty of the United States for them lay not in signaling a new dispensation for the human race, but in offering learned statesmen a fresh opportunity to apply the lessons of the past. Jefferson envisioned something different — the possibility of freeing men from the constricting institutions of the past.

The aristocratic tone of Washington's administration, which Jefferson saw in the provocative elegance of official entertaining, might have remained just a worrisome tendency had not the policies of Alexander Hamilton given substance to his fears that the principles of 1776 were being deliberately abandoned in favor of a British-style government. A bold fiscal innovator, Hamilton put together a package of laws that turned the revolutionary debt into an asset. Brilliantly conceived, his plan converted the ragtag accumulation of state and congressional I.O.U.s into government bonds funded at the original value of the loan, with taxes assigned to pay interest on the new bonds. Those in the know benefited from the sudden rise in the value of the government's debts, and the nation got its first look at a speculative mania in 1791. Highly sensitive to the corrupting influence of such financial derring-do, Jefferson began to articulate a more coherent oppositional picture. Although Washington sought advice from Jefferson, Madison, and Hamilton, it quickly became apparent that Hamilton exercised the greatest influence when Washington signed the capstone of his program, the law incorporating a quasi-public Bank of the United States. Jefferson had insisted that the power of incorporation exceeded the authority that the Constitution gave Congress.

When Washington sought position papers from Jefferson and Hamilton on the question of the constitutionality of the Bank, they produced the classic statements on the strict and loose interpretations of the Constitution, Hamilton arguing that the Constitution's "necessary and proper" clause permitted a broad construction of Congress's grant of power, and Jefferson arguing that only measures that could be shown to be strictly necessary could be added to the enumerated powers. Washington sided with Hamilton, heightening the tension between his two cabinet officers. The outbreak of a general European war and the emergence of the French republic after the execution of the king and queen found them again arrayed on opposite sides, as Hamilton sought to strengthen American ties with Great Britain, and Jefferson identified American libertarian interests with those of France.

By the end of Washington's first term, Jefferson had resigned from the cabinet, finding his official position in an administration he no longer supported untenable and took his concerns out-of-doors to an electorate unused to partisan politics. For the next eight years, Jefferson gave expression to one of the most attractive propositions of the American experiment: the egalitarian notion that political and economic liberty could eradicate the differences imposed by ignorance, superstition, and tyranny. Against the prevailing view that most men were weak, unsteady, and in need of guidance, he offered the Enlightenment conception of the rational, self-improving, independent man who could be counted on to take care of himself and his family if only intrusive institutions were removed. He gave a scientific reading to the natural rights philosophy and mounted a powerful argument against privilege by drawing upon the idea of a dawning new age.

From 1793 until Jefferson's victory in the election of 1800, events conspired to embroil Americans in the most fundamental questions about democratic government, while first Washington and then John Adams struggled to keep the country out of the war that raged among the great European powers. No issue, domestic or diplomatic, escaped contamination from this controversy about privilege and precedent. Not even the enormous prestige of George Washington could persuade the people to return to their personal affairs and let their leaders take care of the general interests.

Looking back on the presidential campaign of 1800 from the safe shore of success, Jefferson reflected on the fundamental differences between the voters who had rallied around him and those who had supported Adams. Writing to Joseph Priestley, the famous English scientist then ensconced in rural Pennsylvania, Jefferson stressed that his opponents had looked "backwards not forwards, for improvement." They favored education, he said, "but it was to be the education of our ancestors"; and he noted ruefully that President Adams had actually told audiences that "we were never to expect to go beyond them in real science."

Yet Jefferson exulted, "We can no longer say there is nothing new under the sun, for this whole chapter in the history of man is new. The great extent of our republic is new. Its sparse habitation is new. The mighty wave of public opinion which has rolled over it is new." Years later, writing to Adams, Jefferson unequivocally characterized their respective parties as "the enemies of reform" and its champions, the two sides splitting on the question of "the improvability of the human mind, in science, in ethics, in government." He lectured Adams, "Those who advocated reformation of institutions, *pari passu*, with the progress of science, maintained that no definite limits could be assigned to that progress."

Accepting the challenge of interpreting natural design, Jefferson turned himself into an agent of change — profound, transformative change in the social relations and political forms of his nation. His opportunity to act upon his ideas as president of his country was unique. Indeed, because his eight years in office were followed by sixteen more from his close associates, James Madison and James Monroe, it is not too much of an exaggeration to say that Jefferson was the nineteenth century's most influential president. The course of American politics had not been set by the war for independence, nor by the ratification of the Constitution. There were many possible futures in the exceptional situation of the new nation. Jefferson articulated one of them, imposing his will upon the federal government and his spirit upon the American electorate.

Before Jefferson could enjoy the fruits of his victory, he had to go through the ordeal of dozens of rounds of balloting in the House of Representatives. He and his running mate, Aaron Burr, had received exactly the same number of votes, leading to the tie for which the Constitution had provided the remedy of a House election, each state having one vote.

Events had clearly already overtaken the Constitution, for its provision for choosing a president suggested the style of an exclusive men's club that arranged to meet quadrennially to perform a solemn duty. After nominations in the states, the House of Representatives, according to the Constitution, would choose the president "from the five highest on the list" forwarded by the state electors. Clearly the Founding Fathers envisioned a decorous process of selecting the president from among a battery of eminent leaders, most of them nominated as favorite sons. They had not anticipated the emergence of an opposition party so disciplined it would produce equal numbers of votes for its number-one and -two candidates. The Twelfth Amendment, ratified in 1804, changed the procedure for electing presidents and vice presidents, but not before Jefferson had to endure two months of suspense while the Federalists, not strong enough to win the election, but with enough votes to block the majority, threatened to rob Jefferson of his victory. On the thirty-sixth House ballot, he was elected president.

In 1801 the State, Treasury, Navy, and War departments composed the executive agencies of the national government. In naming James Madison to head the first and Albert Gallatin to head the second, Jefferson picked nationally known figures long associated with his party. He named Henry Dearborn secretary of war and Levi Lincoln attorney general, a post then without a department. However, the declining appeal of public office became evident when Jefferson was forced to ask four men to serve as secretary of the navy before Robert Smith accepted. The composition of the cabinet was important to Jefferson because he was an attentive administrator who liked to deliberate before acting. He sought advice and took it, even to the point of reading all of his correspondence in search of information and recommendations. With his cabinet members, he maintained an extraordinarily intense and amiable relationship, constantly exchanging memos, canvassing opinions in meetings, and entertaining at weekly White House dinners attended by their wives as well.

Like most newly elected presidents, Jefferson was assailed by would-be officeholders. For him the situation was particularly acute because John Adams had packed the judiciary with last-minute Federalist appointments, the so-called "midnight judges," along with a host of other civil servants. Presiding over the first transfer of contested power under the Constitution, he had a program to implement and immediately sought men who could help him. Claiming the full patronage of the federal executive, he removed a whole cohort of young Federalists from civil and military offices, including John Quincy Adams. Although in his first inaugural address, he had appealed to unity with his statement, "We are all republicans — we are all federalists," in fact he confined his appointments to those who shared his views.

In Congress the Republicans had a strong majority in the Senate and a two-to-one majority in the House. The election of Nathaniel Macon as Speaker of the House guaranteed control for the Republicans of the committees that had come to dominate the proceedings in Congress. When Jefferson persuaded Samuel Harrison Smith to move his paper, the *National Intelligencer,* from Philadelphia to Washington, he had put in place all the institutions that would facilitate his program: a collegial cabinet, a republican civil service, party leadership in Congress, and a newspaper to inform the public of his policies and principles. Only the Supreme Court lay outside his circle of influence, staffed as it was by Federalist appointees, headed by John Marshall.

Adams's last-minute judicial appointments disturbed Jefferson. He ordered Secretary of State Madison not to deliver any more Federalist commissions, including that of William Marbury and three other would-be justices of the peace. Marbury and the others then sought a writ of mandamus to force Madison to give them their commissions. Meanwhile the Republicans in Congress repealed the Judiciary Act of 1801, which permitted the lame-duck Federalist Congress to create a whole new layer of the judiciary with the sixteen new judges that Adams had rushed to pick in the waning days of his administration. Despite Federalist grumbling, the judges accepted the new law, including the elimination of judgeships. Federalist attention turned next to Marbury's suit for the delivery of his commission. Marshall's opinion in *Marbury v. Madison* had two stunning surprises. Marbury and the other expectant justices of the peace, the chief justice said, deserved their commissions, but they could not have them, because of an unconstitutional provision in the Judiciary Act of 1789. Thus, while Marshall denied the Federalists their appointments, he worked out the argument for giving the Supreme Court the much more important power of judicial review of acts of Congress when he deemed unconstitutional key provisions of the earlier Federalist statute. The final round in this highly partisan bickering over the judiciary came when the Republican House impeached the erratic Federalist Associate Justice Samuel Chase, who was then acquitted by the Senate.

Jefferson moved swiftly as president to ensure strict construction of the Constitution, exercise economy in the federal government, move public lands into the private hands of ordinary farmers, and seek freedom of the seas in foreign affairs. In his last days in office, Adams had averted war with France, and a new armistice among European powers gave Jefferson a peaceful world in which to implement his goals. For Jefferson, lowering taxes held the key to a cluster of reforms: a return to republican simplicity in the capital, removal of what he saw as the major source of corruption in government, and tax relief for the vast majority of citizens. By the time of his first annual message he could announce the elimination of all internal taxes along with a decrease of officers in both the civil and military services, moves that made possible substantial reductions in the national debt, which had been the linchpin of Hamilton's fiscal control. Jefferson shrank the size of the federal bureaucracy during his first administration despite considerable growth in both the population and territory of the United States. At the end of his first term, he asked rhetorically, "What farmer, what mechanic, what laborer, ever sees a tax-gatherer of the United States?"

Hamilton had counted on land sales as a source of revenue to help service the national debt, so he had favored payments in cash and tolerated large, speculative purchases through land offices located in the major cities. Jefferson hastened the conveyance of national land to ordinary farmers by increasing the number of land offices and moving them to the frontier. In 1801 Congress set the minimum plot at 320 acres with a price of $2 an acre, and sales increased briskly to half a million acres. Buyers had to pay a quarter of the purchase price within forty days of filing their claim and the remainder in four years. Through this system the government extended close to $750,000 credit annually. After Congress dropped the minimum purchase to 160 acres with an $80 down payment in 1804, upwards of four million acres passed annually into private hands. The west indeed had begun to fulfill its promise of providing an independent citizenry of family-owned farms. In the meantime, Jefferson began the national policy of removal of Native Americans from their ancestral lands to reservations, so convinced was he that the author of progress had not included those Indians blocking the spread of American institutions westward.

More important than direct government credit was the quiet revolution in government fiscal policy that the Jeffersonians effected by allowing Hamilton's bank charter to lapse. Without a central bank to exert some control over private banks, a decentralization of credit and money developed, favorable both to ordinary enterprise and market volatility. In Jefferson's view the absence of government control made the economy more natural and hence more in tune with the fundamental cooperativeness that he believed characterized human relations. In contrast to Hamilton's idea of a perpetual national debt, which could be used to direct the flow of capital, Jefferson retired the debt as fast as he could.

The Constitution had taken away the states' power to issue money, but it soon became apparent that the states would contribute significantly to the country's credit system by chartering their own note-issuing banks. State legislatures, open to voter pressure and eager to give their citizens competitive advantages, fell in line with the demands for more credit and approved new bank charters with amazing swiftness. Stalwarts of the old order decried the leveling tendency in easy money policies, but ordinary people resoundingly approved.

Jefferson had played a key part in the disestablishing of Virginia's Anglican Church in the 1780s. His Bill for Establishing Religious Freedom stands as one of his proudest accomplishments. In his one book, *Notes on the State of Virginia,* he captured this sentiment in a pithy phrase when he wrote that his neighbor did him no injury "to say there are twenty gods or no gods." He maintained with studied indifference to the monotheist views of most of his countrymen, "It neither picks my pocket nor breaks my leg." So it is not surprising that as president, Jefferson chose the occasion of a request for a day of thanksgiving to offer his interpretation of the meaning of the First Amendment's protection of religious freedom by denying the request. He summoned the image of a wall of separation to demark the spheres between church and state.

Not a symbol, a civil servant, or presidential initiative escaped Jefferson's consideration as a tool to use in dismantling the "energetic" government of his predecessors. Naturally averse to personal display, he replaced Federalist formality with a nonchalance in matters of etiquette. He frequently answered the door of the White House himself, and he received guests in his morning robe. Eschewing the elaborate receptions of Washington and Adams he entertained small groups at the White House. Foreign dignitaries were quite amazed at his indifference to protocol, and not a few were outraged by his rule that the person nearest to the dining room entered first. On the Fourth of July, Jefferson received the public in the White House, turning informality into a statement about American political values.

By far the most spectacular opportunity for achievement in Jefferson's first term came in 1802 when he learned that Spain had given the Louisiana Territory to France. Alive to the dangers of having New Orleans fall into the hands of an active European power, Jefferson sent his protégé, James Monroe, to join the minister to France, Robert Livingston, with orders to negotiate a purchase of New Orleans, never imagining that Napoleon would be ready to sell the entire region. An acquisition that doubled the size of the United States, the Louisiana Purchase made tangible Jefferson's dream of an expanding army of peaceful settlers carrying American institutions to the Pacific.

As a strict constructionist, Jefferson feared he was exceeding his powers in buying such an empire, but consummating the deal helped him overcome his scruples. The American government had in fact been entering into land purchase treaties with Native Americans since its inception. Federalists attempted to hoist Jefferson on his own petard of principles, though the sting was taken out of their moral position when some declared a preference for going to war to wrest Louisiana from France. Moving quickly, Jefferson took the lead in establishing the nation's new boundaries and appointed his secretary, Meriwether Lewis, and General William Clark to head up an exploratory trip through the territory. The public was wildly enthusiastic about the purchase and sustained its fervor during the twenty-eight months of the Lewis and Clark expedition.

The singular success of Jefferson's first term in office made it a foregone conclusion that he would be reelected if he ran. Jefferson gave as his reason for not returning to his beloved Monticello the lingering power of the Federalists. Although they never again won the presidency, the Federalists continued to field candidates and win elections. In the event, he and his running mate, George Clinton of New York, carried every state but Connecticut and Delaware.

Unfortunately for Jefferson, the peace he so much cherished was rudely shattered by the resumption of European hostilities. While Americans profited as the neutral carriers for both France and England, the British Orders-in-Council and the Milan decrees made United States merchant ships subject to seizure on the high seas. Jefferson had long favored using commercial discrimination to achieve diplomatic goals, and freedom of the seas was one of his most cherished ones, but he was desperate to keep the country out of war and overconfident of his capacity for economic coercion. At the end of 1807, he declared an embargo, which drastically reduced American trade without having the desired effect of changing the ways of France and Great Britain. Only in his protracted campaign against the Barbary States in North Africa did Jefferson have a diplomatic success in his second term, sending a small American force to stop Tripoli from taking captives and exacting tribute in its patrol of the Mediterranean Sea.

There were to be even more troubles at home. In 1804, Aaron Burr had killed Alexander Hamilton in a duel, an act that led New York State to issue a warrant for the arrest of the vice president of the United States on charges of murder. Although Burr returned to preside over the Senate trial of Justice Chase in the waning days of his term, he became a positive threat to Jefferson's administration when he made overtures to both Spain and England to help him in various schemes to detach parts of the west, including Louisiana, from the Union. Rumors of Burr's appearances in New Orleans and Mobile floated back to the capital with tales of military expeditions and diplomatic treachery. Thoroughly alarmed, Jefferson wanted him arrested and blamed the Federalist judiciary for obstructing justice. In the end Burr was tried for treason and acquitted in the Virginia circuit court, John Marshall presiding.

None of these events improved Jefferson's reputation. The economic coercion he wished to use against France and England had to be applied to American citizens to prevent American ships from slipping through the embargo. Perhaps the only acomplishment of his second term which meant anything to Jefferson was the abolition of the slave trade. Part of a sectional compromise, the Constitution permitted the prohibition of the importation of enslaved men and women twenty years after its ratification, and Jefferson made certain that a bill was ready for a vote when the time came.

Throughout his years as a party leader, Jefferson had been subjected to a level of invective previously unknown in American public life. He hadn't been in office seven months before the *Connecticut Courant* called for his impeachment. His Federalist opponents portrayed him as an atheist and a defiler of all things sacred. They printed doggerel suggesting that he kept a harem of slave women. The more serious, because believable, charge came in 1802 from a disgruntled Republican office-seeker who claimed in the *Richmond Recorder* that Jefferson had sired a number of children with his slave Sally Hemings. These allegations — strongly supported by new DNA evidence — sorely tested Jefferson's commitment to freedom of the press. In part the vicious abuse he suffered was mitigated by the fact that newspapers in the early nineteenth century were conspicuously partisan, and readership divided along party lines. After eight years in office, he was less sanguine about the self-correcting nature of information in a free society, but he never did anything except suggest that Republican governors initiate libel suits against the offending editors. The election of his close friend James Madison to succeed him as president guaranteed that their party would continue to expand the ambit of freedom. Madison, known as the Father of the Constitution, was also the principal author of the Bill of Rights.

Once retired from public office, Jefferson pursued the interests that made him such a remarkable man. He continued to remodel his beloved

Monticello, prepared his extensive library for sale to the Library of Congress, played the violin, mounted and displayed his amazing collection of Indian artifacts, and maintained a staggering international correspondence. Increasingly his thoughts turned to education and the need to provide Southern youth with the level of excellence they sought in Northern colleges. The intensifying sectional strife over the expansion of slavery made Jefferson acutely dissatisfied with the South's dependence upon the North for higher education. From these ideas came his plans for the University of Virginia, which engaged him during the last decade of his life, giving him a chance to display his considerable skill as an architect in the design of the campus buildings that he could see from his hilltop. A widower since 1784, Jefferson devoted a good deal of his time in retirement to his grandchildren, many of whom lived with him at Monticello.

His correspondence reveals that he never stopped thinking about political principles, though old age and near bankruptcy blunted his original optimism a bit. Jefferson placed natural rights at the heart of American identity. He developed a powerful rationale for his liberal reforms with the idea that there was a natural social order open to scientific inquiry. The liberal formula that the government which governs best governs least owed its empirical base to the American experience in the free enterprise economy that the Constitution so well protected. The widening scope of opportunity in the early nineteenth century engaged men's energies and confirmed their expectations about the blessings of liberty. The benign character of the American economy with its owner-run farms and shops and its extraordinarily favorable ratio of natural resources to population seemed to substantiate the conviction that many social goals could be left to "nature" to fulfill, making it credible that natural laws were better than those of legislators.

Successful in perpetuating his pristine republican style of government for a quarter of a century, Jefferson laid the institutional foundation for limited government and the intellectual undergirding for Americans' deep suspicion of political power, even that exercised by the people. Jefferson's emphasis upon the uniformity of nature, however, worked against the acceptance of difference, thus yoking a tradition of discrimination to the nation's highest ideals. These Jeffersonian convictions operated most powerfully in relation to Africans, Indians, and women. Race had long been the defining feature of Anglo-American slavery, and over the years the personal traits slavery produced among African Americans came to be attributed to the genetic inheritance of a particular people, the nurture of enslavement merging imperceptibly with the nature of the enslaved. Instead of recognizing diversity among people or the impact of environment, the Jeffersonian view denied natural rights to those who did not manifest the uniform qualities ascribed to the human race. This confounding of nature and nurture assumed portentous importance when the idea of a uniform human nature was linked to the right to life, liberty, and the pursuit of happiness.

A declared enemy of slavery, Jefferson moved against the South's peculiar institution with caution. He formulated the exclusion that kept slavery out of the Northwest Territory, but he acquiesced in the spread of slavery into the Southwest. As a reviser of Virginia's colonial laws he proposed a scheme for gradual emancipation, but withdrew it from consideration at the last moment. And most importantly his emancipation proposals always contained the provision that freed blacks be carried out of the state and colonized elsewhere. He could tolerate the races living together in slavery; he could not envision a biracial society of freed men and women. When he died with a heavily encumbered estate and two hundred slaves, only five received their freedom; the rest went to the auction block to pay off his debts.

Jefferson relied on nature again in marking out Indians for a subordinate position in white society. For Native Americans, it was not innate inferiority but cultural obstinacy that accounted for their disqualifying differences. He evinced much more sympathy for Native Americans than for Africans. Fascinated by language, he collected Indian vocabularies, but his aversion to those who were not like him overcame even this intellectual curiosity. Writing as governor of Virginia during the Revolution, he proposed the removal or extermination of the Indians because "the same world will scarcely do for them and us" — a not altogether comforting reminder of the ethnic cleansing undertaken by nineteenth-century Americans.

Jefferson said that women had been formed to please men, and few things displeased men more than women talking politics, another appeal to nature. Including women in the Jeffersonian reforms was not inconceivable, for others far less bold than Jefferson recognized their capacity. Treasury Secretary Albert Gallatin offers a case in point. Worried about filling all of the government offices, he suggested naming women to certain posts, a proposal that elicited Jefferson's curt reply: "The appointment of a woman to office is an innovation for which the public is not prepared, nor am I."

The exclusion of women, blacks, and Native Americans from the expansion of free choice and political rights that the Jeffersonians sponsored raises the possibility that perhaps there was something in the Jeffersonian creed which licensed this anomaly. Critics of Jefferson have charged him with hypocrisy, even duplicity, in championing natural rights while continuing to live off slave labor. There is a less morally contentious explanation, if one remembers that the enemies of reform were those who believed that the world had been divided between the few talented enough to govern and to think for the many whose only calling was hard labor. Recorded history, folk wisdom, and religious precepts had long taught that misery and endurance formed the lot of most human beings. In a long public career, Jefferson challenged these morally debilitating propositions. Against them, he pitted the idea of a natural capacity of men to take care of themselves. Only if nature had endowed men with this capacity could the problem of order, which had always been used to justify the power of the few over the many, be solved. During his lifetime, the apparent social characteristics of women, Native Americans, and African Americans seem to undermine this norm of self-sufficiency. In Jefferson's thought, natural difference explained their incapacity and consequently excluded them from participation in a new empire of liberty. A philosophy that taught that nature disclosed the moral ends of human life proved very effective in fighting the venerable arguments for aristocracy, but nature, the mighty liberator in the face of historic privilege, raised its own form of discrimination when it "taught" that some people were naturally inferior or backward or different.

It would be a grave error of historical judgment to underestimate the significance of the Jeffersonians' crusade against the tyrannies of the past. By construing liberty as liberation from historic institutions, it made America the pilot society for the human race. It was not Americans, but all people, who sought freedom from ancient abuses. Jefferson abhorred institutions that exaggerated the frailty of human beings, the better to exercise crippling power over them. His profound antagonism to the debasing effects of tyranny inspired his greatest lines of prose and informed most of his public acts. Indeed, his hostility to authoritarian doctrines, texts, and precedents has turned the historic Jefferson into a recurring principle for those who rise to stand against oppression.

— JOYCE APPLEBY
University of California at Los Angeles

James Madison

1809–1817

b. March 16, 1751
d. June 28, 1836

T HERE IS CONSIDERABLE IRONY in the fact that James Madison was the first American president to ask Congress for a declaration of war. Of all the Founding Fathers, he was the unlikeliest commander in chief. He hardly looked the part. Diminutive in stature and tentative, even timid, in bearing, the scholarly Madison was anything but a commanding presence. As one of his modern biographers has waggishly suggested, he was the kind of man who jumped when a gun was fired. At the peak of his manhood, in his mid-twenties, he had considered himself too physically infirm to fight in America's war for independence. Almost forty years later, as president of the United States, James Madison experienced combat for the first time.

For more than four days in late August 1814 this most cerebral of chief executives virtually lived on horseback, spending more than fifteen hours each day in the saddle. In his capacity as commander in chief, Madison rode among terrified and hopelessly inexperienced American troops, dodged the incendiary Congreve rockets launched by the enemy at the battle of Bladensburg, and gave orders to civilian and military officials alike. It was an onerous and unexpected trial for a sixty-three-year-old man who only the summer before had barely survived a month-long attack of debilitating bilious fever. By all but the most partisan accounts Madison acquitted himself well. He displayed physical courage along with his customary coolheadedness. He was, as always, a steadying influence on others. Only a touching concern for the safety of his wife, also put to flight, occasionally distracted him from his emergency duties as leader of the American republic.

But Madison's heroism as commander in chief — if it can be called that — utterly failed to prevent the enemy from accomplishing its military objectives. On his presidential watch the United States suffered what is doubtless the most humiliating episode in its history. Madison spent the better part of a week on horseback because his home, and the capital city of Washington, were at the mercy of invading British troops who wasted little time registering their opinion of little "Jemmy" Madison (as the British admiral George Cockburn contemptuously referred to him) and his republic. Allegedly retaliating for the behavior of American soldiers the previous year in the Canadian city of York, the British invaders put the torch to the White House and the Capitol, along with most of the federal government's administrative buildings. President Madison may have avoided capture (he had taken a set of dueling pistols with him when he abandoned the White House), but he returned to the smoldering ruins of Washington not triumphant, but homeless. In the haze that lingered after the departing British troops, Madison's fellow citizens had good cause to wonder if there were any government left at all.

For most historians, these dramatic events have marked the nadir of a failed presidency. Conventional wisdom has long been that the United States survived the War of 1812 — its self-proclaimed second war for independence — only through the intervention of several strokes of remarkable good fortune, or put more bluntly, through plain dumb luck. Six months after the disastrous British attack, the Senate hastily ratified a peace treaty that saved America a modicum of face, not to mention its territorial integrity, amid news that American ground troops, after a long string of embarrassing defeats early in the war, had won a remarkable victory. Andrew Jackson's astonishing rout of veteran British troops at New Orleans in January 1815 had no effect on the negotiations that ended the war, but it encouraged Americans to believe that they had won a larger victory. Almost overnight President Madison achieved heroic stature in the eyes of his countrymen as the triumphant leader of a young and muscular republic that had once again tweaked the British lion's tail.

Sober-minded historians have, of course, claimed to know better. Mere survival in the face of chronic disaster was little cause for celebration, and Madison has seemed to most of them sadly miscast in the role of hero. On the contrary, they argue, he was an indecisive bungler, an almost colossally inept president whose naive, amateurish approach to diplomacy and whose lack of leadership over his own cabinet and Congress helped draw the United States into a war for which it was woefully unprepared. And that same timid leadership, tested in the fires of war, brought the fledgling republic to the brink of a military and political catastrophe that Madison's fellow citizens were all too willing, indeed eager, to forget.

This familiar image of Madison's presidency — ubiquitous in the historical literature, including most modern textbooks, often to the point of caricature — can be problematic and potentially misleading. To a remarkable extent, it reflects the enduring influence of Henry Adams's heavily biased treatment of the era in his epic nine-volume history of the United States during the presidencies of Madison and his predecessor Thomas Jefferson, first published in the late nineteenth century. Adams was a brilliant writer and historian. He was also the great-grandson of the second president, John Adams, the Federalist leader whom Jefferson (and Madison) had ousted from office in the election of 1800. Even more important, Henry Adams was a New Englander to the core, culturally undisposed to admire the so-called "Virginia Dynasty" of American presidents; and his harshly critical portrait of Madison's ineptitude, unmistakably reflecting his personal and regional vantage point, set a pattern that succeeding generations of historians were prone to follow. For over a century, with Adams's help, Madison's reputation as president has been held hostage to the national disgrace of Washington in flames.

Yet even Adams appeared to acknowledge, albeit indirectly, a positive side to Madison's presidency. His concluding portrait of the United States after the War of 1812 depicted a robust country that had traveled a remarkably long way toward permanence and respectability from its primitive, tentative condition in 1800, when the voters had unwisely turned his great-grandfather out of office. Even allowing for the great historian's penchant for irony, it is reasonable to assume that he accorded Madison some credit for that progress. Indeed, John Adams, never one to flatter rivals for public esteem, had himself made the point in 1817 on the eve of Madison's retirement from the presidency. Writing to Jefferson, Adams admitted that "notwithstanding a thousand faults and blunders," Madison's administration "has acquired more glory, and established more Union, than all three Predecessors, Washington, Adams and Jefferson put together."

This generous assessment of Madison's leadership was hardly peculiar to John Adams. One modern biographer has even referred to Madison's popularity after the War of 1812, during the remaining two years of his presidency and then during almost twenty years of retirement, in terms of the "adulation" that surrounded him. Of course it would be mistaken, even foolish, to replace the conventionally harsh view of Madison's presidency among modern historians with this partisan view of his contemporaries. Even Madison's modern admirers can easily find (and wince over) the many faults and blunders mentioned by John Adams, and few would deny that Madison lacked the commanding genius of such later wartime presidents as Abraham Lincoln and Franklin Roosevelt. But understanding better the sources and logic of Madison's postwar popularity is one way to avoid the anachronism of imposing on his presidency modern expectations of the office and executive leadership. Adherence to principle can sometimes be mistaken for weakness, and a style of leadership appropriate to an earlier age can seem feeble by the standards of a later one. Above all, understanding Madison's presidency in the context of his larger career, as his contemporaries tended to do, may be the best way of appreciating the vital importance of Madison's personal character to his conduct as president and perhaps the centrality of character to the presidency more generally.

Few presidents seemed better prepared for the office than Madison. His career as a Founding Father was exceptional for both its longevity and its significance, intersecting every major phase of the history of the American Revolution, the adoption of the Constitution, and the early republic. In one sense Madison's personal history cannot be separated from the larger history of the republic that appropriately came to define, in his eyes, the meaning of his own life.

Born a subject of King George II in 1751, James Madison was the first son of the wealthiest planter in Orange County, Virginia, a prosperous but rude provincial outpost of the British Empire noted for its tobacco-growing economy based on African slave labor. Madison's father had both the means and the good judgment to seek for his intellectually gifted son the best education available in such an isolated environment. Five years at a local private school and two more years of tutoring at home prepared the young Madison for his journey in the late 1760s to the College of New Jersey (later Princeton University), where he studied under the exemplary direction of President John Witherspoon, a Scottish minister who introduced his American pupils to the exciting world of the Scottish Enlightenment. Madison studied the writings of David Hume, Adam Ferguson, and Adam Smith, among others, immersing himself in the moral and political philosophy of a group of creative thinkers seeking to develop a science of human nature and social development. After returning home to Orange County in the early 1770s, Madison sorely missed the rich intellectual ambience of his college life in Princeton and suffered through several years of unhappy isolation and indecision concerning his future. Then, abruptly, came the American Revolution — a momentous event that would define not just Madison's subsequent career in politics, but his mature sense of his own identity. For the next sixty years, amid considerable emotional urgency, he zealously pursued the republican dream that gave meaning to his life.

The twentieth-century poet Robert Frost once mused that the best dreamer of the American dream was not a wild-eyed visionary or enthusiast, such as Madison's friend Jefferson, but rather the meticulous, soft-spoken Madison, whose dream, Frost perceptively inferred, was "of a new land to fulfill with people in self-control" — people, Frost might have added, like Madison himself. In character and temperament, the mature Madison who emerged from the Revolution became a model of neoclassical self-command. In time, and especially during his presidency, he was admired as a man whose concern for the public good generally subdued and overrode petty considerations of personal vanity or partisan advantage. Most important, his admirers came to appreciate the extent to which his characteristic modesty, temperance, and perseverance represented a triumph of reason over passion. Along with his contemporaries, Madison knew that unrestrained passion could threaten moral order within individuals as well as in society. He therefore placed the appropriate premium on balance and restraint, not only in his own character and behavior, but in the American republic that he would come to regard as an extension of himself. In so doing he caught perfectly the animating spirit of the enlightened, neoclassical world of the American Revolution. That same rich world of political and cultural experience in turn raised the question and the challenge that became the focus of this provincial Virginian's public life: could people as they actually were in modern commercial society — and not as naive visionaries imagined or wished them to be — govern themselves? More precisely, could popular government succeed in North America?

Madison the republican dreamer is probably best characterized as a conservative optimist and a cautious idealist. He believed that self-government could be made to work in the United States — certainly a radical idea in an age when monarchical or at least aristocratic authority was still widely assumed to be the necessary basis for social order — but only if prudent statesmen took sufficient account of the dangers posed by human nature and also acknowledged the importance of custom and tradition in stabilizing a republican regime. Above all, as Madison wrestled with the problems that arose during the final years of the war for American independence and their aftermath, he came to believe that republicanism and nationality in America were inextricably linked. His own career as a revolutionary perfectly reflected the trajectory of his concern.

He had entered public life as a member of his local county committee, which he represented at the Virginia Convention of 1776 that drafted the state's first republican constitution. After serving in the new state government, he was elected as a delegate to the federal Congress, where he emerged between 1780 and 1783 as a youthful, quiet, but unusually thoughtful and effective defender of his state's interests and the larger needs of a federal union.

By 1787, Madison was once again immersed in the philosophy and history that had been the delight of his college years, only now he was positioned to combine the roles of scholar and statesman in ways that would forever change the course of American history. For the next several years, first as an immensely influential delegate to the Constitutional Convention, then as a tireless and prolific advocate of ratifying the new government, and finally as the key figure in the first Congress's drafting of what has come to be known as the Bill of Rights, Madison earned the renown that he enjoys in scholarly circles today as the principal architect of the United States Constitution. His immense talents as a political thinker were displayed most conspicuously in his contributions to *The Federalist Papers,* which rank among the most original, profound expressions of political theory any American has ever produced and which continue to be scrutinized by scholars for their most subtle nuances and resonances.

Drawing on his experience both as a delegate to the old federal Congress and as a member of Virginia's state assembly during the mid-1780s, the coauthor of *The Federalist* wrestled with the knowledge that passionate, self-interested men and the factions they inevitably formed would always pose a threat to the measure of civilized stability necessary for justice and good government. This danger now seemed emphatically true even in America, where the popular governments created by the new state constitutions, including Virginia's, were trampling on the rights of minorities and ignoring the larger public good by pursuing laws such as those intended to ease the plight of debtors. In Madison's eyes, a revolutionary people were thus failing the rigorous test of virtue (that is, of self-restraint) that he and the other revolutionaries had initially imposed on them. Put simply, he hoped that by extending the sphere of republican government from the state level, where experience now demonstrated that it was not working well, to the federal or national level, the new Constitution might offer a novel remedy for the diseases associated with faction and disorder that were most incident to popular government.

Conventional wisdom suggested that republics should be geographically small in order to remain stable and survive; but Madison, perhaps drawing specifically on the insights of David Hume, now believed to the contrary that an extended republic might offer crucial advantages. By including a greater number and variety of interests, a larger republic would be less vulnerable to the mischief of oppressive majorities, which, with the evidence of the small state republics before him, Madison greatly feared for their tendency to harass minorities and perpetrate injustice. He also hoped that in an extended republic, a filtration of talent, by which the most virtuous, prudent statesmen would be more likely to prevail in the electoral process, might provide the kind of wise, dispassionate leadership that now seemed sorely lacking at the state level and that was necessary for good government. In both respects the Constitution offered a potential vehicle for promoting, at one and the same time, national integration, republican stability, and political justice. The anchor of Madison's enduring republican dream thus became the new and better union of the states forged by the political revolution of the late 1780s.

If the history of the early American republic amply fulfilled Madison's cautious optimism, the road to republican permanence proved far from smooth. The Constitution (and the new union it defined) had to survive several major crises, most notably in the late 1790s and then again during the international havoc wreaked by the climactic phase of the Napoleonic struggle in Europe, when the United States, as a neutral power, found both its foreign commerce and its national honor under siege. In both instances Madison appropriately played an instrumental role in preserving the Constitution: first by organizing and leading the successful opposition to the Federalist faction whose vision and policies lacked broad-based popular support; and then by leading the country, as president, through the extraordinarily trying second war for independence against England. No matter how it is judged, in sum, Madison's presidency is best understood as the culmination of a career rooted in the turbulent matrix of the revolutionary era.

Madison's first term and the first half of his second were dominated by the foreign policy concerns inherited from the administration of his predecessor, Jefferson, which Madison had served for eight years as secretary of state. Madison's and Jefferson's vision of an agrarian republic made up largely of industrious farmers who marketed their burgeoning surpluses abroad depended on a world of free trade, which the restrictive policies of the belligerent European powers, especially Britain, aggressively defied. The failure of Jefferson's great embargo (repealed by Congress just as Madison became president) to establish America's neutral rights through "peaceable coercion," along with the consequent shattering of Republican party unity, meant that Madison faced the challenge of solving the problems that had undone his illustrious predecessor under the most daunting international and political circumstances. That challenge would culminate in the crisis of survival that came with the War of 1812. Only after the successful conclusion of that war would President Madison be able to turn his serious attention, finally, to domestic policy, and even then his legacy would be modest and ultimately ambiguous. During the last two years of his presidency he would give grudging support to the reestablishment of a national bank and to a moderate tariff protecting the native manufactures stimulated by the years of commerical coercion and war; but his veto of an internal improvements bill, on constitutional grounds, shortly before he left office in early 1817 dramatically underscored his concern that a new generation of Americans not lose sight of traditional republican values.

Only by understanding the connection between Madison's ordeal as president and his long-standing republican concerns is it possible to gain a different perspective on the presidential weakness that Henry Adams and other historians have emphasized. Madison often behaved as president in ways that can be attributed not to his mild or timid personality, or to any other alleged deficiencies of character or leadership, but to his principled distrust of executive power and his specific understanding of the office created by the Constitution. For instance, Madison did not seek to impose his views on Congress — or specifically to lead Congress aggressively toward a war that, from the moment he took office, he regarded as the necessary alternative to submission to British control of American commerce — in part because he did not regard such presidential behavior as proper, especially in relation to the vital matter of war and peace. As Madison had himself insisted in the great foreign policy debates of the 1790s, the Constitution gave Congress the sole authority to declare war because history, ancient and modern, amply demonstrated the chronic danger of executive usurpation and corruption when monarchs or their ministers took such matters into their own hands. And once a declaration of war came, President Madison remained just as committed to a posture of presidential restraint vis-à-vis Congress. As always, he communicated his administration's views and preferences on vital policy matters, but he made almost no effort to influence legislative debate or its outcome. This unwillingness to abandon his long-held and deeply rooted views about the limited role of the executive contributed mightily to rampant confusion and inefficiency; but if Madison's restraint is understood in its relevant context, it may also be possible to grasp some of its more salutary consequences.

John Adams's comment that Madison's administration achieved "more Union" than all of his predecessors combined is worth pondering in this connection. Madison's vision of a stable American Union under the Constitution was squarely tied to the agricultural and commercial expansion he saw gravely threatened by British harassment on land and sea during the Napoleonic Wars. But the war fought to preserve that basis for national unity, derisively branded "Mr. Madison's War" by the Federalist opposition concentrated in maritime New England, became itself a supreme test of Union. Partisan opposition to the Madison administration, which at times threatened to take an entire region of the country out of the war, eventually brought extremist New Englanders to the brink of secession. Under these circumstances Madison exercised extraordinary restraint as a wartime leader. Few presidents have been subjected to so much partisan invective and abuse, but he appeared to take little notice. More important, he never hinted at measures abridging freedom of speech or press, even in the face of the rampant obstruction of the country's war effort, which was often difficult to distinguish from treason. Indeed, his administration pursued nothing even remotely akin to the repressive Alien and Sedition Acts of 1798 — the abhorrent badge of the high Federalism that, merely anticipating war, had outlawed virtually any show of opposition to the federal government (and that Madison had so vigorously assailed earlier in his career).

If Madison was determined to prove that a republican government could prosecute war, when necessary, not only according to proper constitutional procedures but without the kind of domestic repression and infringement of civil liberties that would represent a betrayal of its soul, his success in this endeavor became a major cause for postwar celebration, when he was showered with praise for having withstood both a powerful foreign enemy and violent domestic opposition without a single trial for treason or one prosecution for libel. There can be little doubt that this absence of repressive legislation during the war promoted an astonishingly swift healing of what had been near fatal partisan and regional wounds. In this sense Madison deserves more than a little credit for the surge of postwar unity and nationalism during the so-called "era of good feelings" that surrounded his departure from public life. And there is impressive evidence that appreciation of Madison's principled conduct as wartime leader followed him into retirement. Eulogists after his death in 1836 would specifically recall with gratitude "his unswerving protection of civil liberties" at a time when "provocations" had been "greatest for their restraint."

The dividends earned from Madison's principled restraint were truly fundamental. During the War of 1812 some Federalist partisans, opposing increases in the size of the federal army, expressed grave concern about the familiar dangers of executive patronage, executive corruption, and military usurpation — the traditional bane of republican governments. Given Madison's modest bearing and utter lack of military experience, the thought of him becoming a dictator on horseback seems ludicrous; but that he prevented anyone else from assuming that role in the midst of an unprecedented political and military crisis is no joking matter. In an age increasingly dominated by the specter of Napoleon, and in a political culture still very much tied to classical referents, including the danger of "Caesarism," President Madison's executive restraint did more than confirm his principled commitment to, and abiding faith in, the Constitution; it became a symbol for all Americans of their republican identity and a measure of their success in fighting and surviving a second war for independence. Indeed, Americans for at least a generation to come would tie that success specifically to the character and leadership of a president who embodied, much as George Washington had, the values and spirit of their republican revolution.

This unlikely pairing of Madison and Washington, quite common after the War of 1812 but almost inconceivable today, underscores the extent to which two such different figures could represent the central values of a political culture that is now remote from (and even alien to) modern Americans. Specifically, the pairing of the two Virginians reflected a peculiarly eighteenth-century, preromantic, republicanized conception of heroic leadership. So many of the character traits imputed to the decidedly uncharismatic Washington by his adoring contemporaries — modesty, diffidence, self-restraint, patience, steadiness, and perseverance — were precisely the traits Madison displayed in such abundance as chief executive. Indeed, in his stoicism in the face of countless setbacks, not to mention his steady adherence to principle amid alarming confusion and disorder that verged on chaos, Madison, as civilian commander in chief during a second war for independence, offered an example of bravery and self-command reminiscent of the heroic example set on the battlefields of the Revolution. In the eyes of his admiring countrymen, everything that went into Madison's quiet but firm leadership during the war — his unflappable dignity; his refusal to despair; his unflagging confidence in American institutions and the character of the people; and his dogged persistence — somehow overcame his administration's many specific misjudgments and failures. Those close to Madison throughout the crisis emphasized the connections between this brand of leadership and his personal character and temperament. His private secretary in the White House, Edward Coles, later observed that Madison's "persevering and indefatigable efforts to prevent the war" as well as his "manner of carrying it on" were "in perfect keeping with the character of the man, of whom it may be said that no one ever had to a greater extent, firmness, mildness, and self-possession, so happily blended in his character." Coles had watched Madison on a daily basis endure the trials and challenges that would have tested the nerves of the strongest man, and he reported that the president had never been broken. Nothing could "excite or ruffle him"; no matter how vexing the provocation he had remained "collected"; and he had never given way "for one moment to passion or despondency."

The moment of truth had come, of course, during and after the British invasion of Washington, when the very survival of the federal government (and the people's confidence in that government) was hanging in the balance. It was then, especially, that the inspiring example of Madison's strength of character and his abiding faith in his revolutionary dream literally sustained the republic. The president's indefatigable spirit was prominently displayed in his prompt return to the devastated city, indeed his visible presence within hours of the enemy's departure (without knowing, of course, if they would return), and in his immediate order to his cabinet to regroup and convene. One member of that cabinet, James Monroe (who would succeed Madison as chief executive), was convinced that the president's decisive action and reassuring presence saved the republic from what he described as almost unthinkable "degradation." No one at the time could doubt that a different, less happy outcome to the crisis was not only conceivable but likely. And even modern historians critical of Madison's many failures as president have sometimes acknowledged an underlying truth readily grasped by Madison's contemporaries but too easily forgotten by later generations: at the height of the extraordinary challenges posed by the turmoil of the Napoleonic era, when the Constitution was still very much an experiment of uncertain success, the young republic could not have been in better hands than those of the man who would soon be revered, during his retirement, as the Father of the Constitution.

— DREW R. McCOY
Clark University

James Monroe

1817–1825

b. April 28, 1758
d. July 4, 1831

T HE YEARS 1815 to 1825 may be considered the late phase
of the early republic, a time of transition when the memories of the revolutionary and constitutional periods began to
fade into glory, and economic and social change distanced the
country more rapidly from its late-eighteenth-century origins. It was
symbolic that the man who entered the presidency in 1817, James Monroe, was the last hero of the Revolution to occupy that post, the last of
the Virginia Dynasty, and the first to encounter significant — though indirect and short-lived — resistance to the influence of Virginia, and the
slave South generally, in the national government.

Traditionally, historians have portrayed Monroe's two terms as part
of an era when nationalism and sectionalism contended for dominance
of the young nation, with a flush of nationalism after the War of 1812
succumbing in 1819 and 1820 to factional contentions prompted by the
economic depression of 1819 and to sectional acrimony unleashed by
the Missouri Controversy. The age of Monroe's presidency can be more
accurately described, however, as a time when nationalism and sectionalism, along with other forces of social change, remained in uneasy
balance. That tension survived until the Civil War. During Monroe's
presidency, nationalism continued to prevail in some areas, notably
within branches of government least susceptible to popular politics, specifically the Supreme Court and the secretary of state's conduct of foreign policy. Thus, tendencies of both nationalism and sectionalism persisted, with the outcome uncertain.

In 1817 the American economy remained overwhelmingly agricultural, and since the Revolution the lower South had been transformed

into a cotton kingdom, and cotton averaged 39 percent of all U.S. exports from 1815 to 1819. By 1820 westward migration had sent close to two and a half million people to Transappalachia, and the western farming empire reached even across the Mississippi River. A new land law of 1820 reduced the price per acre and the size of a minimum plot so that a settler could buy a sixty-acre farm tract for $100.

At the same time, a transportation revolution and primitive industrialization caused an already vibrant market economy to spread into the nation's vast hinterlands. Since 1790, 213 textile factories had come into being, and in 1822 a group of Boston capitalists would move factory production to a new level by building imposing new structures at Lowell, Massachusetts, along the Merrimack River. By 1820, some four thousand miles of turnpike had been built, and already steamboats and canal barges were supplanting roads as preferred carriers of commercial goods. Indeed, construction of the Erie Canal, that enormous boon to New York's commerce, coincided with the years of Monroe's presidency, and its energizing impact on the Empire State's economy were registered well before its completion in 1825. The New York Stock Exchange had opened in 1817, the year construction on the "big dig" began.

The railroad age, however, lay ahead, and while important mechanical inventions had marked the turnpike and canal era, the communications revolution was still in its early phases. When President Monroe sent Major Stephen H. Long on a scientific expedition up the Missouri River in 1819, for example, the river mud he encountered was a formidable obstacle for the still developing steamboat engine, causing numerous delays.

Commercial institutions surely outpaced political in the new nation, as the size of the national government remained small. The largest executive department, Treasury, grew from 146 employees in 1817 to 165 in 1825; the entire Washington government totaled only 625 four years later. Congress provided the president no funds for a staff, not even for a secretary, and Monroe depended mostly on members of his family to perform clerical duties.

Monroe was only eighteen in 1776 when he fought beside Washington as a junior officer and was badly wounded at Trenton. Four years later he resumed his education as a protégé of Thomas Jefferson, and soon after began a political career that took him through the Virginia legislature, the Confederation Congress, the U.S. Senate, the French ministry (1794 to 1797), the Virginia governorship (1801 to 1802), and several diplomatic missions under Jefferson. He served as secretary of state and acting secretary of war in the years just before his nomination to the presidency in 1816, by the still determinative Republican Caucus — the Republican members of Congress, who functioned as a nominating body. He then received 183 electoral votes to 83 for the Federalist leader Rufus King, though it could not be said that anything like a national campaign or real party contest occurred.

Monroe's advantage over King stemmed from the Federalist party not having been competitive on the national level since 1800. His advantage over potential Republican rivals derived from his Virginia lineage and especially from his Revolutionary War service. Throughout the early national period, the aura of revolutionary patriotism ensured many men long political careers at the state and national levels. This "politics of the revolutionary center" often has flourished in new nations whose citizenry has rewarded military and political leaders of anticolonial wars for national independence. Such men, including some with only a modicum of political ability, often enjoyed long careers in public service. James Monroe, who possessed more than average talent as a "pragmatic politician highly sensitive to contemporary currents," according to a recent biographer, represented the culmination of this tendency in the American presidency.

Monroe's connection to the revolutionary generation influenced not only his ascension to office, but also much of his conduct in the chief executive's office. He was not charismatic, and he did not inspire passionate personal loyalty. But the still vigorous fifty-nine-year-old Monroe recalled the nation's founding in his very appearance, dressed, in the words of Harry Ammon, "in the small clothes of an earlier age — usually a black coat, black knee breeches and black silk hose. On ceremonial occasions he sometimes wore a blue coat and buff knee breeches, an outfit reminiscent of the military uniform of the Revolutionary era."

Although he had shown himself fully capable of partisan activity in the 1790s and later, his early military experience powerfully impressed the young Monroe with the ideal of unity and nonpartisanship. Entering the presidency, he declared that "the Chief Magistrate . . . ought not be the head of a party, but of the nation itself." His exhortations to submerge party differences exceeded Jefferson's "we are all republicans, we are all federalists," so much so that Monroe's conciliatory rhetoric (and nearly unanimous reelection by the electoral college in 1820, with one lone dissident) led historians to dub his White House years the "era of good feelings."

Several causes contributed to this spirit: agricultural prosperity, Federalism's demise, the nationalist disposition abroad following Jackson's victory at New Orleans (ending triumphally an otherwise mostly embarrassing war with England), and Monroe's own antipartisan rhetoric. The president also undertook two extensive tours of the country during his first term, expressly for the purpose of shaking "the foundations of party animosities." His 1817 tour, particularly, which took him to the north and west, drew favorable notice from northern Federalists, some of them hoping for appointive office. But Monroe's rhetoric outran his willingness to allot Federalists any cabinet or even lesser government positions. Yet the warm welcome he received in New England, a region so at odds with Republican national leadership during the War of 1812, did much to nurture the climate of harmony. Monroe also envisioned his 1819 tour — to Georgia, then to the Missouri Territory, then home through Kentucky — as promotion of the western regions of the United States and as further stimulus to nationalist sentiments.

Perhaps the most remarkable aspect of his travels was that in 1817, for example, although Monroe regarded his trips as government business, he paid for them out of his own pocket and was accompanied by only two men (his private secretary and the chief of the Army Corps of Engineers). Equally arresting to the late-twentieth-century observer was his desire to be relatively inconspicuous as a "private citizen," an intention that soon fell victim to fanfares accompanying his progress up the eastern seaboard. The still touchy republicanism of the young nation was exhibited in the slight scolding the fuss provoked from Hezekiah Niles, editor of the *Baltimore Weekly Register:* "We by no means find fault with the marks of respect paid to the chief magistrate on a tour of duty, but think there is more of pomp and parade given to it by the people than the fitness of the thing requires."

Monroe's antipartyism resembled Jefferson's — and Jefferson's generation of Republicans' — in that Monroe expected the Federalists to complete their absorption into one big party, the Republican. He hoped to end partisan divisions also by pursuing national measures enjoying the widest possible agreement. This ambition, however, met with mixed results in domestic policy, in part because of the limits of the legacy of his predecessors and mentors, Jefferson and Madison.

Immediately after the War of 1812, Madison had called for such formerly Federalist measures as improved defenses, a strong navy, a permanent army, a new national bank, protective tariffs, a system of canals and roads for military and commercial purposes, and a national university. With the support of influential southerners such as John Calhoun of South Carolina, Congress in 1816 created a new Bank of the United States and passed a tariff bill designed to protect new industries (over the opposition of most southern and New England members). Because there were as many textile mills in the south as in New England at this point, Calhoun and other southerners assumed that manufacturing would continue to grow in their section, and require protection.

Urging his fellow legislators to "bind the republic together" and "conquer space," Calhoun also pushed through the House a huge internal improvements bill to lay the foundation for a national system. But Madison vetoed the bill just before leaving office because he believed that Congress lacked constitutional power to build roads and canals.

Monroe supported the Bank, tariff, and improved defenses, but followed Madison in rejecting purposeful internal improvements and recommended the need for a constitutional amendment to permit congressional action. The whole subject remained stalled and in something of a muddle during his two terms, with Congress asserting its authority to appropriate funds to the states for, but not itself to build, construction projects, a distinction Monroe seemed to accept. However, in 1822 when Congress passed a bill allowing for the collection of tolls to pay for the maintenance of the national road from Cumberland, Maryland, to Wheeling, Virginia, Monroe promptly vetoed it. Yet he subsequently recommended funding for canals to connect Atlantic Coast rivers with western areas, and approved purchase of stock in a Chesapeake canal company, a general survey bill to project the costs of roads and canals, and a specific $150,000 appropriation for a road from Wheeling to Zanesville, Ohio. Many more appropriations for internal improvements would follow under Adams and Jackson, but Monroe squandered the chance to establish a systematic approach early in the process, during nationalism's honeymoon and before the states' rights backlash set in.

The weakened state of parties and the apparently large Republican majority ironically may have hindered Monroe's presidential leadership. In the previous Congress Republicans outnumbered Federalists 116 to 67; Monroe entered office with the Republican advantage increased to 142 to 40. The absence of a strong opposition may have lessened Monroe's ability to get legislation through Congress. "Everything is scattered," said Justice Joseph Story in 1818. "Republicans . . . are as much divided against themselves as the parties formerly were from each other." Monroe himself complained of his lack of helpers in Congress (such as Jefferson had enjoyed), "except members who occasionally appear as volunteers, and generally without any previous contact with the Executive."

Congress was also becoming more institutionalized, and the man presiding over this process in the House, Speaker Henry Clay of Kentucky, had been offended when Monroe did not appoint him secretary of state.

Clay's resentment turned him into a vigilant and powerful critic of the administration. In 1817 House members were especially susceptible to the crafty Kentuckian's leadership, with 63 percent newly elected, the highest turnover since the first Congress. (Many incumbents had been defeated because the public's wrath had been stirred by passage of a bill in 1816 raising representatives' pay from $6 to $15 per day, which was no more than that of the sergeant at arms or doorkeeper of the House.)

Some observers have characterized the Monroe years as a "sleeping presidency." But this argument can be exaggerated because although Monroe presented few legislative initiatives to Congress, and moreover did not seem as interested as Jefferson had been in leading Congress, in foreign affairs Monroe acted decisively and allowed others — such as General Andrew Jackson and Secretary of State John Quincy Adams — to take aggressive steps in pursuit of the national interest. Monroe's extensive diplomatic experience, and his choice of the nationalist Adams as secretary of state, paid dividends in a climate of better Anglo-American relations.

In 1817 and 1818 the United States and its recent enemy reached important agreements resulting in demilitarizing the Great Lakes and the Canadian border, extending the northern national boundary along the 49th parallel west to the Rocky Mountains, and providing for joint occupation of the Oregon Country. American fishing rights off Newfoundland, granted earlier, were reaffirmed. In 1817, too, Monroe ordered quick naval action against pirates and slave traders operating on islands off the coasts of Georgia and Texas. The president responded aggressively too when in 1820 Adams told Monroe of an expedition of some twenty vessels sailing from New York to the northern section of the Antarctic peninsula on a seal-hunting and whaling voyage. To Adams's concerns regarding a possible confrontation with the British, Monroe requested Adams to ask the secretary of the navy to send a frigate to establish an American station at the site.

The continuing irritant remained Britain's exclusion of American ships from the West Indies trade, minor in commercial significance, large in symbolism as a residue from the two wars with the mother country. Congress, guided entirely by Monroe and Adams, retaliated with its own economic nationalism. The Navigation Acts of 1817 to 1820 imposed a series of increasingly stringent restrictions on West Indian goods and finally banned all British vessels and colonies in the Americas from trade with the United States. Both sides made concessions in 1822 and 1823, but the issue was not resolved until 1830 during Jackson's presidency, and then to the disadvantage of the United States.

Meanwhile, in 1818 and 1819, Andrew Jackson played a controversial role in what became the most spectacular foreign policy triumph of Monroe's first term, the acquisition of Florida from Spain. In the spring of 1818, during a four-month campaign, Jackson embarked on a punitive expedition into Florida against Seminole Indians and fugitive slave settlements. At a minimum, Jackson exceeded his orders, attacking a Spanish post and hanging foreign nationals in the course of sacking the Seminole and fugitive camps. Secretary Adams, however, used the general's demonstration of Spain's impotence in the region to help him negotiate a treaty, known as the Transcontinental, or Adams-Onis, Treaty, by which Spain ceded Florida to the United States in exchange for the United States assuming private American claims against Spain for up to $5 million (the money would not go to Spain, but to American citizens). Spain also agreed to a transcontinental boundary for the Louisiana Purchase running along the Sabine, Red, and Arkansas rivers north and west, thence to the 42d parallel and the Pacific Coast. Spain also abandoned its claims to the Oregon Country (and by a treaty of 1824 Russia accepted 54° 40'N as the southern boundary of its claim). Adams thus sketched in a major part of the geographic blueprint for what later became the "Manifest Destiny" of midcentury expansionists.

Spain's weakening hold on its American empire had made Latin America a source of contention since the start of Monroe's administration. By 1820, all of Spain's former colonies had declared independence except for the islands of Cuba, Puerto Rico, and Santo Domingo. Monroe had begun with a cautious policy of neutrality vis-à-vis Spain and the new republics, especially during the negotiations leading to the Adams-Onis Treaty. Consequently, the administration endured frequent criticism from congressmen, led by Henry Clay, who demanded immediate recognition of the revolutionary governments of Latin America. Not until the spring of 1822, however, did Monroe formally recognize the new republics.

Meanwhile, the crusty secretary of state had been honing the posture and language that would become part of the famous Monroe Doctrine that the president would declare the following year. In negotiations with Spain in 1819, Britain in 1821, and Russia and Britain again in 1823, Adams had bluntly delivered to each in turn words to the effect that they should "leave the rest of this continent to us," and that European powers generally should contemplate no "new . . . colonial establishments." Events in Europe soon led Monroe to broadcast these injunctions to the Atlantic world.

After the Congress of Vienna (1814 to 1815) ended the Napoleonic

Wars, Britain, Prussia, Russia, Austria, and then France set about restoring order and monarchy to the continent. By 1822 French troops, with the Great Powers' blessing, invaded Spain to suppress republican rebels and install the king. Britain balked, however, at the possibility that France would help return to Spain her former American colonies. In August 1823, therefore, the British foreign secretary, George Canning, proposed to Adams a joint declaration warning of renewed European intrusion. While most of Monroe's advisers (including Jefferson and Madison) urged the president to accept the British offer, Adams argued that "it would be more candid, as well as more dignified, to avow our principles explicitly to Russia and France, than to come in as a cockboat in the wake of the British man-of-war."

Adams had in mind the exchange of diplomatic notes, but Monroe incorporated the evolving American position into his annual message to Congress in December and turned it into a general declaration of principles, most notably "that the American continents, by the free and independent condition which they have assumed and maintain, are henceforth not to be considered as subjects for future colonization by any European powers." The United States, he added, intended no interference with remaining European colonies in the Americas or in Europe's own disputes, although Monroe did express his hope that the Greeks would succeed in achieving independence.

Well received by his fellow countrymen and the British foreign office, Monroe's doctrine was ignored by Europe and did not become particularly important until many years later. Its principles had been evolving since Madison's administration, and John Quincy Adams powerfully shaped its content, but James Monroe, who enjoyed mostly good luck as president, eventually would have his name attached to some of the most cherished principles of United States foreign policy.

Although Monroe had little to do with the decisions of the Supreme Court during his tenure, they paralleled the administration's nationalism in foreign affairs. Judicial policy during these years continued the expansion of the Court's power that Chief Justice John Marshall had begun earlier and that had established the supremacy of the federal government over the states. Most notably and controversially, in *McCulloch v. Maryland* (1819), the Court upheld the power of Congress to charter the Bank of the United States and denied the right of any state to tax it. (Ten days earlier the House had rejected a bill to repeal the Bank's charter.) A unanimous Court declared, on the occasion of conflict between a state and federal law, that when the federal government was acting in a constitutional manner *its* laws "form the supreme law of the land." In

1824 (*Gibbons v. Ogden*), similarly, the Marshall court would affirm the federal government's primacy in regulating interstate commerce.

In many ways, 1819 was a pivotal year in American history, if not necessarily for the presidency of James Monroe. Several events that year contributed to awakening a North-South sectionalism that would ebb and flow in the following years. Notably among them were the panic of 1819, the *McCulloch v. Maryland* decision, and the Missouri Controversy. The latter mixed the moral issue of slavery into the rivalries of sections for power, place, and policy preferences.

The panic of 1819 has been called the first major depression in United States history. As measured in sharp declines in export staples, imports, bank notes in circulation, and tobacco, cotton, and rice prices, it was indeed severe. In his annual address to Congress in December, Monroe referred to financial "derangement" and "pecuniary embarrassments," but curiously minimized the importance of the depression and dispensed the conventional wisdom of the day as remedy — what the historian Charles S. Sydnor termed the "time and patience school." The economic downturn seems not to have influenced manuevering for the 1820 presidential nomination, although it seems to have come into factional play *after* Monroe was renominated and nearly unanimously reelected.

The panic did, however, elicit a sharp reaction in the south and west against the Bank of the United States, as many provincial businessmen and political leaders perceived the Bank to have acted selfishly to protect northern and eastern interests during the economic crisis, by contracting credit and calling for repayment of loans. Many editorialists in the sections most aggrieved by the Bank and the panic saw the oppressive policies of the "Money Power" as another facet of the loose constructionism that overrode states' rights in *McCulloch*.

Nothing aroused sectional antagonism along North-South lines, however, more than the passionate controversy that arose in Congress over the attempt to prevent Missouri from coming into the Union with slavery. The conflict in almost all important respects anticipated the slavery-related contests that would increase steadily from the 1830s to 1860; and while the furor over Missouri subsided quickly, it nevertheless elaborated the foundations on which future controversies would ascend. It was, as the historian Glover Moore wrote, "the first full-scale dress rehearsal" for the great sectional division to come.

At bottom, the Virginia dynasty's control of the presidency was very much involved in the genesis of the Missouri debates, as was, more broadly, Southern dominance of national affairs. While slavery-related

THE AMERICAN PRESIDENCY ★ 68

issues had been debated episodically in Congress, the immediate source of the 1819 controversy was the 1816 Republican presidential nomination. In February, the New York legislature adopted a resolution urging the nomination of Daniel Tompkins of New York and pointing out that Virginians had held the presidency for twenty-four of twenty-eight years. Although they fell into line once Monroe had prevailed, on the final caucus ballot all New York congressmen but three had voted for William H. Crawford of Georgia, and one New Yorker after the election published an anonymous pamphlet renewing the call for a "change of dynasty."

In 1818 a Republican backbencher from New York, James Tallmadge, objected to Illinois's admission because its constitution did not explicitly prohibit slavery, and in April a New Hampshire congressman proposed a constitutional amendment to the Missouri enabling bill prohibiting slavery from any state thereafter admitted. In February 1819, Tallmadge attached to the Missouri bill an amendment prohibiting the further introduction of slavery into Missouri and granting freedom at age twenty-five for those slaves born after Missouri's admission. The stormy, engrossing debate that followed, as well as the unprecedented voting along North-South sectional lines, seems to have caught everyone by surprise, including the New York initiators of the move to restrict slavery.

The political power of the Virginians, and Southerners generally, as exercised through the three-fifths clause of the Constitution (granting slave states representation in Congress for three-fifths of their slave populations), seems to have been the lowest common denominator uniting Northerners, including several Federalists who emphasized moral arguments against the spread of slavery. Tallmadge and other Republicans, however, argued that the three-fifths clause, part of the original deal creating the Constitution, should not be extended west of the Mississippi River. Southerners, on their side, generally refused to debate the morality of slavery and stressed rather that Congress did not have the power to dictate terms of admission for new states (conceding in theory that Congress could regulate territories, but not territories becoming states as it were) — probably a superior constitutional position than that of the restrictionists. But the restrictionists had the votes, at least in the House.

In a highly sectional vote, the House approved the Tallmadge amendment 87 to 76, with one Delaware representative voting for both clauses and only one other slave state legislator voting with him for the second clause. The future was forecast in the larger number of free state representatives breaking ranks: ten voted against the first clause, and four others joined them in voting with the South against the second. In the Senate the balance between slave and free states was ostensibly equal, eleven each, but Illinois's delegates and other Northerners held Southern views. There the Tallmadge amendment was easily killed, and Congress adjourned. But with a 105-to-81 majority in the House, the free states seemed capable of blocking Missouri's admission indefinitely.

Before Congress came back in December 1819, various public meetings had been held in the North supporting the restrictionist cause, and the legislatures of New York, Pennsylvania, Delaware, and New Jersey had passed resolutions endorsing Congress's authority to prohibit slavery. But these expressions of restrictionist sentiment in parts of the North failed to prevent the eventually decisive erosion of the Northern majority in the House.

In the renewed congressional debate of early 1820, emotions and rhetoric escalated rapidly, with Southern representatives, especially Virginians, initiating threats of dissolving the Union and even civil war. "By what fatality does it happen," mused John Quincy Adams in his study, "that all the most eloquent orators of the [Congress] are on its slavish side?" Southerners indeed *were* better speakers and parliamentarians, and they possessed the additional advantage of Henry Clay as Speaker of the House. When the district of Maine, once the eastern part of Massachusetts, applied for statehood, the elements of a compromise were in place.

Clay suggested that Maine come in as a free state and Missouri as a slave state, while Senator Jesse B. Thomas of Illinois proposed that slavery be excluded from the rest of the Louisiana Purchase north of 36° 30′N (the line running along Missouri's southern boundary). This idea had earlier been broached when, two days after Tallmadge offered his amendment in 1819, John W. Taylor of New York, an influential Republican, had moved to exclude slavery from the Arkansas Territory. Thomas's proviso would permit slavery there, as well as in Missouri, though it was north of the line. Slavery would be prohibited in the vast expanse west and north of Missouri and popularly thought to be the "Great American Desert."

President Monroe meanwhile acted behind the scenes to preserve slavery in Missouri, urging his son-in-law and political confidant George Hay to disseminate antirestrictionist arguments in the press. He also remained in close contact with Virginia congressmen and with political leaders in Richmond. Monroe believed that the Missouri question was "evidently an effort for power on the part of its authors," and lamented its effect "of alienating all the Eastern members from the Southern." At the same time Monroe did favor some compromise — as long as Missouri entered the Union as a slave state — and on this score risked the wrath of the majority of Virginia's House of Delegates and congressional delegation, both of which threatened to withdraw support from him in 1820 should he "compromise the constitution." Eighteen of the twenty-two Virginia congressmen ultimately voted against the Thomas proviso at the critical point of its passage as an individual part of the compromise package.

In the Senate, Northern senators went on record against the entire compromise by 4 to 18, while Southern senators voted 20 to 2 for it, but, again, even in the Senate, on the key vote on the Thomas proviso Southern senators voted 14 to 8 for it while Northerners went 20 to 2 for the 36° 30' line. The closest and most vital vote came in the House on the original question of Missouri restriction. On March 2, the House rejected what was now the Taylor amendment by the narrow margin of 90 to 87. All seventy-six Southerners voting chose to reject it, while Northerners split 14 against, 87 for, and 4 absentees, for a total of 18 of what Virginia firebrand John Randolph labeled "dough faces" — Northerners who could be counted on to comply with Southern demands.

Two more controversies followed before the matter was settled, the most intense over a clause in the Missouri constitution prohibiting the migration of free blacks or mulattoes into the state, which, because persons of color were citizens in many states, including Tennessee and North Carolina, violated Article IV of the U.S. Constitution extending to citizens of each state the privileges of citizens in all states. Henry Clay then arranged more compromises, getting around Missouri's defiance of Congress (and its disregard for the federal Constitution) by getting its legislature to adopt a circumlocution to the effect that the state's constitution did not mean what it said. Because Northern racial attitudes and practices were hardly benign, restrictionist legislators were more offended by Missouri's contemptuous defiance of *their* sensibilities.

The debates had sectionalized Congress along North-South and slave-free lines as never before. The Second Session of the Sixteenth Congress opened with a contest over the election of House Speaker (Clay had retired) that lasted for twenty-two ballots, before John Taylor was chosen over William Lowndes of South Carolina. Clay later referred to Taylor's election, regretfully, as a sign of "our divisions."

The controversies and compromise virtually created the symbols of North-South sectionalism. The map after 1820 expressed a symbolic cleavage between "us" and "them." The controversy weakened the sense of nationalism just awakening, and in which Southern Republicans had played a conspicuous role. Their capacity to continue in that fashion began to diminish, while westward expansion changed from a nationalizing process to a carrier of symbolic conflict.

Remarkably, James Monroe easily gained reelection, with only one dissenting electoral vote. Still, while there was little immediate carry-over into presidential politics, the damage done to the Southern–middle states alliance, and especially the Virginia–New York axis that initially had put Thomas Jefferson in the White House, shaped the political struggle leading up to the 1824 contest, when the ascendancy of both the Caucus and the Dynasty ended.

After 1821 Monroe's cabinet became a contentious inner arena for presidential aspirants positioning themselves for 1824, particularly Secretary of the Treasury William Crawford of Georgia and Secretary of War Calhoun. Secretary of State Adams, regarded as a likely successor to Monroe by many Northerners, was less active but acutely observant of his rivals' maneuvering. Clay, too, would be a candidate, and outside of Washington the military hero Andrew Jackson aroused popular enthusiasm that caught his insider rivals by surprise.

Initially, Crawford was a formidable candidate, having taken up a Radical states' rights position in opposition to the nationalist policies of Monroe, Adams, and Calhoun, and having exploited his post as head of the government department with the most employees to advance his candidacy. He had also recruited allies in the Land Office and Post Office, as well as Martin Van Buren of New York — the man who would epitomize perhaps more than any other of his generation the party politics looming on the horizon. Crawford and his candidacy, however, were crippled in 1823 by an illness that incapacitated him for months and from which he did not fully recover, though that did not stop his supporters from attacking administration policy, especially any of nationalizing effect, at almost every opportunity.

Crawford's most lasting legacy, however, likely included the Tenure of Office Act, legislation signed apparently unwittingly by Monroe at the eleventh hour in 1820 amid a flurry of other bills. It specified that the term of office for principal officials involved with the collection and payment of money was four years, though they could be reappointed. This harbinger of patronage politics somehow slipped through Congress with no debate and across Monroe's desk without his awareness of its contents. In this still transitional period, neither Monroe nor his successor John Quincy Adams made use of the act but automatically reappointed all officeholders unless they were charged with misconduct. The suicidally apolitical Adams even let political adversaries known to be working against his reelection keep their jobs. Under Jackson and Van Buren, however, the Tenure of Office Act would contribute to the advance of the spoils system, emphasizing again the transitional nature of Monroe's presidency.

The departure of the last revolutionary icon from office did mark the end of an era, as well as the waning of the politics of the revolutionary center and antipartisanship. More importantly in these years, Southern leaders who hitherto had acted as national leaders began to pursue a different, more sectional, and more dangerous course in public life, all the elements of which had been brought into view during the Missouri Controversy. Still, the nationalist trend in foreign policy and the Supreme Court continued after 1821, and for the American people at large, increasingly caught up in material and spiritual "improvement," a sense of optimism and possibilities prevailed.

— RONALD P. FORMISANO
University of Kentucky

John Quincy Adams

1825–1829

b. July 11, 1767
d. February 23, 1848

OHN QUINCY ADAMS always regarded his presidency as a failure, and few historians have disagreed. Although the United States enjoyed peace and prosperity during the four years of his administration, he failed to implement his policy agenda, endured a withering barrage of criticism from Congress and the press, and lost his bid for a second term. Notwithstanding these shortcomings, Adams's presidency casts light on a major turning point in American electoral politics and provides insight into some of the obstacles that stymied presidents who, like Adams, aspired to use the government as an instrument of reform.

The eldest son of President John Adams, John Quincy Adams had been preparing for a life of public service since he had been old enough to read. Born in Braintree, Massachusetts, in 1767, he grew up in a world that had been profoundly shaped by the struggle between the colonies and the Crown. He first learned of the Declaration of Independence from the letters his father wrote his mother from Philadelphia, where the older Adams served, along with Thomas Jefferson, on the committee that prepared the final draft.

Among the decisive influences on the young Adams was his mother, Abigail. Determined that her son would one day become, like her husband, a leader of the new republic, Abigail instilled in him a highly idealized image of his often absent father as a virtuous and selfless republican statesman. It was a lesson John Quincy Adams never forgot. Few Americans can match his extraordinary record of public service. Over a fifty-four-year time span, he held nearly all the principal public offices in the country. The major exception was a seat on the Supreme Court, which, in fact, Adams declined in 1811, after his appointment had been confirmed by the Senate.

Adams's exposure to the world of high politics began in 1778, when, at the age of ten, he accompanied his father on a diplomatic mission to France. Here Adams went to school, met many prominent statesmen — including the aging Benjamin Franklin — and witnessed the peace treaty that ended the American War of Independence. After a brief stint at the University of Leiden, Adams returned to the United States to attend Harvard College, from which he graduated in 1787. Fluent in seven languages and an accomplished classicist, he was broadly conversant with scholarship in almost every field, including mathematics and science. Adams had a special admiration for the writings of Cicero, the Roman orator whose life and speeches he regarded as a model for aspiring statesmen like himself. He even wrote poems. Although Adams conceded that his efforts were uninspired, he doggedly persisted, convinced that writing poetry was among the skills that a cultivated gentleman had an obligation to hone.

Adams entered public service in 1794, at the age of twenty-six, when President George Washington appointed him the American minister to the Netherlands. Three years later, he married Louisa Catherine Johnson, the daughter of the American consul in London. Together they raised three children to adulthood, including Charles Francis Adams, the American minister to Great Britain during the Civil War.

Between 1794 and 1817, Adams filled a string of increasingly prestigious diplomatic posts in Berlin, St. Petersburg, and London. In 1814, he led the special delegation that signed the treaty ending the War of 1812. Three years later, he became secretary of state under President James Monroe. Adams's diplomacy was grounded in a soberly realistic assessment of international relations. It combined a fierce, almost isolationist determination to liberate the United States from neocolonial dependence on European powers with a boldly expansionist vision of the country's future destiny in the Western Hemisphere.

Adams's expansionism fueled his determination to wrest Florida from Spain. Angered by the inability of Spanish authorities to prevent Seminole Indians from mounting hostile raids on American territory, Adams championed Andrew Jackson's controversial invasion of the peninsula, saving the headstrong military leader from a likely court-martial. Adams's bellicosity outraged many at the time (and later, his grandson, the historian Henry Adams). Yet it greatly facilitated the treaty negotiations that ceded Florida to the United States. Among the provisions of this treaty was a clause, which Adams drafted, that specified for the first time that the western boundary of the Louisiana Purchase stretched all the way to the Pacific. With the stroke of a pen, Adams had secured international recognition of the United States as a transcontinental power.

Hostility toward foreign entanglements shaped Adams's diplomatic dealings with Great Britain. In 1823, for example, he pointedly rejected a British offer to issue a joint declaration opposing any European attempt to recolonize the newly independent South American republics. The United States, Adams contended, ought instead unilaterally to proclaim its opposition to future European interference anywhere in North or South America. In December, President Monroe incorporated Adams's position into his annual address; today known as the Monroe Doctrine, it remains a cornerstone of American foreign policy.

Adams's diplomatic achievements were a result of his mastery of the intricate etiquette that undergirded the unabashedly deferential, gentry-based political order of the Founders. By the 1820s, this world was rapidly being supplanted by the ostensibly egalitarian political order that was symbolized by the victory of Andrew Jackson in the election of 1828. Between 1828 and 1840, the new order became solidly entrenched with the emergence of organized competition between the Democrats and the Whigs, the world's first mass political parties.

Adams never entirely reconciled himself to the new political order that the Jacksonians did so much to create. But it would be a mistake to brand him a poor politician. Like most of the statesmen of his generation, Adams was an antiparty politician who deplored partisan electioneering and yearned to be a man of the whole country. His watchword was *union,* an idea that he invested with an almost religious import. "*Union* is to me what *balance* is to you," he explained at one point to his father, "and as without the balance of powers there can be no good government among mankind in any state, so without Union there can be no good government among the people of North America in the state in which God has been pleased to place them."

Nowhere were Adams's political skills more evident than in the months preceding his victory in the presidential election of 1824. At its outset this highly unusual race involved five major candidates, all of whom were at least nominally identified with the Democratic-Republican party of Thomas Jefferson, James Madison, and James Monroe. Each was the natural leader of a geographically based constituency and none enjoyed much support outside of his regional stronghold. Two of Adams's competitors, William H. Crawford and John C. Calhoun, were based in the south; the other two, Andrew Jackson and Henry Clay, in the west. Adams alone represented New England.

None of the candidates secured the requisite majority in the electoral college, throwing the election into the House. Here, in accordance with constitutional mandate, Congress picked the winner from among the three candidates with the largest number of electoral votes. Adams and Jackson were the principal contenders: Clay finished fourth, Calhoun had withdrawn, and Crawford was incapacitated with a paralyzing illness. Voting was by state, with each delegation casting a single ballot. This intricate electoral procedure created enormous opportunities for last-minute maneuvering, which Adams skillfully exploited by discreetly lobbying several state delegations and securing the backing of Clay. Adams's strategy prevailed, the succession crisis was resolved, and he was duly elected on February 9, less than four weeks before the inauguration.

Adams and Clay were logical allies, hailing as they did from different regions and sharing broadly similar views on public policy. Both favored the various government-supported promotional schemes that Clay had dubbed the "American System," and which would soon become a key tenet of the Whig party. But when Adams appointed Clay his secretary of state and Clay accepted, critics charged that they had conspired to deny the presidency to Jackson, the candidate with the largest number of popular votes. It was, plainly, an awkward situation: two-thirds of the American public, Adams dutifully noted in his diary, preferred some candidate other than himself.

Jackson's defeat prompted a small yet determined band of political supporters to harass Adams's administration and champion Jackson as the superior candidate in the next election. Led by the newspaper editors Duff Green, Amos Kendall, and Isaac Hill, these committed Jacksonians became the nucleus of the modern Democratic party. Even had Adams's presidency been otherwise successful, the establishment of this well-organized publicity effort might still have doomed Adams's reelection bid in 1828.

Congressional opposition to Adams's administration was spearheaded by John Randolph of Virginia, who missed few opportunities to excoriate Adams's alleged "corrupt bargain" with Clay. This congressional assault accelerated following the midterm elections, when Adams lost control of both the House and the Senate. Though Congress investigated each of the executive departments, it found few signs of corruption. This is not surprising, since Adams was a gifted administrator and his appointees ran the various executive departments well. Particularly notable was the skillful leadership of Adams's postmaster general, John McLean. By 1828, McLean had not only extended a basic level of postal service to every city, town, and village in the country, but had also established an extensive passenger stagecoach network to serve major commercial centers in the thinly settled south and west.

Ironically, Adams's unblemished administrative record hindered his reelection campaign. Hailing public office as a public trust, he steadfastly opposed the manipulation of government patronage for partisan gain. Adams even refused to dismiss blatantly disloyal subordinates, like McLean, who had plainly gone over to the opposition. Postmasters, customs agents, and other federal public officers calculated, quite sensibly, that if they opposed Adams and Adams won, they might retain their place, while if they supported Adams and Adams lost, they would assuredly be dismissed. In this way, lamented one loyal Adamsite, the president's high-minded appointments policy had inspired the open or covert opposition of three-quarters of the federal officeholders in the country.

Policy setbacks dogged Adams as well. Having long favored legislation to goad Great Britain into opening its markets to American commerce, he now found himself helpless to overturn an unfavorable British ruling that called his bluff and suspended American trade with the British West Indies. Adams's loss of the West Indies trade angered many merchants and farmers and cost him votes in the election of 1828. Just as humiliating, if less controversial outside of Congress, was Adams's inability to persuade the Senate that the United States had an obligation to play a major role in Latin American affairs.

Particularly troubling to present-day sensibilities was Adams's acquiescence in the dispossession of Creeks and Cherokees living within the boundaries of the state of Georgia. Prior to his presidency, Adams had, like most public figures, generally interpreted Indian claims narrowly, fearful that hostile tribes might threaten American sovereignty by allying with Great Britain or Spain. Now that Adams himself was in power, he came to view Indian claims in a more favorable light, convinced that this danger had receded. Adams's solicitude for the Georgia tribes infuriated many southerners, but did little to slow the tribes' forced westward removal across the Mississippi. Only later, after his presidency, would Adams emerge as a consistent champion of Indian rights.

Most frustrating to Adams was his failure to secure congressional support for his domestic agenda. The principal rationale for Adams's program was the desirability of internal improvements, a theme that he considered at length in his remarkable first annual message. Among the improvements that Adams proposed was an ambitious constellation of technological and educational initiatives that ranged from the adoption of the metric system and the launching of expeditions to explore the

Pacific Northwest to the establishment of a national university, a naval academy, and a network of astronomical observatories. The linchpin of Adams's plan was the construction of a comprehensive system of public works to be funded by revenue generated through land sales and the tariff, a venture that bore a distinct resemblance to Henry Clay's American System. Among the projects that Adams supported were a 1,000-mile road between Washington and New Orleans and a 185-mile canal between the Chesapeake Bay and the Ohio River. These public works, Adams predicted, would boost the value of public lands, provide work for the poor, encourage the orderly settlement of the west, and hasten the epochal reorientation of the American economy from the Atlantic seaboard to the transappalachian interior.

Adams defended his domestic agenda as a continuation of a similar policy begun by President Monroe. While partly true, Adams's claim was also somewhat misleading. Unlike Monroe, Adams swept aside as irrelevant all objections to public works spending rooted in constitutional law. Equally innovative was Adams's rationale for government intervention. "Liberty is power," Adams grandly proclaimed, and the spirit of improvement was "abroad upon the earth." Foreign powers might be less blessed with political freedom than the United States, yet they had already made gigantic strides in the "career" of public improvements. So too had the Romans, whose splendid roads and aqueducts remained among the imperishable glories of the ancient republics. How, then, could the American Congress "slumber in indolence" or permit their judgment to be "palsied" by the will of their constituents?

Adams termed his domestic agenda a "perilous experiment" and he was right. Some feared that it would increase the resources available for partisan electioneering. Others rallied behind the familiar banner of states' rights and the strict interpretation of the Constitution. Particularly controversial was Adams's recommendation that Congress ignore objections to those features of his program that provoked the most opposition at the grassroots.

What ordinary Americans thought about John Adams's domestic agenda is hard to gauge. Most well-informed political observers, it is worth remembering, assumed that federal public works spending had broad popular support. Yet Adams's antimajoritarian appeal made an easy target for critics disturbed by his capacious vision of federal power. Lurking behind this critique was the recognition of political insiders like John Randolph that such an expanded administrative apparatus might one day threaten the institution of slavery. If Congress assumed broad powers to build public works, Randolph warned shortly before Adams's election, it could emancipate every slave in the country. Analogous fears undergirded opposition to Adams's domestic agenda, as well as to a controversial and politically damaging tariff bill that he reluctantly signed in 1828. Not until the inauguration of Abraham Lincoln, in 1861, would another president strike such terror into the heart of the South.

Adams always regarded slavery as a great moral evil. So long as he retained presidential aspirations, however, he refrained from identifying himself publicly with the antislavery cause. In his diary, and in candid conversations with public figures like Calhoun, he was decidedly less circumspect. "If the Union must be dissolved" — Adams wrote in his diary in 1820, five years before the start of his presidency — then slavery, the "great and foul stain" upon the country, was the fault line on which it ought to break. And if the Union *were* dissolved, Adams speculated, then it might be reorganized on the "fundamental principle" of emancipation: "The object is vast in its compass, awful in its prospects, sublime and beautiful in its issue. A life devoted to it would be nobly spent or sacrificed."

Adams largely sidestepped the slavery issue during his tenure as president. Determined to be a man of the whole country, he deliberately avoided positions that might antagonize the planters of the South. But it would be a mistake to assume that he ignored the question altogether. Like most public figures, Adams assumed that his domestic agenda would strengthen the power of the central government relative to that of the states, rendering the eventual abolition of slavery far more likely. In addition, and no less importantly, he hoped that his policies would hasten the economic development of the nonslaveholding regions, tip the balance of power within the states from slavery to freedom, and weaken the slaveholding interest as a political force.

It is hard to know whether Adams's policy positions cost him votes in the election of 1828. What is certain is that Adams's feeble campaign effort proved no match for the Jacksonians' well-oiled political machine. Few presidential campaigns in American history have been more scurrilous, mean-spirited, or unfair. Adamsite publicists derided Jackson as a headstrong and unprincipled military chieftain unfit for national office, while the Jacksonian press assailed Adams's administration as corrupt and pledged to restore the republic to the better days of its youth. Adams found the whole process so demoralizing that he refused to attend Jackson's official swearing-in on March 4, 1829, slipping out of Washington just before daylight, just as his father had done twenty-eight years earlier on the eve of Thomas Jefferson's Inauguration in 1801.

Adams returned to public life two years later, in 1831, when he secured a seat in the House of Representatives from the Plymouth district of Massachusetts. Here Adams remained for the rest of his life, serving first as a National Republican and then as a Whig. No longer constrained by national political aspirations, Adams quickly emerged as a leading congressional champion of the antislavery cause.

Adams's long tenure in Congress was the single most important chapter in his long and distinguished public life. For seventeen years he deployed in the House the formidable oratorical stratagems that, decades before, he had taught Harvard undergraduates as a professor of rhetoric. Hailed as "Old Man Eloquent" — and for the first time in his life, a genuinely popular figure — this was Adams's finest hour. Fittingly, Adams died in the Capitol, in 1848, two days after collapsing from a stroke on the floor of the House.

Adams's greatest legislative triumph came four years earlier, in 1844, when, after an eight-year struggle, he persuaded Congress to relax its ban on the free and open discussion of antislavery petitions. No issue more pointedly dramatized the integral relationship that Adams assumed to exist between the perpetuation of slavery and the suppression of constitutionally protected civil rights.

Adams's struggle to uphold the right to petition bolstered his dawning realization that the central government had become subordinated to a malignant and growing "slave power" that was antagonistic to the fundamental principles of the American republic. Emboldened by this insight, Adams abandoned his long-standing support for territorial expansion and publicly castigated as slaveholders' landgrabs the annexation of Texas and the Mexican War. Adams's identification with the antislavery cause intensified in 1841 when he defended before the Supreme Court the cause of fifty-three illegally enslaved Africans who, in a bizarre sequence of events, had commandeered their Spanish captors' ship, the *Amistad*, and landed it in the United States. Adams won the slaves' release, reaffirming, at least temporarily, his wavering faith in American institutions — and, not incidentally, striking yet another blow for American autonomy in international affairs.

Adams's demonization of the slave power led him to reconsider the setbacks that he had suffered during his presidency. The source of all his troubles, Adams came by the mid-1830s to conclude, could be traced to the machinations of a small yet purposeful proslavery cabal. While Adams's analysis was self-serving and a trifle melodramatic, it astutely highlighted the relationship between the rise of the Jacksonians and the demise of Adams's expansive vision of the central government as an agent of change. "When I came to the presidency," Adams explained to a friend in 1837, "the principle of internal improvement was swelling the tide of public prosperity, till the Sable Genius of the South saw the signs of his own inevitable downfall in the unparalleled progress of the general welfare of the North, and fell to cursing the tariff and internal improvement. . . . I fell, and with me fell, I fear never to rise again, certainly never to rise again in my day the system of internal improvement by national means and national energies. The great object of my life therefore as applied to the administration of the government of the United States, has *failed*. The American Union as a moral person in the family of nations, is to live from hand to mouth, to cast away, instead of using the improvement of its own condition, the bounties of Providence."

Adams's reflections on his defeat casts a fresh light on the emergence of political democracy in the United States. In the nineteenth century, American democracy was customarily associated with the negative liberal state that triumphed with the rise of the mass party, a development that Adams always regarded with a suspicion bordering on contempt. In the twentieth century, in contrast, democracy had come to be identified with the positive liberal state, a development that Adams foreshadowed in his prophetic first annual message. Adams always regarded his domestic agenda as broadly democratic in the sense that it was intended to improve the condition of the American people. It is, thus, entirely fitting that he, rather than Jackson, was the first president to term the United States a democracy in a major address.

With the rejection of Adams's domestic agenda, Congress lost one of the few institutional mechanisms that could conceivably have abolished slavery peacefully in the United States, just as Parliament would abolish it in the British West Indies in 1833. Almost overnight, the progressive, developmental state of Adams and Clay had become transmogrified into the reactionary, proslavery state of Jackson and Calhoun. Mindful of this shift, antislavery advocates quickly changed their focus from the promulgation of public policy to the shaping of public opinion. Adams himself came to recognize the merits of this strategy during his post-presidential years in the House. Instead of proposing new legislation, he worked to strengthen the popular identification between antislavery and civil rights. In so doing, he overcame the obstacles that had enfeebled his presidency, and burnished his reputation as one of the greatest statesmen of the age.

— RICHARD R. JOHN
University of Illinois at Chicago

Andrew Jackson

1829–1837

b. March 15, 1767
d. June 8, 1845

NDREW JACKSON, seventh president of the United States, served two terms in the White House, starting in 1829. His administration has traditionally been linked to the dawning of "Jacksonian Democracy," an elusive but attractive concept that implies a celebration of the values of ordinary white men and a demand for majority rule by such voters. The most conspicuous developments of Jackson's administration were probably the president's war on the Bank of the United States, his insistence on the removal of the eastern Indian tribes to reservations west of the Mississippi, his response to the nullification crisis in South Carolina, and his contribution to a permanent system of two-party politics. These events were intimately related, as Jackson built his reputation as a defender of the common man in large part through his attacks on Indians and on the "monster bank," as well as by his defense of a strong presidency and strong Union based on majority rule. He likewise sought to institutionalize this approach to government through the creation of a revitalized Democratic party. In pursuing these objectives, Jackson made a lasting contribution to the structure of U.S. government and the ideology of American democracy.

The major developments of Jackson's presidency unfolded as part of a larger collision between the ideological legacy of America's founding generation and the rapidly expanding economic development of the new nation. The republican principles of the late eighteenth century had stressed an eternal conflict between liberty and power, and called for a balanced and limited government to protect the former against the latter. Eighteenth-century republicans had also believed that the survival of free government depended on the maintenance of popular "virtue," by which they meant a willingness and ability to support the common good above particular, private interests. To possess this sort of virtue, a man must be socially and economically independent, for a dependent person like a tenant or a servant, to say nothing of a child, a pauper, or a woman, would be forced to serve the private interests of those who were more dominant. If political power were widely dispersed among many independent men, and there were few powerful enough to dominate others or too weak to resist their domination, no small group could oppress the others, while the majority would have no incentive to oppress itself. Adverse private interests would not be strong enough to control the state, and no one could advance his own private welfare except through support for the common good. The proper social and political structure could make private interest compatible with public virtue, and liberty might survive. Within this framework, ordinary men of small to medium property and limited education would be expected to choose the wisest and most independent statesmen, probably gentlemen of wealth and standing, to handle the details of governance on behalf of the people as a whole.

The preservation of social and economic arrangements that left most white men free and independent of others was thus an implicit part of the political vision of eighteenth-century republicanism. So too was the expectation that "the common good" would be easily identified by the virtuous and wise and quickly embraced by a consensus among the like-minded populace. In practice, these conditions would be difficult enough to maintain in a society of small, landed farmers with a broad and stable distribution of property and no clashing economic interests. In the rapidly changing economy of the early nineteenth century, protection of these conditions seemed virtually impossible.

Even before the War of 1812 accelerated the process of economic change, the success of the Industrial Revolution in Great Britain had inspired American imitators, and restrictive colonial policies no longer stood in their way. Ambitious entrepreneurs launched successful experiments in factory production, especially of cotton cloth. Inspired by the success of the Erie Canal, begun in 1817 and completed in 1825, a new network of canals reached into the interior and invited farmers to produce more crops for sale to urban and foreign markets. The invention of the steamboat in 1807 and the expansion of the railroad after 1830 created even more opportunities for cheaper transportation and gave a powerful boost to the culture of cotton and use of slave labor. Businessmen eagerly promoted the spread of banks and paper money to finance the new developments, and reformers created public schools and other educational institutions to enable citizens to take full advantage of them. Historians have come to call this period of increased commercial activity the "market revolution."

The prospect of economic development held momentous consequences for newly independent Americans. Abundant new chances for wealth and opportunity offered the possibility of higher standards of living for consumers and producers alike. For some Americans, the stable and deferential villages of the eighteenth century had been stifling traps. Limited economic opportunities kept a few families in security at the top, but a significant number of young people found it harder and harder to improve themselves, or even to replicate the social and economic status of their parents. For these frustrated citizens, the coming of the market revolution and the opening of western lands could offer the chance to cross the mountains for a fresh start, to move to town and learn a remunerative trade, or to cultivate a new crop and ship a profitable surplus to a previously unattainable market. In turn, these new opportunities could upset the frozen social and political hierarchies of the colonial era, giving newly prosperous citizens a chance to claim their own share of social recognition, to win public office perhaps, and to claim the equality with established families that the American Revolution had supposedly promised to everyone.

More ominously, however, the market revolution also presented dangers for Americans, threatening economic bust as well as boom, and fostering the growth of large enterprises that seemed to present far more harm to some citizens than they promised in rewards. A farmer's experiment in market agriculture might fail, leading to the loss of his land and treasured independence. Mass production might destroy the jobs of skilled artisans. In either case, the loss of a farm or workshop could lead to a perceived loss of equality and a social descent into the ranks of permanent wage laborers. The New England farmers' daughters who eagerly took positions in the region's new textile mills may have welcomed the freedom that a regular pay envelope brought them. But men who were brought up to think of independence, property ownership, and manliness as part of an inseparable package were less likely to be satisfied by the new arrangements. A more powerful business cycle, moreover, would create depressions and recessions that could bring bankruptcy and ruin to thousands of families through no fault of their own, while the rise of powerful corporations could overwhelm the political strength of ordinary voters.

In short, the economic developments of the market revolution challenged the political theories of the American Revolution. Beneficiaries

of economic change could argue that development made men more independent and virtuous because nothing made men more dependent
than poverty. Victims of change were likely to retort that development
spread its costs and benefits unequally, making the majority of men dependent upon a few, and thereby undermining the social requirements
of free government. Voting Americans thus experienced the changes of
the market revolution with a mixture of pleasure and anxiety, and they
passed along their ambivalence to their politically elected representatives.

Andrew Jackson, the future president who came to symbolize much of
this conflict, was born in 1767 in the Waxhaws, a frontier region on the
border of North and South Carolina. Jackson himself always believed
that his birthplace was on the South Carolina side of the line, but the
exact spot remains disputed. The future president's parents were
Protestant immigrants from the north of Ireland who settled with relatives in a farming community where life was hard and luxuries were
scarce. Jackson's father died shortly before his wife gave birth to Andrew, the couple's third child. Elizabeth Jackson took refuge with her
sister and brother-in-law, raising her three boys in their home.

As a child, Jackson probably studied in a one-room schoolhouse, followed by a year or two in a local academy. His adult writings reveal that
he learned to express himself with eloquence, but that he never bothered
with the fine points of spelling or grammar. He undoubtedly spent much
time in outdoor sports, riding, hunting, and performing a boy's share of
labor on his uncle's farm. Childhood came to an abrupt end in 1780,
when Jackson turned thirteen and the British army invaded upper
South Carolina in an effort to quell the American Revolution there. The
three Jackson brothers joined the American forces, but Andrew was the
only one to survive the war. In one famous incident, Andrew Jackson
was taken prisoner and ordered to clean his captor's boots. The defiant
youth refused, and the infuriated officer slashed his forehead with a sa-
ber, leaving a lifelong scar.

Mrs. Jackson also died during the conflict, leaving Andrew as the only
member of his immediate family to survive the American Revolution.
After a few desultory years teaching school and experimenting with various other odd jobs, he found an opportunity to "read law" in Salisbury,
North Carolina. Admitted to the bar in 1787, the young lawyer decided
to move west when his friend and fellow student John McNairy won appointment as superior court judge for Davidson County, on the western
side of the Appalachian Mountains. McNairy made his friend the "attorney general," or county prosecutor, and the two soon settled together in

the frontier town of Nashville in the territory that would later become the state of Tennessee.

Once installed in Nashville, Jackson quickly launched a successful law practice and improved his fortune with successful land speculations. He also met Rachel Robards, the estranged wife of one Lewis Robards, and married her when they were informed that Robards had obtained a divorce. The information proved to be incorrect and the embarrassed couple were forced to remarry, an incident that later became the basis for salacious political accusations. To their great disappointment, the Jacksons never had children of their own, but Andrew Jackson served as guardian to many orphaned children of his deceased friends and in-laws. The couple adopted one of Rachel's nephews in 1809 and named him Andrew Jackson, Jr. Rachel Jackson never shared her husband's taste for politics or public life, preferring to stay in Nashville by her own hearth and family, smoking her pipe and reading the Bible.

Jackson had already made a strong mark as a lawyer in the busy and contentious territorial capital. His Tennessee family and business associates had strong political connections, so it was not surprising that local voters chose the promising young attorney to represent them at the state constitutional convention of 1796. A few months later, Tennessee became the sixteenth state to join the Union, and then sent Jackson to the U.S. House of Representatives. The following year, 1797, the legislature elevated him to a vacant seat in the U.S. Senate.

It was a promising start for a thirty-year-old political beginner, but Jackson soon found that he had little taste for the give-and-take of legislative activity. Both in the House and the Senate, he occasionally spoke in defense of the particular interests of Tennesseans, but he did not take up national issues or distinguish himself in debate. Vice President Thomas Jefferson later recalled that Jackson's furious emotions actually choked him with rage when he rose to address the Senate, forcing him to sit down again in silence. Jackson's most memorable action in Congress was to vote against a resolution thanking George Washington for his services as president, on the grounds that the first president, a Federalist, had overstepped the limits of the Constitution. Expressing a strong commitment to states' rights, frugal government, and transappalachian interests, Jackson clearly felt unhappy in Federalist Philadelphia, and after serving one term in the House and six months in the Senate, he gratefully resigned to resume his private life in Nashville.

Financial pressures may have influenced Jackson's decision to leave Congress. A failed land speculation had made him personally responsible for a debt of almost $15,000, and left him with a lasting suspicion of paper credit. Turning his attention to the Hermitage, his plantation outside Nashville, Jackson slowly eased his way back from financial disaster and built up a model estate, with dozens of slaves, ample harvests, and a stable of blooded racehorses. He maintained an appearance of simplicity by living in a modest but substantial house of logs, but finally exchanged these quarters for an imposing two-story mansion in 1819. These years also saw Jackson involved in a series of violent personal outbursts, as he killed one man in a bitter duel, almost fought several others, assisted friends in their own affairs of honor, and even exchanged shots in a free-wheeling barroom gunfight. The puzzling combination of a genial and charming gentleman who was also a bloodthirsty ruffian with a violent temper became a part of Jackson's permanent public image.

While building up the Hermitage, Jackson continued his public services at a reduced level. After leaving Congress, he briefly accepted a position as judge of Tennessee's superior court. In 1802, Jackson won election as major general of the Second Division of the Tennessee state militia. As a prominent westerner, he was naturally sought after by Aaron Burr when the former vice president hatched a hazy plot to erect a personal empire in the Mississippi valley, but Jackson came to suspect Burr's motives and withdrew from the conspiracy before compromising himself. After turning away from this misadventure, Jackson worked hard to improve discipline and preparedness in the militia and longed to lead his men into battle.

Jackson welcomed the beginning of the War of 1812 and saw the hostilities as an opportunity to test his military skill. His first chance came in early 1813, when officials ordered him to take a force of Tennessee militiamen to the defense of New Orleans. After reaching Natchez in midwinter, Jackson received orders to abandon the expedition and disband his army. Fearing that a disorderly mass of soldiers, without weapons, food, or discipline, could never make it back to Tennessee in safety, Jackson decided to violate his orders and keep the men together for an organized retreat. He showed such physical strength and mental resolve on the painful journey back to Tennessee that the men nicknamed their commander "Old Hickory," and the label stuck with Jackson for the remainder of his career.

Later that year, warring Creek Indians killed a large number of white settlers in what is now the state of Alabama and the governor of Tennessee asked Jackson to counterattack. Gathering his forces, Jackson defeated the Creeks in the decisive battle of Horseshoe Bend and later compelled both allied and hostile Creeks to surrender a vast portion of their territory to the United States. After the War of 1812, he used his strength in the area to obtain a series of treaties with the other south-eastern tribes which brought most of their remaining territory into American hands.

President Madison rewarded Jackson's conquests with a commission as major general of the U.S. Army, with command over the southeastern theater. Responding once again to a threatened attack, Jackson took his troops to New Orleans in late 1814 and prepared for a British invasion. Led by General Sir Edward Pakenham, the British advanced on the city and launched their major assault on both sides of the Mississippi on the morning of January 8, 1815. Major British blunders kept the attacking troops pinned down under American guns for much of the morning, and Jackson's forces inflicted heavy casualties. By the end of the day, Pakenham himself was dead and Jackson had gained another triumph, with 2,037 British casualties to 71 Americans.

The battle of New Orleans was the most impressive American victory of the War of 1812, and Jackson was hailed as a national hero who had rescued American pride in a war that had seen humiliating failure. Ironically, the battle had not affected peace negotiations, as commissioners had already signed the Treaty of Ghent before it was fought. The victory sealed American title over the old southwest, however, and brought Jackson's name to national attention for the first time.

A second set of military adventures took Jackson to Spanish Florida in the First Seminole War (1817 to 1818). Jackson's punitive settlement of the Creek War had sent many hostile Creeks into alliance with their Seminole kinfolk, and a series of attacks soon followed on the Georgia-Florida border. President Monroe ordered Jackson to defend the frontier, and Jackson carried out these instructions by invading Florida, which was then a Spanish colony. While there, the impetuous general executed Indians and British agents whom he blamed for the war, and insulted the Spanish authorities. This breach of international law created a diplomatic furor and led to calls for a congressional resolution of censure. That movement failed, but the invasion played a decisive part in Spain's decision to sell Florida to the United States in 1819.

Overshadowing the acquisition of Florida, two major public crises erupted unexpectedly in 1819, leaving deep impressions on American politics and the career of Andrew Jackson. International trade had flourished in the aftermath of the War of 1812, and the high agricultural prices from the era of the Napoleonic Wars continued to stimulate American prosperity. Cotton prices were especially high, and speculation had been rampant in the potential cotton lands of the southwestern frontier. In the spring of 1819, however, cotton prices suddenly broke and many large mercantile and financial houses were dragged down in the so-called "panic of 1819." Led by the Bank of the United States, American banks tightened credit and suspended specie payments, leading to the bankruptcy of thousands of firms and individuals throughout the United States, but especially in the south and west. Unemployment in urban areas was brief though painful, but frontier areas remained distressed for most of the first half of the 1820s. Many settlers lost their homes and farms and developed a lasting hatred of banks and their unreliable paper money.

The second major crisis of 1819 broke out when the territory of Missouri applied for admission to the Union as a slave state. Slavery had once been legal in all the states, but human bondage had never been widespread in most parts of the North, and Northern legislatures had begun to abolish the institution in the generation that followed the American Revolution. The expansion of the cotton frontier gave the institution new vitality, however, and renewed the advantage that slave-owning states enjoyed from the three-fifths clause of the Constitution. Under these circumstances, Representative James Tallmadge of New York responded to Missouri's application with an amendment requiring it to adopt a plan for gradual emancipation as a condition for statehood. Most Northern congressmen came to his support, most Southerners opposed him, and a fierce sectional battle erupted in Washington.

The issue was finally settled in the famous Missouri Compromise of 1820, which admitted Missouri as a slave state, balanced it with the admission of Maine as a free state, and split the remaining portion of the Louisiana Purchase between slavery and freedom. The episode frightened many observers, however, including the aging Thomas Jefferson, who regarded it as "a fire-bell in the night," possibly tolling "the knell of the Union." The Missouri crisis revealed how divisive the issue of slavery could become, especially as it expanded in the western territories, and encouraged politicians to avoid public discussions of this explosive issue. Partly in reaction to the panic and the Missouri crisis, Jackson's presidency would be powerfully influenced by the desire to control the power of the banking industry and by efforts to defuse all controversies arising over the slavery issue.

Despite the tensions aroused by the panic and the Missouri crisis, President James Monroe was virtually unopposed for reelection in 1820, but there was no obvious candidate to succeed him in 1824. Most of the leading contenders were members of his own cabinet: Secretary of the Treasury William H. Crawford, Secretary of State John Quincy Adams, and Secretary of War John C. Calhoun. Speaker of the House Henry Clay also aspired to the honor, and in 1822, Andrew Jackson was nominated by the Tennessee legislature. Crawford was especially popular among those who believed that a congressional caucus should nominate a single candidate to represent the Democratic-Republican party, but because there was no longer any Federalist opposition, other candidates denounced this policy as dictatorial. Secretary Calhoun eventually dropped out of the presidential race, and won the vice presidency with no significant opposition.

Jackson's nomination had grown out of efforts by one faction in Tennessee to get the better of another, but the general himself took it very seriously and the move attracted unexpected attention from local groups around the country who had become disgusted by the unseemly infighting that characterized the campaigns of the other candidates. Jackson corresponded solemnly with inquiring committees, propounding moderate views on the tariff, internal improvements, and related issues of the day, and advocating an old-fashioned set of republican values with dignity and restraint. Jackson's military reputation boosted his campaign, especially among those who believed it demonstrated the strength of his character. His candidacy also appealed to those who believed that "caucus dictation" was outrageous, that official Washington teemed with corruption, and that the popular will ought to have a greater place in the selection of national leaders. Much to the surprise of Washington insiders, the general won the largest number of electoral votes, though no candidate captured a majority.

The election then devolved on the House of Representatives, where Speaker Henry Clay violated his instructions from the Kentucky legislature and threw his support to John Quincy Adams of Massachusetts, the son of the second president, John Adams. Following the election, Adams named Clay secretary of state, provoking charges that Clay had sold the presidency in a "corrupt bargain." Jackson himself was convinced of the charge, and the incident confirmed his preexisting hatred of Clay as an unprincipled opportunist. Almost immediately, Jackson began to think of the coming campaign of 1828 as a battle for personal vindication and for a national assertion of the importance of popular choice in the office of president.

Over the course of the Adams administration, other national politicians hastened to Jackson's standard. In 1825, President John Quincy Adams had outraged congressional opinion by advocating an expensive program of internal improvements and dismissing objections by urging Congress not to be "palsied by the will of our constituents." Senator Martin Van Buren of New York, the leader of the old procaucus faction, urged Crawford supporters to rally in favor of Jackson on the grounds that Adams had embraced the program of his father's Federalist party. It was far better for the country, Van Buren argued, for party lines to be sharply drawn between the equivalents of the old Federalist and Democratic-Republican parties than to risk sectional divisions comparable to the split that had developed over Missouri. Van Buren therefore called for a union of "the planters of the South and the plain republicans of the North" behind Jackson's candidacy. Deferring his own presidential ambitions until after Jackson's anticipated retirement, the now vice president, John C. Calhoun, likewise joined the emerging Jacksonian coalition.

Calling themselves National Republicans to distinguish themselves from the Democratic-Republicans who gravitated to Jackson, Adams's supporters tried in vain to rebut a torrent of demagogic charges that he was profligate, corrupt, aristocratic, Federalist, and even a pimp. Their presses countercharged that Jackson was violent, cruel, reckless, and lawless, and that his marriage to Rachel Jackson was bigamous. These attacks did not succeed, and Jackson triumphed easily in 1828, with overwhelming support from the south and west. At the same time, the voters retained Calhoun in the office of vice president. In the aftermath of the election, Rachel Jackson died suddenly from a probable heart attack, and Jackson took office convinced that his enemies had killed her with their campaign of personal abuse.

The most powerful theme of Jackson's election campaign had been the denunciation of "aristocracy" and "corruption" in the nation's political system. In the broadest sense, Jacksonians had suggested that the republican traditions of the Founders had been corrupted by a failure of civic virtue, as symbolized by the "corrupt bargain" of 1824 to 1825. Jackson also believed that dishonest officials were personally guilty of financial abuses, but this meaning of "corruption" was secondary in his mind to the greater problem of manipulation of the electoral process by an "aristocracy" or privileged and self-perpetuating elite. As a corrective, Jackson favored "equal rights" and strict reliance on the direct preferences of the voters, on the grounds that the people themselves were far more likely to be virtuous than the "aristocrats" whose claims for deference rested on wealth, birth, education, or experience in office. Jackson himself referred to "the first principle of our system — *that the majority is to govern*" in his first annual message, and used it to call for an abolition of the electoral college and the direct election of the president. This proposal attracted little interest, but Jackson had more success with his efforts to revitalize Thomas Jefferson's Democratic-Republican party.

Over the course of his presidency, he consciously sought to build up the Democratic party, as it came to be called, and to use it to marshal the will of the majority and elect reliable Democratic officeholders. That the party could become another form of self-selected elite was an irony that escaped Jackson entirely. The president's hostility to what he regarded as corruption, his emphasis on direct democracy as the cure for it, and his conviction that he himself was the virtuous embodiment of the people's will, were all equal hallmarks of his presidential administration.

Despite these larger preoccupations, the early months of the Jackson administration were taken up with a comic-opera rivalry between the supporters of Van Buren and Calhoun, focused on the personal morals of Mrs. Margaret Eaton, wife of John H. Eaton, secretary of war and ally of Secretary of State Martin Van Buren. Peggy Eaton, as she was known, was the daughter of a Washington innkeeper and the widow of a navy officer who had cut his throat, gossips whispered, in despair over his wife's infidelities with John Eaton. When the Eatons married and Jackson placed John Eaton in the cabinet, other cabinet wives refused to associate with Mrs. Eaton and thereby disrupted the round of official dinners and visits that were an essential part of political life in the nation's capital. The matter took on added political significance because Van Buren's and Calhoun's friends were competing to dominate the administration, and Van Buren sided with the Eatons while many of their critics were allies of Calhoun. Still furious over the recent persecution of his own wife, the president was convinced of Peggy's innocence and decided that Calhoun had cooked up the whole affair in order to embarrass Jackson, force him to discharge Eaton, and dictate his choice of cabinet officers. Distracted by "Eaton malaria," Jackson did not settle into the main business of governing until a cabinet reshuffle that evicted most of the warring parties and sent Van Buren to London as U.S. minister to Britain. In the meantime, Jackson preferred to consult with an informal circle of advisers that critics called his "Kitchen" Cabinet because they seemed to lack distinguished credentials that would qualify them for the official "Parlor" Cabinet.

The administration's first important measure became known to Jacksonians as "rotation in office" and to their enemies as the "spoils system." Jackson had long suspected appointed officeholders of personal corruption and of using the power of their offices to work against his election. He announced a campaign to "reform" the Washington bureaucracy and found several instances of genuine incompetence and embezzlement. His opponents charged, however, that honest and experienced civil servants were swept away along with the guilty, merely to make room for hungry Jacksonian office-seekers. By the standards of the later nineteenth century, Jackson's use of the appointment and removal power to reward his political supporters was rather limited, but it set a precedent for much greater abuses later.

President Jackson also moved quickly to change relations with Native Americans. Dozens of Indian nations still possessed lands east of the Mississippi and in close contact with whites, but the largest were the so-called "Five Civilized Tribes" of the old southwest: Choctaws, Chicka-saws, Cherokees, Creeks, and Seminoles. These peoples had always lived by agriculture as much as hunting and had taken significant steps to adopt white culture and technology. The Cherokees had gone furthest in this regard, replacing their traditional system of governance with a written constitution, adopting a written language, and frequently converting to Christianity. Often descended from white fathers, many Cherokee and other tribal leaders owned slaves and cotton plantations and practiced a lifestyle that was indistinguishable from their white southern neighbors. Under the leadership of Chief John Ross, however, the Cherokees were determined to cede no more land to the whites, but to maintain an independent republic on their remaining traditional territory inside the states of Georgia and Alabama.

White Georgians regarded the Cherokees' aspirations as intolerable. Georgia had ceded its own western territories to the United States in 1802 in exchange for a federal promise to extinguish the aboriginal title to the remaining Indian lands in Georgia as soon as possible, and many years had passed without substantial progress. The majority of Georgia's voters appeared to believe that Indians had no permanent right to lands at all, but were obligated to give way to white people. The abridgement of states' rights and the continuation of nonwhite sovereignty also seemed particularly obnoxious in a state dedicated to the preservation of racial slavery. Georgia's leaders had long catered to popular land hunger by distributing public lands to ordinary families by means of a lottery, and clamoring citizens now demanded that Cherokee land be distributed in the same way. To make matters more urgent, gold had been discovered in the Cherokee domain shortly before Jackson's election, leading to violent clashes between Indians and white intruders.

Georgia had strongly supported Jackson in 1828, confident that this famous Indian fighter would not stand in the way of its campaign against the Cherokees. Soon after the election, Georgia forcefully asserted its own state's rights by abolishing Indian jurisdiction, declaring the end of all tribal governments, placing all Indians exclusively under control of its own laws, and expropriating all lands held under tribal title. It then proceeded to survey these lands in preparation for the lottery. Cherokees who resisted these measures could be assaulted with impunity because Georgia courts would not admit the testimony of Indians against white people.

Jackson shared the Georgians' opinions of Indians' rights, and he determined to use the crisis to force a general solution to what he regarded as the problem of tribal sovereignty inside the states and occupation of desirable lands by eastern Indians. In his first annual message, he proposed legislation that would authorize him to exchange lands beyond the Mississippi for the holdings of all tribes east of that river. Tribes who moved into federal territory would be outside any conflict with the rights of any particular state and could therefore maintain their traditional governments and collective existence. Indians who refused would lose recognition of their tribal identities and be subject to the laws of their states. Individuals would be paid for their improvements to the land, but not for unoccupied hunting grounds. Jackson described the program as voluntary, but he also announced his legal inability to protect Indians from state authorities, clearly implying that states and individual whites would be free to inflict as much violence on resistant Indians as they chose.

In a series of close votes, the Indian Removal Act passed Congress in 1830. To no avail, opponents argued that the program was brutal and unjust, that the rights of the Indians predated the rights of states, and that removal would inevitably inflict serious hardship. After passage of the act, the Choctaws and Chickasaws reluctantly decided to comply and exchanged their lands in Mississippi for territory in the modern state of Oklahoma. Refusing to cooperate, most of the Creeks were eventually rounded up by the army and deported to the same region. The Seminoles fought removal in the Second Seminole War, but most of them were ultimately removed as well. The Cherokees resisted with a series of lawsuits, resulting in two pivotal decisions by the Supreme Court. In *Cherokee Nation v. Georgia,* Chief Justice John Marshall ruled that the Indian tribes lacked the right to sue a sovereign state. In *Worcester v. Georgia,* the Court decided that Georgia's extension of its sovereignty had been unconstitutional, but the decision brought no benefit to the Cherokees because the court lacked power to enforce it. One faction of the Cherokees finally concluded that further resistance was useless and signed a removal treaty in 1836. The tribe's majority rejected this treaty, but the government used the army to force their compliance. Conditions were so harsh on the ensuing "Trail of Tears" to Oklahoma that one-fourth of the Cherokee Nation died en route.

Soon after the adoption of his Indian removal policy in 1830, Jackson signaled the direction of his economic policies by the veto of a bill authorizing a federal purchase of stock in a private Kentucky turnpike called the Maysville Road. Economic change had been rapidly accelerating for decades, fueled in part by developments in transportation technology that supporters called "internal improvements." Supporters of a more highly commercialized economy wanted the federal government to provide capital to public works projects like turnpikes, canals, harbor improvements, and even railroads, but Jackson was reluctant to do so, and he stated his reasons in his veto message. The president acknowledged the benefits of improved transportation, but he worried about the expense and he saw no constitutional authority for federal subsidies. Fundamentally, he feared that a system of federal expenditures for local benefits would corrupt politics by leading congressmen to trade support for one another's favorite projects in a process of logrolling. While internal improvements for military purposes or for projects that were clearly national rather than local in scope might win his approval, Jackson (like Madison and Monroe before him) preferred to see a constitutional amendment to give specific authority for such action. Without directly saying so, the veto message lent support to those who feared that overly rapid commercial change was undermining pure republican principles and promoting forms of economic inequality that were contrary to the spirit of the American Revolution. The Maysville Veto made opposition to federal support of internal improvements a cardinal point in Democratic party doctrine and frustrated backers of a national program of transportation development until after the Civil War.

Jackson's banking policy carried his opposition to federal support for economic development much further. Banks were central to economic change because the government itself issued very little coined money in this period. Instead, banks provided the money and credit to finance the expansion of new ventures. These institutions got started when a group of investors obtained a corporate charter from a state legislature that gave them a legal identity and protected them from the corporation's potential indebtedness. Drawing on the capital the investors paid in, the bank made loans to customers and collected interest from them. Rather than lend out precious specie, bankers issued paper notes that promised to pay a certain amount in coin to the bearer on demand, and these notes began to circulate as money as the borrower spent them. Bankers had a strong incentive to issue as many notes as possible because they collected interest on every loan they made, and it was very rare for all the note holders to descend on the bank at once, all demanding specie for their paper. Banks therefore routinely lent out more in notes than they owned in capital, so times of financial uncertainty could force them to suspend specie payments and refuse to honor the promises on their notes. But Jackson was determined to prevent a repeat of the 1819 panic.

Methodical bankers disliked panics at least as much as Andrew Jackson, but their solution was to form a central authority that would restrain irresponsible banks from issuing too many notes and creating inflationary pressures that could lead to another panic. Under the leadership of its third president, Nicholas Biddle, the Bank of the United States was beginning to serve this function in the 1820s and 1830s. The BUS, as it was called, was a mixed public and private corporation that had been chartered by Congress in 1816 for a period of thirty years. It enjoyed a monopoly of the banking business of the federal government, and its bank notes were the most reliable paper currency in the country. As it collected the notes of other banks in the course of ordinary business, the BUS returned them for redemption in specie and thereby curbed the state banks' propensity to overissue their own notes, and thus exercised a primitive control over the general supply of money in the American economy. Despite its public functions, however, the BUS was fundamentally a private bank with a primary responsibility to provide an ample return to its own stockholders. The conflict between its public and private roles proved damaging to its survival.

Andrew Jackson was far more dubious of the Bank of the United States than its supporters in the business community. Drawing on the ideology of Jeffersonian Republicans, he deeply suspected debt and paper credit, because debt made one man dependent on another, and dependency was inimical to republican government and society. Paper money banking tended to create a financial oligarchy of bankers, note shavers, stockbrokers, and speculators who needed government protection to do business and therefore schemed constantly, he felt, to gain control of the government and its favors. Banking was wrong, in other words, because it would lead to an aristocracy of bankers and their allies and to the virtual overthrow of republican government. In effect, Jackson's attitudes put him in conflict with the central thrust of the market revolution of his own day because it was the ever increasing commercialization of society that had created demands for government-subsidized banks and internal improvements in the first place.

Jackson gave notice of his doubts about the Bank in his first annual message, but avoided a confrontation while Biddle and administration representatives negotiated over possible modifications to the Bank's charter. Led by Henry Clay of Kentucky, however, the president's congressional opponents forced the issue in 1832 by passing a bill to renew the Bank's charter with no more than minor modifications. As the National Republican candidate for president, Clay was running against Jackson in the upcoming election, and he thought the Bank issue would operate to his advantage. Jackson met the recharter bill with a thunderous veto that accused the Bank of unconstitutionality, foreign domination, and violation of the principle of equal rights. His legal and economic arguments were weak, but his egalitarian rhetoric was brilliant and fed the popular impression that this president was truly dedicated to the equality of the common man.

Jackson's Bank veto proved popular with the voters and he handily won reelection over Henry Clay the following fall. Martin Van Buren of New York replaced John C. Calhoun of South Carolina as Jackson's running mate in 1832 and gained a favorable position for his own candidacy as Jackson's successor.

Sometimes to Van Buren's discomfiture, Jackson continued his war against the so-called "monster bank," whose charter did not expire until 1836. Jackson feared that Biddle would use the Bank's resources to sway doubtful congressmen, bribe editors, and influence congressional elections in order to obtain a second recharter bill that could pass Congress by a vetoproof majority. Investigation revealed that the Bank had indeed made questionable loans to influential publicists and officeholders and had secretly used its funds for electoral propaganda. Seizing on this evidence, Jackson decided to remove the government's deposits from the Bank, and thus deprive it of resources for political interference. By the terms of the Bank's charter, however, deposit removal could only be ordered by the secretary of the treasury, not the president. Many believed that the secretary had the responsibility to act independently in such matters, and not simply on the president's orders. When Secretary of the Treasury William Duane refused Jackson's demand to withdraw the deposits, Jackson dismissed him and appointed Attorney General Roger Taney in his place. Taney withdrew the deposits and placed them in selected state institutions that opponents quickly dubbed the "pet banks."

Deposit removal and the dismissal of Duane led to a fierce political uproar. Well-known National Republicans like Clay and Daniel Webster charged Jackson with despotic ambitions, and tepid Jacksonians who had hesitated to break with the president over the Bank veto itself welcomed the opportunity to denounce his apparent defiance of law and the views of Congress. Biddle responded to removal by curtailing the Bank's loans even more than necessary, deliberately inducing a brief, sharp panic, which he hoped would frighten voters into demanding the preservation of the Bank. These measures did lead to a loud outcry against Jackson's policy by urban businessmen, but the president himself was unmoved and loyal Democrats showed greater resentment at Biddle's efforts to manipulate the economy than at Jackson's efforts to curb the Bank. Unable to win a national recharter, the Bank of the United States continued business after 1836 under a state charter from Pennsylvania, but it failed to recover from Jackson's blows and finally collapsed in the early 1840s.

The political consequences of the Bank War were more long lasting. In the spring of 1834, Clay led the Senate to pass resolutions censuring Jackson's conduct, but the president responded with a spirited defense of his prerogatives, including the right to dismiss cabinet officers at will, that has guided the prevailing view of presidential powers ever since. Unsatisfied by this response, Jackson's diverse opponents charged him with monarchical pretensions. Borrowing their name from the Crown's opponents in the American Revolution and from the British party that had traditionally resisted the royal prerogative, Jackson's enemies coalesced under the name "Whigs." The result was a permanent opposition party that far surpassed the feeble organization and narrow voter base of the National Republicans. The Democrats responded by strengthening their own organization, and two-party competition became the norm in every section and at every level of government under the so-called Second American Party System. Two-party competition has been the dominant rule in American politics ever since.

Jackson also strengthened his reputation as a powerful president in a confrontation with South Carolina over the issue of nullification. The growth of northern industry had led to increasing calls for tariff protection to shield the products of American factories and workshops from the competition of more well-established foreign producers. Alarmed by near disaster in the War of 1812 and hoping to build up domestic suppliers of military goods, Congress had responded by enacting a modest protective tariff in 1816. This tariff had won southern support, including a vote from John C. Calhoun, but southerners had complained when Congress raised the tariff, first in 1824 and again in 1828, with minor adjustments in 1832. Protests were loudest in the state of South Carolina, where falling profits had persuaded planters that the tariff destroyed their foreign markets by limiting American imports from Europe, thereby reducing Europe's ability to purchase American cotton at high prices. Surrounded by a large and restive slave population, South Carolina's leaders feared that a weakened slave economy would not survive, and that high tariffs therefore threatened slavery itself. Extremists in the state began to make threats of secession if the protective policy were not repealed, and Calhoun searched for a course that would be militant enough to satisfy his own supporters but flexible enough to preserve the Union and his own presidential ambitions.

Putting together a number of ideas that had been circulating already in antitariff circles, Calhoun formulated the theory of "nullification" in an anonymous essay entitled "The South Carolina Exposition." In it, he argued that the existing tariff was unconstitutional, even though the Constitution clearly gave Congress the power to impose taxes on imports. Calhoun reasoned that the Framers had devised the taxing power for the purpose of raising a government revenue, not of protecting industry. Arguing that Congress had no power to pass a law that violated the *purposes*, as well as the specific wording, of the Constitution, Calhoun concluded that the protective tariff was unconstitutional, but he despaired of obtaining the same interpretation from Congress or the courts. He went on to argue, however, that the Constitution was a compact of equal states and that no entity but the original parties to the compact had the right to judge whether the terms of the compact were being observed. This meant that the states themselves, not the Supreme Court, were the final arbiters of the constitutionality of a given policy. If South Carolina called a special state convention, it could nullify the tariffs of 1828 and 1832 by proclaiming them unconstitutional and unenforceable within its borders. This action would force the other states to call a constitutional convention to clarify Congress's powers. Calhoun hoped that a convention would end tariff protection for the sake of peace, but if it did not, South Carolina could secede from the Union.

Following Jackson's reelection in 1832, South Carolina began to follow Calhoun's scenario. A state convention assembled, nullified the federal tariff, and made it illegal for anyone to attempt to collect it inside the borders of South Carolina. Jackson reacted angrily. As a matter of public policy, he sympathized with the demand for states' rights, but he had endorsed a "judicious tariff" to stimulate production of munitions and to raise revenue to repay the federal debt. More fundamentally, he viewed nullification as a violation of majority rule and thus a serious threat to republican government. On a personal level, he had no intention of being bullied by Calhoun, whom he viewed as an ambitious and disloyal subordinate. He therefore issued a proclamation branding nullification a constitutional absurdity and secession unconstitutional and treasonous. Privately, Jackson even threatened to invade South Carolina to compel its obedience to federal law, but his public posture was more prudent. To avoid any possible clash with state authorities, he arranged to collect the tariff in vessels off Charleston harbor. Early in 1833, Henry Clay joined forces with Calhoun to propose a tariff compromise that gradually reduced protection while leaving federal authority intact, and the crisis passed. Jackson had used the occasion, however, to articulate the doctrine of a permanent and indivisible Union, which later became the basis of Lincoln's actions in the more serious secession crisis that followed a generation later.

In the final years of his second term, President Jackson persisted in his opposition to paper currency and what he called the "credit system." He was particularly concerned about the circulation of small paper notes because these denominations were most often used to pay average workingmen, yet such workers were the least qualified to distinguish sound notes from bogus or devalued ones. Jackson sought to remedy this problem by refusing to accept small notes for the payment of federal obligations, hoping to drive them out of circulation altogether. He also issued his "specie circular" to discourage speculation by requiring payment in coin for purchases of public land.

These efforts were not successful. Generous British credit policies and a dramatic rise in the price of cotton stimulated inflation and speculation far beyond Jackson's power to curb it. Whigs (and some historians) later charged that the pet banks had used the federal deposits to make far more loans than the BUS had made, leading to a dangerous inflation of the money supply and a dizzying boom in land speculation which could not be sustained. Cotton prices broke in the spring of 1837, and a wave of bankruptcies in the leading mercantile houses of New York and New Orleans brought on a financial panic and the suspension of specie payments by the banks. Rejecting the Whigs' analysis of the causes of market collapse, the economic historian Peter Temin has suggested that British lending policies and changes in the international flow of bullion were more responsible for the panic of 1837 than Andrew Jackson's monetary experiments. Whatever the reason, Whigs used the panic to buttress their charges that Jackson's Bank War had ruined the economy, while Democrats used it to prove that banks in general were the real enemy and that the nation must return to an all-metallic currency to avoid the problem of future inflation and deflation. The two parties thus staked out opposite positions in reaction to the credit economy that was transforming business conditions and class structure throughout the country, and maintained their differences for most of the 1840s.

Other noteworthy events of Jackson's second term included the appointment of Roger Taney as chief justice of the Supreme Court, a position he would retain until his death in 1864. A major foreign policy crisis was averted when France finally agreed to pay American spoliation claims dating from the Napoleonic Wars. Jackson closed his administration with a solemn farewell address in which he warned Americans to avoid all forms of subsidy or unequal assistance to corporations and other special interests arising from the market revolution. While he had begun his political career with moderate endorsement of tariffs and internal improvements, he ended it by lumping those policies together with paper money banking and damning them all. Inequality was not only unfair in itself, he said, but it gave rise to struggles for advantage that could undermine the Union and republican government itself. Without directly condemning the economic developments of his own era, he left the impression that government assistance to these changes would destroy the republic. In his prominent vetoes and in this address, Jackson thus reworked the republican doctrines of the revolutionary era to stress equality among all white men and the superior moral authority of democracy (defined as majority rule) over all forms of elitism. Together with his assertion of the power of the presidency itself, this ideological message was in many ways his most powerful historical legacy.

Jackson was likewise harsh on those who endangered the Union, in his opinion, by calling for the abolition of slavery. While he had defended the powers of the federal government in its limited sphere, he insisted that the states must remain free to conduct their own internal affairs, and must not be destabilized by attacks from outside their borders. For the remainder of the antebellum period, even the Northern wing of Jackson's Democratic party was distinctly more proslavery than its Whig opposition. While Jackson had rejected the right of peaceful secession, he had also acknowledged that some unnamed act of tyranny might justify an assertion of the right of revolution by aggrieved states, and this concession would give the Confederate States of America a valid claim on his legacy as well.

With Old Hickory's warm endorsement, Van Buren won the presidential election in 1836. The aging general hastened home to retirement at the Hermitage. From there, he continued to advise Democrats on matters of national policy until his death in 1845 at the age of seventy-eight.

— HARRY L. WATSON
University of North Carolina at Chapel Hill

Jackson's Political Appointments and His "Kitchen Cabinet"

Convinced that a "corrupt bargain" — John Quincy Adams's selection of Henry Clay as secretary of state — had denied him the presidency in 1824, Andrew Jackson entered office determined to make dramatic changes in the process of making government appointments. In his first address to Congress, Jackson affirmed his belief in the rotation of office, declaring that political appointments should be limited to four years. He asserted that long-standing officeholders tended to become indifferent to the needs of the people and were increasingly likely to abuse their authority.

Meanwhile, behind Jackson's maxims, a throng of Democratic party faithful waited anxiously to be rewarded for their political assistance. During his first year in office, Jackson removed more federal officeholders than had all of his predecessors combined and replaced them with his partisans. His opponents, naturally, were shocked, and they denounced the new "spoils system." But in the long run Jackson fell far short of staging a clean sweep of political appointees. During his eight years in office, he replaced only about one-tenth of the total number of federal officers.

Jackson kept a tighter rein over his cabinet. Concerned about the disruptive influence of Vice President Calhoun and disgusted with the treatment of Peggy Eaton, Jackson suspended cabinet meetings altogether for two years. Instead, he relied on an informal group of trusted advisers, including his nephew Andrew J. Donelson, the *Washington Globe* editor Francis P. Blair, Duff Green, Isaac Hill, Amos Kendall, and William B. Lewis, as well as Secretary of State Martin Van Buren and Secretary of War John Eaton. This group, which Jackson's opponents labeled the "Kitchen Cabinet," remained influential until the resignations of Eaton and Van Buren in 1831 prompted Jackson to reorganize and reconstitute the cabinet itself.

Martin Van Buren

1837–1841

b. December 5, 1782
d. July 24, 1862

I FEEL that I belong to a later age," Martin Van Buren declared in his inaugural address, depicting himself as a mere "public servant" treading "in the footsteps of the illustrious men" who had founded the Republic. As Van Buren knew all too well, many in the crowd had come to see one of the last of those illustrious men, the great Andrew Jackson, depart from office. "For once" at an inaugural ceremony, noted Senator Thomas Hart Benton, "the rising was eclipsed by the setting sun."

The first president born after independence, Van Buren lived his entire political life in the shadow of two giants: Jackson and Thomas Jefferson. Meeting Jefferson for the first time in 1824, the rising New York senator was transfixed by "the earnest and impressive manner" of the aged statesman, whom he had long revered as a great political philosopher and the real founder of the Republic. In taking the oath as the eighth president, Van Buren defined his role as one of preservation: "sacredly to uphold those political institutions" created by the Founders and especially to safeguard the hallowed Jeffersonian principles of a limited national government and the liberty and sovereignty of "the people and the States."

If Van Buren's ideology was avowedly traditional, his political accomplishments were not. Known widely as the "Little Magician" and the "Red Fox of Kinderhook," the new president had made his career as a political wizard and innovator — an ambitious, intelligent, and disciplined man who invented the modern American political party and rose to power on its back. Beginning as a state senator and leader of the "Bucktail" Republicans in New York in the 1810s, he won election as a

United States senator (1821–1828) and governor (1829). Then, as the prime architect of the national political coalition that swept Jackson into office, the New Yorker assumed high national positions, serving as secretary of state (1829–1831), minister-designate to Great Britain (1831–1832), and vice president (1833–1837).

It was a remarkable ascent for a diminutive provincial lad whose father was an ordinary farmer turned tavern keeper. Other, arguably superior, men of his generation — Henry Clay, Daniel Webster, John C. Calhoun — sought the presidency in vain. But "Little Van" had a magic touch, eventually using his position as Andrew Jackson's trusted adviser and protégé to rise to the highest political office in the land.

Then, suddenly and unexpectedly, the great prize threatened to dissolve in his grasp. A mere two weeks after Van Buren's Inauguration in March 1837, a financial panic spread across the land, closing hundreds of banks, forcing thousands of businesses into bankruptcy, and throwing tens of thousands of Americans out of work. Simultaneously war with Mexico threatened, over long-standing financial claims and the status of the renegade province of Texas. Further complicating that explosive situation, abolitionists were flooding Congress and the country with antislavery pamphlets and petitions, raising sectional tensions and prompting widespread rioting. And within a few months an anti-British rebellion in Canada and a volatile boundary dispute in Maine raised the prospect of a new war with Great Britain. None of Van Buren's predecessors had faced so many crises in such a short period of time. His would be a crisis-filled presidency that would test the mettle of the man.

Martin Van Buren was the first self-made president. Born in 1782 in Kinderhook, Columbia County, on the east bank of the Hudson River, he grew up among white tenant farmers and smallholders who were dominated by a powerful landed gentry led by the Livingston and Van Ness clans. Shamed as a young man by the refusal of Peter Van Ness to converse with him in public, Van Buren developed a complex relationship with members of that powerful family of notables. He relied on them to get a start in the world, mastering the law as a clerk in the New York City office of William P. Van Ness, but then he repudiated their tutelage and went his own way. Thirty years later, Little Van declared his equality with his onetime patrons, buying Peter Van Ness's mansion and making it his country estate.

This quest for respectability — nay, gentility — was a defining feature of Van Buren's life. While a young senator in Washington, he chose to share quarters with Federalists from distinguished families: Rufus

King, Harrison Gray Otis, and Stephen Van Rensselaer, the greatest landowner in New York. To bolster his standing among such men, Van Buren dressed meticulously and entertained in a lavish fashion. Later, as secretary of state, he threw "aristocratic" parties, a critic complained, and showed scant "taste for republican habits." As minister-designate in Great Britain, Van Buren was most impressed by the British politician he most resembled: Sir Robert Peel, "the son of a cotton-spinner." Reflecting his ambivalent feelings toward the New York gentry, the American minister praised Peel for rising to power as "the pet of the landed aristocracy" and then defying that class by supporting Catholic emancipation and free trade, measures that safeguarded the "welfare of the masses." However, like Peel, Van Buren was never a true tribune of the people; rather, he was an ambitious bourgeois who measured his success in terms of a once dominant and still powerful aristocratic ethos.

Van Buren's climb up the social ladder was the product of decades of political maneuvering and innovation that changed the character of American political life. Beginning in 1817 he created the first statewide political machine, the Bucktails (later known as the Albany Regency). A decade later Van Buren became the architect of the first nationwide political party, the Jacksonian Democrats. In each case the challenge was the same: the overthrow of regimes dominated by men of high status, first De Witt Clinton in New York and then John Quincy Adams in Washington. To subvert these notables, the Little Magician had to conjure up a new political world.

The task was not easy. In the early nineteenth century, American politics was dominated by a phalanx of wealthy merchants, landlords, and slave-owning planters. These notables managed elections by building up a local "interest": lending money to small farmers, patronizing storekeepers and artisans, and treating their workers or tenants to rum at election time. Determined not to become the dependent "tool" of one of these notables, Van Buren set about creating a new political order. First, he attacked traditional republican ideology, which disparaged political parties, and celebrated the role of parties in checking both the government's "disposition to abuse power" and "the passions, the ambition, and the usurpations" of reckless political leaders. Then, to provide himself with the economic means to compete with the notables, Van Buren built his Bucktail faction into a loyal and disciplined party that could capture and control patronage. He recruited ambitious lawyers and journalists from middling backgrounds, men who were part of a new breed of bourgeois Americans pursuing careers "open to talent" (rather than those dependent on aristocratic patronage).

Styling his followers "the plain Republicans of the north," Little Van confronted De Witt Clinton, scion of one of New York's great political families and Federalist candidate for president in 1812. Tall, handsome, self-assured, and college-educated, Clinton was everything that Van Buren was not. And he knew it, condescendingly referring to "the Van Burens and other would-be great men of the day." Moreover, Clinton held different principles. Like John Quincy Adams and Henry Clay, he was a neomercantilist, advocating the active involvement of the government in economic development and winning four terms as governor of New York as a result of his unstinting promotion of the Erie Canal.

To Van Buren, Clinton's status and ambition made him a dangerous man. Using the popular rhetoric of internal improvements, the governor was extracting taxes from the people and using them to enrich wealthy canal contractors and Federalist land speculators. So in 1820 Van Buren challenged Clinton, purchasing a major interest in the *Albany Argus* and using the newspaper to mobilize support for Bucktail candidates. Winning power in the New York assembly, the Bucktails seized control of the Council of Appointment and promptly ousted all the top officers of the state — treasurer, comptroller, attorney general, the chief militia officers, and most county sheriffs. Subsequently the Bucktail council placed its friends in six thousand state offices (carrying salaries and fees worth more than $1 million a year) and elected Van Buren to the United States Senate.

To justify this naked use of patronage, Van Buren redefined republican government as party government. Going beyond James Madison, who had suggested in *The Federalist, Number 10* that differing economic interests would inevitably divide Americans into competing factions, Van Buren argued that coherent "parties would always exist" and celebrated their power, particularly their control over "the selection of candidates for public places." This "spoils system" (as it came to be called) was fair, Van Buren suggested, for it "would operate sometimes in favour of one party, and sometimes of another." And it was thoroughly republican: "That the majority should govern was a fundamental maxim in all free governments."

Even as Van Buren created the modern political party by financing it with government offices and legislative favors, he embraced Adam Smith's philosophy of classical liberalism, opposing public assistance to most economic interests and social groups. His first, and most lasting, bête noire was the nascent banking industry. Fearing that banks would bribe legislators and issue excessive amounts of notes, Van Buren opposed the chartering of new banks in New York. As governor in 1829, he won legislative approval for a safety-fund system imposing strict controls on bank investments and note emissions and requiring banks to create a fund to reimburse the note holders of failed banks.

By this time, Van Buren had extended his small government philosophy to the national level, where he strongly opposed the American system of federally financed internal improvements advocated by Henry Clay and President John Quincy Adams. Declaring his allegiance on "Constitutional questions . . . with the doctrines of the Jefferson School," he voted against federal subsidies for roads and canals and proposed constitutional amendments to limit them.

For political reasons Van Buren initially did not extend this philosophy of limited government to the tariff. Indeed, during the election of 1828 he used the tariff to garner support for Andrew Jackson — masterminding a plan that helped to elect Old Hickory by protecting industries and farmers in New England, New York, Pennsylvania, Kentucky, and Ohio. The *Argus* praised the result as "a national tariff," but most southern leaders condemned the new legislation as a "Tariff of Abominations" that subsidized the rest of the nation at their expense. Following South Carolina's revolt against high tariffs in the Nullification Crisis of 1832, Van Buren began to advocate lower tariffs and free trade in order to preserve southern support for the Democratic party. His political future depended on the party, as did the fate of the Union. Without strong national political organizations, he wrote, there would be nothing to moderate the "prejudices between [the] free and slaveholding states."

Thus, Van Buren entered the presidency not only as the heir to Jackson's policies, Jefferson's ideology of limited government, and Smith's principles of political economy, but also as an accomplished politician with a statesmanlike vision of the dangers facing the nation. This complex heritage would shape the new president's response to the multiple challenges of 1837.

When he entered the White House at the age of fifty-five, Van Buren had lived as a widower for two decades, raising four young sons with the help of relatives and friends. Although popular with women, he seems never to have contemplated remarriage, taking his pleasures instead in good conversation, food, and wine. "Mr. Van Buren is growing inordinately fat," noted John Quincy Adams, in part because of his lavish dinners entertaining political friends and foes. Even as Little Van lost his trim shape and sandy red hair, he kept his sharp intellect — as "penetrating as a mercurial bath," lamented a disappointed office-seeker. However, the crises of 1837 shook his previously imperturbable composure.

The panic of 1837, the most serious crisis, had its origins in a phenomenal economic boom that began around 1830. Huge amounts of British capital flowed into the United States to underwrite state canal projects; by 1836 the aggregate debt owed to foreigners had climbed to $220 million, a threefold increase. Simultaneously American merchants imported ever increasing amounts of British goods, with the annual trade deficit reaching $63 million in 1836 as compared to a surplus of $8 million in 1830. Moreover, the number of state banks more than doubled (from 330 to 778) and, because of changing trade patterns, an unprecedented amount of silver from Mexico — some $35 million — ended up in their vaults. More banks and more specie prompted a striking increase in outstanding bank loans, which rose from $200 million to $525 million, and bank notes, which jumped from $61 million to $149 million.

By 1836 the nation was awash with money and caught up in a spending boom that produced a currency-driven hyperinflation. Many basic commodities tripled in price and, following Jackson's destruction of the Second Bank of the United States, there was no way of stemming the tide. In New York City, the worried banker Philip Hone scribbled in his diary, "a hungry mob charged out of the City Hall Park and raided a warehouse for flour." By the time of Van Buren's inauguration, one of his friends remarked, "The state of the economy" was "damnable," the nation superficially "rich and prosperous and yet on the eve of a general bankruptcy." The long-feared collapse began on March 17, 1837, with the failure of a major New York brokerage house; within a month, one hundred more financial houses had shut their doors.

Events in Britain precipitated the debacle. Late in 1836 the Bank of England raised interest rates in order to encourage investors to pump funds into the faltering British economy. This policy deprived Americans of new supplies of specie and credit. Concurrently, struggling British textile mills reduced their purchases of raw cotton from the south, causing its price to collapse from 20 cents a pound to 10 cents or less.

As American planters, merchants, and canal corporations withdrew specie from banks to pay their foreign loans and commercial debts, they set off a general financial crisis. On May 8, the Dry Dock Bank of New York City closed its doors, and panicked depositors withdrew more than $2 million in coin from other city banks, forcing them to suspend all payments in specie. Within two weeks every bank in the United States had followed suit, shocking high-flying entrepreneurs and ordinary citizens. "It would be difficult . . . to render intelligible in Europe, the stunning effect which this sudden overthrow of the commercial credit and honor of the nation has caused," the British diplomat Henry Fox wrote to the prime minister, Lord Palmerston, "The conquest of this land by a foreign power could hardly have produced a more general sense of humiliation and grief."

As delegations of bankers descended on the White House, Van Buren resorted to a concoction of water, soot, and powdered charcoal to soothe his upset stomach. The president knew that many Americans blamed the panic on Jackson's policies, especially his withdrawal of the national government's funds from the Second Bank of the United States in 1833 and his specie circular of July 1836. The circular required settlers and speculators to pay for western lands in gold or silver coins, which Jackson hoped would cut the money supply and moderate the runaway inflation. But the circular also drew specie away from banks in New York City, where bullion was now desperately needed to satisfy foreign obligations. Nonetheless, Van Buren decided to retain the specie requirement in order to protect the state-chartered deposit banks that now held the government's funds. If the deposit banks were to fail, Van Buren warned, the administration might be unable to resist "the advocates of a [new] United States Bank." He therefore refused demands for emergency legislation and instead called a special session of Congress for September 1837.

The president immediately turned his attention to another crisis. Just before leaving office, Jackson had disregarded Van Buren's wishes and extended diplomatic recognition to the rebellious Mexican province of Texas. By suggesting the prospect of quick annexation, Jackson's action raised the danger of war with Mexico and heightened sectional tensions. New England abolitionists charged that there was a "slaveholding conspiracy" to acquire Texas, and Daniel Webster eloquently denounced annexation.

Boldly reversing Jackson's policies, Van Buren sought peace abroad and harmony at home. He proposed a diplomatic solution to a long-standing financial dispute between American citizens and the Mexican government, rejecting Jackson's threat to settle it by force. With equal firmness, in August 1837 the president rebuffed Texas's formal request to join the United States. A longtime advocate of "mutual forbearance and reciprocal concession" (as he put it during the Nullification Crisis), Van Buren gave a higher priority to sectional harmony than to territorial expansion. As a Texas diplomat reported to his government, the president and many other American politicians feared that annexation would spark a "desperate death-struggle . . . between the North and the South; a struggle involving the probability of a dissolution of the Union."

Having safeguarded the Union by averting a conflict with Mexico, Van Buren presented the special session of Congress with a dramatic financial proposal: the national government should withdraw its funds from the state-chartered banks (just as Jackson had pulled them out of the nationally chartered Second Bank) and keep them in the United States Treasury, thus making the government completely "independent" of the banking industry. Long a critic of unregulated banking, Van Buren charged that the "redundancy of credit and . . . reckless speculation" (and not Jackson's policies) had caused the panic. By cutting the amount of specie held by state banks, he suggested, an Independent Treasury would deter the excessive issue of bank notes and loans, thus preventing future panics.

Despite intricate wheeling and dealing, the Little Magician could not win congressional approval for the Independent Treasury. Aided by his longtime rival John C. Calhoun, Van Buren wriggled the proposal through the Senate by a vote of 25 to 23. But because Calhoun had added a "hard money" amendment requiring the Treasury within four years to accept *only* specie in payment of tariffs and other taxes, the bill lost the support of "soft money" Democrats, who demanded a more flexible monetary system that included the specie-backed notes of state banks. Other Democrats worried that the Independent Treasury would enhance the power of the national government and increase federal patronage, thus undermining sacred Jeffersonian principles. With 23 Democrats defecting or abstaining, the House tabled the bill by a vote of 120 to 107 and, when the measure came up again in early 1838, voted it down.

Ordinary Americans were even less supportive of Van Buren's response to the panic. "The less government interferes with private pursuits the better for the general prosperity," the president had told the special session of Congress, explicitly rejecting financial remedies involving "the aid of legislative grants or regulations by law." Condemned by Webster for his "refusal to prescribe for the sickness and distress of society," Van Buren found his Smithian economic principles repudiated in his home state, where in 1837 the Whigs swept to an overwhelming victory in the assembly elections and in the following year ousted the Democratic governor.

The Red Fox was no longer the master of New York, where yet another crisis threatened his young presidency. In November 1837 British subjects in Lower Canada and Quebec rose in rebellion, protesting their lack of self-government. Sympathetic New Yorkers lent their support to the uprising and, encouraged by cash bounties, hundreds of young "Patriots" prepared to join the rebels. By December one thousand American recruits were camped on Navy Island in the Niagara River. To deter an imminent invasion, British forces crossed to the American bank of the river, where they burned and sank the *Caroline,* a ferryboat being used to supply the recruits. In the melee, one American was killed and others wounded.

To prevent armed conflict with Great Britain, whose financial resources were crucial to the American economy, Van Buren dispatched General Winfield Scott to Buffalo. Drawing upon his military reputation and considerable political skills, Scott persuaded the American Patriots to disband. For his part the president warned that "no aid or countenance" would be given to Americans who crossed the border and persuaded Congress to enact a tough neutrality act. With Canada, as with Mexico, Van Buren was determined to avoid war.

This pacific policy was dictated in part by the woeful condition of American military forces. The navy sailed antiquated ships, and the regular army consisted of a mere eight thousand poorly equipped men. Since early 1836 half the army had been in Florida, trying to force four thousand Seminole Indians and their fifteen hundred African American allies (including many fugitive slaves) to resettle in Indian territory west of the Mississippi River. Unable to secure a military victory, in October 1837 the army resorted to deceit, flagrantly violating a flag of truce to capture Osceola, the leading Seminole warrior. Soon the army had removed half the Indians to Oklahoma, but the war dragged on and Van Buren refused to compromise. The Seminoles had to be "totally expelled," he declared, so as not to set an "evil example" for other tribes facing resettlement. Despite mounting pressure from leading Whigs, the president was determined to carry out Jackson's policy of Indian removal. In May 1838 he ordered General Scott to expel fifteen thousand Cherokees from their ancestral lands in Georgia and neighboring states, setting in motion that Indian nation's tragic journey to Oklahoma on the "Trail of Tears." Even as removal devastated many Indian peoples, its huge cost — some $50 million during Van Buren's presidency — stimulated the American economy. By the early summer of 1838 the financial crisis had ended and most banks resumed specie payments.

After a trying year in office, the president took a well-deserved vacation to Virginia Springs. While fellow Democrats had criticized Van Buren for lacking "energy & vigor" and for being "willing to let things

THE AMERICAN PRESIDENCY ★ 112

take their own course," he had in fact pursued a purposeful and successful set of policies averting military conflicts with Mexico and British
Canada and maintaining harmony between north and south. Moreover,
he had upheld Jeffersonian principles and Jacksonian policies by restoring fiscal stability and the specie standard without recourse to a national
bank.

As 1839 dawned, Van Buren began to think about a second presidential term. In February he averted another diplomatic crisis with Britain
by repudiating the aggressive military tactics of the Democratic governor of Maine in an ongoing boundary dispute with the Canadian
province of New Brunswick. And in June, the president defied tradition
by campaigning for reelection, traveling to Kinderhook by a circuitous
route and delivering "nonpolitical" speeches in Maryland, New Jersey,
and Pennsylvania. Partisanship came to the fore in New York City, where
Democrats welcomed Van Buren with a great parade and thousands of
workers voiced strong support for hard money and the Independent
Treasury. The president's optimism grew when the fall elections maintained the party's majorities in Congress.

Then, late in the year, disaster struck in the form of a second financial
deluge. Once again the Bank of England precipitated the collapse by
doubling interest rates, thus cutting the flow of capital and credit to the
United States. In December cotton prices plunged and nearly half the
nation's banks again suspended specie payments. This time Van Buren
took the offensive, blaming British policies for the collapse and demanding establishment of an Independent Treasury to escape "the control of a
foreign moneyed interest." Departing from traditional Democratic doctrine and his previous views, Van Buren also urged Congress "to cooperate with the States" in regulating bank notes and loans. Although the
southern-dominated legislature rejected the president's call for national
supervision of state-chartered banks, it finally passed the Independent
Treasury Bill. Calling the legislation a "Second Declaration of Independence" from Britain, Van Buren signed it into law on July 4, 1840.

Although Van Buren had a strong grasp of the workings of international finance, he was mistaken in thinking that the Independent Treasury would free the American economy from British influence. In fact,
by following Jeffersonian principles and repudiating high tariffs and
federally financed internal improvements, Van Buren (and Jackson) had
encouraged Americans to rely on British imports and capital, with results they publicly deplored. In 1832 Jackson had lashed out at the
stock-owning "foreigners" who had invested in the Second Bank and
would profit from its rechartering; and late in 1839 Van Buren had con-

MARTIN VAN BUREN ★ 113

demned "money power in Great Britain" that precipitated the two financial panics. Because they rejected the neomercantilist program of national banking, high protective tariffs, and nationally financed roads
and canals advocated first by Alexander Hamilton and later by Henry
Clay, the two Democratic presidents were unable to moderate the mad
economic boom of the early 1830s and the dramatic busts that inexorably followed. For Jackson and Van Buren, these Federalist and Whig
policies had too many defects: a "consolidated" government, a lower
standard of living for ordinary Americans, and — perhaps the crucial
factor — the loss of southern support for the Democratic party.

In the event, Van Buren would not carry the south — or the nation —
ever again. In 1840 the Whigs exploited the president's lifelong quest for
gentility, depicting him as a "lily-fingered aristocrat," and falsely portraying his opponent, General William Henry Harrison, as a simple farmer
living in a spartan log cabin. But Van Buren's defeat in the election was
less the result of Whig propaganda than of the swiftly deteriorating
economy. As imports of goods and capital plunged, thousands of businesses failed and many thousands of workers lost their jobs. Roused
by hard times and a superbly organized Whig party, 80 percent of the eligible voters went to the polls in 1840 (up from 58 percent in 1836).
Nonetheless, Harrison won only by a narrow margin; a shift of 8,088
votes in four northern states would have returned Van Buren to the
White House.

The Little Magician had lost his powers and would never fully regain
them. In 1844 southern Democrats denied Van Buren the party's presidential nomination because he continued to oppose the annexation of
Texas. Four years later, following the Mexican War and bitter sectional
strife over the Wilmot Proviso, the New Yorker took his revenge. Mobilizing a radical segment of the "plain Republicans of the North," Van
Buren ran as the candidate of the Free Soil party, winning 10 percent of
the popular vote and garnering enough support in New York to cost the
Democrats that state and the presidency.

As an elder statesman in the 1850s, Van Buren gravitated back to the
Democratic party and, in the crucial election of 1860, voted for Stephen
Douglas, hoping that he could unite the divided nation. But, following
the secession of South Carolina, Van Buren declared that the Constitution was "a perpetual and irrevocable compact" and called upon the Democracy of New York to support President Lincoln's use of force to preserve the Union. As Northern and Southern armies clashed in Virginia
in July 1862, Van Buren died at his New York estate.

The wheel had come full circle. As president in 1837 Van Buren had preserved his party and the Union by refusing to annex Texas. But his cautious statesmanship ultimately had not been able to contain the passions unleashed by Northern abolitionists, slavery expansionists, and the prophets of Manifest Destiny. Indeed, the robust democratic polity that Van Buren had helped to create as a young politician had eventually done him in. "Behold," Philip Hone had astutely exclaimed upon Van Buren's defeat in the spectacular "Log Cabin" campaign of 1840, "old things are passed away and all things have become new."

— JAMES A. HENRETTA
University of Maryland

William Henry Harrison

1841

b. February 9, 1773
d. April 4, 1841

B Y THE USUAL MEASURES, William Henry Harrison accomplished little during his one month as president of the United States. He took the oath of office, delivered a lengthy and tedious inaugural address, appointed many men to office, called a special session of Congress, became ill — and abruptly died. If Harrison's presidency was insignificant, however, his pursuit of the office transformed the history of American presidential elections. Until Harrison, men rarely engaged in electioneering to obtain the nation's highest office; Harrison sought it openly and vigorously. Although opponents grumbled that Harrison was debasing himself, his efforts stimulated new ways of thinking about presidential elections. Public campaigning for the presidency, although not embraced by candidates as the norm until the end of the nineteenth century, effectively began in 1836.

Born in 1773 in Charles County, Virginia, the third son of Benjamin Harrison, a Virginia signer of the Declaration of Independence, William Henry Harrison began adulthood intending to become a physician. After the death of his father, however, Harrison abandoned Dr. Benjamin Rush's famous medical school in Philadelphia. Instead, like many a younger son before and after, Harrison joined the military and headed toward the frontier. An officer in the army from 1792 to 1798, he parlayed military service and his family connections into a series of public offices — secretary of the Northwest Territory from 1798 to 1799, the territory's first congressional delegate to the U.S. Congress from 1799 to 1800, and (when the Northwest Territory was divided in two in 1800) the governorship of the Indiana Territory, a position he held until 1812.

In 1811, as governor living in the territorial capital of Vincennes, he claimed a small corner in the recesses of the public imagination. Like other territorial governors, Harrison supported the expansion of white settlement and removed Native Americans from their land. But he also repeatedly expressed sincere regrets about unjust confiscation. Invariably, despite genuine sympathy, lust for land won out. During his gubernatorial term, he gained control of millions of acres for the Indiana Territory. His most formidable Native American opponents lived just to the north of Vincennes, at the junction of the Tippecanoe and Wabash rivers, in Prophetstown. The village was the center of a pan-Indian religious revival, led by the prophet Tenskwatawa, and of a political effort to forge intertribal unity, led by the prophet's brother, Tecumseh. For several years, Harrison and the Indian leaders negotiated, threatened, and deceived one another. Finally convinced that only military force would resolve conflict, Harrison determined to attack Prophetstown while Tecumseh was away seeking allies among the southern tribes. For his part, Tenskwatawa ordered his warriors to assault Harrison's troops the night before a formal meeting of the two leaders, but the Indian forces were repulsed, with substantial casualties on both sides. Harrison's victory shattered Tenskwatawa's reputation as a seer and religious leader, crippled Tecumseh's intertribal alliance, and established Harrison as a minor military hero.

During the War of 1812, Harrison resigned his governorship to take the field and rose to the rank of major general. He led American forces in the battle of the Thames as they defeated the British and their Indian allies, killed Tecumseh, and secured control of Detroit and the surrounding region. Although questions about his leadership, judgment, and bravery led to a congressional investigation, the committee absolved him of all charges.

Harrison returned briefly to farming, but financial reversals in the speculative postwar economy and a desire for public life led him to seek election to Congress. After representing Ohio for several years, he won a seat in the Ohio state senate. Thereafter, he was defeated in attempts to return to the state senate and the House of Representatives as well as to gain election to the U.S. Senate. But persistence paid off. The legislature elected him to the U.S. Senate in 1825. He resigned from that office in 1828 to accept an appointment as ambassador to Colombia, a post he held for little over a year. Harrison's public support for Colombian republicanism, at a time when President Simón Bolívar was contemplating the reestablishment of monarchy and crushing a republican rebellion, enmeshed Harrison in Colombia's domestic politics and nearly led to his arrest. A timely recall by the new Jackson administration spared him that embarrassment.

Harrison came home to Ohio, and once again suffered a sharp reversal of his personal fortunes. The death of one son and the neglectful business practices of another left Harrison deeply in debt and responsible for the support of a large extended family. He repaid most of the debt by selling land and earned a regular income as clerk of the court of common pleas in Hamilton County, Ohio.

Yet the now aging Harrison burst back into national politics in 1834. As Democrats contemplated a successor to the incumbent, some latched onto Kentucky's popular Richard Johnson. They promoted him as a war hero responsible for the American victory at the battle of the Thames (though at the time he held the rank of colonel) and as the slayer of Tecumseh. In the Democratic party's literature, Harrison's role in the war and reputation shriveled as Johnson's swelled. Harrison's former aides, and then the former general himself, challenged the claim that Johnson had slain Tecumseh and ridiculed the idea that Johnson led American forces at the battle of the Thames. Instead, they asserted Harrison's own preeminence as an Indian fighter and military commander, first at the battle of Tippecanoe and then at the battle of the Thames.

Opponents of the Jackson administration, who were now calling themselves Whigs, recognized in Harrison a war hero who might attract broad popular support for the presidency. As the incredulous Harrison remarked to a friend: "Some folks are silly enough to have formed a plan to make a President of the United States out of this *Clerk* and Clodhopper!" Nevertheless, the clerk quickly became a presidential candidate. Soon after the tiff with Johnson, delegates at a Whig and Anti-Masonic party meeting in Harrisburg, Pennsylvania, nominated him for the presidency.

Although Harrison had a substantial record of military and political accomplishment, the kinds of achievements traditionally expected of presidential candidates, he was laboring in an obscure local office and was largely unknown outside of Ohio and Indiana. In order to become a candidate for the office he had to alert politicians and other politically active citizens of his interest in the presidency, become better known to the public, and prove his viability as a candidate. And because of his advanced age (he turned sixty-two in 1835), he needed to dispel concerns about his physical capacity.

In response, Harrison toured Kentucky and Indiana. The gatherings, ostensibly called to celebrate the former general's military exploits, actually promoted his candidacy. During his travels, Harrison denounced

Democratic partisanship and the supposedly unconstitutional behavior of Andrew Jackson. His campaign met with favor. Citizens at local political meetings in the northwest endorsed his candidacy. Then, in December, he gained the crucial support of the Whig and Anti-Masonic parties of Pennsylvania. Those endorsements stimulated support from anti-Jackson leaders throughout the north, and soon Jackson's opponents throughout the north and west nominated him for the presidency.

Running against the Democratic nominee, Martin Van Buren, as one of three opposition candidates — Hugh Lawson White of Tennessee ran primarily in the south and Daniel Webster mainly in Massachusetts — Harrison blithely disregarded custom and campaigned actively for the presidency. His lengthy speeches emphasized traditional subjects — his military service, the economy, and presidential abuses of power — yet their novelty captured the public imagination. Harrison's victory in seven states confirmed the value of this new approach.

The narrowness of Martin Van Buren's victory and the strong showing of Harrison and White combined with the panic of 1837 and the subsequent depression to spur Whig party development. Confident that a well-organized Whig party could defeat an incumbent in the midst of a depression, leaders called for the party's first national convention in Harrisburg, the state capital of Pennsylvania. Despite the efforts of the perennial presidential hopeful Henry Clay, Harrison easily gained the nomination. Convention delegates then chose Virginia's John Tyler, a Clay supporter, as Harrison's running mate. Initially, Harrison confined most of his campaign activities to writing platitudinous letters and to delivering a handful of innocuous speeches in Ohio. In those speeches, he embraced the traditional notions that a president should be chosen based upon a man's lifetime of service to his country, not his stand upon particular issues. As the historian M. J. Heale observes, "Harrison . . . resorted to new modes of electioneering in order to broadcast a very old-fashioned message."

The Whig's unprecedented log cabin campaign transformed the nature of presidential races. When Democrats derided Harrison as a man who lived in a log cabin and drank hard cider, Whigs responded by embracing the idea that Harrison was like ordinary Americans with a taste for drink. The party held parades, and Whig politicians gave speeches from the rooftops of log cabins to enormous crowds. Whigs joked, sang campaign songs, marched around the local log cabin, and also drank hard cider. Well organized, well financed, and active at the national level, Whigs carried out a remarkably thorough, enormously effective campaign.

Harrison increasingly joined in the effort. After bowing briefly to the tradition of reclusive and silent presidential candidates, the irrepressible Harrison took to the stump during the summer and fall of 1840 and delivered more than twenty meandering, extemporaneous speeches to huge crowds. He disclaimed all interest in a second term in office, pledged to relinquish the veto power in all but the most extraordinary cases, and promised to restore republican simplicity to Van Buren's monarchical White House. The rousing, highly competitive campaign ended with Harrison winning in nineteen of twenty-six states for an overwhelming 236 electoral votes, though he garnered only 52.9 percent of the popular vote. Even more remarkable, an extraordinary 80.2 percent of the eligible electorate (virtually all of whom were white male adults) turned out to vote.

Immediately after the election, Harrison was besieged by office-seekers deprived of patronage during the presidencies of Andrew Jackson and Martin Van Buren. The demands grew so insistent that Harrison felt compelled to travel to Washington weeks before his inauguration. There he chose his cabinet. He accommodated his former rival, Henry Clay, by appointing Clay's lieutenant, John J. Crittenden, as attorney general, but — sensing that Clay expected to become the real power behind the presidency — refused Clay's other requests with the sharp admonition, "Mr. Clay, you forget that I am the President." In the same fashion in which he had built broad public support, Harrison dispersed cabinet appointments widely across the country.

On Inauguration day, March 4, Whigs mounted a fantastic parade, filled with rolling log cabins, bands, Tippecanoe clubs, and militia companies. Harrison, to prove that he was indeed like ordinary Americans, eschewed a carriage in favor of a horse. After taking the oath of office on that bitter, windy day, Harrison spoke for two long hours. He reiterated Whig doctrines, especially his commitment to a weak presidency and to a harmonious relationship between the branches of the federal government, and promised to retain all competent public officials (in implicit contrast to partisan patronage appointments by the Democrats). He appeared at virtually every inaugural ball held in the city. Besieged by office-seekers, Harrison spent most of the first days of his administration handing out jobs.

The new president immediately faced an economic crisis. In a Whig administration, Congress would be expected to be the center of economic policy making, and Clay, as the party leader in the Senate and its chief spokesman, would be expected to play a leading role in shaping that policy. Given Whig assumptions about the respective roles of Congress and the president, it was unsurprising that the first major issue confronted by the Harrison administration was whether the new president should convene an extra session of Congress. A week after the Inauguration, the cabinet split evenly on the question. Unconvinced that a special session was necessary, Harrison broke the tie by voting against the summons.

Less than a week later, he changed his mind. Concerned about the financial condition of the federal government, he asked Secretary of the Treasury Thomas Ewing to evaluate the situation. Ewing reported that the treasury would not have adequate revenue to sustain the government until the next regular meeting of Congress in December. In order to avert federal bankruptcy, President Harrison called for a special session of Congress to meet on May 31.

Whigs believed a weakened president would not only transfer political authority to Congress, but also would share substantial executive authority with his cabinet. They assumed that cabinet members would do more than advise the president, but also vote on executive decisions. The president, as in the initial decision to refrain from calling a special session, would cast one vote, just like members of the cabinet; presumably the president would be expected to abide by any decision made by a majority of the group. Although cabinet members had wielded great power in the years before Jackson's election, Old Hickory had granted cabinet members a distinctly subordinate role. But Whigs, anxious to rein in presidential power, accorded much more authority to the unelected cabinet. Because appointees often owed allegiance to a different political benefactor, moreover, President Harrison could not expect unanimous support for his own point of view. Despite some grumbling and occasional outright resistance, Harrison accepted his reduced role.

After his Inauguration, Harrison was ever present on the Washington social and political scene. He met with office-seekers and officeholders, with happy Whigs like Webster and William Seward and unhappy Whigs like Henry Clay, and even with ousted Democrats. Apparently quite vigorous, Harrison walked everywhere. But in late March, he developed a severe cold — which may have begun on his blustery Inauguration day — that turned into pneumonia. On April 4, before politicians in Washington had an opportunity to adjust to the new president's style of governance, William Henry Harrison lay dead, having served just one month of his forty-eight-month term.

— MARC W. KRUMAN
Wayne State University

John Tyler

1841–1845

b. March 29, 1790
d. January 18, 1862

THE ADMINISTRATION OF JOHN TYLER was filled with surprises for the president, for his fellow Whigs, and for other citizens. When John Tyler was inducted as vice president of the United States on March 4, 1841, neither he nor the men who elected him expected him to become president of the United States. Indeed, like virtually all vice presidents before him, Tyler could anticipate four quiet, inconsequential years in office. He would preside over the United States Senate as its president, yet with a small but comfortable Whig majority, he would have few opportunities to break a tie vote.

A month later, on April 4, the death of President William Henry Harrison abruptly ended John Tyler's vice presidency and placed him in official limbo. The U.S. Constitution was vague about succession, and no president had ever before died in office, so there was no model for Tyler to follow. Would the vice president become acting president or the actual president of the United States? Many in the Harrison cabinet and in Congress expected him to assume the role of acting president, but Tyler recognized that an acting presidency would compromise his ability to lead the country. He therefore boldly took the presidential oath of office and established a custom that would govern presidential succession until 1967, when the Twenty-fifth Amendment formalized the practice. Tyler's decision assured the smooth transition of full presidential power to the vice president upon the death of the president.

In normal times, President Tyler might have taken several months to learn his new responsibilities, but Harrison's call for a special session of Congress required an immediate response. Whigs like Henry Clay assumed automatic cooperation from the new president; after all, Tyler had been nominated for the vice presidency because he supported Clay. As a Whig presumably averse to the exercise of substantial presidential power, moreover, Tyler could be expected to eschew the veto and leave policy making to the Whig Congress.

In early conversations, Tyler seemed to confirm Whig expectations. Welcoming members of Congress in 1841, Tyler and Clay predicted a harmonious special session. Congress indeed moved quickly to pass new bankruptcy legislation and laws facilitating the sale of federal lands to settlers. Clay and other Whig leaders, acting upon what they considered to be the Whig mandate, sought the reestablishment of the National Bank. Clay assumed that because the Harrison-Tyler ticket had swept to an easy victory and because Whigs had gained majorities in the House and Senate, the public embraced all aspects of Whig economic policy. Yet as other Whigs (including the president) and the Democrats recognized, the party rarely had discussed the bank or specific economic policies during the presidential campaign. For many years, moreover, Tyler had opposed the reestablishment of a truly national bank. He instead proposed the creation of a central bank in Washington, D.C., that could set up branches in other states only with the permission of the legislatures of those states. But Clay insisted upon a charter for a National Bank in which directors could establish branches in any state. Congress sent Clay's bill to the president on August 6; he vetoed it ten days later.

Congressional Whigs made superficial changes that accommodated the president while ensuring further conflict. They renamed and recast the bank bill as the Fiscal Corporation Act, in deference to Tyler's views, but authorized the fiscal corporation to create branches in any state, more power than the president would concede. Tyler vetoed the second bank bill on September 9. Within days, five of the cabinet's six members resigned. Only Secretary of State Daniel Webster remained at his post. After Congress adjourned on September 13, Whig congressmen ousted Tyler from their party after denouncing him for usurping the powers of Congress.

Now a president without a party, Tyler faced a nearly empty treasury when the regular session of Congress opened in December 1841. He urged Congress to increase tariff duties. He also proposed that Congress end the practice of distributing the proceeds of public land sales to the states and apply those revenues to a budgetary deficit. Congressional Whigs twice passed legislation increasing the tariff while maintaining distribution; Tyler vetoed them. Whig congressmen, fearful of voter retribution for government bankruptcy, uncoupled the issues and passed both bills. Tyler quickly signed the tariff bill while rejecting distribution.

But this was a pyrrhic victory. The uncoupling infuriated most Whigs, the new tariff bill antagonized Democrats, and the distribution bill veto angered Southern Whigs.

Tyler faced difficulties not only in Washington with politicians, but in New England with rebellions and border disputes. In 1842, Rhode Island's government still functioned with a constitutional structure established in 1663. Unlike elsewhere in the country, white male suffrage in Rhode Island was severely circumscribed. Reformers organized an extralegal constitutional convention, followed by a legal one authorized by the legitimate government. Asked by both sides to intervene, President Tyler helped to defuse the crisis. He declined to send the army to Rhode Island, urged compromise by the two parties, and called upon the rebels to disperse. After a brief skirmish, the so-called Dorr Rebellion ended. The rebels went home and accepted an expansion of the suffrage that still fell short of universal white manhood suffrage.

The second New England problem was the persistent border dispute between Maine and Canada. Political leaders in Maine, supported by Anglophobes everywhere, staunchly opposed any land concessions to Great Britain; the British insisted upon a compromise. Secretary of State Daniel Webster's astute political maneuvering secured crucial popular support in Maine and elsewhere for his negotiations with the British ambassador, Lord Ashburton, and for the Webster-Ashburton Treaty of 1842, which resolved all disputes over the Canadian–United States boundary from the Great Lakes eastward. Negotiators also settled the case of the American ship *Creole*. American slaves had gained control of the ship and had steered it to the British possession of Nassau, where slavery was illegal. The British freed the slaves but agreed to compensate the American slave owners. The two countries also agreed to cooperate further to stop the slave trade off the coast of West Africa.

After Webster concluded the negotiations, he resigned his office and moved to repair his relationship with other Whig party leaders, who had scorned him for remaining in the cabinet. President Tyler replaced Webster with the Virginian Abel Upshur, who had contributed significantly to the modernization of the navy during his brief tenure as secretary of the navy. In that role, he had sought to transform the navy by modernizing its command structure and encouraging the use of steam-powered ships.

As secretary of state, Upshur pressed hard for the annexation of Texas — ultimately, the most important and the most controversial act of the Tyler administration, and an act that he and the president hoped would benefit the South. In 1836, American settlers in Texas had gained their independence from Mexico in the expectation that they would become part of the United States. But Presidents Andrew Jackson and Martin Van Buren feared that any attempt to add another slave state to the Union would arouse antislavery feeling in the North, intensify sectional conflict over slavery's expansion, and lead to war with Mexico. Tyler and Upshur felt no such qualms. Annexation, they believed, was good for the South in particular and the nation in general. They assumed that Southern interests and American national interests were synonymous and that annexation extended American freedom (including the freedom to own slaves) while benefiting the regional and national economies. With many other Southerners, then, the two Virginians sought to expand American freedom by expanding American slavery.

President Tyler also thought that Texas might invigorate his own political career. Ousted from the Whig party, he was loathed by most Democrats for having abandoned *their* party and President Andrew Jackson a decade before. Nevertheless, Tyler hoped to establish himself as a candidate for reelection and vindicate his reputation. Texas, he believed, might propel him to the leadership of a new third party or of the Southern-dominated Democratic party.

As a consequence, in 1843, Tyler and Upshur urgently sought negotiations on annexation in response to rumors that England was seeking to help abolish slavery in Texas. For their information, they relied upon Duff Green, special commercial envoy to England, longtime supporter of John Tyler, and son-in-law of the South Carolina political leader and Southern rights advocate John C. Calhoun. Green inferred from the English press and parliamentary debates that most Englishmen believed British emancipation had undermined the economies of their colonies in the Caribbean. At the same time, he learned that the British foreign minister, Lord Aberdeen, had engaged in some general discussions with two American abolitionists about British financial support to end slavery in Texas. Green concluded, in an urgent report, that the British had conspired with American and British abolitionists to do so. If slavery were abolished in Texas, inevitably that republic would become a refuge for runaway slaves and a base for antislavery military assaults on the southwestern part of the United States. The British goal, Green was convinced, was to abolish slavery in the United States and around the world in order to restore the competitive position of the British colonies.

Green found a receptive audience in the new secretary of state. Upshur, a proslavery ideologue and an antidemocratic proponent of property qualifications for voting, had prophesied several months earlier that the expansion of abolitionism and democracy would undermine the South and slavery. The assault would begin in Texas, where the English would agree to protect Texas independence in return for emancipation. If Texans abolished slavery, Upshur feared, the rest of the Southern states would not be far behind. Green's report confirmed Upshur's worst fears. The secretary soon learned that Lord Aberdeen had squelched notions of immediate intervention in Texas, but had held out the possibility of future British financial support for abolition there. Further, Aberdeen proposed to Mexico that it recognize Texas independence in return for the republic's termination of slavery.

Aberdeen soon confirmed his proposal on the floor of the House of Lords. Without waiting to learn if Mexico or Texas accepted the proposal, Upshur and President Tyler concluded that only annexation would save slavery in Texas and, subsequently, the South. They also recognized that only a united South, one that transcended allegiance to the Whig or Democratic parties, could demand Texas annexation. Upshur sought to initiate negotiations with Texas in October 1843, but his overtures were rejected on the grounds that Tyler and Upshur could not promise ratification of an annexation treaty. But when Texas President Sam Houston's mentor and friend, the former president Andrew Jackson, urged him to reconsider on the grounds that ratification seemed likely, Houston relented.

While awaiting Houston's response, Upshur celebrated one of the great triumphs of his naval career; the inaugural voyage of the new battleship, the *Princeton*, took place on February 27, 1844. Among the guests were the president and many members of the cabinet. In addition to taking his guests on a cruise on the Potomac, the navy captain Robert Stockton displayed the power of the ship's cannon, the "Peacemaker." After several successful tries, and with members of the cabinet looking on, Stockton fired the cannon once again. This time the ball imploded and killed Upshur and Secretary of the Navy Thomas Gilmer, who were standing nearby.

Upshur's death left formal Texas negotiations to his successor, John C. Calhoun, who assumed the office of secretary of state on March 29. Yet Upshur and the Texas ambassador, Isaac Van Zandt, had worked out the agreement in such detail that Calhoun's only task was to formalize the agreement and prepare a defense of annexation. Negotiators signed and announced the treaty on April 12; Secretary Calhoun delivered it to the Senate on April 22, along with Calhoun's defense of the treaty (contained in a letter to Richard Pakenham, the British ambassador to the United States).

In that letter and in a second one written to Pakenham after the Senate began its deliberations, Calhoun denounced Britain's supposed attempt to absorb Texas and abolish slavery there. He claimed that ending slavery in Texas would erode the institution throughout the South. Because slavery was, he argued, both morally correct and profitable, any attempt to destroy it would meet vigorous American resistance. Therefore, in order to protect slavery, the United States government had sought, and reached an agreement on, annexation. In the Pakenham letter, the secretary of state pursued a strategy he had followed since the Nullification Crisis and would follow until his death in 1850. He tried to alert white Southerners to the danger that an independent Texas posed to slavery; Southerners then, he hoped, would join a Southern rights party and shatter national party loyalties. Finally, a united South would compel Northerners to bend to the Southern will for the sake of union.

Tyler and Calhoun's campaign for Texas gathered even greater force when former president Andrew Jackson embraced annexation as crucial to the interests of the Union and to the defense of slavery. Responding to the urgent request of Sam Houston, Jackson roused himself from his deathbed to write numerous letters contending that, if Texas were not annexed, it would seek another protector, England; in return, England would insist on abolition. A slaveless Texas occupied by the British army would endanger both the military security and the slave economies of the southwest.

By reintroducing Texas annexation to the political stage, President Tyler profoundly influenced the future of American politics. Tyler's supporters planned to nominate him for reelection in May. He hoped to use Texas to win substantial support in the South, especially because the two leading candidates for the presidency, the Democrat Martin Van Buren and the Whig Henry Clay, opposed immediate annexation. Apprehensive Southern Democrats quickly abandoned Van Buren and demanded that the Democratic party nominate a pro-annexation candidate. The party's nomination of the expansionist James K. Polk of Tennessee on May 29 satisfied Southern demands and crushed Tyler's plans for reelection.

But the president was determined that no one crush his plans for annexation. On June 8, after three weeks of debate, senators met in secret session and defeated the treaty by the overwhelming margin of 35 to 16. Despite Calhoun's efforts to sectionalize the issue, only one Southern Whig supported annexation. Initially on June 10, and later when Congress assembled in December, President Tyler asked congressmen to ignore the treaty and grant Texas statehood by means of a joint resolution.

Tyler's proposal gained political momentum from the presidential campaign. Polk, who believed that annexation would benefit the entire country, transformed Texas from a Southern issue into a national one. He and the Democratic party reinforced this view by demanding the acquisition of the Oregon Country. Polk's victory — he won the popular and electoral college votes in the North and South — made Texas's annexation seem inevitable.

As Congress assembled for its final lame-duck session in December, pressure for annexation built. Basing his demands on the election mandate and on the threat of English interference, Tyler pushed Congress to adopt a joint resolution. The government of Texas reinforced the president's sense of urgency by expressing disinterest in annexation. Texas's new president, Anson Jones, for one, ignored the annexation issue in his inaugural address. In response, on January 25, the House passed a resolution annexing Texas as a state. A Tennessee Whig, Milton Brown, introduced an amendment allowing the creation of up to four additional states from Texas with state legislative approval. In an apparently nationalist spirit, the proposal required that any new states north of the Missouri Compromise line of 36°30' would be free; those to the south would be open to slavery. But Texas legislators could spin off four new states to the south of the compromise line, adding a total of five slave states to the Union. Nevertheless, House members passed the bill.

The Senate Foreign Relations Committee tried to dash annexationist hopes, but Senator Thomas Hart Benton of Missouri, the only Southern Democrat to oppose annexation, and President-elect Polk revived them. On February 4, the committee voted in the negative on the joint resolution; the next day, Benton proposed admission of the Republic of Texas as a state once a new treaty — to clarify Texas's boundaries, debt, and public lands — was approved by the two parties. President-elect Polk came to Washington intending to unite his party behind a compromise that would lead to immediate annexation, and he succeeded. Mississippi's Senator Robert Walker proposed that the joint resolution allow the president to accept either the House resolution or the Benton proposal. Both houses supported the Walker resolution. On March 1, President Tyler signed the resolution. Although congressmen intended to leave the decision to Polk, the resolution allowed Tyler to conclude annexation. Pressured by Calhoun, Tyler polled his cabinet on March 2; it voted unanimously for him to proceed. He accepted the House version of the resolution and, on March 3, sent instructions to his representative in Texas, Andrew Jackson Donelson, to announce annexation. The next day, John Tyler left office, but the Texas issue played out as he had hoped. After a brief period of indecision, President Polk ordered

Donelson to carry out Tyler's instructions. Soon after, the envoy proclaimed annexation.

John Tyler, the president without a party, opposed alike by Whigs and Democrats, had succeeded in acquiring Texas. Annexation was part of a pattern that persisted throughout the Tyler administration — diplomatic triumph combined with personal political defeat. Tyler's persistent states' rights opposition to the re-creation of a National Bank, combined with Henry Clay's intransigent refusal to accept anything less than a national charter, guaranteed conflict with Whig congressmen, and ultimately led the Whigs to expel Tyler from the party. Yet during his term in office, "His Accidency" presided over the resolution of crucial boundary disputes with British Canada and the annexation of Texas, in the process altering the contours of nineteenth-century political and constitutional history.

— MARC W. KRUMAN
Wayne State University

James K. Polk

1845–1849

b. November 2, 1795
d. June 15, 1849

T HE PRESIDENCY OF JAMES KNOX POLK presents a paradox. Polk was far more successful than most presidents in defining and achieving an ambitious agenda. Few presidents have worked so hard at the job, demonstrated such detailed managerial capacity, or kept their administrations so untainted by corruption. Few have bequeathed the nation such an enduring legacy, for under Polk the United States almost doubled in size with the acquisition of Texas in 1845, Oregon in 1846, and the Mexican Cession in 1848. Despite his commitment to public rectitude, however, Polk seemed duplicitous to fellow party members. A devoted Democrat who hated his Whig foes, he seriously split his own party and provided Whigs with ammunition to recapture the House of Representatives in 1846 and the White House itself in 1848. And to his consternation, territorial expansion, his crowning nationalistic accomplishment, pushed Americans down the road to civil war.

Polk's political inexperience or naivete hardly explain these paradoxes. Often labeled the nation's first "dark horse" presidential candidate because he was not expected to be the Democratic nominee in 1844, Polk was mercilessly mocked by Whigs as a political pygmy compared to their own candidate, Henry Clay. Yet Polk was a savvy advocate of Democratic principles with substantial experience in state and national office. His past record helped him win his party's nomination and shaped his presidential agenda.

Born in North Carolina in 1795, Polk grew up in Tennessee after moving there with his parents in 1806. After four years' experience in the Tennessee legislature, he was elected in 1825 to the House of Representatives and served continuously from December 1825 to March 1839. An early acolyte of Andrew Jackson, Polk acted as House floor manager for Jackson's Bank War in the early 1830s and was elected Speaker from December 1835 to March 1839, primarily because of his skilled championship of Democratic economic policies against Whig detractors. In 1839 he won the governorship of Tennessee, but he narrowly lost gubernatorial races in 1841 and 1843.

Those two losses largely account for Whig jeers that he was a nonentity, yet his three gubernatorial campaigns profoundly influenced his nomination, his election, and his presidency. The energy and physical stamina that Polk displayed stumping across Tennessee three times accurately previewed the literally killing pace he would set as an administrator. More important, Polk focused all three campaigns on the sharp distinctions between Whigs and Democrats on national economic policy. Especially in 1841 and 1843 he vigorously attacked programs that Whigs, who won the White House and Congress in 1840, had enacted or might enact should Clay, their legislative leader, be elected president in 1844. Widely reprinted in Democratic newspapers across the country, those speeches cemented his public association with Martin Van Buren's hard-money wing of the party.

Orthodox Jacksonian Democrats like Polk and Van Buren believed that any positive governmental action to promote economic growth inevitably created privilege for some that violated the rights of others. Thus they opposed Whig demands for protective tariffs. And they fought subsidies for internal improvements, whether they were to come from the federal government or indirectly from sharing federal land revenues with the states. Of greatest concern to them during the depression following the financial panic of 1837, however, were Whig calls to expand currency and credit through creation of state-chartered banks and a new national bank. Jacksonians like Polk condemned paper money as fraudulent and argued that "hard money," government-minted gold and silver specie, must serve as the country's primary currency supply, a program Whigs charged would dry up credit and stultify economic growth.

To reduce the circulation of paper banknotes, Jackson's successor, Van Buren, now leader of the Democrats' antibanking wing, urged Congress in 1837 to pass his Independent Treasury plan. This sought to remove federal funds from private state banks, where they might be used to support the issuing of paper money, and to deposit them instead in government subtreasuries. The bill required the government to accept and pay out only specie. It was finally enacted in July 1840, but Whigs immediately repealed it in 1841, after the election of William Henry Harrison;

and in 1842 Whigs also enacted a high protective tariff. Only vetoes by John Tyler, who succeeded Harrison in April 1841 and who opposed much of the Whig program, prevented Clay's congressional Whig allies from chartering a new national bank and distributing federal land revenues to the states. This uncompleted Whig program explains why Polk feared Clay's election and why he made tariff reduction and reenactment of the Independent Treasury two of his priorities as president.

Had the 1844 election focused solely on national economic policy, however, Van Buren, not Polk, would have been the Democratic nominee. Polk himself favored Van Buren initially, and until the eve of the 1844 Democratic National Convention he angled instead for the vice-presidential nomination. Van Buren was opposed by soft-money pro-banking Democrats; by John C. Calhoun, who had loathed him since Jackson's first administration; and by men who considered him a loser. But without a new issue to divert attention from economic questions, they could not stop Van Buren.

The issue that derailed Van Buren and provided the most famous part of Polk's presidential agenda was territorial expansion. Ostracized by the Whigs, President Tyler sought to boost his chances of reelection (as an independent or as the Democratic nominee) by annexing the proslavery Republic of Texas in 1844. Jackson ardently endorsed the project, as did Polk in a public letter. But when Tyler sent his treaty of annexation to the Senate in April, both Clay and Van Buren, fearful of antagonizing Northerners who opposed the expansion of slavery, came out in public opposition.

Because annexation enthralled most Democrats, that stand allowed Van Buren's Democratic enemies to upset his bandwagon. After furiously agitating for Texas annexation in the weeks before, they scored an early victory at the Democratic Convention by pushing through a new rule requiring the party's presidential nominee to receive the votes of two-thirds of the delegates, thus negating Van Buren's majority. Meanwhile, prior to the convention, Jackson had called Polk to his homestead, the Hermitage, announced that Van Buren would not do, and anointed Polk as his favorite for the nomination. Polk astutely instructed his convention managers to stick by Van Buren so long as the New Yorker had a chance, thereby assuring Van Burenite support for Polk as a compromise choice should a deadlock develop, as it did. Polk was nominated with the aid of Van Burenites who saw him as the only pro-Texas candidate committed to Jacksonian economic orthodoxies. Aside from traditional Democratic economic dogmas, the platform committed the party to "the reoccupation of Oregon and the reannexation of Texas, at the earliest practicable period." There had been little interest in Oregon before the convention; it was added to the platform to give sectional balance to the party's expansionist designs. But it, too, became one of Polk's top priorities.

Martin Van Buren's furious Northern supporters blamed their defeat on pro-Texas agitators, vindictive Calhounites, and soft-money men, not on Polk, for whom they loyally campaigned. But Van Burenite Democratic senators helped Whigs resoundingly reject ratification of Tyler's annexation treaty in June, and their adamant hostility to Texas would have important postelection ramifications. Nonetheless, the pro-expansion platform measurably helped Polk in the South and some midwestern states, but not in critical New York and Pennsylvania, where the race was decided.

Polk relied on the disappointed Van Burenites to carry New York for him. To reassure them and proponents of other candidates, he pledged in his letter of acceptance to serve only a single term. More important, to help carry Pennsylvania, where protectionist sentiment was strong among Democrats, he wrote a letter for publication in which he seemed to promise to keep a protective tariff, even as he privately assured Southerners that as president he would immediately insist that Congress lower duties. Brandishing this letter, Pennsylvania's Democrats brazenly campaigned under the banner "Polk, Dallas, Shunk [the Democratic gubernatorial candidate], and the Democratic Tariff of 1842. We dare the Whigs to repeal it!" Their sense of betrayal when Polk did what he had always intended explains some of the bitter discord that Polk's presidency provoked among Democrats.

Clay's clumsy handling of the Texas annexation issue helped cause ardent antislavery men to cast protest votes for a Liberty party candidate. The Whigs' nativist stigma drove new immigrant voters into the Democratic column. Both contributed to Polk's victory in key Northern states. The popular vote was exceptionally close, and a small number of additional Whig votes in New York or in Pennsylvania and Indiana would have given the prize to Clay. Polk won the electoral vote 170 to 105, carrying fifteen states to Clay's eleven.

The support of Van Burenites had been crucial to Polk's triumph in New York, but Polk alienated them even before he assumed office. Rejecting their suggestions, he appointed their factional enemy William L. Marcy as secretary of war, and he outraged them by naming as treasury secretary Robert J. Walker, the man most directly responsible for derailing Van Buren's nomination with the Texas issue. More important, in the winter of 1844 to 1845, lame duck President Tyler pressed Congress to

offer Texas admission as a state by joint resolution. Van Burenite Democrats, who along with Whigs could have blocked the proposal in the Senate, agreed only after adding an amendment. It gave the president a
choice between offering the huge Texas Republic admission as a single
state or renegotiating the terms to admit only part of Texas immediately
as a slave state, keeping the remainder a free-soil territory that would
presumably produce free states later on. Van Burenite senators assumed
Polk would take the second option. On his last day in office, Tyler chose
the first option. Polk not only refused to reverse that decision after entering office, but worked vigorously during 1845 to get Texas to accept
Tyler's offer. At Polk's urging the Democratic majorities in both houses
of Congress admitted Texas as a state as soon as Congress met in December 1845. For the remainder of Polk's presidency, however, furious
Van Burenites considered him an untrustworthy liar. They would not be
the only Democrats to think so.

Shortly after his Inauguration, Polk told Navy Secretary George
Bancroft that he had four goals as president: reducing tariffs to a level
that would produce only necessary government revenue; reestablish-
ing the Independent Treasury; settling the Oregon boundary with England; and acquiring California, then a Mexican possession. At Polk's
instigation Democratic congressional majorities passed the first two
measures in 1846. But the Walker tariff, with its low rates on manufactured goods and high rates on imported raw materials used by manufacturers, angered Pennsylvania's betrayed Democrats. All but one Pennsylvania Democrat in Congress voted against the Walker tariff, forcing Polk
to depend on midwestern and Southern Democrats to push it through.
Midwesterners delivered those votes in return for support from Polk
and Southern Democrats in Congress for a measure lowering western
land prices. As soon as the new tariff passed, however, Polk and Southerners abandoned the land bill, which quickly died. That same summer Polk vetoed a large rivers-and-harbors bill eagerly sought by midwestern Democrats. These actions swelled the ranks of disillusioned
Democrats.

Polk's handling of the Oregon question especially infuriated midwestern Democrats. Since 1818 the entire Oregon Country, from the 42d parallel in the south to the 54° 40' line in the north, had been opened to
joint settlement by British and American citizens with the stipulation
that either England or the United States could end this arrangement by
giving the other one year's notice. Americans wanted a deep-water port
on the Pacific coast to facilitate trade with China. Because only Puget
Sound met that requirement, they had wanted to divide Oregon at the

49th parallel for years. But virtually all American settlers in Oregon lived south of the Columbia River, whose shallow mouth prevented it from serving as a port. So the British had countered that the Columbia should be the dividing line, thus excluding Puget Sound from the American portion.

Then, in 1845, the British suggested that the matter be submitted to international arbitration. Since Polk feared arbitrators would fix on a line south of Puget Sound, he rejected the offer. Instead, he decided to bluff the British into accepting the 49th parallel as a compromise by calling on Congress in December 1845 in his annual message to give England the one-year notice ending joint occupation, appropriate money to build a line of fortifications in Oregon, and extend American authority over the entire Oregon Country up to 54° 40'. In short, Polk seemed prepared to fight England to get all of Oregon, and for months war seemed imminent. Ultimately, however, Polk's bluff worked. In June 1846 the English agreed on the 49th parallel, with a deviation of the boundary through the Fuca Straits to give British Columbia all of Vancouver Island.

The political problem for Polk was that some midwestern Democratic congressmen never knew or refused to believe that Polk was running a bluff. Belligerent Democratic senators led by Lewis Cass of Michigan, Edward Hannegan of Indiana, and William Allen of Ohio responded to his tough message by loudly raising the cry "Fifty-four forty, or fight!" during the winter and spring of 1846. So even though Polk had always intended to agree to the 49th parallel, midwestern Democrats believed that he had intentionally duped them. What was especially galling, they complained, was that they had loyally helped the slaveholder Polk acquire Texas for other slaveholders, but that he and a minority of Southern Democratic congressmen led by Calhoun had then refused to acquire all of Oregon for the free states. This false belief that Polk was the puppet of the Southern slave power would have serious consequences for his presidency.

The Mexican War provided the chance for various Northern Democrats to vent their accumulated grievances against Polk. War with Mexico was a direct result of Polk's fourth objective for his administration — acquisition of California. The Democratic platform of 1844 committed him to Texas and Oregon, but only his personal vision of extending the United States across the entire North American continent explains his obsession with bringing the entire Pacific coast into American hands. The Mexicans, however, showed no disposition whatsoever to give California to him, and from the start Polk treated them with arrogant contempt.

He was determined to get California by hook or by crook. Even before Texas accepted the offer of statehood in July 1845, he tried to intimidate the Mexicans into ceding California by a show of American military might in the Gulf of Mexico and along the disputed Texas-Mexico border. Twice that year, he ordered American naval officers in the Pacific to seize San Francisco and other California ports should war break out. He tried to incite Californians to revolt against Mexico on the assumption that they would later seek incorporation into the United States as Texans had done. He attempted to bully Mexico into selling California for $25 million and United States payment of all claims against the Mexican government by private American citizens. He intrigued with exiled Mexican president Santa Anna to restore him to power in return for acceding to Polk's heavy-handed demands. When all these forays failed, he resorted to naked force. Altogether, Polk's dealings with Mexico constitute one of the shabbiest episodes in American diplomatic history.

As early as January 1846, Polk began planning to ask Congress to declare war against Mexico. War was necessary, he insisted, to avenge what he called Mexico's insulting refusal to meet with an American minister sent to negotiate payment of Americans' often spurious monetary claims against it, claims he knew could only be settled by a territorial cession that would topple the weak Mexican government from power. In the end, however, the pretext for war became the disputed Texas-Mexico boundary. When Texas was a Mexican state, Mexicans had always considered the Nueces River its southern and western boundary. Thus they were incensed by Texas's claim, backed aggressively by Polk, that the border was farther south, at the Rio Grande. Only about one hundred barren miles separated the two rivers in the southeast where they met the Gulf of Mexico and where war would start. But the Rio Grande flowed hundreds of miles farther west than the Nueces, *west* indeed of Santa Fe, cutting the Mexican province of New Mexico almost in half. Accession to this grandiose claim, in short, meant the dismemberment of Mexico.

Polk had ordered troops under Zachary Taylor to Corpus Christi on the southern or western bank of the Nueces as soon as his administration took power, and in early 1846 he sent Taylor to the banks of the Rio Grande itself. When Mexican troops attacked a small unit of Taylor's soldiers on the northern (or eastern) side of the Rio Grande, Polk unblushingly told Congress in May that war existed because Mexico "has invaded our territory, and shed American blood upon American soil." To the fury of Whigs and a few Southern Democrats, led again by Calhoun, Polk did not ask Congress to declare war against Mexico as required by the Constitution. Instead Democratic majorities, on party-line votes, adopted a carefully worded preamble asserting that "by the act of the Republic of Mexico, a state of war exists between that government and the United States" and attached it to a measure requisitioning troops and supplies. Outraged Whigs responded accurately that "the river Nueces is the true western boundary of Texas" and that "it is our own President who began this war," but once Democratic votes had forced adoption of the noxious preamble, they had little choice but to vote for the war bill. For the remainder of the contest the same fear of appearing unpatriotic caused them continually to vote for men and supplies.

Polk finally had the war he wanted, and his aims were reflected in his initial moves. Insisting upon absolute authority over military affairs as commander in chief, Polk meticulously managed the war effort from start to finish. At first, he made no attempt to reinforce Taylor's small army at the mouth of the Rio Grande. Instead, he dispatched troops to seize Santa Fe and the rest of New Mexico; and he repeated his orders to the Pacific squadron to seize San Francisco, San Diego, and other California ports. He wanted the Americans to have military control of both territories when the overmatched Mexicans sued for peace, as Polk expected them to do. He could then demand New Mexico and California as an indemnity for the costs of the war to America. Polk, in fact, expected the Mexicans to give up by the summer of 1846. But despite a series of stunning American victories, first by Taylor in northern Mexico in 1846 and early 1847 and then by General Winfield Scott, who captured Mexico City in September 1847, the stubborn Mexicans did not agree to Polk's insulting peace terms until early 1848. Ratified by the Senate on March 10, 1848, the Treaty of Guadalupe Hidalgo fixed the Rio Grande as the Texas-Mexico boundary and gave the United States a huge tract of Mexican territory encompassing the modern-day states of New Mexico, Arizona, California, Nevada, Utah, and part of Colorado, in return for $15 million and government payment of private American claims against Mexico. Polk had fulfilled his continental vision.

Yet the unexpectedly long Mexican War had three untoward consequences for Polk. First, he wore himself out with his obsessive attention to every detail. He would die on June 15, 1849, barely three months after he left office. Second, the war gave his hated Whig rivals a unifying and politically potent issue. On the other hand, the war's two greatest heroes,

Taylor and Scott, were Whigs who publicly quarreled with the administration. To Polk's great vexation, he could neither replace them nor stem
their growing popularity, and Taylor would win the presidency for the
Whigs in 1848. On the other hand, although Whigs dared not vote
against men and supplies, they gained great political mileage from denouncing the war itself as an immoral land grab from a weaker neighbor. They called Polk's management of it partisan and inept, and they
denounced his rigid insistence upon a huge territorial indemnity as dangerous to North and South alike.

Polk's territorial demands, indeed, became the war's third and most
important problem for him. Polk saw territorial expansion as fulfilling
the entire nation's manifest destiny. He did not see it in sectional terms.
Yet from the war's start, Northern Whigs portrayed it as an attempt to
get more land for slavery. Already furious at Polk and Southern Democrats for a variety of reasons, Northern Democrats refused to defend the
president against those charges. To protect themselves at home and
avenge themselves against Polk, they insisted that territorial expansion
must not be accompanied by slavery extension. When Polk asked Congress in August 1846 for an appropriation as a down payment on Mexican territory to induce the Mexicans to surrender, a Pennsylvania Democrat named David Wilmot offered an amendment, thereafter known as
the Wilmot Proviso, barring slavery from any territory taken from Mexico as a result of the war. This proviso never passed Congress, but it instantly and repeatedly pitted Northern Whigs and Democrats against
their Southern counterparts.

By 1847 and 1848 fourteen of fifteen Northern state legislatures instructed their senators to impose the proviso on any territorial acquisition, while all Southerners denounced it and some threatened secession
should Congress pass it. Both parties frantically searched for compromise solutions even before the Mexican Cession became a reality. Personally uninterested in slavery extension and appalled that the divisive
question tarnished his achievement, Polk initially favored extension of
the Missouri Compromise line to the Pacific coast. The preferred Democratic dodge, known as "popular sovereignty," would have removed the
decision from Congress altogether and allow residents of the territories
themselves to decide whether slavery should exist within them.

When the primary champion of that formula, Lewis Cass, won the
Democrats' presidential nomination in 1848, Northern Van Burenites,
who had long hated Cass and fumed at Polk for four years, helped form
the Free Soil party. Running Van Buren himself on a platform demand-

ing congressional prohibition of slavery from all federal territories, it achieved their goal of siphoning enough Democratic votes from Cass in New York and other Northern states to assure Taylor's triumph.

Polk's greatest legacy to the nation was territorial expansion. But that expansion brought with it a bitter sectional dispute over the spread of slavery. The dispute delayed formal territorial organization of even Oregon until the summer of 1848. No solution for the Mexican Cession was found until 1850, after Polk's death. The bitter controversies ignited by Wilmot and other Northern Democrats as retaliation against Polk would not go away. Only a bloody civil war would resolve them.

— MICHAEL F. HOLT
University of Virginia

Zachary Taylor

1849–1850

b. November 24, 1784
d. July 9, 1850

I
N SOME WAYS, Zachary Taylor's name seems more likely to appear
as a "Jeopardy!" question than on any list of presidential greats.
Zachary Taylor's sixteen months in office were marked by clashes
over sectionalism, the California gold rush, and divisive debates
over the Compromise of 1850.

Taylor, the twelfth president, like six of the men who served before
him, was born in Virginia. He grew up outside Louisville, Kentucky,
wearing a coonskin cap and leather moccasins, a son of the frontier, but
also the son of a slave owner. An uneducated young man, Taylor was
offered a commission as a lieutenant in the U.S. Army in May 1808 by
his cousin James Madison. In 1811 Taylor led troops to the Indiana frontier, supporting General William Henry Harrison's campaign against
Tecumseh at the battle of Tippecanoe.

Taylor served in the War of 1812, was promoted to captain, and continued to rise through the ranks, earning his nickname "Old Rough and
Ready" during campaigns in the Black Hawk War and the Seminole
War. After his service in Florida, Taylor requested an assignment in the
southwest, and in 1841 was put in charge of all forces in southern Louisiana. He settled his family at Cypress Grove, a plantation near Baton
Rouge, where he invested heavily in land speculation and slave-owning.
He was given a series of military promotions and assignments, all in the
southwest.

Troubles along America's border with Mexico created tensions between the United States and Spain, especially over the inflammatory issue of the Republic of Texas, which had declared its independence in
1836. Following years of campaigning for annexation, Texas was admit-

ted to the Union in December 1845. General Zachary Taylor was dispatched to the Rio Grande to stand guard.

When the Mexican War commenced following an ambush or skirmish in early May 1846, Taylor led his troops against the Spanish at Palo Alto and Resaca de la Palma (May 8 and 9, 1846), and won easy victories. His capture of Monterey (September 2 through 25, 1846) and triumph at the battle of Buena Vista (February 22 and 23, 1847) catapulted Taylor to national prominence. At Buena Vista, Taylor's men were outnumbered three to one, yet American forces dealt General Santa Anna a stunning defeat. Taylor's American rival for headlines, General Winfield Scott (by contrast nicknamed "Old Fuss and Feathers") captured Vera Cruz and marched on to Mexico City. But it was Taylor who had captured the national imagination and who emerged a national hero.

Taylor has been caricatured as someone with little interest in politics, someone who had never even bothered to vote in presidential contests before his own candidacy. This image considerably distorts Taylor's views, as he was far from indifferent to national affairs.

Zachary Taylor believed military duty required that he hold himself above politics. But privately, he repeatedly revealed political views: "I am & always have been a democrat of the Jeffersonian school, which embodies very many of the principles of the Whigs of present day." When Taylor was being courted to run for high office in 1847, he expressed his admiration for Henry Clay, the thrice-defeated presidential candidate for the Whig party. Taylor confessed he would have voted for Clay in 1844 "had I voted at all, which I have never done for any one of our chief magistrates since I entered the army or before, which is near forty years." The fact that Taylor never voted was misinterpreted by contemporaries and subsequently by historians as a sign that he was apolitical. Rather, Taylor clearly *declined* to vote, as a matter of principle.

Whig political kingmaker Thurlow Weed struggled to woo this maverick general because the nation was so divided over slavery, the tariff, and other sectional issues that he believed only a transcendent patriotic figure could rise above the fray and scrape together enough votes to win. Ironically, a Louisiana slave owner was perceived as a vital center to hold the country together at a time when disunion over slavery was tearing the nation apart — simply because of his reputation as a war hero.

When Taylor was first approached to run for the presidency, he dismissed the idea out of hand. Then, slowly, it began to appeal to him. He wanted to transcend party and still get elected. He believed in 1847 that he should refuse any offer from the two major rival parties because "I would become the slave of a party instead of the chief magistrate of the nation should I be elected." His political naivete was legendary. Despite his lofty hopes, Taylor succumbed to Whig party blandishments.

When the Whigs nominated Taylor, party leaders sent a letter to his Louisiana home on June 18, 1848. They were alarmed when, weeks later, Taylor had not responded. One complained, "The delay is ominous." Finally, Taylor's acceptance arrived on July 15, with his apologies for tardiness: the invitation had been held in the dead letter office, as the frugal Taylor had instructed his local postmaster not to deliver mail that had not been prepaid. The Whigs used this story for political capital, but it underscores Taylor's deliberate and studied indifference.

Perhaps Taylor was more shrewd than he appeared. He confessed in 1847: "If I am elected at all, it will be by a union of a portion of Whigs, Democrats & native votes." This from a man allegedly with no interest in politics.

His professed goal was to foil presidential hopeful Winfield Scott. As Taylor wrote in 1847: "Between ourselves, General Scott would stoop to anything however low & contemptible as any man in the nation, to obtain power of place . . . I look on him as heartless & insincere an individual as exists. I would undergo political martyrdom rather than see Gen. Scott or Cass elected." Taylor thwarted Scott's attempts to win favor with the Whigs, and captured the nomination for himself. Nominee Taylor eventually beat Democratic candidate Lewis Cass and Free Soil candidate Martin Van Buren as well.

His election was even more remarkable, as Taylor was willing to place his name in the ring but refused to campaign. The Whig candidate was unusually reticent and made only a handful of public appearances in Mississippi and Louisiana. Despite this handicap, his bland reassurances struck a chord: "I hope some compromise will be entered into between the two parties slavery & antislavery which will have the effect of allaying violent passions on both sides." Taylor's soothing message seemed to douse sectional tensions threatening to ignite. Taylor appealed as a moderate, but a strong military man who might quell divisive forces erupting within the national political arena.

Most divisive was the question of slavery in the territories. President Polk had wanted Congress to extend the Missouri Compromise westward, a measure that passed in the Senate but was voted down by the House of Representatives. Michigan Senator Lewis Cass proposed that settlers should determine for themselves a state's policy on slavery. In 1848 the Democrats, in an attempt to offend no one, refused to adopt a position on this question of "popular sovereignty," and rejected both the Wilmot Proviso of 1846 (stating that no territory acquired through the

Mexican War would allow slavery) and John C. Calhoun's resolution to condemn the proviso in 1847.

The "Conscience Whigs," as exemplified by Charles Sumner of Massachusetts, had bolted the party over the slavery issue, while the "Cotton Whigs" (known as such because many were textile magnates) had opposed the Mexican War but were willing to ride the general's coattails into the White House. The alliance of the "lords of the loom" and the "lords of the lash" proved a winning combination.

In letters prominently circulated by Whig supporters, Zachary Taylor had pledged not to veto whatever Congress determined was best concerning slavery in the territories. Southerners believed Taylor was their ally: when he was asked during the campaign if his election would enhance or diminish slave property, Taylor replied he owned three hundred slaves and thought no further answer necessary. At the same time, Northern Whigs and free-soil advocates such as Abraham Lincoln took Taylor at his word and believed his promise to let Congress decide. Taylor won eight of the fifteen slave states and seven of the fifteen free states, accumulating 163 electoral votes to Cass's 127 on America's first national election day. (A bill passed in 1845 had established a uniform day for presidential balloting.)

Taylor offered the shortest inaugural talk to date, heralding "attachment to the Union" and pledging "whatever dangers may threaten it, I shall stand by it and maintain it in its integrity to the full extent of the obligations imposed and the powers conferred upon me by the Constitution." Taylor intended for Congress to control the country's destiny, and declared he would not exercise his veto power. He offered pablum when he argued: "We should abstain from the introduction of those exciting topics of a sectional character which have hitherto produced painful apprehension in the public mind."

But the gold nuggets discovered at Sutter's Mill in California in January 1848 contributed significantly to the sectionalist miasma consuming Congress. Although the government wanted to include this Pacific Coast territory as a state as settlers poured into the region, slavery remained a stumbling block. Antislavery politicians blocked every effort to allow territories that permitted slavery to join the Union, while Southern statesmen vetoed any move to admit a region that would prohibit slavery. Congressional debates were disintegrating into fistfights — and no one was sure how long the center would hold.

Following elections in 1849, the Free Soil party sent twelve men to the House of Representatives. Political troubles were painfully evident when it took three weeks and more than sixty ballots before a Speaker of the House was selected, signaling the growing political stalemate within Congress.

Taylor was increasingly disturbed by the "intolerant" and aggressive attitude of Southern "fire-eaters" — as secessionist firebrands were known. In October 1849 John C. Calhoun had called for a group of slave-state congressional delegates to convene in Jackson, Mississippi, to discuss partisan issues and announced plans for another Southern secessionist caucus the following June. Taylor was particularly aggrieved by the participation in this movement of his former son-in-law, Mississippi Senator Jefferson Davis (Davis's first wife, Sarah Knox Taylor, had died after less than a year of marriage). Taylor condemned Davis as a disloyal conspirator and wanted to take action to stem political treachery.

In his message to Congress in January 1850, President Taylor endorsed the admission of California as a (free) state immediately, and the admission of New Mexico as a state in the near future. By the end of January 1850, Henry Clay offered a series of resolutions, nearly forty in number, designed to satisfy both North and South. This omnibus bill included statehood for California and division of New Mexico into two territories (with no mention of slavery attached to either proposal), abolition of the slave trade in D.C., and a strengthening of the fugitive slave law in the North and South.

Taylor actually opposed Clay's plan, arguing that California's admission as a state should not hinge on all these tangential issues. Vigorous debate followed and stimulated oratorical flourish as Daniel Webster, William Seward, and others made famous speeches. John C. Calhoun died three days after his own rousing address, and, indeed, before issues could be resolved. So did President Taylor.

Taylor fell ill on July 4, 1850, after participation in Independence Day festivities in the broiling heat. Crowds gathered around the White House as the president lay sick in bed, stricken with horrible stomach pains. Hourly bulletins updated his condition, until his death on July 9.

One of his last official acts was to sign the Clayton-Bulwer Treaty (an agreement to settle disputes between the United States and Britain over a canal project in Central America), which addressed the only significant foreign policy issue during Taylor's presidency.

The historian David Potter argued that Taylor's death was "one of those extraneous events which suddenly and in an irrational way alter the course of history." Taylor's vice president, Millard Fillmore, took the oath of office and replaced the entire cabinet (the first and only time such a clean sweep has taken place). Championing Clay's Compromise of 1850, Fillmore stewarded the measures, piece by piece, through Congress in the autumn following Taylor's death.

A small number of scholars have argued that if Taylor had lived, the Civil War would have come sooner. And an even smaller number allege that Taylor may have been murdered by proslavery advocates.

This theory recently gained attention when a scholar argued that it was not cold buttermilk and raw cherries (what the president allegedly consumed on the raging hot day of the Fourth of July celebration) that contributed to Taylor's untimely death, but arsenic. Following the scholar's demands for exhumation, 150 spectators gathered in Louisville, Kentucky, in July 1991 to watch Taylor's body, draped in a flag, get carried off for testing. Since no traces of poison were detected, this conspiracy theory, like Taylor's remains, was finally laid to rest. Horace Mann's impression in 1850 provides a fitting epitaph: "He has the least show or pretension about him of any man I ever saw; talks artlessly as a child about affairs of State, and does not pretend to a knowledge of anything of which he is ignorant. He is a remarkable man in some respects and it is remarkable that such a man should be President of the United States."

— CATHERINE CLINTON
Baruch College

Millard Fillmore

1850–1853

b. January 7, 1800
d. March 8, 1874

"God save us from Whig vice presidents," cried an anguished Ohio Whig in November 1850. He was referring to John Tyler, who still epitomized apostasy to Whigs because of his party-wrecking betrayals in the early 1840s after he succeeded William Henry Harrison, and to Millard Fillmore, who assumed the presidency upon the death of Zachary Taylor on July 9, 1850. The Whigs should have known that Tyler, a Virginia slaveholder, strict constructionist ideologue, and ex-Democrat, had little commitment to their party and its principles. But Fillmore, a New Yorker and a loyal Whig from the party's founding in 1834, had seemed an eminently safe choice for vice president in 1848. Indeed, he was added to the Whig ticket precisely to appease Northern Whigs; they had opposed Taylor's nomination because he was a Southerner, a large slaveholder, and a man of little commitment to Whig programs. He, not Fillmore, seemed likely to become another John Tyler. But Fillmore quickly antagonized Northern Whigs at least as much as Taylor once had. He did so because of the ultimate irreconcilability of his two main objectives as president: "to save the country [and] the Whig party, if possible."

Fillmore had spent his early years trying to elevate himself from the poverty of hardscrabble farming into which he had been born in 1800. A genuinely self-made man, he had by dint of hard work, self-education, and diligent persistence become a successful lawyer and civic leader in Buffalo by the late 1820s. Strikingly handsome, robust in physique, impeccable in dress, and stolid in temperament, he was elected to the legislature in 1828 and 1830 and to Congress in 1832, 1836, 1838, and 1840. Intelligent, if never flamboyant, Fillmore developed expertise in Whig programs to promote economic development. He pushed improvement of the Erie Canal system as a legislator, and later, as state comptroller in 1847 and 1848, he brilliantly revised New York's free banking system, making it the model for the later National Banking System. As chairman of the House Ways and Means Committee between 1841 and 1843, Fillmore wrote the Whig Tariff of 1842 as well as two earlier tariff bills that the hated Tyler vetoed.

This record made Fillmore a leading contender for the Whigs' vice-presidential nomination in 1844. But Thurlow Weed, boss of the state Whig organization, insisted that he run for governor instead, a race Fillmore narrowly lost. Two years later, when Whig prospects were considerably brighter, Weed blocked Fillmore's renomination for governor. By then Fillmore had aligned himself with a faction of self-defined "conservative" Whigs in the state party who bitterly resented its domination by Weed and his favorite disciple, William Henry Seward. Conservative Whigs also disliked the agenda of Weed's majority wing: state constitutional revision, an outreach to immigrants and Catholics and a simultaneous rejection of potential nativist allies, and strenuous opposition to slavery and slavery extension in order to win over third-party antislavery men. Ever intensifying, this rancorous factional feud among New York's Whigs decisively shaped Fillmore's remaining political career.

Fillmore's conservative allies, indeed, secured his vice-presidential nomination in 1848 primarily to prevent Weed from getting that post or a subsequent position in Taylor's cabinet for Seward. To keep Fillmore from brokering federal patronage for New York, Weed then secured Seward's election to the Senate, launching an instant rivalry between the two for the administration's favor.

That rivalry made Fillmore's experience as vice president a humiliating nightmare. The clever, energetic Seward ran rings around the dignified, seemingly torpid Fillmore. Weed's men, not conservatives, got the state's juiciest federal jobs, even in Buffalo itself. Seward, not Fillmore, attended cabinet meetings and advised the administration on policy. "My recommendations in my own State and even in my own city have been disregarded," the mortified Fillmore complained in July 1849. "My advice has neither been sought nor given as to the policy to be pursued." Nor had that freeze-out thawed a year later when Taylor suddenly died.

When Millard Fillmore became president in July 1850, therefore, he had ample incentive to seek revenge against the scornful Sewardites. Nor was he alone. Party regulars from other states, who had been shortchanged by Taylor's cabinet in the distribution of jobs and who particularly resented Taylor's calculated — and politically foolish — retention of Democrats in many positions, howled for a massive purge of Taylor's appointees as soon as Fillmore took office. Fillmore's beloved party, in short, had been shattered by patronage quarrels, and restoring harmony to it became a top priority for him. He immediately replaced the widely reviled cabinet with different Whigs who represented aggrieved factions. But he shunned the advice of his most vindictive allies for a massive bloodletting, which he realized would be counterproductive. In 1850 and 1851 he removed only a very few Sewardites from jobs in New York and other states and allowed the vast majority to keep them. To balance the books by rewarding Whigs who had been ignored by Taylor, he relied instead primarily on appointing them to the jobs Taylor had allowed Democrats to keep. The most vindictive of the Whig regulars continued to call for a purge of their Whig enemies, but Fillmore's handling of the vexatious patronage question was fair, adroit, and constructive.

To help "save" the fragmented Whig party, Fillmore also tried to draw a line between Democrats and Whigs in his annual messages to Congress in December 1850 and 1851 by reviving the economic issues that had long divided the two parties. He requested a new, higher tariff with specific duties to replace the low Democratic Walker tariff of 1846, and he made an impassioned case for the necessity and constitutionality of federal subsidies for internal improvements. Yet these pleas got nowhere with a Congress controlled by Democrats, and even Fillmore knew his proposals were merely expedients. What divided and threatened his party even more than grievances about federal jobs was the sectional crisis over the issue of slavery expansion which he inherited from Taylor, who in turn had inherited it from James K. Polk.

Because that crisis also threatened to disrupt the Union, it put Fillmore's twin objectives of saving the country and saving the Whig party into conflict with each other. Though he would try, he could not please both Northern and Southern Whigs. And when he sided with Southerners in order to stop secession, furious Northern Whigs would ask God to save them from yet another Whig vice president.

When Zachary Taylor died, the divisive problem of what to do about slavery in the lands acquired from Mexico in 1848 — lands known as the Mexican Cession — remained unresolved. Some Northerners continued to call for enactment of the Wilmot Proviso, which would bar slavery from all the new territory. Some Southerners threatened secession should the proviso pass. But since January 1850, congressmen had bickered primarily less over the Wilmot Proviso than over the alternative plans offered by Taylor and Henry Clay, both of which sought to defuse the explosive issue. Taylor insisted that Congress admit California and New Mexico as free states and pass no other legislation for the cession whatsoever. Clay wanted Congress to admit California as a free state; organize formal governments for the other territories without any restrictions on slavery; and fix the disputed boundary between Texas and New Mexico. In addition, Clay advocated two measures that Taylor had ignored — a new, considerably harsher federal fugitive slave law and the abolition of public slave auctions (but not slavery itself) in Washington, D.C. Taylor's plan lacked the votes to pass; but its supporters, along with Southern Democrats who hated both proposals, had stopped Clay's compromise package, even though it had been changed considerably by Democratic amendments since January to make it more palatable to them.

By July 1850, the impasse had escalated to dangerous levels. Taylor and Clay openly vilified each other and their respective plans. Southerners threatened secession should California be admitted as a free state without concessions to slaveholders in the remainder of the Mexican Cession. Yet northern Whig supporters of Taylor's plan, who far outnumbered their Southern colleagues in the House, vowed never to allow the organization of territorial governments without the Wilmot Proviso or to allow Texas an inch of land west of the Nueces River in its boundary dispute with New Mexico. More serious, military conflict appeared imminent at Santa Fe between United States troops and the militia of the state of Texas, which claimed all the land east of the Rio Grande as Texas soil. Taylor sought reinforcements to block the Texans, Southern Whig congressmen threatened to impeach him if he sent them, and, across Dixie, Southerners vowed to send militiamen to aid Texans against federal troops should fighting break out.

This was the inflammatory situation that Fillmore inherited. He and his closest Buffalo advisers had ardently supported Taylor's plan since the summer of 1849, but by early July he recognized that Clay's compromise package had the best chance to pass and to alleviate the sectional crisis. Several days after becoming president, moreover, he apparently saw for the first time a letter from Texas Governor Peter H. Bell to Taylor, threatening to ask the Texas legislature for militia to march on Santa Fe when it met on August 13, 1850. Convinced of the urgent necessity of settling the boundary dispute, something only Clay's plan could do, the new president filled his cabinet with procompromise men, most importantly Daniel Webster, who became secretary of state.

In August, Fillmore and Webster prodded Congress to resolve the Texas–New Mexico boundary dispute and pass the other compromise measures that had seemed doomed after a series of Senate votes on July

31. Most Southern Whigs and Northern Democrats already supported the compromise. Passage in both houses of Congress now required winning support from a considerable fraction of the Northern Whigs, who had backed Taylor's plan and believed the compromise would violate their repeated pledges to keep slavery out of any part of the Mexican Cession. Fillmore and Webster provided Northern Whigs with a modicum of political cover by stoutly rejecting Texas's claims to eastern New Mexico in a masterly message to Congress on August 6 which broke the logjam. And they twisted the arms of recalcitrant Northern Whigs by threatening to withhold patronage nominations from them unless they went along. Through this skillful use of the carrot and the stick, Fillmore and Webster persuaded enough Northern Whigs to vote for, or at least not vote against, the pro-Southern parts of the compromise — the new Texas boundary far west of the Nueces, the new Fugitive Slave Act, and the organization of Utah and New Mexico with a promise to admit them as slave states should they so choose — to secure their passage.

Fillmore, in sum, played an indispensable role in securing passage of the Compromise of 1850, which immediately (if only temporarily) soothed frayed sectional tempers. As Fillmore well knew, however, it did not restore sectional harmony to his party or avert the danger of secession. In at least four Deep South states, hotheads called the compromise a complete Southern defeat and demanded secession. They were stopped by new Union parties, which, like Whigs in other slave states, insisted that the new Fugitive Slave Act be rigorously enforced by Fillmore's administration and that Northern Whigs and Free Soilers never attempt to change it or impose the Wilmot Proviso on federal territories. They insisted, in sum, that anticompromise Northerners acknowledge the compromise as a final settlement of all slavery questions.

Fillmore and Webster certainly considered it final. Fillmore said so in his annual messages of 1850 and 1851, and within a month of the compromise's passage Webster told a Massachusetts lieutenant that "the present administration will not recognize one set of Whig principles for the North and another for the South." But Northern and Southern Whigs had always taken different positions on slavery, and most Northern Whigs simply refused to have Fillmore's administration impose the straitjacket of finality on them. They continued to demand imposition of the proviso on Utah and New Mexico and revision of the Fugitive Slave Law's harshest provisions. Throughout the remainder of 1850 and 1851, therefore, brutal battles in Northern states occurred between proadministration and anticompromise Whigs over platforms and nominations. When those divisions contributed to massive Whig defeats,

Northern anticompromise Whigs did more than call on God to save them from Whig vice presidents. They also began to promote Winfield Scott for the Whigs' presidential nomination in 1852 so that neither Fillmore nor Webster could get it.

Southern Whigs, in contrast, almost unanimously sought Fillmore's nomination. He had little interest in running in 1852, but he was determined to restore harmony among feuding Northern Whigs and between Northern and Southern Whigs. The biggest obstacle to that goal was implementation of the new Fugitive Slave Act, a sine qua non to Southern Whigs and Northern administration men. Most Northerners, in fact, grudgingly obeyed the law, but there were a few spectacular "rescues" of captured fugitive slaves by antislavery forces which deeply embarrassed Fillmore's administration. His problem, therefore, was how to implement the law without appearing blatantly to favor the South at the expense of Northerners' sensitivities.

His solution was ingenious, if barely appreciated by his Northern Whig opponents or by later historians. The administration was obliged to enforce neutrality laws against Southern filibusterers who were illegally seeking to invade Cuba so as to annex it and add it to slavery's realm. For two years, therefore, Fillmore pursued a determinedly evenhanded course of law enforcement to show he played no sectional favorites. Almost every time he issued a proclamation denouncing or calling out troops to quell Northern violators of the Fugitive Slave Law, he issued within days — indeed, often on the very same day — similar orders for the use of federal authority against Southern violators of the neutrality laws.

The biggest threat to American neutrality, however, did not come from Southern filibusterers, and squelching that threat constituted one of the signal, if unsung, achievements of Fillmore's administration. In December 1851, Louis (Lajos) Kossuth, the exiled leader of a failed Hungarian revolution against the Austro-Hungarian Empire, landed in New York. Greeted everywhere he went by huge and adoring throngs, Kossuth presumptuously demanded that the United States abandon its policy of nonintervention in European affairs: by immediately recognizing Hungary's independence; by raising arms to renew the rebellion; by warning Russia, whose intervention in 1849 had helped Austrians crush the revolt, that the United States would send troops to aid Hungarians if it again did so; and by sending an American fleet to the eastern Mediterranean to give teeth to that warning. Because Kossuth's defiant appeal was especially popular among new German immigrants, whom Whigs and Democrats alike expected to vote in unprecedented numbers in

1852, denying his demands had a real political cost, especially as Democratic aspirants for the presidency recklessly endorsed them. Yet Fillmore resolutely refused to change American policy, and at the Whig national convention in 1852 his backers would insist upon a platform plank that reiterated America's commitment to nonintervention.

None of this averted a divisive intraparty battle before and during the Whigs' 1852 national convention between Scott's anticompromise supporters and proponents of Webster and Fillmore. In the end, Whigs could only paper over their deep fissures. With a majority of delegates, the Webster and Fillmore forces imposed a procompromise platform on the party. But because Webster's men stubbornly refused to combine with the far more numerous Fillmore delegates, Scott secured the nomination. Many Northern Whigs openly repudiated the platform; many Southerners openly repudiated the candidate. With the Whig vote decimated in the Deep South and fractured in the North, Scott carried only 4 of 31 states against his Democratic opponent Franklin Pierce. Democrats also won huge majorities in Congress. It would be the last presidential race the Whig party would ever run. Fillmore had helped save the country from disunion, at least temporarily, but he had manifestly failed to save his beloved Whig party from division and defeat.

Fillmore and his closest administration advisers acknowledged as much. Within months of his leaving the White House they plotted to form a bipartisan Union party of compromise supporters to replace the Whigs. For two years those efforts came to naught. But the apparently final sectional rupture of the Whig party by the Kansas-Nebraska Act in 1854 and the stunning success of the anti-Catholic, anti-immigrant, and virulently anti-Seward Know-Nothing party in that year's elections gave conservative Whigs who admired Fillmore one last chance to smite his longtime enemies and to save the country from disunion. They joined the Know-Nothings and made Fillmore the nativist party's presidential nominee in 1856. He carried only Maryland and garnered only 21 percent of the popular vote. By then, the political future lay with the new Republican party, which Sewardites and the vast majority of Northern Whigs had joined. Fillmore's active political career was over. He would live until 1874, long enough to witness a four-year sectional bloodbath that he had worked to avert during his presidency. The man who had achieved so much through indefatigable striving to remake himself could not, in the end, achieve his most cherished goals.

— MICHAEL F. HOLT
University of Virginia

Franklin Pierce

1853–1857

b. November 23, 1804
d. October 8, 1869

F RANKLIN PIERCE easily won the presidential election of 1852, defeating his Whig rival, General Winfield Scott, resoundingly, despite Scott's significant accomplishments during the Mexican War. The youngest man elected president to that time, Pierce had not been as prominent a national figure as most earlier presidents. He had no credentials as either a major political figure or as a statesman. And he was not a military hero, although he had also served in the Mexican War. His tenure in legislative offices had been comparatively brief. Clearly a dark horse candidate at the outset of the race, he was his party's compromise choice, emerging out of a group of other, quite similar candidates, although most of those in the Democratic field were more nationally prominent than he was. But, despite his relative lack of renown, Pierce was not an unusual selection as the Democrats' nominee.

The son of a leading New Hampshire politician, Pierce graduated from Bowdoin College and then studied law. He quickly entered local politics and rose to a dominant position in the Democratic party of New Hampshire, as a member of the Concord Regency leadership group at the state capital. He served in the state legislature while his father was governor, then as a member of Congress for four years, and finally as United States senator from 1837 to 1842. He resigned from the Senate to practice law back in Concord before volunteering for Mexican War service. He was well known enough in the mid-forties to have been offered the attorney generalship by President James K. Polk in 1846.

In the run up to the presidential nomination in 1852, Pierce drew support not only because of his long career as a dedicated Democrat activist, but also for his unblinking support of the party's policy commitments. It was recognized and celebrated that Pierce's political life had been imbedded in the partisan world of Jacksonian democracy. He was an unswerving acolyte of the Democratic ideological heritage, following along the pathways first laid out by Old Hickory, Andrew Jackson, almost thirty years before: that limited government secured freedom, and powerful government threatened it; that states' rights and local control of one's destiny were preferable to any centralizing impositions by the national government; and that the nation's leaders must be forever militantly alert to threats to the nation from unfriendly foreign powers or divisive groups on the domestic scene. As Pierce put it in his inaugural address, "if the Federal Government will confine itself to the exercise of power clearly granted by the Constitution, it can hardly happen that its action upon any question should endanger the institutions of the States or interfere with their right to manage matters strictly domestic according to the will of their own people."

The forty-eight-year-old party stalwart's partisan dedication was considered a prime political virtue in this "party period" of American history. Partisan activists like Pierce were the interchangeable foot soldiers and officers of a major movement dedicated to expanding and defending American liberties from the threats they faced. "Young Hickory of the Granite Hills," as he was known, was able to settle into the footsteps of his illustrious party ancestors because he believed as they did, and was as committed to the Jacksonian cause as they were, and because other Democrats saw the virtues of such qualities.

It was not all smooth sailing. Pierce's personal and family life imposed a significant burden on him. His wife, Jane, described by one of Pierce's biographers as "shy, retiring, frail and tubercular," had never been happy in her husband's political world and hated living in Washington. The Pierces had lost two sons to early deaths. Their only surviving child, Benjamin, eleven years old, was killed before his parents' eyes in a train wreck just before the family moved to Washington. As a result, Mrs. Pierce, overcome with melancholia, was largely a recluse during her husband's presidential years. Pierce himself already had a reputation as a heavy drinker, a reputation that caused some concern among those close to him politically, although how much that affected subsequent events is unknowable.

Pierce's one term in the White House was marked by the reopening and deepening of the nation's sectional wounds because of the controversy over the Kansas-Nebraska Act. That this erupted as it did was a surprise to Pierce and his colleagues. Party leaders had spent a great deal of effort dampening down the difficult sectional tension that had so strained the party since 1846, and they believed that they had succeeded, thanks to the legislative compromise of 1850 and the subsequent declaration by both the Democrats and Whigs that each considered that legislation a final settlement of all outstanding sectional questions.

But they were very wrong. A major political explosion occurred in 1854 and grew powerful enough thereafter to alter fundamentally the existing trajectory of American politics and promote and accelerate the forces that would culminate in civil war. Pierce presided over these momentous events unable, in the eyes both of many of his contemporaries and of historians since, to prevent them from worsening, or to manage them effectively once they did. He was to become better remembered for his lack of grip in a very difficult situation than for anything else.

Yet this all lay in the future when Pierce traveled from New Hampshire to the White House in the late winter of 1853. Immediately before him were the selection of his advisers and the refinement of a political agenda. At the outset, there was the usual scramble for appointments to office, combined with the usual disappointments and anger that always accompanied patronage activities. Pierce worked very hard in choosing people to assuage old party feuds, the embers of which still glowed in many places. He made it clear that he wanted no section of the party proscribed due to past actions. In the words of his outstanding biographer, "schism was to be banished."

His cabinet selections tried to balance the various party and regional groups that had claims to office, and included New York's William Marcy as secretary of state, and Jefferson Davis of Mississippi as secretary of war. Nevertheless, there was grumbling about some of Pierce's other choices and his general philosophy of reconciliation, particularly about his decision to readmit the Northern Free Soil Democrats to party councils despite their destructive bolt from the party four years before. Critics thought that the president went much too far in trying to mollify all factions. But, like many patronage situations, much of this seemed to go beyond one person's capabilities to settle quietly and without hurt feelings and anger. It was a no-win matter for the president, as it had been for others before him, whatever his well-intentioned notions of reconciliation.

Pierce brought to his new office a mindset about its role and powers originally formulated by his Democratic predecessors, Jackson and Polk, both of whom had exercised strong executive authority in what still remained a legislatively dominated political system. Jackson had defied a hostile Congress and faced down extremist Southerners in South Carolina. Polk had worked with his fellow Democrats on Capitol Hill to enact a wide-ranging legislative program imbedding the party's conceptions of appropriate financial, tariff, and land policies into law and had faced down dissident Democrats in Congress when they went against the party's ideals and commitments.

Pierce had every reason and inclination to follow Polk's example. When he took the oath of office in early March 1853, both the House and Senate were safely Democratic. Although there was some disagreement among party members about policy priorities and appropriate initiatives, all Democrats were united, as they had been eight years earlier, behind a notion of eternal vigilance against persistent Whig efforts to expand and alter the nature of the American government, to the country's detriment. Pierce and his colleagues expected trouble from their opponents and were ready for it.

But it was not from the Whigs that trouble emerged. The ambitious Democratic leader of the Senate, Stephen A. Douglas of Illinois, precipitated the events that have forever defined Pierce's years in office. The chair of the Senate Committee on the Territories, Douglas was determined to organize the areas west of the Mississippi River into federal territories for a number of reasons. Normal settler pressures to enter ostensibly unoccupied farmable land, boom times that further invigorated those pressures, and rivalries over the future route of a transcontinental railroad all played a role in Douglas's determination to move to organize the "Nebraska country," the areas across the river from the states of Missouri and Iowa.

The bill was introduced into what quickly became a political cauldron, one which neither Pierce nor Douglas could control. Although the Thirty-third Congress had Democratic majorities, the potential for fault lines to appear among them was always present, and Douglas's actions powerfully exacerbated them. In the early stages, Douglas realized that he needed Pierce's support as party leader to assure the necessary votes, support that the latter was willing to give for ideological and partisan reasons. All his political instincts led the president to seek to work with his party colleagues and accept the fact that differences were always present, but to strive to overcome them and to involve all groups in the party.

Pierce also had practical and personal reasons to join with Douglas: he was already facing a number of difficulties with senators over his patronage policies and needed to work with them, a need that fit comfortably with his political approach as well. Whatever the primary reasons,

Pierce was, in all that followed, loyal to his ideology and to his party as its embodiment. He fell into line with Douglas and supported what the Illinoisan was attempting to accomplish. And that support had enormous consequences for everyone involved.

The question of slavery emerged immediately although it was not a prime concern for Douglas, a fact that led to a series of missteps on his part — missteps that dragged Pierce into the situation directly as well. At the center was the question of whether slavery could be brought into the territories in question. Under existing law, it could not be, because by the terms of the Missouri Compromise legislation of 1820, its introduction was forbidden in that part of the Louisiana Purchase north of longitude 36° 30', that is, into the very Nebraska country that Douglas was seeking to organize.

But Southerners, led by a number of Missourians, many of whose constituents had already drifted into the portion of the territory that is now Kansas, bringing their slaves with them, could no longer accept the Missouri restriction. Southern senators put a great deal of pressure on Douglas to modify his bill accordingly. It was here that Pierce and Douglas ran up against the power of the Democratic nabobs in the Senate, particularly the Southern-dominated "F Street Mess," Senators Mason and Hunter of Virginia, Butler of South Carolina, and Atchison of Missouri. The president and Douglas had to deal with them if they hoped to get the bill through, and the price of doing so was high. Douglas agreed to the Mess's terms because he wanted the bill, because he was sensitive to the need for Southern votes, because he was relatively indifferent about the spread of slavery in the west, and because he believed that the Compromise of 1850 had introduced a more "Democratic" way of deciding on whether slavery should exist in a particular place, that is, by the vote of the people there, not by the dictates of the central government.

"Popular sovereignty" had been placed in the Utah and New Mexico territorial bills in 1850, and Douglas now inserted it into what was the Kansas-Nebraska Bill as well. But the Southerners demanded — and got — more. They wanted the bill *explicitly* to repeal the previous restriction from the 1820 law. Douglas complied, although he feared that such overt action would kick up a serious political storm. It did. The Southerners had overstepped themselves by these demands and dragged everyone else along with them toward disaster.

Pierce, like other Democratic leaders from the North, was always quite alert to the political pressures emanating from the party's important Southern wing. He had been a strong supporter of the compromise measures of 1850, including the controversial Fugitive Slave Law, which continued to ruffle many Northerners. Most Northern Democrats were also strongly committed against any further agitation of the slavery question. Such roiling of the waters, they saw, stemmed from a few very mischievous fanatics intent on hurting both the Union and the Democratic party. That attitude shaped a great deal of their support for Douglas and his bill in the battle that followed.

The problem was that despite all of their efforts, the sectional divisions of the forties still plagued the party. Other Northern Democrats quickly came out in opposition to Douglas as they had against the expansion of slavery into the Mexican Cession a decade before. Pierce, however, made support of the bill, no matter how locally controversial in many Northern constituencies or politically dangerous to some representatives and senators, a test of party loyalty. This strategy was initially successful. The bill passed both Houses of Congress by early May, relatively easily in the Senate but with much difficulty in the House, ultimately needing some Southern Whig votes.

But, by then, the issue had blown up beyond the normal give-and-take of legislative deal making largely confined to Washington, into something quite a bit more profound. The repeal of the Missouri Compromise line provoked an unprecedented uproar, first among congressional Free Soilers, then in the Northern states where many loud and denunciatory anti-Nebraska meetings convened to express shock and outrage. And, as a result of their constituents' reaction, Democrats, both in Congress and among leaders and activists on the local scene in the Northern states, split with the administration.

No one expected the anti-Nebraska protests to be as potent as they were. What no one could know was that the political system was already under great stress, which added fuel to the flames. The appearance and spread of a powerful backlash against a recent surge of immigration, largely Irish and Catholic, into the Northern states, was affecting the loyalty of some Democrats and building into a quite disruptive force against the usual two-party politics of the day. Hundreds of thousands of voters, particularly in the North, were angry enough, and fearful enough, about the alien hordes, to demand a political solution to the immense threat posed to their well-being.

The old parties, however, were, in the nativists' view, unwilling or unable to do anything. The Democrats were too dependent on Catholic voters, while the Whigs had recently been courting them as well. The nativist reaction to these insults began seriously to undermine the existing two-party-based political stability even before Douglas's bill had exploded into controversy. In local elections in 1853 and 1854, nativist electoral tickets won widespread victories, a pattern that continued into the more significant fall elections of 1854.

The Kansas-Nebraska matter fed additional discord into this political discontent, even after Douglas's bill had moved successfully through Congress. Pierce and his party colleagues believed that the worst was over with the passage and signing of the Kansas-Nebraska Bill into law, that political attention would turn to other matters that were nonsectional in nature. But things went from very bad to much worse. The congressional and state elections of 1854 were a disaster for the Democrats, as their nativist Know-Nothing and their new Free Soil Republican opponents surged forward. The next House of Representatives would find the Democrats in the minority.

At the same time, Kansas itself exploded into violence, which persisted, with only occasional respite, for the rest of Pierce's administration (and beyond), as violent proponents of slavery restriction, and of its expansion, battled to win the war on the scene by establishing themselves as the popular majority there and by discouraging their opponents from coming into the territory or staying if they did. In Congress as well, the rhetorical war between the free-state and slave-state forces continued, culminating in the violent caning of the antislavery firebrand Senator Charles Sumner of Massachusetts, by an infuriated Southern congressman.

The winner in all of this was the new Republican party. Built on the energy unleashed by the protests against the Kansas-Nebraska Bill, its claims that the Pierce administration was pro-Southern, under the thumb of an avaricious, unyielding, aggressive slavocracy, in fact, and that the Whigs were no better, attracted many Northern Whigs and former Free Soilers to its ranks behind its sectional platform. It gained additional strength from the Know-Nothings, disgruntled Democrats, and new voters determined to halt the march of Southern aggression in the territories. The power of that imagery affected everything — the process of political realignment had begun, a process further fostered by the continued festering of the Kansas issue.

Worst of all, Pierce seemed unable to do anything to alleviate the situation. Each of the four governors of the territory whom he appointed in rapid succession were failures, seeming only to make matters worse when they were on the scene. Pierce had no wish to use armed force to suppress the war in the territory, and he was unable to think of any alternative, which made him seem weak and feeble in the face of the violence. The continued bloodshed in the west, the weak responses, and the inability of the administration to get out of the mess it was in, sapped confidence in Pierce, damaged his reputation, and ended any hope that he might have nursed for renomination by his party. He remained caught in a situation only partially of his own making. He had created little of the conditions that had led to political implosion, but he had proved unable to do anything to stop what was occurring. Further electoral losses for his party followed as the situation worsened. It was an appalling time for every Democrat.

Not everything that occurred during Pierce's presidency occurred on the plains of Kansas. Other legislative battles and policy differences emerged as expected, given the long-standing pattern of confrontation between the national political parties. In the domestic realm, the Polk administration in the previous decade had secured a large part of the traditional Democratic agenda. It had established an independent treasury to hold the federal monies received from customs duties and land sales, and significantly lowered the tariff. As a result, Pierce and his colleagues pushed few new initiatives, focusing, rather, on the need to protect what was already in place from any renewed Whig challenges.

There was one domestic matter that splintered the Democrats in an unusual manner for this era. American expansion westward provoked repeated demands for federal aid for internal improvements projects: the deepening, clearing, and maintaining of rivers and harbors, and the projected Pacific railroad and other routes, were all seen as a major promoter of settlement and the prosperity of the western regions. The propensity of Congress, with its local orientation, to appropriate money for such projects was a reality that Democratic presidents, going back to Andrew Jackson, had had to be alert to, and ready to turn back via the presidential veto, unless they could be persuaded that a particular project was national, not local, in character. Pierce, faced with such legislation, vetoed nine such bills as not meeting Democratic party commitments against what was considered to be the profligate and unconstitutional use of federal resources. Every one of his vetoes was bitterly challenged, by Whigs and by members of his own party, who were more than willing to push aside the traditional Democratic commitment on the issue in favor of some immediate local advantage.

In the foreign policy realm, Pierce demonstrated a traditional Democratic assertiveness. When the president and his colleagues came into office, they faced a number of lingering disagreements and ongoing problems. There were significant tensions with a weak Spain, a reclusive Japan, and a quite powerful, and often hostile, Britain, busily dabbling in Central America, despite the Monroe Doctrine. More critically, the American expansionist thrust to acquire new lands in the Western Hemisphere, to develop commercial interests overseas, and to protect the nation from enduring threats so that it could remain the only secure republic in a world of monarchies, all came alive again after a ten-year hiatus.

There appeared to be more intensity about these matters than had been true for some time. The emerging Young America movement, largely made up of Democratic editors, officeholders, and other party supporters, aggressively promoted its own expansionist inclinations, focusing particularly on the seizure of Cuba, further adventures in Mexico, and the acquisition of territory in Central America. The essential statement signaling a more rambunctious America on the world scene was the Ostend Manifesto, which, in 1854, declared the nation's intention, and right, to acquire Cuba from Spain. The fact that it was issued by three American ministers in Europe gave it an official quality that provoked an uproar both overseas and at home. The result was more tension and divisiveness on the political scene, adding to the already deafening tumult.

There were successes in the foreign field: the purchase of land from Mexico to facilitate the building of a Pacific railroad, the "opening" of Japan to American commercial interests, the settlement of a number of the outstanding disputes with Great Britain, and the avoidance of war despite some of the more aggressive activities of American expansionists. But these did not help when the reckoning was done. Secretary of State Marcy was the administration's point man in foreign policy, and it is not clear how deeply involved Pierce was in the enterprises occurring under his name. Nevertheless, his leadership formed the central focus of attention, as it did in domestic affairs. And in foreign policy as in the Kansas imbroglio, Pierce seemed, to many, to be overwhelmed by forces he could not control.

Pierce probably wanted to be renominated by his party in 1856, but it was not to be. He believed that he had done reasonably effectively in the face of what he considered parochial political considerations and extremist sectional agitators intent on bringing him down. But by 1856, his reputation as a leader and as a vote getter was in shreds among the reeling, distraught Democrats. He was succeeded by another party member remarkably like himself. James Buchanan came out of the same Jacksonian–Van Buren Democratic political tradition that had shaped Pierce. He was, like Pierce, a strong promoter of states' rights, local control, and unswerving partisan commitment. He, too, was hostile to extremist sectionalists in the North, sympathetic to the South, and very wary of doing anything to split the Democratic party further. And he ultimately did no better than Pierce — worse, in fact, as the domestic situation degenerated into political chaos, and then civil war.

Pierce had saved a portion of his salary while president, and friends had successfully invested it for him. That made possible a period immediately after he left office of idle, aimless retirement marked by long stretches of residence in Europe. The Pierces finally settled down back in Concord in 1860. The ex-president played little role in politics, speaking out occasionally as the nation moved closer and closer to civil war. In the crisis years of 1860 and 1861, there were futile efforts to involve him and other former presidents in a mediating role between North and South. Little came of them.

During the war itself, Pierce expressed bitter, if usually private, hostility toward the Lincoln administration, especially over its emancipation policies. When he appeared in public he exuded despair, echoing other aggressive naysayers among his party colleagues who saw no possible success coming from Lincoln's use of military force to restore the Union. Accused of being a Confederate sympathizer, even a traitor, Pierce vigorously denied such calumnies but refused to make any gesture to prove his loyalty. It did not really matter, except for occasional local flare-ups against him in Concord, because he was largely ignored and forgotten. Jane Pierce died in 1863, the former president in 1869.

Historians have largely shared the views of Pierce's contemporaries: one scholar refers to him as "a weak and incompetent President," an affable weakling, afraid of confrontation and unwilling to take strong stands against powerful colleagues even when he might have done so with some effect. In their occasional rankings of American presidents, Pierce has usually nested at the bottom of the list among others labeled as failures. Such evaluations seem very near the mark. Of course, Pierce lived in extraordinarily difficult times when a host of problems rose to challenge and overmatch the stability of the political arrangements in which he had grown up and prospered. He could neither understand nor manage the challenges, and he made many mistakes in trying to deal with them.

It is not certain that any other Democrat of his time and lineage could have functioned any better than he. Still, whatever the circumstances beyond his control, the country paid a high price for his and his party's limitations at a time when affairs and institutions were wracked and twisted by the pressures of unyielding, powerful forces that demanded unusual political talents.

— JOEL H. SILBEY
Cornell University

James Buchanan

1857–1861

b. April 23, 1791
d. June 1, 1868

JUST BEFORE NOON on March 4, 1861, James Buchanan got into a waiting carriage and rode to Willard's Hotel, where he picked up president-elect Abraham Lincoln. Armed troops lined the streets and sharpshooters peered from rooftops as the two presidents, one the representative of the long dominant Democratic party, the other of the newly ascendant Republican party, rode together down Pennsylvania Avenue to the Inauguration ceremonies at the Capitol. The unusual security measures revealed that this Inauguration represented something more than the normal transfer of power from one party to another. Relieved to be finally surrendering the burdens of the presidency, Buchanan remarked to his successor, "If you are as happy in entering the White House as I shall feel on returning [home] to Wheatland, you are a happy man indeed."

Few presidents have assumed the office with a more distinguished record of public service than James Buchanan. And yet, paradoxically, few have experienced such devastating political defeats or adopted policies that were so self-defeating. Buchanan left office abandoned by his friends and scorned by his enemies, and with the nation on the brink of civil war.

Born near Mercersburg, Pennsylvania, on April 23, 1791, Buchanan graduated from Dickinson College in 1809, and after studying law in nearby Lancaster was admitted to the bar in 1812. Intelligent and hard-working, he quickly became a prominent and successful attorney and began to accumulate a sizable estate that made him financially comfortable for the rest of his life.

Buchanan's interest in oratory soon led him into politics. While he was serving as a Federalist member of the legislature, his fiancée suddenly died shortly after breaking their engagement. Seeking sympathy, Buchanan created a romantic myth of blighted love, vowed never to marry, and despite a number of flirtations remained a lifelong bachelor. He turned instead to public life for social companionship, and from then on his closest and warmest friendships were in the all-male world of politics.

Buchanan was elected in 1820 to the U.S. House of Representatives, where he served for a decade and became an important member of the new Democratic party that rallied around Andrew Jackson. In 1831 he accepted an appointment as minister to Russia but soon resigned. Upon his return, he was elected in 1834 to the U.S. Senate and was reelected in 1837 and 1843. He was essentially a political trimmer, opposing both a protective tariff and free trade and advocating both hard money and expansion of the currency. As a loyal party man, however, he faithfully supported the policies of Jackson and his successor, Martin Van Buren.

On the slavery issue, Buchanan took a more decisive position and displayed strong pro-Southern sympathies. While he opposed slavery in the abstract, he never held strong moral feelings about the institution and bitterly opposed the new abolitionist movement. In the congressional battles over slavery, Buchanan supported the right of Southern postmasters to exclude abolitionist material from the U.S. mails, played a key role in the adoption of the so-called gag rule that automatically rejected antislavery petitions submitted to Congress, opposed the abolition of slavery in the District of Columbia, and favored the acquisition of the slaveholding Republic of Texas.

After Buchanan unsuccessfully sought the 1844 Democratic presidential nomination, James K. Polk named him secretary of state. Polk soon concluded, however, that Buchanan was too erratic and timid and conducted foreign relations himself. Buchanan was "an able man," the president noted in his diary, "but in small matters without judgment and sometimes acts like an old maid." One policy that Buchanan consistently advocated was the acquisition of Cuba from Spain, a goal he promoted for the duration of his career. After failing to win the 1848 Democratic presidential nomination, he retired from office the following year.

Buchanan made another bid for his party's nomination in 1852, but lost to Franklin Pierce, who, following his election, offered Buchanan the ministry to Great Britain. Buchanan's tenure in London was uneventful except for his signing of the Ostend Manifesto, which urged that the United States use force if necessary to acquire Cuba. The ensuing public outcry checked the administration's expansionist schemes.

Buchanan returned to the United States in the spring of 1856. His absence abroad proved to be a political godsend, as he had been out of the country in 1854 when Congress passed the controversial Kansas-Nebraska Act, which enacted the principle of popular sovereignty, by which the residents of a territory were to decide the status of slavery for themselves. Fearful that the resulting disorder in Kansas would defeat the party, Democratic leaders turned to the cautious and experienced Buchanan as the party's presidential nominee. Although Buchanan was elected in November, the most striking feature of the election was the strength of the sectional Republican party, which nearly won despite having virtually no support in the South. For the first time in the country's history, a Northern antislavery party posed a legitimate threat to win national power.

During the recent campaign, Southern leaders had warned that the South would secede if the Republicans carried the election. Taking these threats seriously, Buchanan announced that his goals as president would be "to arrest, if possible, the agitation of the slavery question at the North, and to destroy sectional parties," as well as to acquire Cuba.

For all his extensive experience, Buchanan was ill equipped to deal with the brewing sectional storm. Hailing from one of the most conservative Northern states, he had little understanding of the antislavery movement and lumped abolitionism together with more moderate forms of antislavery sentiment. His absence from the country in 1854 and 1855 was an additional handicap, for he had not witnessed firsthand the angry sectional passions aroused by the Kansas-Nebraska Bill, did not comprehend the situation in Kansas, and failed to grasp the reasons for the Republican party's success. A product of the Jacksonian world, he looked back to an earlier time when the antislavery movement was not respectable and sectionalism was not condemned, and he did not appreciate how different conditions now were.

The tall, somewhat stout new president was sixty-five years of age when he assumed office. His meticulous if old-fashioned clothing, white hair, excessive dignity, and courtly manners all reinforced his image as an elder statesman. An eye defect caused him to tilt his head when in conversation, which gave the impression of deference and enhanced his reputation for tact and conciliation. While he was diligent and methodical and paid close attention to details, he was also so fussy and meddlesome that John W. Forney, his 1856 campaign manager, described him as "a sort of masculine Miss Fribble." Cautious and unimaginative, he took a narrow legalistic view of problems, and throughout his long career displayed no talent for writing eloquent prose or uttering memorable phrases.

Still, he knew how to have a good time. He enjoyed fine food and cigars, was fond of dancing, and delighted in social occasions. He was famous for being able to hold his liquor and spent long evenings drinking with cronies. After the dreary Pierce years, Buchanan and his advisers cut a brilliant swath in Washington society. Yet there was an emotional reserve about the chief executive that discouraged familiarity, prompting members of the cabinet privately to refer to him as the "Squire."

Buchanan disliked confrontation and intended to make the cabinet the center of his social activities. Thus, rather than including all party factions, he selected men he was comfortable with personally. Its leading members were Howell Cobb of Georgia, the treasury secretary, whose company Buchanan especially delighted in; Secretary of the Interior Jacob Thompson of Mississippi, and Attorney General Jeremiah S. Black of Pennsylvania. In organizing his administration, Buchanan deliberately excluded any supporter of Senator Stephen A. Douglas of Illinois, whom he heartily disliked.

Buchanan's leadership style was to seek consensus among his advisers before adopting a policy. Nevertheless, he was not simply a tool of the cabinet, as throughout most of his term he fully agreed with his advisers on policy matters. Still, he was remarkably hesitant to lead or forcefully voice his opinion. Henry S. Wise, a longtime friend, recalled that "even among close friends he very rarely expressed his opinions at all upon disputed questions, except in language especially marked with a cautious circumspection almost amounting to timidity." Yet he could be quite stubborn when his authority was challenged.

Buchanan's inaugural address was uninspired and unremarkable. He announced that he would serve only one term, promised to uphold the Union, called for fiscal restraint, and advocated geographic expansion. The most important section, however, dealt with the slavery issue. He denounced the continuing agitation against slavery as dangerous to the nation and voiced the fervent hope that it would end. As for the question of slavery's expansion, he endorsed the principle of popular sovereignty but voiced the opinion that, as most Southerners argued, the people of a territory could determine slavery's status only at the time they drafted a state constitution. He added that the Supreme Court would soon rule on the status of slavery in the territories and, feigning ignorance of its deliberations, pledged that he would cheerfully abide by the Court's decision.

The Court's opinion in the Dred Scott case, issued two days later, gave

Buchanan's presidency a serious jolt just as it got under way. By a 7 to 2 vote, the Court ruled that Congress could not prohibit slavery from a territory, which in effect declared the platform of the Republican party unconstitutional. Naively believing that this decision would settle the issue of slavery's expansion, Buchanan had secretly intervened prior to the Inauguration to get a Pennsylvania justice to vote with the Southern majority. Instead, the ruling precipitated a storm of protest in the North, badly inflamed sectional feeling, and was a serious setback to his goal of calming sectional tensions. Buchanan's role in the matter provided stark evidence of how little he comprehended the sectional crisis or Northern feeling.

Buchanan's presidency suffered another reverse when an economic panic broke out in August 1857. As the party in power, the Democrats were badly hurt by the ensuing depression, which lasted until 1861. At the same time, hard times reinvigorated traditional economic issues and increasingly pitted the North against the South. Efforts to raise the tariff failed because of strong Southern opposition, and sectional rivalries prevented any agreement on the route for a transcontinental railroad. In addition, at the behest of Southerners, Buchanan vetoed several bills that would have aided the North, including a homestead bill, a bill to improve navigation of the Great Lakes, and a bill to grant public land to support agricultural colleges. These developments all strengthened the Republican party.

Buchanan was more effective in dealing with a crisis in the Utah Territory. When federal officials, charging Mormon harassment, fled the territory, the president took vigorous action to put down alleged Mormon defiance. Naming a new set of appointees, he dispatched an army of 2,500 men to the territory in 1857 to sustain federal authority, but conflict was averted when Mormon leaders agreed to settlement before the troops arrived.

The most important problem confronting Buchanan, however, was the situation in Kansas. For the critical post of territorial governor, Buchanan turned to Robert Walker, a skillful Democratic politician with ties to both sections. In reluctantly accepting the post, Walker insisted that the president publicly endorse the policy to allow the residents of Kansas to vote on their new state constitution. "On the question of submitting the constitution to the bona fide resident settlers of Kansas," Buchanan affirmed, "I am willing to stand or fall."

Walker went to Kansas intending to implement this policy, but he soon came into conflict with the proslavery faction in the territory. Because antislavery voters refused to participate, proslavery elements won control of a constitutional convention that met in the town of Lecompton in the summer of 1857 and proceeded to draft a constitution that recognized the legality of slavery. Despite Walker's earlier pledge, the delegates refused to submit the constitution to popular ratification; instead, they stipulated that voters were to decide only if additional slaves could be brought into the state. The electors could not approve the constitution without slavery, and they could not reject the constitution entirely. Moreover, the proposed constitution prohibited any amendments for seven years. With free-state men again boycotting the election, voters approved the Lecompton constitution with the right to import new slaves into Kansas.

Aware that a large majority of the residents of Kansas opposed the Lecompton constitution, Walker denounced it as a fraud and urged Buchanan to reject it. Buchanan, however, was under intense Southern pressure to make Kansas a slave state, and in the end he endorsed the document. In response, Stephen A. Douglas broke with the administration, proclaiming that the survival of the Democratic party in the North was at stake. Rejecting any compromise, Buchanan offered patronage positions, lucrative contracts, and even cash to force the Lecompton constitution through Congress. A Democratic insider later claimed that the administration's expenditures totaled $30,000 to $40,000. But after one of the fiercest struggles in American political history, the House rejected the Lecompton constitution. Democratic leaders eventually fashioned a compromise that by indirect means sent the constitution back to Kansas for a new vote. In August, with both factions participating, the residents of Kansas overwhelmingly rejected admission under the Lecompton constitution. This outcome marked the end of the Kansas issue in national politics and meant that Kansas was certain to enter the Union as a free state (as it did in 1861).

The Lecompton imbroglio was a political disaster for the Democratic party. Not only had the administration suffered a humiliating defeat when Congress rejected the Lecompton constitution, but the fall elections in the North were a Democratic debacle. Almost half the party's Northern seats in the House were lost and with them control of the new Congress. The state elections were equally discouraging, as the Republicans made significant gains in key states. Moreover, in yet another blow to the administration's prestige, Douglas won reelection to the Senate, further evidence that Northern Democrats backed him over the president.

At this point, it was imperative that Buchanan heal the breach with Douglas and restore party unity. Instead, seething with hatred for the

Illinois senator, he refused to make any overtures to estranged North-ern Democrats and commenced the systematic removal of anti-Lecomptonites from federal office, while abetting the successful move-ment to strip Douglas of his committee chairmanship in the Senate. This steadily worsening division left the Democratic party badly crip-pled as the 1860 election approached.

In contrast to the domestic situation, Buchanan achieved some suc-cess in foreign affairs. He worked out a new arrangement with Great Britain in Latin America, concluded a treaty with China, and estab-lished relations with Japan. In the face of growing Northern opposition, however, he failed to acquire Cuba, his major foreign policy objective, and his attempt to establish a military protectorate over Mexico's north-ern provinces, which opponents feared was a guise for annexation, was rejected by the Senate. While he acted more vigorously than critics ac-knowledged to halt Southern-backed filibustering expeditions of mili-tary adventurers who invaded various Latin American countries hoping to add these regions to the United States, he achieved only limited suc-cess in this endeavor.

Buchanan was further weakened politically by a series of scandals that wracked his administration. A congressional investigating commit-tee uncovered that, among other things, government money had been diverted to Democratic candidates, that public printing contracts in-volved kickbacks and bribes, that party contributors had been recom-mended for lucrative government contracts, and that the administration had offered outright bribes to congressmen if they would support the Lecompton constitution. The committee's report, which Republicans widely distributed as a campaign document in 1860, established that Buchanan presided over the most corrupt administration in American history before the Civil War.

These developments served as a backdrop for the 1860 presidential election. Buchanan's main objective when he assumed office was to pre-vent the Republican party from gaining national power, but once again his personal dislike of Douglas and his pro-Southern sympathies caused him to adopt policies that thwarted his larger goal. At the 1860 Demo-cratic National Convention, Buchanan's advisers aided the Southern ef-fort to block Douglas's nomination. In the end, the party was unable to agree on either a platform or a candidate, and eventually the Northern wing of the party nominated Douglas on a platform endorsing popular sovereignty, while most of the Southern delegates, joined by the Bu-chanan men, nominated Vice President John C. Breckinridge and de-manded enactment of a congressional slave code. A majority of rank-

and-file Democrats backed Douglas, but Buchanan refused to support his hated rival, and all efforts to heal the Democratic breach failed.

In November, Abraham Lincoln, the Republican candidate, swept to an easy victory in the electoral college, though he garnered less than 40 percent of the popular vote. The very outcome that Buchanan had assumed office determined to prevent — the election of a Republican president in 1860 — had become a reality.

Lincoln's election precipitated a secession movement in the South. The first state to act, South Carolina, seceded on December 20. One by one the other states of the Deep South followed, so that by February 1, when Texas seceded, the entire lower South had left the Union. Buchanan spent his final months in office, vainly trying to deal with this crisis.

The secession movement disrupted Buchanan's once united cabinet, as its previous harmony dissolved in the bitter passions that secession aroused. Buchanan was paralyzed by the deep divisions in his official family. In the course of one month, four members resigned and two assumed new posts; the most notable departures were Cobb and Thompson, the two leading disunionists in the group. In the process, the cabinet came under the firm control of Unionists, headed by Jeremiah Black, who became secretary of state. He was backed by Edwin Stanton, the new attorney general, and Joseph Holt, the postmaster general, both staunch Unionists. They were soon joined by John A. Dix, who assumed Cobb's place in the Treasury Department. The most important change, however, was in the War Department, where John Floyd of Virginia was in charge. Despite some indecision, Floyd had resisted secession, but it was now revealed that he was implicated in the theft of $870,000 in federal bonds by a kinsman. Buchanan's unwillingness to fire Floyd when the secretary refused to resign demonstrated his basic weakness. After much public blustering and a belated conversion to secession, Floyd left the cabinet at the end of December and was replaced by Holt. By mid-January, Buchanan's cabinet presented an unbroken Unionist front.

Buchanan was in an unenviable position. A lame duck, he had little personal prestige or political power: Northern Democrats repudiated him, Republicans denounced him as a weakling and demanded his impeachment, and his closest Southern friends now abandoned him. Even a sympathetic supporter conceded that the harried president was "execrated now by four-fifths of the people of all parties." Beset by pressure from all sides, his health broke down, he ceased his accustomed daily walk, and he was frequently too sick to leave his study. "He looks badly," the wife of one senator reported. "His face indicates much unhappiness."

Nevertheless, despite the incredible public abuse he suffered, his commitment to the Union never wavered, prompting Dix to observe, "There is no warmer friend of the Union."

Buchanan addressed the country's situation in his annual message to Congress in December. Though he forcefully rejected the legality of secession, he undercut this position by insisting that, as president, he had no power to prevent it, because only Congress could "authorize the employment of military force." He also blamed the current crisis entirely on the North, tracing it to "the long-continued and intemperate interference of the Northern people with the question of slavery in the Southern states." This one-sided analysis, which was indicative of his long-standing view of the antislavery movement, destroyed what little influence he still retained over Northern public opinion.

Terrified at the prospect of disunion and civil war, Buchanan lacked the will to lead. Although he later complained that Congress refused to pass any legislation to call up volunteers or employ force, in reality he did not desire such legislation. Instead, his goal was simply to hang on until his term ended, doing nothing that would either recognize the legality of secession or start a war. He blamed the existing crisis on Lincoln and the Republicans and thus believed that Lincoln should resolve it. "His policy," Holt explained, was "to preserve the peace if possible and hand over the Government intact to his successor."

In late December Major Robert Anderson, hoping to reduce friction with the city's residents, moved the Union garrison in Charleston from Fort Moultrie on shore to Fort Sumter, which was under construction on an artificial island in the harbor and was thus more secure militarily. Anderson's action was unauthorized, but after wavering Buchanan finally endorsed it. Shortly thereafter, Buchanan authorized a relief expedition to reinforce Fort Sumter, but it failed when shore batteries drove the unarmed ship away. Badly shaken by this narrow escape from war, the president accepted an informal truce in Charleston, under which the government agreed not to send supplies to Fort Sumter unless Anderson specifically requested them; he was nevertheless careful to avoid any action that would imply acceptance of the legality of secession. This stalemate continued throughout the final weeks of his presidency, and it was with enormous relief that he turned the reigns of power over to his successor in early March.

When the Civil War began, the former president endorsed the use of force to preserve the Union but generally avoided the political arena because of the controversy that surrounded him. He was denounced as a traitor even in his hometown, and fellow Masons, fearful for his safety, for a time guarded his house. He spent these years writing his memoirs, which were published in 1866. Shortly before Buchanan died on June 1, 1868, he told a friend, "I have always felt and still feel that I discharged every public duty imposed on me conscientiously. I have no regret for any public act of my life, and history will vindicate my memory."

Contrary to his expectation, historians have not been generous in their evaluation, and he has consistently ranked low in polls rating presidential performance. To be sure, he inherited a difficult situation as president, but he also pursued policies in office that were destructive of the ends he sought. His unwillingness to compromise on the Lecompton constitution and his unrelenting animosity toward Douglas were especially harmful. While well intentioned, he surrounded himself with a narrow set of advisers, shut out dissenting viewpoints, adopted policies that were excessively pro-Southern, was insensitive to Northern public opinion, failed to assess accurately the political consequences of his actions, and evidenced no ability to shape or mobilize public opinion. These manifold shortcomings ultimately wrecked both his presidency and his party.

— WILLIAM E. GIENAPP
Harvard University

Abraham Lincoln

1861–1865

b. February 12, 1809
d. April 15, 1865

O N NOVEMBER 6, 1860, Abraham Lincoln won the American presidency in a unique election that foreshadowed the division of the nation. Four candidates, including two Democrats, had contested what became a referendum on sectional allegiances. Most Southerners voted for the Kentuckian John Breckinridge, the former vice president in the Buchanan administration who ran on a platform of congressional protection of slavery in the territories and acquiring Cuba. With a numerical power that threatened the South, Northerners supported Lincoln. And so Lincoln won, with a substantial majority of electoral votes, a plurality of 40 percent of the popular vote, and 54 percent of the North's votes. He had received less than 3 percent of the South's ballots. The Illinoisan was the first member of the recently organized Republican party to win the presidency, and he would be the only president-elect in American history whose victory, even before his Inauguration, would significantly affect national events.

While Southern secessionists cheered the possibilities that the election of a despised, incorrectly supposed abolitionist afforded them, Lincoln said little during the four-month interregnum between his election and Inauguration in March 1861. He did write a few private letters to friends and party leaders, reaffirming his commitment to what was for him the central issue of this election — keeping slavery out of the territories. By nature the sixteenth president was a moderate who believed that Congress, not the president, held primacy in directing the nation's affairs.

As a Whig until the late 1850s, Lincoln had absorbed that party's doctrine opposing, as he wrote in 1861, "a very free use of any . . . means, by the Executive, to control the legislation of the country. As a rule I think it is better that Congress should originate as well as perfect its measures without bias." In fact he had made his slender reputation as a Whig congressman by attacking President James Polk's use of his war powers to maneuver the United States into a conflict with Mexico in 1846.

There were some Americans who worried that the new president was undermatched for the challenges that awaited him in Washington. To those eastern leaders with years of formal education in the classics and a lifetime of informal schooling in manners and high culture, the Illinoisan appeared as a poorly dressed country bumpkin whose habit of using homely metaphors and rural syntax seemed ill suited to the presidency. Yet for most of his life Abraham Lincoln had been a public man, a lawyer who delighted in party politics and discussions of contemporary affairs as well as complicated legal cases. Still it had been ten years since he had held any public office, and Lincoln had hardly been a notable congressman during his term in the Thirtieth Congress from 1847 to 1849.

Certainly he was well known in Illinois after serving four terms in the state legislature during the 1830s. He had gained more recognition after his debates in 1858 with Democrat Stephen Douglas. But he had lost this race for the U.S. Senate, just as he had lost a previous election for the same office in 1855. He had failed as well to gain a much-sought patronage appointment from President Zachary Taylor. In May 1860 there were better-known Republican aspirants such as William Henry Seward, Edward Bates, and Salmon Chase to be considered by the party's presidential nominating convention, where Lincoln nevertheless emerged, after the early roll calls, as a popular second choice. After his nomination on the third ballot he told the author of his campaign biography that there was not "much of me." It was a judgment with which many of his prominent contemporaries agreed.

Nor was there, at least in recent practice, much to the office to which he came. The writers of the Constitution had placed the article defining the executive power second, following that which organized, and gave extensive powers under the necessary and proper clause, to Congress. With the exception of an implied authority over foreign affairs and an explicit authorization of the president as commander in chief of the army and navy and the state militias, the United States Constitution extended few specific powers to the president. While there had been examples of expansive, imaginative uses of executive authority in earlier administrations, particularly during times of war and over issues of national expansion, the presidency in 1860 was like an underinflated balloon. In fact, as was intended by the framers who were preoccupied with their memories of the English monarch George III, the national government had developed into a congressional republic, not an executive one.

Lincoln's immediate predecessors — Democrats Franklin Pierce and James Buchanan — had set the pattern for a weak executive, conceiving their roles as little more than clerks and national caretakers who either approved or disapproved legislation developed from Congress's agenda. Buchanan did not consider adopting the presumptive, presidential function of shaping opinion as the republic's only nationally elected official. Throughout his term, the nation's capital simmered with disturbing sectional disagreements over the return of fugitive slaves, the Dred Scott decision, the future of Kansas, and the status of slavery in the territories.

Still, Southerners were sufficiently convinced of the presidency's potential to prevent the expansion of slavery into the territories that they began the process of secession immediately after Lincoln's election. The hated Lincoln, a representative of what Southerners labeled a sectional party, provided a convenient target for the mobilization of Confederate nationalism. By the first of February, seven Southern states had seceded, and before Lincoln's Inauguration, the Confederate States of America had organized as an independent nation. The result was to afflict Lincoln, when he took the presidential oath to "preserve, protect, and defend the Constitution of the United States," with unprecedented challenges to the survival of the American republic on a level that no president before or since has faced.

But it was still peacetime, and in Lincoln's restrained reading of the Constitution the presidency held few proactive powers. The sixteenth president's understanding of his office's prerogatives lay undeveloped, as he sought to reassure his fellow countrymen and to prevent more states from seceding. In his inaugural address delivered on March 4, 1861, he offered few policy dictates, but rather asked questions (more than twenty in all) that were to be answered not by him, but by his "dissatisfied" countrymen living in the South and the border states. Although Lincoln made clear that he believed secession illegal, he also guaranteed enforcement of the much loathed (in the North) fugitive slave law. And he quoted his party's platform to the effect that the domestic institution of slavery as it currently existed in the states was protected within the federal system. As president, he insisted, he would do no more than "cheerfully" abide by the laws and the Constitution.

In an address peppered with subjunctives and hypothetical instances, the new president seemed to foreshadow another feeble executive presence. For example, he would not construe the Constitution by any "hy-percritical rules"; he would deliver the mails if he could, although he might forgo any "irritating" or "obnoxious" exercise of this process. He would allow the Confederacy to replace federal officers, and crucially he would use the national power given to him to "hold, occupy and possess" government property. This he would somehow accomplish without using federal force or invading the South.

Accordingly it was the Confederate attack on Fort Sumter, the fortification in the middle of Charleston Harbor and by April 1861 one of the last remaining forts in federal hands, that transformed Lincoln's understanding of his powers. Confronted the night of his inauguration with the need either to surrender Fort Sumter or to resupply the garrison under Major Robert Anderson, Lincoln ordered a supply ship to Charleston, advising the South Carolina governor of its peaceful intention. But just as Lincoln could not surrender this last symbol of federal sovereignty in the Confederacy, so the South could not permit the continuation of national authority within sight of the most fertile fields of secessionism. On the night of April 12, 1861, Confederate batteries in Charleston Harbor opened fire on Fort Sumter. The next day Anderson surrendered the fort. The war that some Americans had looked forward to, that more had hoped to avoid, that a few had anticipated since the founding of the Republic, and that Lincoln despised but if it must come wanted the Confederacy to initiate, had begun. Not the least of its effects would be the transformation of the presidency as interpreted by Abraham Lincoln, who became not just a war president, but the Civil War president, for the remainder of his presidential tenure.

As a neophyte administrator who lacked even an organized executive branch, Lincoln was uncertain about how to implement his directives. Later he admitted his lack of knowledge about "the most basic manner of doing business in the executive office." There had been no declaration of war by Congress, which was not in session at the time and would not, in the normal course of its schedule, return until December. Instead Lincoln, absorbing both congressional and judicial powers, undertook an extraordinary series of unilateral actions, which lasted until July 4, 1861, the date Lincoln chose to call Congress back into session.

First the president announced an insurrection undertaken by "combinations too powerful to be suppressed by the ordinary course of judicial proceedings"; next, in this same proclamation, he called up and nationalized seventy-five thousand members of state militias under powers clearly extended to the executive in the Constitution. But other actions taken by the president were by no means understood to be powers of the executive, and thus from the moment the war began the Lincoln presidency shattered all precedents.

Four days after his first proclamation, Lincoln announced a blockade of Southern ports, and in so doing employed a war power without a declaration of war. Under international law any legitimate blockade established a condition of war between belligerent powers that Lincoln's description of insurrection denied. Then in a series of proclamations the president increased the size of the army, calling for forty-two thousand volunteers and additional regiments in the regular army and navy without either appropriations or congressional authority. Faced with a menacing accumulation of Confederate troops easily observable from the roof of the White House and operating as yet without a substantial Union force in Washington, he suspended the writ of habeas corpus, the ancient common law protection of individuals from the government, in areas along the railroad line from Washington to Philadelphia.

Lincoln justified this precedent-setting expansion of executive power (the constitutional discussion of the writ appears in Article 1, which pertains to the legislature) by invoking what became a familiar refrain during his presidency — he was constitutionally designated as commander in chief and the military peril to the Union made such actions necessary. In July the president sought retroactive sanction from Congress for his actions.

By this time Lincoln had already clashed with Supreme Court Justice Roger Taney, who, sitting on circuit in Baltimore as a federal district judge, held that Lincoln's suspension of the writ was illegal. The case involved the Marylander John Merryman who had been arrested on May 25, 1861, for sabotaging railroad lines used by Washington-bound troops. Merryman had also been drilling a contingent of troops for Confederate service. When General George Cadwalader, the military commander at Fort McHenry, acted under Lincoln's suspension order and refused to honor the justice's order to produce Merryman in civil court, Taney challenged the president on the grounds that the power to suspend the writ lay with Congress and not with the president.

Several weeks later Lincoln argued to Congress that whether "strictly legal or not," his actions were undertaken in the name of "popular demand and public necessity." The latter interpretation became a staple explanation of his other dilations of the presidential prerogative, the latter an elusive interpretation of the office as imbued with an inherent national power. The theory of executive prerogative was not Lincoln's creation, but could be traced to John Locke. The English political philosopher argued in the seventeenth century that in emergencies laws must give way to a discretionary executive power exerted as a fundamental law of nature.

It was the appeal to the people's will that expressed Lincoln's war-borne, modern understanding that his presidential office represented the sovereign will of the people. As the only nationally elected official in the government, he, not Congress, and certainly not the courts, embodied the people's will. In six other suspensions of the writ during his four years in office — in Missouri, in Kentucky, and once across the entire United States, as well as in his other stretches of the executive power — Lincoln relied on the same reasoning. Would he not have broken his oath of presidential office, asked the president, had he allowed the scrupulous observance of one law to stand in the way of preserving all the others and thereby permitted the government to go to pieces?

Although Democratic party and border state critics were already using the words *dictatorial* and *tyrannical* to describe these unilateral actions, Lincoln had in fact absorbed already defined authority extended to Congress and, in the case of habeas corpus, the courts. His assumption of power during an undeniable crisis was not extraconstitutional, but rather represented a funneling of powers, delivered to other branches of the government in peacetime, into the presidential office in wartime. He was centralizing authority. The issue, as the president often repeated, was time and the necessity of acting quickly to mobilize an army, to prevent the Confederates from attacking Washington, and to stop saboteurs and dissidents from overthrowing established governments in Missouri and Maryland. Instinctively Lincoln understood that the presidency carried with it a unity of office which allowed for swift action impossible for the legislature. Congress agreed. In August the U.S. Senate and House retroactively endorsed his actions, declaring them "legalized and made valid," as if Congress itself had enacted them. In 1863 in the Prize Cases arising from the blockade, the Supreme Court also ratified his actions on the grounds that the president was bound to meet the challenges of the Civil War without waiting for Congress to baptize it with a name.

In the remaining years of his term, Lincoln continued to act expeditiously on other matters. Always he sought legitimacy for his activism. Whatever his critics said, Lincoln and his innovations to the presidency did not represent the pursuit of power for personal or political aggrandizement. Rather they were conscientious efforts to prevent the most serious threat to any republic — the loss of its national integrity. Few of his actions were taken without some disclaimer about how sparingly he had used his authority. In less challenging circumstances Lincoln might well have carried on the traditional passive nineteenth-century presidency, although he supported a more active government than did most Democrats.

From 1861 until his assassination in April 1865, Lincoln also relied on his oath of office as a lever of influence. He believed this simple declaration extended authority to him to defend the laws of the United States currently violated by the Confederacy, and he referred to Section 3 of the executive power to "take care" that the laws were faithfully executed, sometimes even quoting the phrase "take care." To this he added an expanded capacity as commander in chief of the Army and Navy of the United States, as he breathed executive life into a previously inert presidential balloon.

Such a reading of the Constitution underwrote the sixteenth president's instructions to the Treasury Department to pay two secret agents two million dollars to buy military supplies before the requisition system was regularized. Such an arrangement violated the provision that expenditures could be made only when carrying out specific congressional appropriations. Albeit grudgingly, in 1863 Lincoln accepted an ersatz government in West Virginia as fulfilling the constitutional mandate that existing state governments — in this case Virginia — agree to the creation of new states established within their territory.

Infringing on First Amendment privileges, he closed down newspapers not because they criticized him — for Lincoln remained remarkably thick-skinned to personal abuse — but rather because they repeatedly opposed the war effort, discouraged volunteering, and published grossly misleading information that undermined popular support for the war. Such was the case in 1864 when the *New York World* printed a fraudulent, demoralizing story that the president had called for four hundred thousand additional soldiers. Following orders signed by the president, General John Dix, the commanding general in New York, shut down the paper and took the managers of the *World* into custody — but for only three days. Never did the president ask Congress for legislation establishing a sedition or treason act.

Sometimes Lincoln was embarrassed by the excesses of his military commanders, some of whom, like General Ambrose Burnside, moved far too harshly against Southern sympathizers and war protesters. In the case of Ohio Democrat Clement Vallandigham, who had been arrested and sentenced to prison by a military commission after an inflammatory antigovernment and draft speech in 1863, Lincoln commuted the sentence to exile. Ultimately Lincoln was responsible for the military arrest of two thousand civilians, though few of these were political prisoners.

Most, as recent historical research has shown, were legitimately arrested as draft evaders, ex-Confederate soldiers, and smugglers. In no instance was Lincoln's expansion of his presidential authority and his justification for executive prerogatives more starkly presented than in his treatment of emancipation. Fearful of shallow Union allegiances in the slave border states of Missouri, Maryland, Delaware, and Kentucky, and the effect that a clumsy, overly swift emancipation would have on public opinion there, in the spring of 1862 Lincoln suggested to congressional representatives of these states that they accept a plan of compensated emancipation. He was rebuffed, but such a proposal from the president suggested an activist leader in an area previously closed to such intrusions.

Refusing to delegate his authority over the critical issue of emancipation, Lincoln also rescinded the directives of two military commanders, Generals David Hunter and John Frémont, who had unilaterally announced a policy of emancipation within their military districts. The president believed, as he wrote in a famous letter to the influential editor of the *New York Tribune,* Horace Greeley, in August 1863, that freeing the slaves was less important than maintaining the Union. The former might well imperil the latter, according to the president.

No integrationist or believer in racial equality, Lincoln instead supported unrealistic constitutional amendments that would pay for the expenses of colonizing blacks overseas; he even presented to Congress an amendment that would begin an agonizingly long process of gradual emancipation ending only in 1900. What Lincoln wanted was to remove contention over the slave issue from the politics of the border Union states, where, in any case, his interpretation of his wartime powers denied him authority to do so.

The president, however, did not live in a vacuum. He existed in a wartime community and observed not just military parades and drills, but the abysmal condition of the black refugee camps located near the U.S. Capitol. The war, which like most Americans he thought would be brief, festered on with more casualties, more army hospitals to visit, and more bad news from the Virginia front, especially after the battle of Fredericksburg in December 1862. Increasingly the president was assailed by antislavery congressmen who, ahead of him on the need to free the slaves, had begun to write their own antislavery agenda. The Republican program included ending slavery in the territories, killing the fugitive slave law, freeing slaves in the District of Columbia, and considering slaves as property that might be confiscated from disloyal owners. About this last legislation, Lincoln noted his reservations in a threatened veto, but he eventually signed both the First and Second Confiscation Acts.

Influenced by the events which he said controlled him, Lincoln gradually modified his ideas about the relationship of slavery to the Union. By 1863, to end slavery, in his mind, was to help the war effort by removing an important source of Confederate labor. And the final Emancipation Proclamation sanctioned the double benefit of accepting blacks into the army, although in limited roles.

Thus it was in his capacity as commander in chief that Lincoln took steps toward freeing Southern slaves, first in the preliminary order that warned the rebels to end slavery by January 1863 or face emancipation. (Presumably the Confederate states could have laid down their arms and returned to the Union with slavery intact. To be sure, it defied the logic of the situation that they would follow such a course.) Accordingly January 1863 marked the end of slavery in the Confederate states — in those areas not under Union control. Observers, then and now, have complained that Lincoln emancipated slaves only where he could not reach them.

But such criticism misses the point of Lincoln's justification, made on the basis of his military powers. It also overlooks the implications of the order. Not only had the president committed the nation to slave liberation; not only was the Union army now consigned to freeing slaves as a war goal; but as his generals moved into the Confederacy, they implemented his orders. Slaves fled their masters and rushed to the protection of the Union Army.

A month earlier in a startling display of his tender caution about the limits of his prerogatives, Lincoln, in his annual message to Congress in December 1862 delivered just weeks before his final Emancipation Proclamation, proposed a constitutional amendment to Congress. It was an unusual display of executive direction of Congress, but a conservative way to end slavery: Lincoln's proposition was that Congress provide bonds to pay slave states that abolished slavery before 1900. He also proposed that Congress appropriate money to pay for the colonization of free African Americans who would emigrate. Thus, even as he transformed himself into the most powerful president in the nineteenth century, Lincoln remembered that he was a civil war president harnessed to a constitution that gave him no jurisdiction over Union states and their domestic property of slavery.

That aspect of emancipation he intended to share with Congress. Lincoln never changed his mind on this point, even lobbying members in Congress as well as Republican leaders for passage of what became the Thirteenth Amendment in 1865 prohibiting slavery in the United States. Amending the Constitution was the preserve of Congress and state legislatures. It was in his expanding role as a national leader shaping public opinion that Lincoln forcefully communicated his support of the amendment. The president made certain as well that the 1864 Republican party platform included a resolution calling for emancipation, believing, as he had come to understand during his own journey from antislaveryism to abolitionism, that slavery was the cause, the comfort, and the power of the rebellion.

While Lincoln derived his powers to emancipate slaves in the Confederacy from his military authority and while he believed that he must share this power to end slavery in loyal states with Congress, he showed no hesitation about running the army and navy unilaterally. Out of necessity, given the ineptitude and passivity of generals in chief Winfield Scott and Henry Halleck, the president did almost everything but lead troops into battle, especially in the first two years of the war. He selected commanders, sometimes choosing those whose martial talents seemed less developed than their ability to represent various political constituencies. He removed them, replaced them, and sometimes, as was the case with General George McClellan, reinstalled them in a search for competent generals which lasted until he appointed Ulysses Grant a lieutenant general and general in chief of the Union Army in March 1864. He wrote endless letters to his generals, inquiring where they were, what they had done, and crucially what they intended. He visited the front eleven times. Lincoln even delivered general war orders more appropriate for eighteenth-century commanders than a president running a modern war.

Lacking any military experience save his short service in the Illinois militia during the Black Hawk War in 1832, Lincoln nonetheless actively shaped strategy and, occasionally, tactics. Ever an autodidact who was confident after less than a year of formal schooling that he could teach himself the intricacies of maneuver, flanking operations, and enfilading an enemy's lines, Lincoln undertook a crash course in military tactics. He read the manuals of Napoleon's strategist Henri Jomini and West Point instructor Dennis Hart Mahan. Throughout the war he spent hours with maps and telegraphs from the front.

The most famous instance of his intervention in strategic affairs occurred when, to General McClellan's everlasting exculpatory chagrin, the president diverted General Irvin McDowell's corps during the Peninsular Campaign to cover the capital from a possible attack by Stonewall Jackson's troops. Earlier in the first of his General Orders, the president had instructed the lethargic McClellan, who since his appointment as commander of the Army of the Potomac had preferred organizing his troops over engaging them in conflict, to move his army against "insurgent forces." But McClellan presented his plan for the complex flanking operation of the ill-starred Peninsular Campaign, and Lincoln agreed to a delay. Still even after his approval the president issued instructions on details of military organization, troop dispositions, and the timing of the offensive.

The specifics of fighting the war continued to absorb a major proportion of the president's time and energy, but by 1863 Lincoln's personal involvement lessened. In part this was the result of the successful administration of the War Department by his second secretary of war, Edwin Stanton. By this time the size of the Union armies and navies and their deployment on a front including seventeen military departments and stretching from Virginia to Texas precluded any intimate involvement by the president. Also the commander in chief had located a general who shared his convictions that the war could be won only by hammering the Confederate armies with the superior resources of the North and that the war in the west must be a critical feature of Union strategy. But before Grant became commanding general, he too had felt the sting of an activist president who opposed the general's complex tactical maneuvers at Vicksburg. Later, after Vicksburg fell in July 1863, Lincoln acknowledged that Grant had been right.

To some extent the president shared the running of the army and navy with a restive Congress. Determined to play a role in the war and endlessly suspicious of Democratic generals, the Republican-controlled Congress organized the Committee on the Conduct of the War to investigate delinquent commanders and their behavior on the battlefield. Like Lincoln, Congress employed its powers of publicity, but the president never tried to impede the committee's self-assigned watchdog role during the Civil War.

Congress and the president also contested an entirely new area of jurisdiction — forming new governments in those states whose territory had been substantially wrested from the Confederacy. By 1864 the Union, requiring an unconditional surrender or military control over all seceded territory to achieve its war goals, had occupied portions of the Confederacy. Under a formula of his own, which did not establish black voting, Lincoln had set in motion governments in Louisiana, Arkansas, and Tennessee. To Congress in his annual message in December 1863, the president noted his authority to initiate the Reconstruction process as commander in chief of the military forces that were the first agents restoring national authority.

Sensitive to congressional prerogatives given that body's control over its membership, the president invited Congress to participate. In May 1864, Maryland's Congressman Henry Winter Davis and Ohio's Senator Benjamin Wade wrote legislation that established tighter guarantees of white loyalty and more insurance for black freedom than had Lincoln's plan. In one of his few confrontations with Congress, Lincoln, in an assertion of his executive authority, pocket vetoed the Wade-Davis Bill; in response Wade and Davis issued a manifesto denouncing Lincoln's "plenary dictatorial power." With neither precedent, constitutional or statutory direction, the process of Reconstruction remained a source of friction between the executive and legislative branches of government, especially when the war ended and the president could no longer assert his power as military commander. But by this time Lincoln was dead, and a less competent successor grappled with the issue.

Despite this exception, during his administration Lincoln maintained generally amicable relations with a Congress dominated by his party, in part because he initiated no legislation. In the tradition of nineteenth-century presidents, he had no interest in shaping a domestic agenda. On the contrary Congress sought to establish its own dominance over the president. Not only did individual congressmen attempt to extend the legislative body's control over Lincoln's cabinet, but several senators argued that they should serve as his advisers during this emergency.

Congress also created various select committees to investigate military contracts and indulged in harsh attacks on the president delivered in Congress. Freed from the negative votes of Southerners, the wartime Congresses, especially the Thirty-seventh, displayed their own power as they wrote through their nonmilitary legislation what one historian has called "a blueprint for modern America." During the war Congress passed homestead legislation and the Pacific Railway Act, established land-grant colleges, raised tariffs, and initiated an income tax and National Banking System that helped to nationalize a previously decentralized economic system.

This was not Lincoln's agenda for which he labored as a legislative leader. Rather it was the Republican program, delineated in the party's progressive 1860 platform. It would be hard to argue that Lincoln, who extended the presidential leadership in so many areas, exerted any leadership over domestic legislation, with the exception of emancipation, which, of course, he viewed as a military matter.

Nor did Lincoln pay much attention to his cabinet. He had chosen a distinguished group of seven men who represented different factions and coalitions within the Republican party. The exception, his first secretary of war, Simon Cameron, proved corrupt and ineffective and was gone by early 1862. Secretary of State William Seward and Secretary of the Treasury Salmon Chase were political rivals whose challenges to the president were somewhat neutralized when they accepted a cabinet office in his administration; others, like Secretary of the Navy Gideon Welles, were former Democrats.

Clearly Lincoln viewed the cabinet more as a political mechanism and administrative apparatus to run departments than as a consultative body that, as had been the case in Buchanan's administration, shaped policy. Usually Lincoln kept his own counsel, and Lincoln's cabinet officers found that meetings were infrequent affairs without specific agendas during which the president often told jokes. Their function, as the astute Chase recognized, was to administer their departments, not to shape executive policy. Certainly this was the case with Chase, to whom the president relegated the immense challenge of running the economy and trying to pay for the war.

Lincoln, like most presidents, understood himself to be the head of his party. At first he had tried to rally a wartime coalition of former Democrats, Whigs, Know-Nothings, and other remnants of prewar politics around a new Union party. But such efforts floundered on natural instincts and political habit. By 1862 the Democrats had revived their organization and became a persistent minority for the remainder of the war.

As a good partisan who had absorbed the culture of a time and place in which politics was in the air everywhere, Lincoln had systematically removed nearly all the Democrats from the 1,520 presidential appointments he controlled. During the war his largesse increased as the number of civilian jobs in the federal service expanded five times over prewar levels. By 1865 nearly 195,000 civilians worked for the federal government, many in new positions created in the War Department's Quartermaster Corps. Often these positions were low-paying. Still the opportunities to bind a follower to the young Republican party through a salary gratefully received during a period of general economic dislocation were never lost on Lincoln. As the needy and greedy, the worthy and the demanding, flooded the halls of the White House second floor to press their claims, Lincoln spent hours on this party function. By the time of his renomination in the summer of 1864, what Lincoln's wife, Mary Todd Lincoln, had once characterized as our "Lincoln party" indeed existed.

It was this Lincoln party that filled the seats of Baltimore's Front Street Theater on June 7, 1864, to renominate the incumbent on the first ballot. There had been challenges to the president from Secretary Chase and from dissident Republicans in Congress. In fact Frémont, the party's first presidential candidate and a former Union general, had organized a separate faction. But the effective control of Lincoln forced these political threats outside the mainstream Republican party. They frittered away during the spring and summer of 1864, as Lincoln became the principal opponent of the Democrat McClellan. For a time the gloomy Lincoln believed he might lose this election — at least until General Sherman's September success in Georgia.

This wartime election of 1864 provides an observation point for viewing Lincoln's presidency. One of the best ways to evaluate his precedent-setting uses of power is to think in terms of what he did not do as chief executive. Principal among the assertions of power that might have been exerted but were not was interfering in the political process by postponing the presidential election. Too much the democrat to use even the false news reports in Northern newspapers as a justification for tampering in electoral arrangements, Lincoln believed that free government was impossible without elections. "If the rebellion could force us to forego, or postpone a national election, it might fairly claim to have already conquered and ruined us," said the president, who was returned to office for a second term in November with a majority of four hundred thousand votes cast and 55 percent of the popular vote.

Like most presidents, after both his elections Lincoln assumed a nonpartisan cast and sought to be the leader of all the people, an especially critical task during the Civil War. His eloquent presidential messages, especially the Gettysburg Address and his second inaugural, were masterly presentations of the meaning of the war and the values of the nation, framed in simple and compelling language. Lincoln believed that the implications of this American Civil War reached beyond the United States to affect the "family of man" and the future of self-government everywhere. He therefore provided stirring global claims to those caught in an internal, parochial war that seemingly would never end. The United States, as he put it in his message to Congress in December 1862 during the darkest days of the war, was the "last, best hope of earth."

Lincoln's commonplace images to explain complex issues such as his wartime expansion of the executive power helped an often discouraged population to understand what the Union was trying to preserve and what it must change. Characteristically, the president, explaining his trespasses on individual liberties, used a comparison that all Americans of this era could understand. "A limb," he said, "must be amputated to save a life, but a life is never wisely given to save a limb." Widely circulated, such messages cast Lincoln as the head of a government educating the people in the necessary sacrifices of wartime. Always he tried to boost morale, often asking his audiences to give three cheers for Grant and all the armies under his command. In this novel capacity, Lincoln also came to think of himself in a new presidential role as the embodiment of the American people.

As the only Civil War president in American history, Lincoln offered no valid precedents for other presidents, although his actions and words would be cited by other wartime chief executives. Viewed from the perspective of his own time, his presidency was a paradox and its own thing. Lincoln greatly inflated some presidential functions but treated others, especially the initiation of legislation and his authority over foreign affairs, in a conventional way. He developed neither new forms of institutional machinery nor bureaucratic innovations which a successor might inherit to puff up his powers. Lacking an imperial style, Lincoln was self-consciously aware of what he was doing. Thirty years earlier the French visitor to the United States Alexis de Tocqueville noted that circumstances kept American presidents weak, not the laws or the constitutional arrangements. From 1861 to 1865 public necessity during the only war ever fought on continental American soil changed these circumstances and so temporarily did the presidency.

— JEAN H. BAKER
Goucher College

Ascending the Stage

Born in Kentucky, Lincoln was the first president to be raised outside of the original thirteen states. His family was poor, and as a boy Lincoln experienced the deprivation and instability of frontier life as his family moved to Indiana and then to Illinois. Determined to rise, Lincoln worked as a riverboatman, store manager, surveyor, and postmaster. His ambition and interest lay in public affairs, however, and these led him to the study of law and a seat in the Illinois state legislature.

By the early 1840s he had become a leader of the Whig party in Illinois and a prominent lawyer in Springfield. After serving one term in the U.S. House of Representatives, however, Lincoln withdrew from politics, disappointed that President Zachary Taylor had not appointed him to a prestigious position in his administration. Lincoln then devoted the next five years to his successful law practice.

The issue of slavery revived Lincoln's political activism. He opposed slavery and hoped that it would slowly die if confined to the South. In 1854, however, Illinois Senator Stephen Douglas sponsored the Kansas-Nebraska Act, which replaced the limits on slavery set by the Missouri Compromise with the principle of "popular sovereignty," permitting territories to determine by vote whether they would be slave or free. With the disintegration of the Whig party into sectional factions, Lincoln in 1856 joined the new Republican party, which sought to check the expansion of slavery.

In 1858 Lincoln ran against Douglas for the U.S. Senate, and the two candidates faced off in the most famous political debate in American history. Lincoln displayed superb oratorical skill, relentlessly attacking Douglas on the incongruity between popular sovereignty and the immorality of slavery. Douglas won the election, but Lincoln's performance drew national recognition. Two years later Lincoln defeated Douglas in the presidential election.

Andrew Johnson

1865–1869

b. December 29, 1808
d. July 31, 1875

THE FIRST PRESIDENT ever impeached by the House of Representatives and tried before the Senate, Andrew Johnson was one of the least successful American chief executives. Succeeding to office after the assassination of Abraham Lincoln, Johnson confronted one of the gravest crises in American history — reuniting the nation after the Civil War and coming to terms with the abolition of slavery. His years in office, 1865 to 1869, were marked by a tumultuous conflict between president and Congress. Johnson's personal stubbornness, racism, doctrinaire belief in states' rights, and inability to gauge the currents of Northern public opinion contributed mightily to the triumph of his political opponents, and did much to destroy his presidency.

Born on December 29, 1808, into the poorest circumstances of any man who ever reached the White House, Johnson was the son of Jacob Johnson, a landless laborer in Raleigh, North Carolina, and Mary McDonough Johnson, a seamstress and laundress. Apprenticed to a local tailor as a youth, he absconded at age fifteen to Laurens, South Carolina, and in 1826 moved to Greeneville, Tennessee, a town of fewer than five hundred residents, where he opened his own tailor shop. Johnson's life, as his later political rival Oliver P. Temple wrote, was "one intense, unceasing, desperate upward struggle." Like Lincoln and many other contemporaries, Johnson turned to politics as an avenue of social advancement. He achieved remarkable success. Beginning as an alderman in Greeneville in 1829, he held office almost continuously for the remainder of his life. He was elected mayor in 1834, and later served as a member of the state legislature (from 1835 to 1837 and 1839 to

1841), as a Democratic congressman (1843 to 1853), as governor for two terms (1853 to 1857), and as U.S. senator, beginning in 1857. Johnson achieved a statewide reputation as a brilliant stump speaker and impromptu debater. He also prospered economically, acquiring substantial landholdings. In 1860, he owned five slaves, a couple and their three children. But Johnson remained a lonely, self-absorbed man, who, as Gideon Welles, later his secretary of the navy, remarked, "has no confidants and seeks none." His major decisions, Welles added, seemed to have been made without consultation with "anyone whatever."

Despite his success, Johnson had nothing in common with the planter class that, as in other Southern states, dominated Tennessee society and politics. Indeed, he rose to office as a self-proclaimed tribune of the state's yeomanry, and in campaign speeches thundered against the state's slavocracy, which he called a "bloated, corrupted aristocracy." His speeches never failed to express pride in his plebeian origins. In the legislature, he proposed that congressional district lines be drawn according to the white population, without considering three-fifths of the slaves, as the federal Constitution required, thereby reducing the power of the plantation counties to the advantage of the small farming counties of the eastern part of the state. Johnson defended slavery and condemned the abolitionists, but he devoted most of his energies to measures that would uplift poorer whites. He championed tax-supported public education, a reform enacted into law during his term as governor, and the granting of free land to settlers in the west, which he promoted in Congress as a measure that would "make many a poor man's heart rejoice." He was also a fanatical advocate of economy in government, opposing nearly every proposal for federal expenditures. As senator, he voted against the construction of an aqueduct and other public works in Washington, D.C., and he demanded that the capital's citizens pay for their own police force and schools.

A staunch supporter of the Union, Johnson was the only senator from a Southern state to campaign aggressively against secession and then remain at his post in Washington when his state joined the Confederacy in 1861. Like Lincoln, he saw the secession crisis as a test of "man's capability to govern himself," and he believed that a breakup of the Union would suggest that democracy in America had failed. He thus became a national symbol of what both he and Northern Republicans imagined to be a legion of white Southerners who had remained loyal to the Union. After the federal capture of Nashville in February 1862, Lincoln appointed him military governor of Tennessee, a post he occupied almost until the end of the war. His administration was nothing if not vigorous.

Even as Lincoln sought to conciliate Southern planters in occupied Louisiana if they promised to resume their loyalty to the Union, Johnson seized the Bank of Tennessee, suspended hostile newspapers, arrested opponents, ousted elected officeholders, and took to using the phrase that would win him an undeserved reputation for radicalism: "Treason must be made odious and traitors punished."

Before the war, Johnson had offered no criticism of slavery, even though he had condemned the planters for oppressing poorer whites. During the conflict, he remarked to a Union general, "Damn the Negroes, I am fighting those traitorous aristocrats, their masters." Yet once the Lincoln administration adopted emancipation as a war aim, Johnson went along. The Emancipation Proclamation did not apply to Tennessee, but in October 1864, Johnson unilaterally proclaimed the end of slavery in the state, and announced to a black gathering, "I will indeed be your Moses, and lead you through the Red Sea of war and bondage to a fairer future of liberty and peace."

By this time, Johnson had been nominated as Lincoln's running mate in the election of 1864, replacing Hannibal Hamlin, Lincoln's ineffective first vice president. The circumstances of his nomination remain somewhat obscure, but essentially Johnson was chosen, with Lincoln's acquiescence and perhaps design, as a symbol of the Republican party's determination to extend its influence into the South once peace had returned. The Inauguration, on March 4, 1865, was a disaster for Johnson. Prior to the ceremony he imbibed a considerable amount of whiskey to fortify himself against a case of typhoid fever. His speech, one contemporary reported, became a "rambling and strange harangue, which was listened to with pain and mortification," an unfortunate prelude to Lincoln's memorable second inaugural address. The press ridiculed Johnson as a "drunken clown."

Six weeks later, on April 15, 1865, Johnson was president. Less than a week had passed since Robert E. Lee surrendered the Army of Northern Virginia, effectively ending the Civil War and inaugurating the era of Reconstruction. Johnson's past career led some observers to suppose that he favored sweeping changes in Southern society. Some Radical Republicans, advocates of immediate suffrage for Southern blacks and punishment for "rebels," rejoiced at Johnson's accession. "Johnson, we have faith in you," declared Senator Benjamin F. Wade of Ohio. "By the Gods, there will be no trouble in running the government."

Despite his comments about punishing treason, however, Johnson had little in common with Northern Radicals. He saw Reconstruction as a way of empowering loyal yeomen, not former slaves, and he did not share the Radicals' expansive view of federal power. Indeed, Johnson's Reconstruction policies began with the premise that secession was illegal and that the states had therefore never left the Union. Individual "traitors" should be punished, but the states retained all their constitutional rights and could not be dictated to by the federal government. Johnson did plan to appoint new governors and impose conditions for the restoration of civil government in the South, implying that the states had surrendered at least some of their traditional powers. Unlike the Radicals, however, he did not believe blacks had any role to play in Reconstruction. Indeed, he harbored deeply racist sentiments toward the former slaves. His private secretary, Colonel William G. Moore, recorded in his diary that Johnson at times "exhibited a morbid distress and feeling against the negroes." In his December 1867 message to Congress, Johnson would insist that blacks possessed less "capacity for government than any other race of people," and when left to themselves showed a "constant tendency to relapse into barbarism." A few months after assuming the presidency, he did propose to enfranchise propertied blacks in Tennessee, as a tactic, he explained, to foil more sweeping Radical demands. But he never wavered from the conviction that the federal government lacked the authority to impose such a policy on the states.

On May 29, 1865, Johnson issued two proclamations that announced his plan of Reconstruction. The first conferred amnesty and pardon, including the restoration of property rights (except for slaves) to Confederates who took an oath of future loyalty to the Union and support for emancipation. Certain groups of Southerners, however, were excluded, and required to request individual presidential pardons. Most notable were owners of property valued at over $20,000, in other words the bulk of the planter class. The second proclamation appointed William H. Holden provisional governor of North Carolina and instructed him to call a convention to amend the state's prewar constitution so as to enable the state to regain its place in the Union. Those who had not been pardoned could not vote; but otherwise, prewar voting qualifications remained in effect, leaving blacks with no role to play in the new postwar system. Similar proclamations for the other states soon followed.

Johnson always claimed that he was simply implementing Lincoln's plan of Reconstruction, but in many ways his proclamations represented his own point of view. Lincoln had never proposed anything as draconian as excluding all wealthy Confederates from participation in government, and near the end of his life had publicly called for enfranchising educated blacks and black soldiers. Johnson's plan seemed to rest on the assumption that with planters and blacks excluded from politics, power would flow to the loyal white yeomanry of the South. Alternatively, if planters applied for individual pardons, essential for regaining control of their lands, Johnson could assure himself of their support (including support for a reelection bid in 1868) by granting their request. And, indeed, most wealthy Southerners did ask for, and soon received, presidential pardons.

Thus was launched the era of presidential Reconstruction (1865 to 1867). The terms Johnson laid down for the South's readmission were amazingly lenient, especially for a man who had spoken so insistently of punishing treason. Apart from the requirement that the former Confederate states abolish slavery and repudiate secession and the Confederate debt — all inescapable consequences of Southern defeat — Johnson gave the new governments a free hand in managing the transition from slavery to freedom. Their conduct turned much of the North against Johnson's policies. Contrary to his expectations, the Southern electorate returned former Confederate leaders (many still unpardoned), not honest yeomen, to power. The new legislatures enacted the Black Codes, which severely restricted the rights of the former slaves and sought, through vagrancy and breach of contract laws, to compel them to return to work on the plantations. Meanwhile, in the summer of 1865, Johnson ordered the Freedmen's Bureau, a federal agency established earlier that year to assist the freed people, to return abandoned plantation lands to their former owners, including parcels that had already been divided among blacks in Virginia, South Carolina, and Louisiana. Clearly, Johnson envisioned no sweeping social revolution as a sequel to emancipation.

Given the political disaster that soon overtook him, Johnson's course in tying his administration to the planter class and the policies of the new Southern governments seems the height of folly. Because he rarely wrote letters, kept no diary, and had few confidants, his motives will always remain something of a mystery. But it seems that as 1865 progressed, Johnson concluded that his best chance for future political success was to forge a new political alliance that would occupy the center of the political spectrum, excluding the Radicals on one side and the extreme secessionists on the other. Moreover, the unexpected militancy of former slaves, who mobilized in 1865 to demand civil rights, the suffrage, and land, alarmed Johnson and led him to conclude that only the planters possessed the experience and authority to reassert racial control in the South. Thus he gradually abandoned the idea of transferring power to the region's poor whites. That most Northerners initially seemed willing to give Johnson's plan a chance to succeed persuaded him that he was following the correct course.

Johnson's annual message to the new Congress that convened in December 1865 insisted that the "work of restoration" was now complete. All that remained was for Congress to readmit Southern representatives. At the same time, he expressed paternal concern for the freedmen's future. When it came to substantive measures to improve their condition, however, he proved unyielding. And it was already clear that the majority of Northern Republicans had serious misgivings about Johnson's program. Widespread reports of a return to local power by "rebels" and mistreatment of the former slaves and white Unionists persuaded Congress that Johnson's plan needed amending. While Radicals such as Thaddeus Stevens, congressman from Pennsylvania, and Charles Sumner, senator from Massachusetts, proposed that new governments be established based on universal manhood suffrage, moderate Republicans proposed to work with Johnson while modifying his program. In the meanwhile, Congress refused to seat the South's new representatives and senators.

Early in 1866, Congress enacted two bills reported to the Senate floor by Lyman Trumbull of Illinois, one of the most influential Republicans in Washington. The first extended the life of the Freedmen's Bureau, which was due to expire shortly. The second was the Civil Rights Bill, intended to overturn the Black Codes and grant the former slaves the same legal protections as white citizens, short of the suffrage. The bill declared all persons born in the United States citizens (thus abrogating the Dred Scott decision of 1857, which ruled that no black person could be a citizen of the country) and spelled out the rights they were to enjoy regardless of race, including making contracts, bringing lawsuits, and enjoying the equal benefit of the law. The first statutory definition of citizenship, the bill embodied a profound change in federal-state relations. Traditionally, citizens' rights had been delineated and protected by the states. The bill empowered the federal courts to hear cases involving civil rights and invalidated scores of discriminatory state laws, Northern as well as Southern.

Most Republicans assumed that Johnson would sign the two bills. Had he done so, he would have avoided many of his future problems. But he vetoed both. His messages insisted that Congress should not legislate while eleven states remained unrepresented, and they rejected the idea of federal protection of blacks' civil rights and the broad conception of national power which lay behind it. States' rights and racism merged in his messages. "The Government," he declared in the Freedmen's Bureau veto, "has never felt itself authorized to expend the public money for the thousands, not to say millions, of the white race who are honestly toiling from day to day for their subsistence." Thus, it had no obligation to assist the former slaves. The Civil Rights Bill, he claimed, violated "all our experience as a people," and represented a "stride toward centralization." Moreover, Johnson had somehow convinced himself that clothing blacks with the privileges of citizenship discriminated against white people — "the distinction of race and color," he wrote, "is by the bill made to operate in favor of the colored and against the white race." He even charged that the bill threatened to encourage interracial marriage. He also offended Republican congressmen by a spontaneous, intemperate speech at the White House on Washington's Birthday, 1866, in which he implied that Radical Republicans were plotting his assassination. The two vetoes alienated Johnson from virtually the entire Republican party. Congress failed by a single vote to override the Freedmen's Bureau veto, although it later succeeded in extending the agency's life. But in April 1866, it reenacted the Civil Rights Bill, the first important piece of legislation in American history to become law over a president's veto.

The breach with Congress became complete when Johnson opposed ratification of the Fourteenth Amendment, passed by Congress in June. This amendment fixed in the Constitution the principle of birthright citizenship and forbade states from depriving any citizen of the "equal protection of the laws." It constitutionalized the vast shift of power from the states to the federal government, which the Civil War had accomplished, authorizing Washington to overturn state laws that deprived any person of "life, liberty, or property" without "due process of law." To prevent ratification and save his faltering administration, Johnson moved to organize a new political alliance, holding a National Union Convention to bring together Democrats and Republicans who supported his Reconstruction program.

The congressional elections of 1866 became a referendum on Reconstruction, especially the Fourteenth Amendment. Late in August, Johnson, breaking with tradition, embarked on a speaking tour of the North to generate support for candidates who supported his position. The "swing around the circle," as the tour was called, became a political disaster. Interrupted by hecklers, Johnson could not resist responding in kind, exchanging epithets with his audience and launching tirades against his opponents. He compared himself with Jesus Christ and at one point suggested that divine intervention had removed Lincoln to elevate Johnson himself to the White House. His behavior thoroughly alienated Northern voters and contributed to a massive Republican victory in the fall. In the next Congress, Republicans would outnumber Johnson's supporters by well above the two-thirds majority necessary to override vetoes.

Some Southern leaders now concluded that their region must accept Congress's terms and ratify the Fourteenth Amendment. Johnson, however, remained adamant in opposition, and one by one the Southern legislatures rejected the amendment. (The lone exception was Tennessee, which ratified it and regained its representation in Congress.) Completely bereft of influence in Congress, Johnson in January and February 1867 could only stand by helplessly as Republicans dismantled presidential Reconstruction. In the Reconstruction Act, it temporarily divided the South into five military districts and outlined how new governments, based on manhood suffrage, were to be established. Believing, as he told Charles Nordhoff, managing editor of the *New York Evening Post*, that "the people of the South, poor, quiet, unoffending, harmless, are to be trodden under foot to protect niggers," Johnson returned the bill with a veto, which Congress immediately overrode. Thus began the period of congressional or Radical Reconstruction (1867 to 1877), which brought to power new governments in the South in America's first experiment in interracial democracy.

Among the anomalies of congressional policy in 1867 was that it left the implementation of Reconstruction in the hands of an institution — the army — controlled by its greatest adversary, President Johnson. While a few Radicals were already calling for Johnson's impeachment, the majority of Republicans preferred to shield Reconstruction from presidential interference by requiring that all orders to army commanders pass through the general of the army (Ulysses S. Grant), and, in the Tenure of Office Act, authorizing officials who had been appointed with the consent of the Senate to remain in office during the term of the president who had appointed them, until the Senate had approved a successor. One intent was to have the Republican majority in the Senate protect Secretary of War Edwin Stanton, an ally of Congress, from removal by Johnson.

Johnson, however, was not easily cast aside. The Tenure of Office Act allowed him to suspend Stanton while Congress was not in session, and in the summer of 1867 he did so. Soon he was encouraging Southern whites to oppose the new Reconstruction policy, and he replaced several Republican military commanders with officers sympathetic to himself. Emboldened by Republican losses in the fall elections of 1867, Johnson determined to remove Stanton permanently. In February 1868, after the

Senate refused to concur in Stanton's removal, Johnson dismissed him, appointing as his successor General Lorenzo Thomas.

Johnson contended that he was not obligated to abide by the Tenure of Office Act, which he deemed flagrantly unconstitutional. (Alternatively, he also suggested that whether constitutional or not, it did not apply to Stanton, who had been appointed to office by Lincoln, not himself.) But to Congress, Johnson's violation of the law was the last straw. With unanimous Republican support, the House of Representatives voted to impeach the president. There followed the unprecedented drama of his trial before the Senate, charged with "high crimes and misdemeanors."

Despite having alienated the entire Republican party, Johnson's conviction was far from certain. For one thing, the constitutionality of the Tenure of Office Act was certainly open to question, and nine of the eleven articles of impeachment hinged on Johnson's violation of that law. Two others charged the president with denying the authority of Congress and attempting to bring it "into disgrace." The problem was that the real reasons Republicans wished to remove Johnson — his political outlook, the way he had administered Reconstruction, and his sheer incompetence — seemed to some senators not impeachable offenses. Many Republicans, moreover, were unhappy at the prospect of Benjamin F. Wade, president *pro tem* of the Senate, succeeding to the White House and, presumably, to the next Republican nomination. Not only was he a Radical on Reconstruction, but Wade's high-tariff, soft-money economic outlook alienated many Republican businessmen.

Presiding over the trial was Chief Justice Salmon P. Chase, himself a perennial presidential aspirant who probably preferred to have Johnson finish out his term rather than see Wade, long a rival in Ohio politics, become president. Chase steered the trial in a narrowly legalistic direction, making it impossible for the impeachment managers to raise broad questions about Johnson's leadership. It soon became apparent that an influential group of moderate Republicans were uneasy with the prospect of removing Johnson, both because they disliked Wade and because they wished to preserve the constitutional separation of powers. During the trial, moreover, Johnson's lead attorney, William M. Evarts, assured wavering Republicans that if acquitted, Johnson would cease his efforts to obstruct Reconstruction policy. When the vote was taken in May 1868, thirty-five senators voted for conviction, one short of the necessary two-thirds. Seven Republicans, including Trumbull, voted to acquit. Contrary to later myth, they were not read out of the party, and all campaigned in the fall for the Republican nominee, Ulysses S. Grant.

Johnson continued to harbor hopes that the Democrats, who had supported him on Reconstruction, would nominate him for a second term. "Why should they not take me up?" he asked as the Democratic Convention gathered in July. "They profess to accept my measures; they say I have stood by the Constitution and made a noble struggle." He was bitterly disappointed when the nod went to New York's governor, Horatio Seymour.

With Grant's election, Johnson's presidency came to an end and he returned to private life in Tennessee. Meanwhile, Reconstruction followed its turbulent course. The new Southern governments enacted pioneering legislation protecting civil rights, promoting economic development, and establishing tax-supported public schools in the Southern states that lacked them before the Civil War (all but Tennessee, where Johnson himself had spearheaded the drive for public schooling years before). Southern whites, who could not accept the idea of former slaves voting and holding office and who opposed the egalitarian policies adopted by the new governments, used every method at their disposal to restore white supremacy. In some states, such as Tennessee, the white voting majority soon coalesced against Reconstruction and restored the Democratic party to power. Where Republicans maintained a majority, Reconstruction's opponents turned to violence. The Ku Klux Klan launched a campaign of terror that decimated the Republican organization in many localities.

By 1875, when Johnson returned once more to Washington, elected to the Senate by the Tennessee legislature, Reconstruction was already on the wane. A few months later, on July 31, he died at his home in Tennessee. By then, Democrats had already taken control of a majority of the Southern states, a process that would be completed in 1877.

When he died, Johnson's reputation was at a low ebb. In the early twentieth century, however, as historians adopted an interpretation of Reconstruction that saw black suffrage as a colossal mistake and Southern governments of the era as travesties of democratic self-rule, Johnson was portrayed as a courageous defender of the Constitution, whose policies were far wiser than those of his congressional opponents. In *The Tragic Era,* a best-selling combination of mythmaking and history published in 1929, Claude G. Bowers wrote that Johnson "fought the bravest battle for constitutional liberty and for the preservation of our institutions ever waged by an Executive," but was overwhelmed by the "poisonous propaganda" of the Radicals. As a result, Southern whites were "literally put to the torture" by "emissaries of hate" from the North, who manipulated the "simple-minded" freedmen. More recently, as Reconstruction has come to be seen as a time of significant progress for black Americans and as the nation's first effort to establish the principle of racial equality, Johnson has been condemned by historians for racism and lack of statesmanship. Today, his presidency is generally regarded as a failure.

— Eric Foner
Columbia University

Ulysses S. Grant

1869–1877

b. April 27, 1822
d. July 23, 1885

U LYSSES S. GRANT is a study in paradox. The definition of the ordinary American, he achieved greatness. Totally in command as a general, he was unable to take commanding leadership as president of the United States. If history has judged him a failure as president, it is usually for the wrong reason. Grantism is seen by some as a synonym for corruption, but corruption by members of his administration did not distinguish Grant from other presidents. If we judge him as a failure, it must be for not securing the safety and full citizenship of the former slaves for whom his victory in the Civil War achieved freedom. Regarded by Henry Adams as "simple minded," Grant wrote one of the towering works of American literature.

The future president was born in Point Pleasant, Ohio, on April 27, 1822, to Hannah Simpson Grant and Jesse Root Grant, who named him Hiram Ulysses (pronounced *U*-liss-is) because his maternal grandmother had been reading Racine. When the boy was still an infant the family moved inland to Georgetown. Just across the road from their house was the tannery that Jesse Grant operated. From his earliest days, the boy was repelled by the shrieks of the aged horses slaughtered for their hides and the stink of the tanning process.

Hannah was a deeply religious, often forbidding woman who in later years pulled inward, refusing to communicate with many outside of the family. From her, Grant seems to have derived his own undemonstrative demeanor. His father was the quintessential self-made man. Ambitious and hardworking, the elder Grant, given up for adoption by his widowed and alcoholic father, built his tanning business into a success. Over time, he owned both the tannery and profitable harness shops.

Jesse Grant saw to it that his children received an education at good schools, but in little else did he and his eldest son see eye to eye. The father despaired of making a businessman of Ulysses and bringing him into the family enterprise. The boy seemed interested only in horses — he became an excellent, confident horseman — and his father thought of the army as the only place where Ulysses might make a living. Jesse lobbied the local congressman and gained an appointment to the United States Military Academy for his son.

In 1839, Grant, still a boy five feet one in height and weighing 117 pounds, arrived at West Point. The congressman, not recalling Ulysses's full name, but mindful of his mother's maiden name, sent in the appointment for Ulysses S. Grant, as he has been known ever since.

The cadet's performance at the academy was unspectacular; he did well in mathematics, but the only course that truly engaged him was the drawing class designed to prepare officers to visualize and sketch vistas of fields on which they were to fight. A well-rendered sketch by Grant of a Native American woman cradling a baby survives. He had little patience with scientific studies of earlier wars and did not join the Napoleon Society. He detested the little warrior as much, perhaps, for his similar stature as for the worship of vainglory indulged in by Napoleon's admirers.

Commissioned a second lieutenant in 1843, Grant requested a cavalry assignment. Instead, he was appointed to the Fourth Infantry and posted to a base outside St. Louis. There he courted his West Point roommate's sister, Julia Dent. Their engagement was a long one; Grant was transferred to posts first in Louisiana and then in Texas.

He was near what is now Brownsville, Texas, when President Polk's pressure on Mexico resulted in the Mexican War. When Grant later wrote about the war, not only did he chronicle the battles well, but he was perceptive about the politics leading to the war and sensitive to the plight of Mexican peasants caught between two marauding armies. The future general also took notice of many fellow officers he would fight with and against in that war. But as a first lieutenant when the war was over, he was about as noticeable as he had been at West Point.

Grant, however, was far from simply an observer assigned to the quartermaster corps, but a young officer eager for action. In the battle of Monterey, he seized his opportunity and participated in the house-to-house fighting; at one time clinging to the side of a horse opposite the firing Mexican soldiers, he galloped out to secure reinforcements. For this, he earned mention in official dispatches for bravery under fire.

Back in St. Louis after the war, Grant married Julia on August 22,

1848. Raised to be a belle by a beautiful mother, the unglamorous Julia proved a determined and strong mate. Their marriage was one of life-long affection. The couple's first son, Frederick Dent Grant, was born in 1850. When their second, Ulysses S. Grant, Jr., was born in 1852, Grant had left the family behind to report for duty in the remote Pacific north-west. He was crossing the Isthmus of Panama when the baby was born and had to wait for six months to learn, when the mail finally arrived, that Julia had had a safe delivery.

Absence from the family and mistreatment by a martinet of a com-manding officer at the remote Fort Humbolt in California contributed to, perhaps even caused, a severe depression to which Grant responded with too much alcohol. His letters to Julia during the lowest chapter in his life portray a deeply troubled man. Finally, in 1854, now a captain, he was forced to leave the army because of his drinking. After a greatly dis-piriting trip, he reached St. Louis and was reunited with Julia and the boys. Exactly nine months later, on the Fourth of July 1855 Nellie Grant was born.

Grant struggled to find a place for himself in the civilian world, a struggle remarkable for its lack of success. Over the next six years, on land provided by his father-in-law, he tried his hand at farming; the crops failed. Then Julia turned to a cousin for help, and Ulysses be-came a bill collector (it would be hard to imagine anyone less suited to that calling). In time, the poverty of which Grant, recalling his desti-tute grandfather, had a lifelong fear, seemed to defeat all his efforts to achieve comfortable independence. The former captain in the United States Army was even driven to cutting firewood and hauling it into St. Louis to sell from a street corner.

Desperation drove Grant to swallow his pride and ask his father for a job. He was given one, working as a clerk in the Grant harness shop in Galena, Illinois, under a younger brother. At thirty-eight, he appeared to have settled into a life of obscurity. Only the family sustained him. There are accounts of Ulysses romping affectionately on the floor with the chil-dren; there now were four — Jesse Root Grant had been born in 1858. But at work, Grant appeared downcast to his friends. He climbed the steep steps up the bluff to his house at the end of the day with slumped shoulders.

All this changed abruptly in the spring of 1861 with the firing on Fort Sumter and President Lincoln's call on the governors for troops at the start of the Civil War. Grant worked to organize a company of men at Galena and, with all graduates of West Point in demand to train the new armies of civilian recruits, expected to be called back into the regular army as an officer. In anticipation, he went to the state capital, Springfield, and drilled volunteers, but the commission was slow in coming and a discouraged Ulysses Grant considered going home. Finally, at the urging of Congressman Elihu Washburne of Galena, Governor Richard Yates promoted Grant to colonel of the Twenty-first Illinois Volunteers.

Grant's spirits rose immediately. Full of resolve, he trained and led his regiment into Missouri to secure that slaveholding border state for the Union. For the whole of the terrible war ahead, Grant's confidence seldom failed him. He was alive in the war as he never had been — or would be again, save in the last year of his life. Successful in his first, limited battles, Grant was promoted to the rank of brigadier general.

In January 1862 Grant, stationed now in Cairo, Illinois, convinced his commanding officer, Henry Halleck, to order an assault on Fort Henry on the Tennessee River and on nearby Fort Donelson on the Cumberland. The two forts at the Kentucky-Tennessee border blocked the entry of Union armies into Confederate Tennessee. Grant took Fort Henry and, to an offer of a truce at Fort Donelson, replied, "No terms except unconditional and immediate surrender can be accepted." Grant was on record that there must be total victory in the war.

Pleased with the first notable victory since the humiliation of the battle of Bull Run, President Lincoln spotted an officer who could win and promoted Grant to the rank of major general. After a brief reassignment by General Halleck, who was both displeased by Grant's seemingly insubordinate independence and jealous of his quick ascendancy, Grant regained command of his army and led it down the Tennessee River to Pittsburg Landing and Shiloh, Tennessee.

There, the Confederate General Albert Sydney Johnston, marching north from Corinth, Mississippi, on April 6, 1862, attacked the poorly defended position of Grant's advance force under General William Tecumseh Sherman. During the first day of the battle, the Union forces were badly battered and nearly driven into the river. Grant slept that night in the rain rather than endure the blood and screams in the building where amputations were taking place. In the morning, rather than call a retreat — and aided by reinforcements — Grant called for an attack.

General Johnston had died of a wound; with General G. P. T. Beauregard in command, the Confederate army was driven from the field and retreated to Corinth. This sense that, no matter how bad the situation, he would press forward was to prove characteristic of Grant throughout the war. Shiloh, judged a victory in Washington, was won at a hideous cost in lives. But Shiloh was only the first of the huge and terrible battles of the war.

If Shiloh ended with questions about Grant's abilities, the long but successful Vicksburg campaign dispelled those criticisms. Ultimately, it was hoped that by driving the Confederate forces from the town, with its commanding position on a bluff above the Mississippi River, the Union would control the river's length and split the Confederacy in two. After a series of failures, Grant's force, bypassing the city from the north, took up a position to its south. In a daring move in which he abandoned his line of communication, Grant placed his armies behind the city and between those of the encroaching General Joseph E. Johnston and General John Pemberton in Vicksburg itself.

With the city under siege, the inhabitants were starving; finally, on July 4, 1863, Pemberton surrendered. At the same time, General George Meade won the battle of Gettysburg, but it was Grant, whose quiet manner — at five feet seven and about 140 pounds, he was never imposing — and iron fist had captured the attention of the press and the public. He was now a national hero.

Later that year, with Union forces encircled in Chattanooga, Tennessee, Secretary of War Edwin M. Stanton came to Louisville, Kentucky, and placed Grant in command of all the armies of the west. Despite an injured leg, Grant crossed the mountain over the only trail still open to the east Tennessee city to try to break the Confederate siege. In an example of another of his traits, he sized up the conflicting reports of the generals in command of the units under him, and with a remarkable ability to visualize a whole theater of the war, wrote, in his own hand, orders moving all of the forces in the west.

Shortly, with the aid of General George Thomas, Grant broke through the Confederate lines and, at the close of November 1863, drove their armies from Chattanooga and out of the southeast corner of Tennessee into Georgia.

As Grant's victories caught the public's eye — his picture appeared on magazine covers — he entered the national political conversation. As early as the battle of Shiloh in 1862, a newspaper speculated that Grant might be elected president; after Vicksburg and Chattanooga, the idea surfaced frequently.

Grant had always understood, as many other generals did not, that his job was to fight and that the policy making in Washington was, for better or worse, not to be complained about. Wisely, he took no note of the political speculation. Lincoln knew that more than one of his generals made no secret that he thought he could do a better job as president. Grant seemed one who would stick to his knitting.

On March 8, 1864, Grant, ordered to Washington, arrived at Willard's Hotel with his young son Fred. He registered as "U. S. Grant and son, Galena, Ill." That evening Grant walked over to the White House, where President and Mrs. Lincoln were holding a reception. As the guests gathered around Grant, who stood modestly at the door, the president came over and introduced himself.

From that moment, the two midwesterners got on well, a simple fact of great importance to the winning of the war. This rapport was sustained largely because Grant never forgot that the president and not the general was the boss. But the fact remained that Grant's entry into the Washington world could not have been more politically astute.

Grant had been summoned to the city to be commissioned lieutenant general, a permanent rank held previously only by George Washington, and put in command of all of the nation's armed forces. He did not linger in the capital. He established his headquarters in Virginia and ordered another assault on the Confederate capital, Richmond. All of the other Union commanders who had set out for the city had been repelled with terrible losses. Now there would be no turning back no matter how serious the casualties were.

As the supreme commander, Grant adopted a simple but merciless approach to the conflict, one that Lincoln shared. The North had far more industrial capacity than the South, but more crucial was the fact that the North's population greatly exceeded the South's. When the Confederacy had run out of men (or boys) of fighting age, there would still be plenty of Union men to handle a gun. This formulation was cold and ruthless; all that was necessary was not to lose any more men in battles than the enemy lost.

Union armies crossed the Rapidan on May 4, 1864, and began the Wilderness Campaign in which the losses would indeed be great. Nonetheless, with Grant in complete command, Northerners were optimistic that this time Richmond would fall. Had Grant succeeded in taking the Confederate capital that spring, he might have become the seventeenth, rather than the eighteenth, president.

The historian James Ford Rhodes later wrote that Grant "was now by all odds the most popular man in the United States." The National Union party — as the Republicans, in a wartime coalition with border-state Unionists, styled themselves — met in June in Baltimore to nominate a candidate for the fall presidential election. Victory in the war seemed to elude the country, and Lincoln was by no means assured of the renomination he sought.

Declining an invitation to a carefully timed rally just prior to the convention to express "gratitude" to Grant, Lincoln noted that the general already had a job to do and did not need another. And Grant wisely gave no puff to the trial balloon. Richmond wasn't taken, and the convention did not stampede to Grant when the Missouri delegation cast its 22 votes for the general.

In Virginia, Grant confronted General Robert E. Lee. For weeks, Confederates and Union men fought with terrible savagery. At one point in the battle of the Wilderness the tangled secondgrowth thicket was set on fire by ceaseless, immense gunfire and hundreds of wounded men suffocated or were burned to death. Relentless, Grant, his eye on Richmond, ordered attack after attack. Again and again, Lee blocked him.

At Cold Harbor, Grant faced an entrenched Confederate force. Believing that to breech Lee's center would mean the capture of Richmond, Grant ordered a suicidal frontal assault and lost six thousand men in one hour. The horrible losses were made worse when Lee and Grant engaged in a macabre game of generalship, refusing to let either side have the cease-fire long enough to allow the wounded to be cleared from the sun-parched field.

Grant abandoned the attempt to take Richmond from the north and moved his army east and south of the Confederate capital. At Petersburg, Grant, facing Lee's firmly entrenched forces, laid siege for nine months. It seemed a stalemate. But while Lee remained on the defensive (and never again mounted an offensive), General Sherman led his troops across Georgia and north through the Carolinas, General Philip Sheridan laid waste the Shenandoah Valley in Virginia, and General Thomas blocked the last Confederate attempt to invade northern Tennessee in December 1864.

A deadly war of attrition was nearly over. On April 2, 1865, Grant broke the Confederate line, forcing Lee to abandon Richmond. With Grant in pursuit and the Confederate army badly depleted, Lee agreed to terms of a military surrender of his army at Appomattox Court House on April 9, 1865.

The war won, Ulysses S. Grant moved into the politics of constructing a peace. The assassination of Abraham Lincoln (a fate that Grant escaped by declining an invitation to join the president at the theater that night) meant that the commanding general of the army had to negotiate between the Radical Republicans in Congress who were sympathetic to the needs of the freed people and the new president, Andrew Johnson, who was not.

State governments sanctioned by Johnson then passed the Black Codes, which severely restricted the rights of the freed people, who were also subject to countless acts of outlaw violence. This the Congress was determined to end. Over Johnson's veto, Congress passed the Radical Reconstruction Acts, designed to rescind state restrictions on the freed people and to have the army ensure their safety, as well as to register the men to vote.

In 1866, Congress revived the rank of full general, believing that Johnson would nominate Ulysses Grant for the post, which he did. Johnson lobbied strenuously to keep Grant in his camp, going so far as to take the general on his ill-fated 1866 "swing around the circle" to rally public support for his plan for Reconstruction.

Grant's appearances with the president on the extensive trip had fueled speculation that he was Johnson's man. In truth, Grant was nobody's man but his own. Grant's shrewdly political stance was that of an apolitical soldier above the political battle, but, at the center of a tremendous battle over what should be the fate of the former slaves, such detachment was difficult to achieve.

With their congressional majorities sustained in the election of 1866, the Republicans, again over Johnson's veto, passed the Radical Reconstruction Acts. As general of the army, Grant was obliged to enforce the acts. Johnson, hoping that he would do so with leniency toward the restored white state governments, fired Secretary of War Stanton, who would not, and named Grant to the post. Johnson considered himself betrayed when Stanton, claiming he could not be fired under the recently enacted Tenure of Office Act, sought to reenter his office and Grant, expected to lock him out, allowed him to enter and once again assume the authority of his cabinet post.

This charade over occupancy was good Washington comedy, which diverted Congress from the unfunny problems of facing the freed people. The Republicans, angered by Johnson's frustration of their Reconstruction policies, chose the relatively minor matter of the president's defiance of the Tenure of Office Act to bring a bill of impeachment.

The impeachment trial of President Johnson, which discredited him even as it failed to remove him from office, did not bring great esteem to the Radicals who challenged him. A war hero seemed necessary to mend the badly divided nation. Grant was the choice of the Republican party as their candidate for president in 1868; Speaker of the House Schuyler Colfax was nominated for the vice presidency.

Contrary to campaign rhetoric, Grant was no Cincinnatus, reluctantly called back from a postwar plow to take up the rigors of rule. Grant had known obscurity and poverty, and he had no urge to confront either again. Quietly returning to Galena and doing no public campaigning, he seemed the reluctant warrior awaiting a call to duty. In point of fact, he stayed close to the telegraph line watching every move of his supporters and offering shrewd advice.

The Republican strategy was simple: accuse the Democrats of causing the rebellion and wave the "bloody shirt" of war to demonstrate their own valor in putting it down. The Democrats, by no means powerless, nominated the governor of New York, Horatio Seymour, for president and the Union General Francis P. Blair for vice president. Grant received 214 electoral votes to Seymour's 80, and 52.7 percent of the popular vote. The 1868 national election was the first in which black men, including the recently enfranchised former slaves, voted in significant numbers. Grant would have won without their votes, but he would have had only a minority of the popular vote. He owed them a political debt.

There were several major achievements of the Grant administration. One still much valued was the establishment of the National Park System. A vast, beautiful stretch of the present states of Wyoming, Idaho, and Montana was set aside as Yellowstone National Park. It was to be free of commercial or residential development and forever available for recreation and for the contemplation of nature. Yellowstone was the forerunner of the vital National Park System so cherished, and so endangered, at the close of the twentieth century.

Another achievement was the settlement of a long-simmering quarrel with Great Britain. During the Civil War, Britain had permitted its shipbuilders to supply the Confederacy with warships that had done great damage to Union cargo ships. The most famous of these Confederate raiders was the *Alabama*, and the claims of the United States government against Britain were known, collectively, as the Alabama Claims.

In the face of increasingly belligerent rhetoric from Congress — demands for huge sums and suggestions to annex Canada — Grant's aristocratic secretary of state, Hamilton Fish, conducted gentlemanly negotiations with the British minister to Washington. Under the 1871 Treaty of Washington, the claims were referred to a board of international arbitration. Charles Francis Adams argued the case before the board, which determined that Britain should apologize and pay the United States $15.5 million to cover the direct claims. Britain did so and a major crisis in Anglo-American relations was resolved while setting an important precedent for the resolution of international disputes by arbitration, rather than war.

A start was made on establishing a professional civil service rather than one based solely on patronage. Grant appointed George William Curtis, a reformer, to head the new Civil Service Commission. But when the Republicans split with the reformers in 1872, the Congress, with Grant's approval, refused to fund the commission. Reform had to wait until another administration.

A major event of the Grant administration was the passage in February 1869 and ratification in February 1870 of the Fifteenth Amendment to the Constitution, stating that "the rights of any citizen of the United States to vote shall not be denied or abridged by the United States or any State on account of race, color, or previous condition of servitude." As a candidate for president, Grant had pledged support of the amendment, and he now hailed its ratification.

Unforeseen were the various ways the amendment would be scorned. Opponents of black voters claimed that people were citizens of the states when it came to deciding who should vote. This concept was upheld in a series of findings of the Supreme Court, beginning with the Slaughter-House Cases of 1873. Legislators employed ingenious means of blocking the vote of African Americans, such as grandfather clauses (one could vote only if one's grandfather could vote, and no slave grandfather could) and poll taxes that the poor could not afford.

But in 1870 the Fifteenth Amendment was, with the other two Reconstruction amendments to the Constitution in place — the Thirteenth ending slavery and involuntary servitude and the Fourteenth establishing the civil rights of citizens — optimistically thought to have completed the work of Reconstruction. Then Congress created the Justice Department, and the attorney general, once simply the government's lawyer, now had vastly increased responsibilities overseeing United States district attorneys bringing cases under the amendments in federal district courts around the country. In addition the new post of solicitor general was created to bring the government's cases to the Supreme Court. The once promising process of establishing the concept of national rather than state-defined rights of citizens was begun.

Under Senator Charles Sumner's leadership, the Civil Rights Act of 1875, forbidding discrimination in places of public assembly such as theaters and restaurants, was passed. The act was ruled unconstitutional in 1883 but was the forerunner of the Civil Rights Act of 1964, sustained by a later Supreme Court.

Less salutary was Grant's almost obsessive desire to annex the Dominican Republic. Caribbean expansion was out of character for him; he seemed less interested in imperial notions than in providing a safe haven for America's own black citizens whom he found himself increasingly unable to protect at home.

Secretary of State Fish, never particularly conscious of domestic racial troubles, was not enthusiastic about annexation, but the leader of the successful opposition was Charles Sumner, the abolitionist chairman of the Senate's Foreign Relations Committee, who vigorously fought annexation. Instead, he wanted enforcement of African Americans' rights. Sumner's break with the president and many of his black supporters such as Frederick Douglass caused a breach in the ranks of those striving to aid the freed people.

Such assistance seemed guaranteed by the election of the victor of the war. Grant had great personal authority in the nation; a hero in the North, he was held in some respect in the South. Grant was never despised in the former Confederacy in the way General Sherman or General Benjamin Butler were. President Lincoln had appointed a majority of the justices of the Supreme Court, and Republicans, deemed the friends of the freed people, were in the majority in both houses of Congress.

Grant was positioned to carry out a Reconstruction program that would ensure the safety and rights of the nation's new black citizens. The first undercutting of that effort was a personal loss, the death, not a year after the election, of Secretary of War John A. Rawlins, the only true champion of black rights in the original Grant cabinet. Rawlins, Grant's chief of staff during the war, was a friend not afraid to criticize his commander. With his death in 1869, Grant lost not only his wisest counselor on public issues, but also an intimate friend who might have helped him achieve a greater sense of purpose with regard to not only Reconstruction issues, but all others.

The second was the massive effort by a significant proportion of the white South to reassert white dominance in the region. Violence against the freed people, common since the wartime ending of their bondage, had intensified. Since its founding at Pulaski, Tennessee, in 1866, the regionally organized terrorist organization, the Ku Klux Klan, had worked to intimidate and suppress the freed people. The tortured corpse of a man who had been active politically or was succeeding economically was a grim warning to aspiring citizens of a whole county or, indeed, region.

Grant's first attorney general, E. R. Hoar, had not favored aggressive federal action in the civil rights field, but his successor, Amos T.

Akerman, appointed in 1870, did. Southern Republicans, consisting of local white citizens who accepted that the war had truly changed their region; Northerners who had moved south since the war; and newly enfranchised African Americans urged the appointment.

Akerman was a native of New Hampshire who had long lived in the South and had been a colonel in the Confederate army. He knew the Klan firsthand in his home country of north Georgia, as did Congress after extensive hearings. In the spring of 1871, with the support of President Grant, Congress enacted a strong Ku Klux Klan Act designed specifically to bring those who attacked black people into the federal courts. Akerman, assisted by United States attorneys in the South, directed the prosecution of cases in South Carolina, a state with a large black population that was making significant advances. These sped the Klan into action.

So extreme was the violence that Grant, as a military man himself, reluctant to send troops to carry out civilian policy, was willing to use the courts. He suspended the writ of habeas corpus, enabling those accused to be held by federal marshals, without possibility of relief, until indicted. Cases were heard by conscientious federal judges, but brutal intimidation of witnesses made prosecutions difficult. Nevertheless, there were six hundred convictions. This vigorous federal action curtailed the Klan's effectiveness, and white supremacists turned to other methods of achieving their goal.

As for Amos Akerman, his zealous pursuit of justice ended with his dismissal. Opponents of his civil rights policies and railroad men angry over his stern scrutiny of their activities lobbied strenuously, and Grant forced Akerman to resign. His successor did continue the prosecutions but was far less interested in the issue than Akerman had been; Grant's two succeeding attorneys general were still less so.

In differing ways, white supremacists successfully moved to reassert their power in each of the states of the former Confederacy. In Mississippi, for example, in 1875, the Republican governor was Adelbert Ames, a Union general from Maine, while one senator, James Alcorn, was a native Mississippian and the other, Blanche K. Bruce, was African American, as were many local and state elected officials. White supremacists, utilizing the Democratic party, turned to politics to defeat Reconstruction government and were frank about their goal. The "white liners," as they termed themselves, printed on newspapers across the state: "Carry the election peaceably if we can, forcibly if we must."

The Democrats organized in a disciplined way and not only put forward their white line candidates forcefully, but demonstrated what their opponents could expect if they continued to resist. The candidate for county treasurer assassinated in Macon, Mississippi, was but one of many African Americans murdered in order to intimidate other candidates and voters. The survivors of a bloody assault on a Republican meeting in Clinton fled to Jackson and begged Governor Ames to protect them.

When Ames petitioned the Grant administration for federal troops, yet another of Grant's attorneys general, Edwards Pierrepont, replied: "The whole public are tired of these autumnal outbursts in the South." Ames was told to handle it on his own. When election day came, there was no violence. There was no need for any; the intimidation worked. The Democrats swept into office with huge majorities.

By 1877, all of the states of the former Confederacy were governed by men firmly committed to white supremacy. The section had been, in their parlance, redeemed; Reconstruction was over. Grant had presided over the federal apparatus at the moment when it seemed to be in the best position to substantiate the gains the Civil War victory had pointed to. But, as president, he had not used his considerable powers to the full, and the dream of a truly reconstructed nation proved a dream deferred.

Native Americans commanded attention too. Grant had been sensitive to the destitution of displaced Indians when stationed on the Columbia River in the Pacific Northwest. As president, he saw the relentless expansion of white America into the Native American reaches of the west. To respond to this dilemma, he appointed Eli S. Parker, a Seneca, commissioner of Indian affairs, but they were faced with generals like Sherman who explicitly used the word *extermination* as the inevitable solution to the Indian *problem* and by the white settlers who demanded protection from attacking warriors defending their lands.

To thwart Sherman's lethal approach, while depriving the Native Americans of an ability to retaliate, Grant listened to Christian reformers. Under the peace policy various Christian denominations were assigned to reservations on which Indians were required to live and were given the job of assimilating the Indians into the ways of white America. The forced change in the way of life of a people had grave consequences. Problems of economic and cultural dislocation were not addressed and still remain.

The Grant administration was also confronted with financial crises. Financially conservative measures, including the demonetization of silver and rushed reduction in the national debt, curbed economic growth. In 1869, a plot to corner the gold market caused many business failures, and, in 1873, at the beginning of Grant's second term, Jay Cooke and

Company failed. The banker who had financed the war effort had over-extended his bank in his efforts to sell the securities of the Northern Pacific Railway. The panic that ensued exposed the weakness in the whole economy. The resulting depression lasted six years, or, by some reckonings, was the beginning of the economic troubles that beset the nation until the end of the century.

Grant's reelection in 1872 was widely expected and the election remarkable if only because Grant's strongest challenge came from dissidents in his own party. The Liberal Republican movement was led by the genteel reform element within the party who believed the Grant regime to be corrupt. As they looked around they saw a lack of intellectual force in the government and blamed it for corruption that was reaching into city halls and state legislative chambers. Joining forces with the Democrats behind the presidential candidate Horace Greeley, the well-known editor of the *New York Tribune,* they challenged Grant.

Leaving nothing to chance, the "tanner" and the "cobbler" — Grant's running mate, Senator Henry Wilson, had once worked in a shoe factory — reached out to working-class voters with considerable success. Aided by loyal black voters, Grant won 55.6 percent of the popular vote.

The Democrats were not long in recovering. The country's economic problems, coupled with the continuing reports of scandals within the ranks of the Grant administration, enabled them to regain control of the House of Representatives in the election of 1874. As the new majority took its seats, investigations were launched that soon implicated many of Grant's cabinet ministers. The most notable scandal involved Secretary of War William Belknap, who blatantly accepted regular kickbacks from agents running Indian trading posts. Belknap was impeached and tried by the Senate and would have been found guilty if not for the technicality that he no longer held the office. Grant had allowed him to resign.

One scandal hit close to Grant personally. His private secretary, Orville Babcock, was shown to have been directly involved in widespread and lucrative schemes to defraud the government on taxes on the production of whiskey. Even when Secretary of the Treasury Benjamin Bristow confronted Grant with damning evidence, Grant refused to fire Babcock. At Babcock's trial, Grant provided a deposition that won his aide's acquittal.

Grant was still popular with the people and considered running for a third term. But the scandals had taken their toll, and others in the Republican party sought power. He was denied the nomination. The eight years in the White House had been the longest stay in one residence of the Grants' marriage and they left feeling like "waifs." Their solution was to set out on a two-and-a-half-year world tour that became a triumph. In northern England, thousands of workers, committed to a labor movement, turned out to celebrate the "Emancipator," whom they saw as having freed American workers who were as oppressed as they were. In Japan, Grant became the first mortal with whom an emperor shook hands.

Returning home, there was another try, in 1880, for a third term, but it too failed. At a loss, the Grants moved to New York City and the general went into the investment business with his son and a speculator named Fernando Ward. Grant was about as well suited to Wall Street finance as he had been to bill collecting, and when Ward defrauded the firm, the former president of the United States again faced the poverty that he had so long feared.

To provide for his wife, Grant undertook to write articles about Civil War battles for *Century Magazine,* which then led him to consider a book. His friend Mark Twain, warning him against a poor contract, offered Grant a contract with his own publishing house. When the book was published, Julia Grant received a record royalty check. But for Grant, the book represented far more than money.

At just about the same time, late summer 1884, that he began work on his book, he developed the first symptoms of cancer of the esophagus. Thousands and thousands of cigars had taken their toll. When he died on July 23, 1885, only two weeks after his final work on proofs of his memoir, he had, in ten months, completed a great book. No other presidential memoir can match it in literary quality; critics as diverse as Gertrude Stein and Edmund Wilson have hailed it. The *Personal Memoirs of U. S. Grant,* which gives scant attention to his presidency, is a classic account of the Civil War. Ulysses S. Grant, as a writer, brought to bear the same force of will that had been his as a warrior and which, in the main, eluded him as president.

— WILLIAM McFEELY

Rutherford B. Hayes

1877–1881

b. October 4, 1822
d. January 17, 1893

A T FIRST GLANCE, the presidency of Rutherford Birchard Hayes might not seem worth a second glance. The novelist Thomas Wolfe wrote that Hayes and the Gilded Age presidents who followed "swam together in the sea-depths of a past intangible, immeasurable, and unknowable. . . . Which had the whiskers, which the burnsides: which was which?" Historians in recent years, however, have restored Hayes to deserved prominence. It was Hayes, after all, who rehabilitated the presidency after the impeachment and near conviction of Andrew Johnson and the scandal-ridden term of Ulysses S. Grant. During the Hayes presidency, the United States drew closer to becoming a world power: sectional animosities were tempered, the civil service was reformed, and domestic and foreign commerce thrived. Yet the period was also marked by ethnic and class conflict. Rather than attend to the distinctive needs of lower-class laborers and ethnic minorities, the Hayes administration promoted the development of a homogeneous national culture based largely on middle-class economic and political values. For Hayes, the key to creating homogeneity and unity was education, so that every American could learn to engage with the market economy and to act responsibly in the political system. Until all were educated, Hayes once wrote (at a time when he had in mind southern blacks), the elite must "forget to drive and learn to lead the ignorant masses around them."

Like many nineteenth-century presidents, Hayes was a war hero, but he had earned his fame more through his political than his military experience. Born in Delaware, Ohio, on October 4, 1822, Hayes was raised by his mother, Sophia Birchard Hayes, and his uncle, Sardis

Birchard; his father, Rutherford Hayes, a farmer, died before he was born. Educated at preparatory schools in Ohio and Connecticut, and then at Kenyon College and Harvard Law School, Hayes practiced law in Lower Sandusky (later Fremont), Ohio, from 1845 to 1849, and then in 1850 settled in Cincinnati, where he made a significant reputation in the courts. In 1852 he married Lucy Ware Webb. The couple would have six children. A Whig during the 1840s and early 1850s, Hayes began attending Republican functions in 1855 and in 1858 successfully ran as a Republican for solicitor of Cincinnati, his first political office. After the Civil War broke out, Hayes was commissioned as a major in the Twenty-third Ohio Volunteer Infantry and eventually attained the rank of brevet major general. In 1864, while still in the military, Hayes was elected to the Thirty-ninth Congress, where he served one term. Elected as governor of Ohio in 1867, he retired in 1870 after two terms but was elected again in 1875. This last campaign, in which Hayes successfully united the rival factions of his party, won him national attention.

Hayes's reputation as a party healer and his popularity in Ohio, a politically important state, made him a potential presidential candidate in 1876. In a field of candidates known as machine politicians, lackeys of big business, or zealots for reform, Hayes had the advantage of being something of an outsider, neither tainted by scandal nor fixated on reform. When James G. Blaine, the Speaker of the House of Representatives, failed to secure an early nomination at the convention, and none of the other candidates could gain a majority, the anti-Blaine delegates joined to nominate Hayes. The convention then nominated the New York Congressman William A. Wheeler for vice president. Hayes's Democratic opponent was Samuel J. Tilden, the governor of New York and smasher of the notorious Tweed Ring of Tammany Hall. As was the custom, Hayes did not campaign for the presidency, but he did make two public appearances, both at the Philadelphia Centennial Exhibition. Initially hoping that Republicans could win the election on the issues of hard money and civil service reform, Hayes gave way to the more traditional "bloody shirt" tactics of the Republicans by encouraging campaigners to stress "the danger of giving the Rebels the Government." On election day, Hayes predicted defeat, and his prediction seemed to come true. Tilden received a plurality of 250,000 votes and won 203 electoral votes when only 185 were needed for victory.

But the election was far from over. Republicans disputed the returns from Louisiana, Florida, and South Carolina. If Hayes received the electoral votes from these states, where Republicans controlled the returns, he would win. By disqualifying enough Democratic ballots, the boards in these states declared a victory for Hayes and sent their results to Congress. Democrats saw fraud — a reasonable charge, but one counterbalanced by the white intimidation and violence that had kept from the polls thousands of blacks who would have voted for Hayes — and they sent to Congress their own sets of returns showing a Democratic victory. Oregon also sent in two sets of returns because the Democrats there claimed that one of the Republican electors should be disqualified. As the Constitution required, a joint session of Congress assembled in which the president of the Senate was to count the electoral votes. Congress was immediately deadlocked, however, because the Democratic majority disagreed with the Republican president of the Senate over which sets of ballots should be counted. To resolve the crisis, Congress created an electoral commission composed of five senators, five representatives, and five Supreme Court justices. The commission was supposed to be politically balanced, with seven Democrats, seven Republicans, and one independent — ostensibly Justice David Davis. But almost immediately Davis removed himself because he was elected to the Senate, and he was replaced by Justice Joseph P. Bradley, a Republican. With an 8 to 7 majority on the commission, the Republicans resolved every disputed ballot in favor of Hayes, giving him the 185 electoral votes needed for election.

The Democrats refused to turn away empty-handed. They could still filibuster, creating enough delay to prevent Hayes's inauguration and cause a constitutional crisis, but few wanted to play this card. Some were willing, however, to use the bluff to gain concessions from the Republicans. In particular, Democrats wanted the removal of federal troops that still supported the governments of Louisiana and South Carolina, the last Republican regimes in the south. They also hoped to gain some federal appointments and federal subsidies for a southern railroad. Hayes promised nothing, but some of his associates, including Grant, privately pledged a withdrawal of the troops if the southern governments promised to uphold freed people's rights. During secret negotiations at the Wormley Hotel in Washington, which Hayes did not attend, other agreements may have been worked out, but probably none of these affected Hayes's policy once he was inaugurated. Indeed, much that Democrats believed had been conceded to them was already fated. Even before the election, Hayes had hoped that, without damaging the rights of African Americans, federal troops could be removed from the south in order to promote national unity and attract southern whites into the Republican party. As part of this goodwill strategy, Hayes was destined to support federal subsidies for southern railroads and to appoint white southerners to federal office. Despite Hayes's prior intentions, southern Democrats treated Republican pledges as concessions, not predictable policies, so they believed a bargain had been struck. Northern Democrats, in the meantime, had no desire to bring on a constitutional crisis that might create a financial panic, so they too agreed to let the contest go to Hayes, who was declared elected on March 2, 1877.

Three days later at the inaugural, Hayes made a dual pledge "to protect the rights of all by every constitutional means" and to restore "honest and efficient local self-government." He did not yet see that the two objectives were mutually exclusive. Making good on his promise, Hayes removed federal troops from the capitals of South Carolina and Louisiana, where Republican and Democratic governments vied for power. He thus assured the collapse of Republican rule in these states, but only after receiving promises from southern Democratic leaders that black rights would be respected. A month after his inaugural, South Carolina was restored to Democratic rule; three weeks after that, a commission appointed by him upheld the Democrats in Louisiana. To nurture an alliance with white southern elites, Hayes also appointed as postmaster general a southerner, David M. Key of Tennessee, who used his patronage powers to try to draw southern whites into the Republican fold. None of Hayes's efforts bore fruit. The Republican party in the south fizzled with the restoration of home rule, and African Americans there suffered unprecedented abuses, especially during political campaigns. In November 1878, Hayes privately admitted that "the experiment was a failure." The end of Republican rule in the south was not the only factor threatening the livelihood of southern African Americans. The southern economy's overreliance on cotton, combined with harsh terms of credit for agricultural laborers, made it nearly impossible for poorer southern blacks and whites to become independent landowners. In 1879, tens of thousands of African Americans gave up on the South — and on Hayes — and moved west.

Hayes never meant to abandon African Americans. He valued the opinions of black leaders and appointed some to important posts — Frederick Douglass, for example, became marshal of the District of Columbia. During the last two years of his presidency, Hayes did his best to enforce the Reconstruction amendments, which he believed guaranteed political but not social equality. As a result of the election of 1878, the Democrats, who already controlled the House of Representatives, gained a majority in the Senate. Seven times in the next Congress, Democrats attached to appropriation bills "riders" that would have restored civil rights to ex-Confederates and repealed the Enforcement Acts of the early 1870s, which authorized federal troops to ensure a fair vote. Despite the united opposition of congressional Republicans, each of these bills passed, and Hayes vetoed every one. While the policy of enforcement was sustained, the practice of enforcement remained weak, in part because of a diminished and underfunded military. Also, southern state governments devised more subtle methods of disenfranchisement, such as the poll tax initiated by Georgia in 1877. African Americans did not simply disappear from southern politics, however. They still occupied important roles in states such as Virginia, where the biracial Readjuster party ruled from 1879 to 1883, and they still turned out for elections. In the presidential election of 1880, for example, 50 to 80 percent of eligible African Americans voted in each southern state. But Hayes's plan of enlisting southern whites to end partisan and racial division remained stillborn, and as his term ended he still embraced the one plank of his southern policy that he would never drop: public education for African Americans.

Hayes made this plank central to his policy for Native Americans as well. Under Hayes and Secretary of the Interior Carl Schurz, a "permanent civilization fund" created Indian boarding schools. Designed to integrate Native Americans into Anglo-American society, the schools usually isolated pupils from their tribes, forced them to learn English, and trained them in housework and industrial arts. As described by Richard Henry Pratt, head of the Carlisle Indian School in Pennsylvania, the goal was to "kill the Indian and save the man." Literal killing of Indians also marked the Hayes years. In the summer of 1877, war broke out in the northwest with the Nez Percé, who defeated local and federal troops at the battle of White Bird and then conducted a fighting retreat hundreds of miles to Bear Paw Mountain, in present-day Montana, where the bulk of the tribe surrendered and Chief Joseph uttered his famous words: "I will fight no more forever." Although the federal commander who accepted the surrender promised the Nez Percé safe passage to their reservation, General William T. Sherman annulled the agreement, and after a series of forced migrations in the Indian territory of present-day Oklahoma, most of the tribe perished. A new crisis arose in 1879, when a small band of Poncas tried to move from the Indian territory to their original home in the Dakotas. After the Poncas were arrested in Nebraska, a federal district judge set them free, at which point the leader of the Ponca band and two Omahas went on a speaking tour that aroused popular sympathies for the mistreatment of the Poncas and all Native Americans. The sensational tour spurred Helen Hunt Jackson to write *A Century of Dishonor,* published in 1881, which condemned United

States Indian policy, and it led Hayes and Schurz to urge Congress to grant the Poncas financial compensation, to allow them to live where they wished, and to give them title to their lands in severalty — that is, as individuals rather than as tribes. The recommendations conformed to Hayes's general Indian policy, for the granting of lands in severalty would undermine tribal life and accelerate the integration of Native Americans, as individuals, into the market economy. Congress agreed to compensate the Poncas, but it did not pass a severalty plan; that would come — for all Native Americans — in the Dawes Act of 1887.

In Hayes's ideal republic, citizens were to be educated and economically independent, and civil servants were to earn their positions through merit, not political connections. He had supported civil service reform as a congressman and had granted appointments to worthy Democrats as governor, but as president he faced a special challenge. As the federal bureaucracy had grown from fifty-one thousand employees in 1871 to a hundred thousand in 1877, the abuse of patronage had become more deeply rooted. After the scandals of the Grant administration and his own controversial election (critics called him "Rutherfraud"), Hayes thought it particularly important to establish a principled government service. He began by selecting cabinet members who had little connection to the previous administration and who had not been contenders for the presidency. The most influential of his officers were Secretary of State William Max Evarts, Secretary of Treasury John Sherman, and Schurz, who, as a leader of the "liberal" Republicans, championed civil service reform. The assault on machine politics began with Hayes's order of June 22, 1877, prohibiting political assessments of civil servants and civil servant participation in political campaigns. Next Hayes took aim at the New York customhouse, which was headed by Chester A. Arthur, the lieutenant of Senator Roscoe Conkling, the chief of the Republican antireform "Stalwarts." Eventually defeating Conkling by suspending Arthur and applying the merit system to the customhouse, Hayes went on to implement strict rules for appointment at all customhouses. His reforms alienated many Republicans, however, and were the primary cause of the Republican defeats in the congressional elections of 1878. With continued opposition from Republican spoilsmen and the Democratic majority in Congress, Hayes's effort to create a civil service modeled on the British system was doomed. Yet he increased the efficiency of the customhouses and paved the way for the Civil Service Reform Act of 1883. Hayes was not above using political appointments for political ends, and he allowed civil servants to make voluntary contributions to political campaigns, but he was definitely less friendly to the spoils system than any president had been since John Quincy Adams.

Hayes also shared with John Quincy Adams a reputation as a puritan. It was true that Hayes was pro-temperance, but it was "Lemonade Lucy" who was most responsible for the policy of keeping the White House alcohol-free. Lucy was also a much more devout Christian than her husband, who attended Methodist services with his wife but never joined a church. If there was a puritan in the administration, it was not Hayes but Anthony Comstock, the post office investigator who arrested such supposed purveyors of obscenity as Ezra H. Heywood, cofounder with his wife of the New England Free Love League. Hayes pardoned Heywood but upheld other convictions resulting from Comstock's campaign. On the issue of women's suffrage, however, Hayes was consistently conservative. He approved of women working outside the home — under his administration, many women were employed in unskilled jobs in the civil service — but both he and Lucy thought that maternal duties were incompatible with political participation. They frowned upon the campaign for women's suffrage, which at the time was particularly active but unsuccessful in the west.

Hayes was also a poor friend to organized labor. Membership in labor unions declined rapidly after the onset of economic depression in 1873, but organized labor did not disappear. The Knights of Labor, founded in 1869, became particularly popular during the Hayes years, in part because of the leadership of Terence V. Powderly, who headed the Knights after 1879. Laborers were vulnerable to the whim of unregulated limited liability corporations, most notably the railroads, that cooperated to fix wages and prices. In July 1877, wage cuts on eastern lines led to the Great Railroad Strike, which spread from Martinsburg, West Virginia, east to Philadelphia and west to Chicago and St. Louis. For two weeks, as nonrailroad workers joined with conductors, brakemen, and porters to cripple transportation, a public increasingly hostile to monopolies lent its support. Hayes used the army to quell the violence in some communities, making him the first president to send federal troops against a labor uprising. But despite the requests of some business leaders and local officials, he did not use the government to operate the railroads or to suppress the entire strike. In the west, Anglo-American labor focused its discontent on Chinese immigrants, whose low wages represented a threat. Influenced by Denis Kearney's Workingmen's party, the California constitutional convention of 1878 denied Chinese immigrants access to state jobs and prohibited them from voting. Congress followed with a bill restricting Chinese immigration. Hayes vetoed the bill, but only because it violated the open immigration clauses of the 1868 Burlingame Treaty with China. That treaty was renegotiated, with Hayes's encouragement, to allow the exclusion of Chinese. Future conflict could also be averted, thought Hayes, through "education of the strikers" and "judicious control of the capitalists." By again assuming that ignorance lay at the root of conflict, Hayes ignored the deeper cause of discontent: concentration of wealth in the hands of the few. Disparity in wealth helped make a bestseller out of Henry George's 1879 *Progress and Poverty*, which argued for a single tax to undercut the unnatural accumulation of capital through land ownership.

Hard currency, not wealth redistribution, was the hallmark of Hayes's economic policy. Determined to restore confidence in the American economy by retiring the wartime debt without inflating the currency, Hayes and John Sherman refinanced much of the debt and resumed specie payments in 1879, as called for in the Resumption Act of 1875. Against the administration, which favored a gold-backed currency, stood western silver producers and southern and midwestern small farmers. These farmers, who tended to be heavily in debt, preferred inflation to contraction and therefore supported the issuing of "greenbacks," currency not backed by gold. Silver producers, who still bristled from the demonetization of silver in 1873 ("the crime of '73"), favored "free silver," the unlimited coinage of silver dollars, which were intrinsically less valuable than gold dollars. Advocates of soft money seemed to triumph in 1878. In that year, Congress overrode Hayes's veto to carry the Bland-Allison Act, which required the government to purchase and coin an annual minimum of $2 million in silver. But the act fell short of the goal of unlimited coinage, and the administration never purchased more than the minimum silver required. The other apparent victory for soft money in 1878 was the success of the Greenback Labor party, which won fourteen congressional seats. However, the election results were probably due less to popular support for soft money than to widespread sympathy for labor, particularly after the Great Strike of 1877. Hayes eventually yielded to limited silver coinage once the economy showed signs of recovery.

Restoring the national economy was, for Hayes and many Americans, part of a strategy to establish the United States as a legitimate world power. Eager to embrace symbols of national progress and technological superiority, Americans could be proud of the telephone that was installed in the Hayes White House, although the new machine was rarely used. The Hayes administration found it difficult to translate the ideal of

American dominance into a strong foreign policy, however, because of the tiny size of the army and navy. Cautious diplomacy nonetheless advanced American commercial interests. A hard-line refusal to recognize the new Porfirio Díaz regime in Mexico because of border raids by Mexican bandits gave way to cooperative efforts by the two countries to suppress the raids, followed in 1878 with official recognition. Peace on the border more than doubled yearly exports to Mexico. Again with commerce foremost in mind, Hayes denounced the efforts of the private French citizen Ferdinand Marie de Lesseps to build a canal across Panama. Although nothing came of the plan, Hayes seized the opportunity to invoke the Monroe Doctrine against ventures by foreign citizens — not just foreign governments — in the Western Hemisphere. Meanwhile, under Evarts's guidance, the State Department negotiated a new commercial treaty with China (in conjunction with the new immigration treaty). It also secured Pago Pago as a naval station, and it expanded commercial rights in Samoa, Hawaii, and Korea.

Hayes had announced in his inaugural address that he would serve only one term — a pledge he honored — and that "He serves his party best who serves the country best." By the end of his administration, he had served his country well, but his party had reaped only modest benefits. Badly divided between "Stalwarts" and "Half-Breeds," the Republicans in 1880 ran a dark horse candidate, James A. Garfield, and nearly lost the election. Hayes, who lent little assistance to the Garfield campaign, retired to Spiegel Grove, his estate in Fremont. There he sustained his commitment to education through service to the Peabody and Slater funds. As president of the National Prison Association from 1883 until his death, Hayes argued that even the most deviant person could be reformed. It was a position consistent with his goal as American president: unity and peace achieved by the eradication of ignorance. On January 17, 1893, three days after an attack of angina, Hayes died at home.

— MICHAEL VORENBERG
Brown University

James A. Garfield

1881

b. November 19, 1831
d. September 19, 1881

T HE PRESIDENTS of the late nineteenth century make inviting targets. Their critics charge that they provided little leadership in the face of major challenges — that they ignored
the plight of former slaves and the continuing racial tensions
throughout the country; that they remained passive in response to the
emergence of giant corporations, the rise of a new and largely immigrant industrial working class, the explosive growth of cities, and the
widening economic disparities among regions; and that they did little
about rampant corruption in the political system.

Yet the federal government was not in a very good position to do much
about these matters in those years, and neither were the men who occupied the White House. Although federal revenues were rising and the
number of government employees was growing rapidly in the 1870s and
1880s, this increase reflected population and economic growth more
than a broadening of government activity or authority. Most federal employees worked for the Post Office, delivering the mail, or the Treasury
Department, collecting customs duties. Most Americans expected Congress — in session only about half of each year — to provide policy leadership. They looked to the president for efficient administration of the
government and management of foreign policy and to serve as a solid,
dignified symbol of the nation.

The brief administration of James A. Garfield should be assessed in
this context. He was elected on a platform that promised a continuation
of many of the policies of his predecessor, friend, and fellow Ohio Republican, Rutherford B. Hayes. He proposed modest reform of the civil
service, with term limits for junior and midlevel positions. He directed

the postmaster general to ferret out the details of an emerging scandal in the assignment of independent postal routes — a scandal that, incidentally, implicated Garfield's own campaign managers. He arranged for the government to pay off its debt in specie. He promoted the "Industrial Revolution" (a term he used decades before it became popular) but was prepared to consider some form of regulation of the railroads. He supported universal suffrage but believed that the lot of African Americans in the South would improve in the long run only with better public education. In foreign policy, the president sought to limit Chinese immigration and to minimize European influence in the Americas. To help build solidarity in the hemisphere, he approved Secretary of State James G. Blaine's plans to host a conference of American states in Washington in 1882.

The most important promise Garfield brought to the presidency was not new policies or innovative solutions to the ills of the nation, but rather improved relations between the White House and Congress, which had been bitterly contentious since the early days of Reconstruction. Garfield was a man of consensus, a "pacificator" who labored to overcome divisions within his party and the Congress. He was also a man of the Congress, where he had spent seventeen years, and he brought deep knowledge of its personalities and workings to the presidency. By temperament and experience, he seemed well suited to govern in a spirit of compromise. Typically, he took an evenhanded view of presidential power. The job of the president, he declared, was "to recommend to Congress such measures as he considers for the public good," not to "dictate to Congress the policy of the government, nor use the power of his great office to force upon Congress his own peculiar views of legislation." On the other hand, Garfield saw a need to restore authority to the office of president. He agreed with Hayes that the presidency had been severely weakened during Reconstruction and needed fortification, especially in the president's ability to make appointments.

The question facing the Garfield administration in 1881 was whether, in an era of aggressive partisan politics and lingering rancor from the Civil War, a government based on compromise and consensus could make much progress. The new president himself expressed private reservations, calling the presidency "the bleak mountain" and wondering about his suitability for the position. "I love to deal with doctrines and events," he confided to his diary soon after taking office. "The contests of men about men I greatly dislike."

*   *   *

His inner doubts aside, the forty-nine-year-old Garfield had seemed tailor-made for the presidency — as biographers, including Horatio Alger, had emphasized during the campaign. He was born in a log cabin (the last president of whom this would ever be said) to poor but hardworking parents in a rural part of northeastern Ohio on November 19, 1831. His father died when he was eighteen months old. Young James rose from a childhood of material deprivation through industry and high native intelligence. As a teenager, he spent several months as a boat driver and handyman on Ohio canals. When he was seventeen, a district school-teacher recognized his potential and arranged for him to enter a local academy. That boost, and a religious conversion experience a year later, propelled him on his way. He attended the Western Reserve Eclectic Institute (subsequently Hiram College) while teaching school and preaching in local churches and then finished his education at Williams College.

From there, Garfield's rise was extraordinarily swift. At age twenty-four, he was a professor of ancient languages and literature at his alma mater in Hiram; at twenty-five, the college's acting president; at twenty-eight, a Republican state senator (and ardent abolitionist); at thirty-one, a major general in the Union Army (where he distinguished himself especially in staff assignments during the early years of the Civil War); and at thirty-two, a Republican member of Congress. By then, he was married to the former Lucretia Rudolph (affectionately known as "Crete"). The couple would have five children.

In Congress, Garfield established a reputation as a hard worker, a competent legislator, and a speaker who delivered well-crafted and ornate orations. He endorsed the impeachment of Andrew Johnson and the congressional program of reconstruction. In the 1870s, he supported the laissez-faire economic policies of the Grant and Hayes administrations. Although he won acclaim for his defense of hard currency — an unpopular stance in Ohio and other western states — he usually devoted himself to less controversial subjects: the reform of the Bureau of the Census, the establishment of the Department (later Bureau) of Education, and the creation of the U.S. Geological Survey. His reliability and conservatism on economic issues won him the chairmanship of the Committee on Appropriations. When Blaine moved up to the Senate in 1876, Garfield became the leading Republican in the House and the party's candidate for Speaker. He served on the special commission that settled the 1876 presidential election in favor of Hayes.

Yet, for all of his accomplishments, Garfield was a diffident politician who prided himself on scholarly pursuits and enjoyed the life of a man of letters. He read voraciously and was often found in the Library of Congress. During one six-week period, while recuperating from a minor affliction, he worked his way through General Sherman's two-volume memoirs, a twelve-volume history of England, several plays of Shakespeare, and new works by Thackeray and Tennyson, all the while taking copious notes. He numbered the writers William Dean Howells and Henry Wadsworth Longfellow among his friends and enjoyed hearing Charles Dickens when he lectured in the United States. Garfield occasionally wrote essays on political and economic topics, but most of his publications were based on his speeches, over which he labored long and hard, fine-tuning them until the moment of delivery. He also maintained a voluminous correspondence, and from adolescence until the night before his shooting he kept a diary that is now an invaluable source for learning about American social and political life during the middle part of the nineteenth century.

Garfield's indifferent skills as a politician became manifest in a series of questionable dealings in the 1870s. Although these episodes seemed to his enemies evidence of deep-seated corruption, they more accurately indicated clumsiness, poor judgment, and a conviction that his political success had earned him a right to make some money. In one instance, Garfield accepted a $5,000 fee for little work in advising a paving contractor who was seeking (and who later won) lucrative public business in Washington, D.C. Garfield led the Appropriations Committee, which was responsible for public works in the capital at the time, and he should have understood the impropriety of his actions. But the furor over the matter led him to wonder instead "whether a member of Congress has any rights."

Another damaging affair stemmed from his relationship with fellow U.S. Representative Oakes Ames, who presided over the Crédit Mobilier, the corrupt entity that helped to finance construction of the Union Pacific Railroad. As details of Ames's machinations came to light, he listed Garfield among many congressmen who had profited from the Crédit Mobilier. The facts of the case were muddy and Garfield's testimony did little to clear them up. It appears, however, that he accepted stock in the company but never took possession of it. He apparently allowed dividends to pay for the purchase price and then collected a dividend of $329. Garfield recalled only arranging a loan from Ames and repeatedly denied owning stock. A congressional inquiry did not answer the question of stock ownership but did find that he had received the dividend. But the inquiry did not portray this payment as a bribe and produced no evidence that Garfield had acted improperly. (Garfield escaped censure, but the number 329 would come back to haunt him during his presidential campaign, when it showed up everywhere, on barns, doors, walls, streets, sidewalks, and even the steps of Garfield's home in Washington.)

Garfield prospered in politics not because he was corrupt, but because he was competent. He also avoided extreme positions and seemed able to transcend factional differences within the Republican party. He once wrote that "the law of my life [is] never to ask for an office." He attributed his own rise to hard work and providence, but others construed his ostensible lack of ambition as an indication that he would be an acceptable compromise candidate. He was first nominated to run for the state senate, then Congress, then the U.S. Senate, and ultimately the presidency, all without campaigning, because party leaders considered incumbents or leading contenders worse alternatives. During the late 1870s, when the Stalwarts (as advocates of a third term for Ulysses S. Grant were known) and Half-Breeds (as the opponents to the Stalwarts were known) sparred for control of the Republican party, Garfield sought to remain above the fray. His admirers saw this independence as a strength, while his detractors portrayed him as a weakling. Grant said that Garfield lacked the backbone of an angleworm. Even his friends, such as John Hay, worried that Garfield tried too hard to please everyone.

On the eve of the Republican Convention in June 1880, the leading candidates for the party's nomination were Grant and Blaine, with some sentiment for Secretary of the Treasury John S. Sherman. Some party figures whispered that Garfield might be a good candidate to unify the party, but he had not agreed to run and lacked formal support. He had recently been elected to the U.S. Senate, with a term due to begin in 1881, and his energies were directed there. Indeed, he made no plans even to attend the convention until May, when fellow Ohioan Sherman asked him to serve as his floor manager and give the nominating speech.

The first ballot at the convention revealed the relative strengths of the candidates: with 379 votes needed to secure the nomination, Grant drew a count of 304, Blaine 284, and Sherman 93. These positions changed little over the next day and a half during thirty-three ballots, although Blaine and Sherman each lost a sprinkling of votes to dark horses, favorite sons, and potential compromise candidates, including

Garfield. Grant's supporters continued to hover around 300, while the other delegates could find little common ground besides opposition to Grant.

On the thirty-fourth ballot, the situation began to change. Garfield, who had attracted no more than two votes in preceding tallies, surged to 17, as the Wisconsin delegation broke from Blaine and a dark horse candidate. On the next ballot, Garfield attracted 50 votes, drawing from both Blaine's and Sherman's supporters. Finally, on the thirty-sixth ballot, the deadlock gave way, as the anti-Grant delegates committed en masse for Garfield. With 399 votes, he won the nomination. Although he had not pursued the honor, he admitted that "a nomination coming unsought and unexpected like this will be the crowning gratification of my life."

Having narrowly won the nomination, Garfield narrowly won the presidency. The Democrats controlled both houses of Congress, and Republicans of all stripes were anathema in the south. Garfield knew the election would turn on a few hotly contested states, especially New York. With this concern in mind, the convention chose Chester A. Arthur, a New York Stalwart and associate of the powerful Senator Roscoe A. Conkling, as the vice-presidential candidate.

During the campaign, Garfield remained on his farm near Mentor, Ohio, where he received the customary stream of delegations and visitors who came to pay their respects. Only once did he venture outside of Ohio, traveling to New York City to shore up support in the state. The local politicians there were seeking to restore their role in making federal appointments after the dry spell of the Hayes administration, and Garfield agreed to give them a voice. According to his diary, that is all he promised, however, and he returned home believing that "no serious mistake had been made and probably much good had been done. No trades, no shackles, and as well fitted for defeat or victory as ever."

The standard-bearer for the Democrats, Winfield Scott Hancock of Pennsylvania, was a sort of mirror image of Garfield: a former Civil War general who was moderate, competent, and colorless; a compromise candidate nominated at a divided convention. To the extent that the election turned on matters of policy, the principal difference was the tariff. To Hancock, it was "a local issue"; to Garfield, it was the key to protecting American workers. In the end, Garfield eked out the narrowest of victories — a plurality of 9,464 votes out of more than nine million cast. The outcome in the electoral college was more decisive — 214 votes for Garfield to 155 for Hancock. Each man carried nineteen states, but Garfield won New York and its 35 electoral votes — which provided him

THE AMERICAN PRESIDENCY ★ 230

his margin of victory. Meanwhile, the Republicans regained a small majority in the House, with the Senate evenly divided.

Garfield viewed the prospect of assuming office with a kind of dread. Days before the Inauguration, at a farewell gathering at Hiram College, he noted the solemnity of this good-bye, comparing it to how he had felt in 1861 when he left to fight in the Civil War.

His principal problem involved making appointments, and he set about the process determined to bridge factions in the party and balance sectional interests. Almost immediately, however, he was drawn into a bitter fight with Conkling, the leader of the Stalwarts. Garfield felt an obligation to offer the first major appointments to the independent and Half-Breed Republicans who had thrown him their support, and he designated Blaine as secretary of state. This action infuriated Conkling, and very little that Garfield could do afterward seemed to mollify him.

Garfield eventually named two Stalwarts to the cabinet (Robert Todd Lincoln as secretary of war and New Yorker Thomas L. James as postmaster general) and accepted Conkling's advice about several key appointments in New York State. But after that, he drew the line, nominating William H. Robertson, a key Garfield supporter in the presidential convention and Conkling's enemy, to the vital position as collector of the New York customhouse (a position once held by Chester Arthur). Conkling took this news as a personal affront and tried various stratagems to defeat the appointment, all but paralyzing the nomination process for several months. He first appealed to the Republican Caucus to force Garfield to withdraw Robertson's name. Garfield refused, seeing principle at stake: it was time, he said, to determine once and for all "whether the President is registering clerk of the Senate or the Executive of the Nation."

Conkling and fellow New York Senator Thomas C. Platt next sought to have the Senate confirm Garfield's uncontested nominations and adjourn without acting on Robertson. Garfield countered by withdrawing all nominations except Roberston's — a step that meant that the senators would have to confirm him or sacrifice the appointments of their friends. In a last act of desperation, on May 16, Conkling and Platt resigned their Senate seats. The gesture was ostentatious and theatrical, as the two men believed that the state legislature would immediately reelect them. That proved an enormous miscalculation, however. New York Republicans, fed up with the whole affair and Conkling's years of bullying, declined to return Conkling and Platt to office. By late June, it appeared that Garfield had won a decisive victory over his most nettlesome antagonist in the Republican party. It also seemed likely that when Congress gathered in the fall, the president would command newfound respect.

During the first weeks of his presidency, Garfield settled into a daily routine. He rose early to read the papers and attend to correspondence. At 10:00 A.M., he opened his office to virtually anyone who wanted to see him — the first president since before the Civil War to make himself so accessible. After lunch, he spent most of the rest of his day meeting with office-seekers or with members of the cabinet. He spent the hours from late afternoon until the middle of the evening with his family, and then returned to work, especially to read. He also found time to read Carlyle's *Reminiscences* and the new novel *Ben-Hur*, which he greatly admired. (He had already named its author, General Lew Wallace, U.S. ambassador to Paraguay. After reading the novel, Garfield changed the posting to Constantinople in hopes that Wallace would "draw inspiration from the modern East for future literary work.")

In May, Lucretia Garfield contracted malaria, and the president arranged for her to convalesce on the New Jersey shore. The coming weeks proved a dreary time for him. He missed his wife and thoroughly disliked "the personal aspects of the presidency" and the fact that "almost every one who comes to me wants something which he thinks I can and ought to give him." By late June, however, the First Lady's health improved enough for her to join the president on a planned swing through New England, which was to include a speech at his alma mater, Williams College. Buoyed by his apparent defeat of Conkling, Garfield seemed in high spirits. On July 2, as he was chatting with Blaine at the Baltimore and Potomac railroad station before embarking on the New England trip, he was shot twice from behind by Charles J. Guiteau, a delusional religious fanatic who was frustrated by efforts to secure a federal post. One bullet grazed the president's arm. The other penetrated his back.

Garfield's wounds, although serious, were not at first believed to be fatal. The major problem was that the bullet that had entered his back could not be removed. Indeed it could not even be found. Doctors tried repeatedly and unsuccessfully to locate it. Given the unsanitary nature of their probing, they undoubtedly worsened the president's condition. At one point, Alexander Graham Bell was brought in to look for the bullet with a new electrical device that he had designed specifically for the purpose. That effort, too, failed.

Meanwhile, with Congress away and only routine business for the administration to carry out, the government went on. Vice President Arthur refused entreaties to become acting president until Garfield's fate was clear. The president remained in the White House, bedridden and suffering from high fevers and constant pain, until early September, when he was moved to the Jersey shore to escape the oppressive heat in Washington. The change in location did not help, and on September 19, 1881, he died. During the eighty days between his shooting and his death, the president's lone official act was to sign an extradition paper.

Despite his obvious insanity, Guiteau was tried, convicted, and sentenced to death. He was hanged on June 30, 1882. The assassin was portrayed, conveniently, as an embittered office-seeker, and his crime gave momentum to the movement to reform the civil service, culminating in the Pendleton Act of 1883. The problem posed by Garfield's extended convalescence was not addressed until much later, in the aftermath of the Kennedy assassination. In 1967, the Twenty-fifth Amendment to the Constitution provided for an orderly succession when the nation's chief executive was found to be incapacitated.

— DAVIS DYER
The Winthrop Group

Chester Arthur

1881–1885

b. October 5, 1829
d. November 18, 1886

"I AM A STALWART, and Arthur will be President!" exclaimed Charles Guiteau after he fired the bullets that mortally wounded President James A. Garfield in July 1881. Guiteau, certainly insane, was one of the few Americans who welcomed the prospect of a Chester Arthur presidency. Before the Republicans nominated him to be Garfield's running mate in 1880, Arthur's public career consisted of two administrative posts that he landed through patronage. After Garfield's assassination, the *New York Times* concluded that "Arthur is about the last man who would be considered eligible to that position, did the choice depend on the voice of either a majority of his own party or of a majority of the people of the United States."

Chester Arthur is the most obscure of what now seems like an interchangeable group of unrenowned and bewhiskered men who occupied the White House in the late nineteenth century. If the period's other presidents do not deserve such derision — in their day they inspired passionate popular followings and achieved in office about all they could — Arthur comes closer to the dismissive stereotype. Contemporaries knew him as a pleasant, gracious, tactful man skilled at raising money for his party and pacifying egos within it. His political career and unlikely presidency were the products of nineteenth-century ideas about gentility and the power of political parties. His easy movement among men of means fitted him well for the services he performed for his party, especially raising money by (politely) raking government employees and businessmen for contributions. His status as a gentleman allowed for his career as a spoilsman and administrator. That career made up the slim record that brought him the vice-presidency and presidency.

In the mid-1820s, William Arthur, the future president's father, attended a revival, and the fervor stirred the Scotch-Irish immigrant teacher to convert to the Free-Will Baptists and preach the Word himself. In the hardscrabble rural town of Fairfield, Vermont, where he and his wife, Malvina, settled, William headed a congregation of forty-six and taught school to help support his family. Chester Alan, the couple's fifth child, was born in October 1829. The family moved frequently in search of a stable living, and during their stay in Schenectady, New York, Arthur attended an academy and, in 1845, entered Union College. After graduation, he taught school, studied law, and moved to New York City to open a practice.

Arthur ran from his upbringing by fashioning himself into an urban gentleman. Polished manners and a knowledge of the classics marked a midcentury gentleman, as did refined feelings and a taste for lavish food, drink, and furnishings. Like other Americans who strove for genteel respectability by rejecting the hellfire sermons, stern God, and plain architecture of Baptist and Congregational churches, Arthur abandoned the religion of his parents in favor of the more fashionable Episcopal Church. It offered attractive surroundings and ritual, and did not condemn fashionable dress and amusements, all of which Arthur's parents interpreted as a backward step for their son, rather than a move forward.

Yet, according to a friend, Arthur took "great interest in matters of dress" and was "always neat and tasteful in his attire." As soon as his income allowed, he spent considerable sums on clothes, trying on as many as twenty pairs of custom-made trousers before finding the perfect fit and fabric. He wore his brown hair fashionably long as a young lawyer, and grew full "whiskers of the Burnside variety, invariably trimmed to the perfection point." And Arthur, a slim six feet two inches tall, had a good start on a gentleman's bearing. His later paunchiness registered his fondness for the good life. "He loved the pleasures of the table," a friend recalled, "and could carry a great deal of wine and liquor without any manifest effect other than a greater vivacity of speech." Later, in the White House, Arthur's staff uncorked French wines to accompany fourteen- to twenty-one-course meals. Mrs. James G. Blaine described "the flowers, the damask, the silver, the attendants, all showing the latest style and an abandon in expense and taste." Arthur undertook substantial renovations of the White House, hired the decorators, chose the colors and moldings, and made some major purchases. He found decorating at least as engrossing as the affairs of the federal government.

Even critics of Arthur's politics praised his tact and deportment. From his days as a college student, Arthur impressed those with whom he came in contact with his ease with a story and his impeccable manners. According to a campaign biographer, he possessed "manners of the utmost geniality." Wrote one observer, "There is nothing about him of the politician. He does not talk in offensive accents, his voice is low and gentlemanly." His graciousness helped in putting wealthy men and high-strung politicians at ease, his most important asset in politics.

By 1860, Arthur had the education expected of gentlemen, good looks, a socially ambitious and attractive wife, and a profession. What Arthur needed was sufficient income. Politics would provide that.

In politics as in religion, Arthur fastened himself to respectability and power. He put himself at the service of the most conservative wing of the Republican party, the political equivalent of Episcopalianism, and to it Arthur owed his first substantial paycheck. Arthur had worked hard for Edwin D. Morgan's election as governor of New York, and for his troubles he gained an appointment to Morgan's staff. When the Civil War broke out, Arthur managed the chaotic paperwork that covered everything from seeing to it that the right number of socks got to a regiment to the letting of contracts for supplies. Morgan appointed him quartermaster general in 1862 (he retained the honorific title of General until his presidency), and he accomplished his work without notable corruption. When New York voters turned the Republicans out of office in 1863, the new Democratic administration removed Arthur from his post.

Roscoe Conkling gave Arthur his next break. In the Republican state convention in 1867, Conkling, New York's new senator, turned to Arthur and the New York City conservatives to help him oust Reuben Fenton for control of the party. Conkling was a politician of tremendous gifts and edgy temperament. He cared little for propriety: with his wife back in Utica, he carried on a scarcely concealed affair with a colleague's wife while in the Senate. A powerful orator in the florid nineteenth-century style, Conkling belittled his opponents savagely. He aimed to dazzle with his looks as well as his words. A trim six feet three inches, Conkling wore such items as green trousers, striped shirts, polka-dotted ascots, purple coats, and mauve waistcoats, fortunately not together. A perfect reddish blond curl hung across his wide forehead. Conkling was a rising star in the Senate and the party.

Conkling's machine rested on federal patronage and the campaign funds and workers it provided. Thousands of postal workers were political appointees, as were the more than one thousand employees of the New York customhouse, the largest single source of patronage in the nation. Federal workers defrayed the campaign costs by kicking back as much as 6 percent of their salaries. Parties also assessed candidates for the privilege of running for office. Firms that depended on customhouse workers to process their goods paid bribes and considered it the cost of doing business. Corporations, especially the railroads, made campaign contributions in exchange for legislators not acting on proposed regulations. Companies that sought government contracts also offered donations.

Arthur's chance to shine as a fundraiser came in 1871, when Conkling vanquished Fenton's machine. The customhouse collector, Tom Murphy, resigned that year and recommended Arthur as his successor. Congress (and Conkling) approved the choice. Arthur spent only a few hours each day at the customhouse because his real work was political. He calmed those whom Conkling had insulted, and he made sure that worthy partisans received their just reward and that a contingent of party workers loyal to Conkling attended state conventions. Most important, Arthur raised money. A customhouse worker remembered that "Arthur had been prominent in the large collections . . . made in 1864 and 1868 and he was anxious that the Customs House should show under his reign no decline in productiveness." Arthur denied knowledge of assessments, but customhouse workers were asked to make "voluntary" contributions and to work for Ulysses S. Grant's reelection.

Arthur's position was safe with Grant in office because he was part of the party faction that included Grant's strongest supporters. The Stalwarts, as they would be labeled by 1880, supported the toughest policy possible for the defeated south and opposed civil service reform if directed at their followers. Conkling, one of the leading Stalwarts, sought the presidential nomination in 1876, as did James G. Blaine, who led a faction that would become known as the Half-Breeds. The Half-Breeds took a more conciliatory view of the south and made gestures at civil service reform, especially when it affected the Stalwarts' methods of raising campaign funds. Both Conkling and Blaine had passionate support and equally dedicated foes. The nomination went to Ohio's governor, Rutherford B. Hayes, the man "nobody could object to."

When Hayes took office in 1877, he announced new civil service policies. While an investigation of the largest customhouses went forward, he ordered Arthur to reduce the size of the staff in New York and to for-

bid customhouse employees from engaging in partisan work. He mandated that "no assessments for political purposes of officers or subordinates should be allowed," although Arthur had raised money for Hayes's
campaign in exactly that way. Conkling ridiculed the "old woman policy
of Granny Hayes" and "snivel service" reform, but Hayes had the last
word. In October, he asked the Senate to name Theodore Roosevelt, Sr.,
to the New York customhouse to replace Arthur. The Senate rejected the
nomination. In July 1878, Hayes removed Arthur. Tired of the issue and
Conkling's performances, the Senate finally approved the nomination of
Roosevelt.

Conkling attempted to recover his political future by working for a
third term for Grant. When it was clear that neither Blaine nor Grant
had enough support to win the nomination, the party turned to James
A. Garfield, an uncontroversial Half-Breed. In the interest of party harmony, Garfield's representatives sounded out Stalwarts for the vice presidency. When the banker Levi P. Morton declined on Conkling's advice,
Arthur came next. Conkling assumed Arthur would also reject the offer.
But Arthur, feeling that "the office of the Vice-presidency is a greater
honor than I ever dreamed of attaining," accepted.

Arthur coordinated the campaign in New York State and the assessment of officeholders. Federal employees, from judges to letter carriers,
received requests for "donations" of 3 percent of their salaries. State
workers, down to night watchmen, also received carefully worded letters. Contributions were "voluntary," but if anyone missed the hint, a
more pointed letter followed. Arthur raised an estimated $125,000 from
federal workers in New York State. With Morton, he tapped New York
City businessmen for contributions totaling as much as $400,000. The
infusion of money and the Republicans' late focus on the tariff helped
turn the competitive midwestern states to Garfield.

Conkling believed the Stalwarts' efforts on behalf of the ticket had
earned him the right to control federal patronage in New York. Garfield
thought otherwise. On Secretary of State Blaine's advice, he went out of
his way to offend Conkling by proposing that William H. Robertson, a
Half-Breed, serve as collector of the New York customhouse. Arthur and
other Stalwarts signed a petition protesting the appointment; many
senators were alarmed at this show of executive power. Garfield persisted, and eventually the Senate relented. As a dramatic protest, Senators Conkling and Thomas Collier Platt resigned their seats in May 1881,
perhaps hoping that the New York Assembly would return them in vindication. Guiteau's profession of loyalty to the Stalwarts and the growing strength of the Half-Breeds in New York upset any possibility that

Conkling and Platt would return to the Senate. The assembly replaced them with two Half-Breeds.

The morning Garfield was shot, Arthur was in New York City to meet with Conkling about the Senate seat. He took the next train to Washington, amid rumors that his faction was responsible for the assassination. Aside from offering condolences to Garfield's family, Arthur steered clear of the White House and remained out of sight. His obvious grief upon hearing of Garfield's death helped silence suspicions that Arthur had encouraged Guiteau. The whole of his political life had consisted of managing the New York machine; he had absolutely no experience in national politics and no feel for popular leadership. When Arthur was inaugurated on September 22, he quickly concluded that he needed to be president, not the avenger of the Stalwarts. He did nothing to resurrect the fortunes of his faction. Arthur retained Garfield's cabinet, including Blaine. To the delight of good government reformers, he vetoed a rivers and harbors bill on the grounds that it provided for nothing but wasteful local projects. When Congress turned to tariff revisions in 1883, he argued for an expert commission, which, following business principles, could rationally revise tariffs downward.

Arthur offered no legislative direction — few expected that of any president in the late nineteenth century — but significant measures became law during his term. White westerners, especially Californians, had agitated against Chinese immigration since the 1870s. Mobs in western states had attacked Chinese workers. They would work so cheaply, white organized labor argued, that they would drive down "Americans'" wages. They could never be assimilated, echoed ministers, professionals, and politicians. Western congressmen spearheaded a drive to restrict Chinese immigration. Arthur vetoed a law barring Chinese immigration for twenty years. Congress sustained the veto, but passed new legislation with a ten-year ban, which Arthur signed in 1882. For the first time, the United States had restricted immigration on the basis of nationality or race.

In early 1883, Arthur signed into law another landmark piece of legislation — the Pendleton Civil Service Act. Reformers had pressed the cause of civil service reform for decades, supported by politicians who complained about the nuisance of filling large numbers of patronage jobs from a much larger pool of supplicants and businessmen who called for more professional government services. The widely held idea that Guiteau was a "disappointed office-seeker," driven mad by the patronage system, rather than merely a madman, gave new life to the cause of reform: the absence of a civil service system had killed a president. Arthur proposed civil service reform in his annual address in 1883. The Ohio Democrat George Pendleton offered legislation, reasoning that his party would be hurt in 1884 if the issue went to the Republicans. His bill established a three-member Civil Service Commission, and brought into the civil service system employees of large customhouses, Washington clerks, and some postal workers — about 10 percent of the federal workforce by 1885. Arthur, who built his career on appointing and assessing federal employees, signed the bill.

As Arthur's term neared its end, appraisals divided between faint praise and mild derision. Mrs. James Blaine claimed Arthur had a talent for "seeming to do things, while never putting his hands or his mind near them." A contemporary concluded he "has done well . . . by not doing anything bad." His term seemed most remarkable for Washington's improved social life and the languid pace with which he pursued official business. More than sluggishness dictated his light schedule: in 1882 or 1883, his doctors determined he suffered from Bright's disease, a liver ailment that included lethargy among its symptoms. It is likely he had suffered from Bright's for a number of years before the diagnosis.

Although in poor health, Arthur tried for the Republican nomination in 1884. But without a base in his party or a popular following (he lowered postage rates in an inept effort to cultivate it), Arthur's halfhearted campaign was doomed. Arthur intended to return to his law practice, but failing health prevented that. He died on November 18, 1885, eight months after leaving office, of a cerebral hemorrhage.

As a gentleman and party functionary, Arthur followed nineteenth-century mores. By the 1890s, both Arthur's personal style and political techniques had fallen out of fashion. His version of gentility was ridiculed as bloodless and artificial. The assessment of government workers lingered into the twentieth century, but mainly at the local level. Corporate contributions, significant in Arthur's day, became the essential source of money for increasingly expensive national races in the late 1880s and 1890s. As patient managers replaced buccaneers like Conkling, new connections between money and politics generated new sources of corruption.

— PAULA BAKER
University of Pittsburgh

Grover Cleveland

1885–1889, 1893–1897

b. March 18, 1837
d. June 24, 1908

STEPHEN GROVER CLEVELAND occupies a distinct position in the history of the American presidency as the only individual to serve two nonconsecutive terms in the nation's highest office. He has also received many positive historical evaluations of his record as president for his ostensible courage and resolve in resisting monetary panaceas during the depression of the 1890s. More recent assessments have subjected Cleveland to more criticism for his failure to deal effectively with the economic and political disasters of his second term. He is now generally considered as the last of the old-style presidents of the nineteenth century and someone whose lack of skill as a leader contributed to the decline of the Democratic party in the 1890s. Grover Cleveland's ineptitude, historians now claim, helped the Republicans achieve their electoral majority at the end of the Gilded Age.

He was born on March 18, 1837, in Caldwell, New Jersey, the son of a Presbyterian minister who died when Grover was sixteen. The youth moved to Buffalo, New York, and became an attorney. In the Civil War, using a provision of the draft laws that even then was controversial, he hired a substitute to take his place. Cleveland was a hardworking plodder who spent a good deal of his leisure time with male friends in the saloons of Buffalo. He had many female acquaintances, one of whom, Maria Halpin, would later figure in his first race for the presidency. As Cleveland grew older, he put on a good deal of weight. Members of his family called him "Uncle Jumbo."

A loyal Democrat with some interest in city politics, Cleveland held several local offices during the 1860s. In 1870 he was elected sheriff of Buffalo County, a post he held for one term. His political destiny changed in 1881 when his party selected him to run for mayor of Buffalo. Once in office he called for thrifty, efficient government, a position that suited the conservative men who had supported his election.

Soon Cleveland was mentioned as a possible 1882 gubernatorial candidate. A split among the Republicans that year made the Democratic nomination an attractive prize. Because New York State commanded the largest bloc of electoral votes, Cleveland's friends saw the race as a stepping-stone to the White House. Cleveland won a decisive victory in the fall elections.

As governor, Cleveland relied on the veto to hold back legislation that, in his view, unduly expanded the range of government power. He opposed bills that would have reduced transit fares in New York City. While he came into conflict with the New York Democratic machine, Tammany Hall, he earned a national reputation as the embodiment of austere, economical government in which honesty was the essential virtue.

There hadn't been a Democrat in the White House since 1860, and the party was looking for a fresh face for its 1884 national campaign. Cleveland was a new personality who could carry New York State and run well in the northeast and Middle Atlantic states. Any Democrat was assured of the votes of the Solid South in those years, and Cleveland thus promised a winning coalition for his party in the general election. He received the Democratic nomination on the second ballot on July 11, 1884. Thomas A. Hendricks of Indiana was selected as his running mate.

Cleveland went into the election campaign against the Republican nominee, James G. Blaine, with the support of dissenting members of the G.O.P. who called themselves Mugwumps (from a Native American word for "big chief") and who liked Cleveland's honesty. Although Blaine was very popular, there were questions about his honesty and financial affairs. But soon Cleveland faced an apparent scandal of his own. A newspaper in Buffalo revealed that he had been romantically involved with Maria Halpin, who had a ten-year-old son. She claimed that Cleveland was the father. He accepted responsibility and supported the boy while he was in an orphanage.

At first Cleveland's reform friends were shocked and the Republicans were gleeful. But Cleveland responded with a telegram urging a simple course on his supporters: "Tell the Truth." He stressed his honorable action in paying for the child's upbringing. Because the scandal came out early in the race, it did not have a decisive impact on the voters.

Presidential candidates did not campaign much in the late nineteenth century, and Cleveland adhered to that pattern. He delivered a speech on July 29 and issued a letter of acceptance that became the main text for his candidacy. While he made a few appearances at the end of the campaign, he placed his main reliance on the Democratic organization to win the election for him.

The election of 1884 has become famous for the incidents that bedeviled Blaine as the voting neared. Blaine attended a meeting with Protestant clergymen just before the election, at which one of the ministers, the Reverend Samuel Burchard, called the Democrats the party of "Rum, Romanism, and Rebellion." Burchard's anti-Catholic slurs, which Blaine did not contradict at the time, aroused much controversy, but in the end did not cost the Republican either the votes of New York State or the presidency.

Circumstances favored Cleveland in 1884. The Republicans had held power for more than twenty years, and the country was in the mood for a change of government. A recession in the early 1880s helped the Democrats' chances. So did the electoral arithmetic that gave them 140 electoral votes from the south to start. If Cleveland carried New York, New Jersey, and a midwestern state, he would win. That was what happened. He won Indiana, Connecticut, Delaware, New Jersey, New York, and the south, garnering 219 electoral votes to Blaine's 182. The Democrat won 48.5 percent of the popular vote to 48.2 percent for Blaine. It was a very close contest, but Grover Cleveland was the new president.

Cleveland took office on March 4, 1885. In his inaugural address, he pledged to provide "reform in the administration of government, and the application of business principles to public affairs." His cabinet reflected his commitments to business leaders and conservative Democrats. During his politically eventful first term, Cleveland championed reform of the civil service as a way of reducing the influence of politicians in government.

In one of his more celebrated policies, the new president struck out against the elaborate and expensive system of pensions for Civil War veterans with which Republicans had become identified in the 1870s and 1880s. The G.O.P. maintained that government expenditures for the brave men who had saved the Union were justified. Southern Democrats, whose constituents received no compensation for their war service in the Confederacy, opposed pensions. Cleveland worked late into the night going over individual pension bills that Congress sent to him for special cases of allegedly needy or infirm veterans. In many instances, he argued that the claims were fraudulent and he vetoed those cases as bogus or unworthy. As one unsympathetic Democrat said of the president, "Cleveland delighted in the little and would labor pantingly at the windlass of small things. It was this bent of the infinitesimal that led him to put in hours darkly arranging a reason to shatter some old woman's pension with the bludgeon of his veto."

The president used his veto pen as well to teach the nation what he considered the important lessons of personal self-reliance. When Congress appropriated $10,000 for seed for drought-stricken Texas counties in 1886, the president struck down that law because "federal aid in such cases encourages the expectation of paternal care on the part of the government, and weakens the sturdiness of our national character, while it prevents the indulgence among our people of that kindly sentiment and conduct which strengthens the bonds of a common brotherhood."

During Cleveland's first term, Congress produced some notable legislation such as the 1887 Dawes Severalty Act, which sought to assimilate Native Americans into American society by dividing up their tribal land into individual plots. That same year the Interstate Commerce Act created the first national regulatory agency to oversee a major industry. While these laws occurred during Cleveland's presidency, his connection with them was minimal. His main concerns remained honesty in pension legislation and Democratic control of appointments and patronage.

But Cleveland alienated many segments of his party during his first term with his inept handling of the patronage appointment system. In an era of intense partisanship, the issue of selecting party members for federal offices commanded national attention. Cleveland often ignored the advice of the Democratic leaders in key states and selected men who had impressed him as honest and trustworthy. The result was friction among Democrats as party leaders became angry with Cleveland. As one Democrat acidly remarked: "Faithlessness is a passport to recognition by this mass of presidential fat."

The most popular public gesture of Cleveland's first term was his marriage in 1886 to the young and beautiful Frances Folsom, the daughter of his former law partner. The bride was in her early twenties, and their romance inspired a wave of national attention to the wedding on June 2. Reporters tracked the couple to their honeymoon spot, and from a distance flooded newspapers with stories about the newlyweds. The episode indicated how much presidents and their families had become celebrities. It was a process that the intensely private Cleveland found highly distasteful.

The results in the 1886 congressional elections produced some Republican gains, and political observers anticipated a close race for Cleveland's reelection in 1888. To seize the initiative in the upcoming campaign, the president devoted his annual message in December 1887 to a call for a reduction in the protective tariff. The Republicans reacted with pleasure to a contest based on the issue that held their party together. James G. Blaine argued that keeping the Democrats in power would be a menace to prosperity.

Cleveland had made a bold start to his reelection bid, but then he backed away from tariff reform as a major issue during the rest of 1888. He did not pressure congressional Democrats to enact his ideas, and in the House his party produced a bill that did not embody reform principles. The Republican Senate wrote their own bill, and Congress adjourned with nothing accomplished.

The Republicans selected Benjamin Harrison of Indiana to oppose Cleveland. He proved to be a very effective campaigner who made deft, persuasive speeches on behalf of the protective policy. A united G.O.P. rallied behind its standard-bearer and the idea of a higher tariff. On the other hand, Cleveland accepted the tradition that an incumbent president should not campaign for reelection. He did not exercise decisive leadership over the Democratic campaign either. The party played down the tariff, which divided their ranks.

Cleveland won a plurality of the popular vote in the general election, largely because of Democratic strength in the south, where African Americans faced institutional and extralegal barriers to voting. But Harrison won a 233 to 168 victory in the electoral college and defeated Cleveland.

A number of elements contributed to Cleveland's loss. The Democrats waged a lackluster canvass; the Republicans were resolute and effective. Harrison campaigned well; Cleveland was passive. At a time of stalemated national politics in which party strength was evenly divided, a slight change in voting could change the results. And in 1888, the pendulum swung back to the Republicans. Despite his loss, Cleveland remained the most visible national Democrat and the party's most popular figure. The story that his wife predicted a return to the White House in four years is a myth, but many Democrats were forecasting the same thing when Cleveland left Washington as Harrison was inaugurated on March 4, 1889.

From 1889 to 1893, Cleveland practiced law in New York City and watched national politics closely. His schedule was not demanding, and he had plenty of time to ponder his next move in the presidential sweepstakes. The good fortune that had marked his ascendancy to the White House continued to favor his return. The Republicans controlled both houses of Congress after 1888, and they proceeded to enact an ambitious program of economic legislation that led to a voter backlash in the 1890 elections. The Democrats made striking gains. They resumed control of the House and narrowed the G.O.P.'s hold on the Senate. The Harrison administration found itself with a divided and unhappy party as many Republicans looked for an alternative to the president in 1892. Adding to the volatility of politics was the emergence of the agrarian-based People's party, which made significant gains in the midwest and south in the 1890 contest. Prospects seemed good for a Democratic rebound in 1892.

As the major national leader of his party, Cleveland became the focus of much speculation about his possible renomination. He maneuvered skillfully to placate the warring factions within the party who disagreed about the inflationary doctrine of free silver as an answer to the falling commodity prices that plagued cotton and wheat farmers. Although Cleveland believed firmly in the gold standard and abhorred silver, he obscured the intensity of his feelings throughout 1892 and sought party unity. He easily defeated his only major rival for the nomination, Governor David B. Hill of New York. At the Democratic Convention in Chicago, Cleveland was selected on the first ballot. The delegates chose Adlai E. Stevenson of Illinois as the vice-presidential nominee. The Democrats' main theme was lowering the tariff and a restoration of Democratic economy and thrift after the excesses and spending of the Republican years.

The Republicans renominated President Harrison, and the People's party selected James B. Weaver. The Republican campaign lacked the fervor and cohesion of 1888, and the president found the tide of events running against him. Cleveland earned a decisive victory with 277 electoral votes to 145 for Harrison and a margin of 400,000 in the popular vote. The Populists carried several states to reap 22 electoral votes and earned more than a million popular votes, but the national strength of the Democrats won them both houses of Congress and a second term for Cleveland. Happy party members chanted: "Grover, Grover, four more years."

In his second term, Cleveland kept the presidential office simple and uncomplicated. He had a single secretary, Henry T. Thurber, and a very small staff. His cabinet had the same conservative tilt that had marked his first term. Cleveland made no effort to reach out to the Washington press corps. In fact, as time passed he became secluded in a well-guarded and inaccessible executive mansion. One reporter said that finding out about the inner workings of the administration was done "much after the fashion in which highwaymen rob a stage coach."

The president did not travel around the country to promote his programs, and he resumed his habits of working steadily at his desk in isolation.

These attributes did not serve Cleveland well in his troubled second term. Within a month of his return to power, the economy slid into the depression that was known as the panic of 1893. The downturn lasted four years and was one of the most severe periods of hard times that the nation experienced during the nineteenth century. The conventional economic wisdom taught that there was little government could do to improve matters. Cleveland shared that belief and stood firmly against efforts to deal with the growing number of unemployed in the nation's cities and farms.

The president did have a remedy for the economic slowdown, and he threw the weight of his office behind that proposal. Cleveland believed that the Sherman Silver Purchase Act of 1890 had undermined confidence in the monetary system. That measure specified that the government should buy a fixed amount of silver each month. Although the metal was not coined into money, the subsidy program, so Cleveland contended, was a sign of government sympathy for inflation. That, in turn, had led to the business closings and other manifestations of the panic. An indication of this problem was the fall in the nation's gold reserve, which hovered around the psychologically important level of $100 million. Cleveland's reaction was simple and direct: repeal the Sherman Act in a special session of Congress, which he summoned to meet on August 7, 1893.

The problem for Cleveland, however, was that his own party contained deep divisions over the currency issue. Eastern Democrats wanted to maintain the gold standard. Western and southern Democrats, now tinged with Populism, believed that inflation through the coinage of silver would raise prices and ease their burden of debt. So when Cleveland insisted on repeal of the Sherman Act, he drove a wedge into his party which was not closed for five years.

Unknown to the public, the president also experienced a medical emergency during the summer of 1893. A heavy cigar smoker, Cleveland felt a rough spot on the roof of his mouth which doctors agreed was cancerous. In a secret operation, the president had part of his jaw removed. The public was told nothing of the procedure because Vice President Stevenson was in favor of silver and the possibility of his becoming president might unnerve the financial markets.

Using the full powers of his position to drive the repeal measure through Congress, Cleveland succeeded in attaining the policy change he wanted by October 1893. The price of victory was high. The Democratic party became even more disunited as the two factions debated the issue of silver. Because Cleveland had promised that repeal of the Sherman Act would end the depression, he suffered the political consequences when the hard times did not improve in 1894 and 1895.

The next issue to arise was tariff reform, on which most Democrats agreed. The party's control of the Senate was so narrow, however, that it proved impossible to pass a measure that lowered rates. A compromise emerged, called the Wilson-Gorman tariff, which cleared Congress in the late summer of 1894. In a gesture of party harmony, Cleveland might have signed the bill as the best that the Democrats could do. Or, in furtherance of his commitment to reform, he might have vetoed it to show that the Democrats were true to principle. Instead, he let it become law without his signature, a gesture that left the Democrats with the worst of all political worlds.

Contributing to the difficulties of the Democrats as the 1894 elections approached was the White House attitude toward the social unrest that marked the mid-1890s. Cleveland and his administration displayed distaste for the efforts of the unemployed, led by Jacob S. Coxey, to march to Washington to seek congressional aid. More damaging still was the president's response to the nationwide rail strike that grew out of a walkout at the Pullman Palace Car Company in Chicago in mid-1894. When Eugene V. Debs and the American Railway Union added their support with a broader boycott, Cleveland's attorney general, Richard Olney, obtained court injunctions and used the army to disperse the strikers. Labor voters turned against the Democrats in many parts of the country.

The midterm elections of 1894 brought disaster for Cleveland and the Democrats. The Republicans gained 117 seats in the House, the largest reversal of congressional strength in the nation's history. The Populists won some support too, but failed to establish themselves as a credible alternative to the major parties. Although his presidency had received a dramatic repudiation, Cleveland remained true to his belief in the gold standard and governmental economy during his last two years in office.

Despite all of the president's actions, the gold reserve was still in danger. To stave off disaster, the White House called on the New York banking community for assistance. J. P. Morgan and his colleagues purchased government bonds and exercised their power to see that the reserve remained intact, making substantial profits in the process. Democrats in the west and south reacted with hostility to this commingling of government policy with the power of eastern capital.

Amid all the domestic problems of his second term, Cleveland also encountered foreign policy difficulties that foreshadowed the era of American imperialism. The Harrison administration had laid the foundation for the annexation of the Hawaiian Islands. Cleveland repudiated the treaty that would have completed the process and forestalled annexation until the McKinley years. The revolution that began in Cuba against Spanish rule in February 1895 did not receive Cleveland's support. He allowed Madrid time to put down the uprising, and he resisted congressional moves to promote American intervention. His pro-Spanish policy contributed to the decline of Cleveland's popularity.

The most sensational foreign policy episode of Cleveland's second term occurred when a controversy between Great Britain and Venezuela about British Guiana came to a head over the discovery of gold in the region. Cleveland and his second secretary of state (and the former attorney general), Richard Olney, asserted the Monroe Doctrine and challenged the British role in the dispute. The threat of possible war loomed at the end of 1895, but the British opted for arbitration and the crisis eased. Cleveland's actions represented an affirmation of the Monroe Doctrine that gained him a momentary rise in his popular standing.

During the last two years of his presidency, Cleveland became a political liability for the Democrats as everyone looked ahead to the next presidential election. The correspondence coming to the White House amounted to a fraction of what a chief executive usually received. Hate mail denounced the president's policies and contributed to his conviction that the White House needed additional guards. The president seemed "constantly more hedged in and mysterious" in his virtual isolation.

In the 1896 election the Democrats repudiated Cleveland and nominated William Jennings Bryan on a pro-silver platform. At the party's national convention, Cleveland's supporters represented a minority of the Democratic rank and file. The victory of the Republican candidate, William McKinley, attested to the decline of Cleveland's party from the optimism of the 1892 elections. In retirement in Princeton, New Jersey, Cleveland remained a controversial figure. Even the mention of his name as a presidential possibility in 1904 stirred anger among his Democratic enemies from the 1890s. Cleveland died on June 24, 1908. For many years, the Republicans ran against the memories of the 1890s and the Cleveland administration, just as the Democrats were to do in their turn with Herbert Hoover in the 1930s. Cleveland did not live to see his reputation rebound during the Great Depression, when his resolve in defending the gold standard was depicted as the epitome of political courage. In the 1960s, a newer generation of historians questioned whether Cleveland's insistence on having his own way had served the long-term interests of the Democrats or the nation. Honest and hardworking, Cleveland was also an inept party leader whose stubbornness and dogmatism contributed to the erosion of Democratic electoral fortunes. His individualistic presidential style soon gave way to the more bureaucratic, publicity-oriented, and activist methods of his immediate successors, William McKinley and Theodore Roosevelt.

— Lewis L. Gould
Eugene C. Barker Professor Emeritus
in American History,
University of Texas at Austin

Benjamin Harrison

1889–1893

b. August 20, 1833
d. March 13, 1901

PRESIDENT BENJAMIN HARRISON disliked shaking hands. He thought it a foolish and demeaning custom that may well have hastened the death, in 1841, of his grandfather, President William Henry Harrison. When the grandson resolved "to be president to my finger-tips," he meant something different from the clasping, grasping politics of the day. When a member of the hale-fellow, high-rolling Fifty-first Congress — the "Billion-Dollar Congress" — lingered too long on a political errand, Harrison was given to drumming his fingers on his desk, though it is unlikely he meant that either. Harrison's reserve — he allowed he was "not a gusher" — inspired many nicknames. Republican National Chairman Clarkson dubbed him "refrigerator"; House Speaker Reed and detractors in Indiana knew him as "the human iceberg"; Mrs. Blaine, wife of the secretary of state, called Harrison "the oracle in the White House." By the time he left office, Washington City was in a chronic bad mood; Henry Adams described it as a "torpor."

That so unconventional a politician could be selected as a presidential candidate is surprising, except when one remembers the bitter factions of both major parties in the 1880s. Harrison was chosen Republican nominee through a many-sided play on the rule the-enemy-of-my-enemy-is-my-friend. He had votes from Blaine's supporters to derail Sherman; from western delegates to defeat DePew; from DePew's people to forestall Iowa's Senator Allison; from Mark Hanna to prevent a groundswell for Blaine. He himself aimed to spoil the plans of his Hoosier nemesis, Walter Gresham. Harrison was tinged with neither gold nor bimetallism, scandal nor surpassing zeal. Such a procedure might be expected to result in a nominee without a strong following of his own: in the popular vote, Harrison trailed Cleveland, who barely campaigned; only through special Indiana methods did Harrison carry the state where he was best known.

Indirection characterized Harrison's political career throughout. He took up his first federal position, Mississippi River commissioner, after President Hayes announced it, unaware Harrison had already declined. He was buoyed in his move to be named U.S. senator in 1881 by information that Gresham was plotting his exile to President-elect Garfield's cabinet. He ran for reelection as president to squelch a boomlet for Blaine. Harrison claimed that he had wanted only one office in his life, reporter of the Indiana Supreme Court, for the pay. Informed of his re-nomination in 1892, he told those gathered at the White House he had never suggested, much less demanded, personal loyalty from anyone. He asked only that public officers faithfully perform their duty.

Harrison's remark might sound disingenuous, coming from one just selected on the first ballot by what reformers and bosses alike contemptuously dubbed "the postmasters' convention." But the presidency, at least in peacetime, had changed little from an era when his grandfather's friends spoke in those terms. New national undertakings after the Civil War proceeded upon enlarged authority in the Congress and a widened jurisdiction for the federal courts, both under the aegis of a militant Republican party. No activity or doctrine or party had yet uprooted the original idea of the president as public officer preeminent, chief magistrate, with constitutional — that is to say, legal — duties and privileges, fixed to restrain as much as to enable their holder. By training and temperament, Harrison was the consummate lawyer.

Formality in the presidency may have hardened over time. Harrison's descriptions convey a stiffness more commonly identified with monarchy than democracy or even the pretensions of the Gilded Age: "The President and the ex-President again take their places in the carriage — the ex-President now on the left — and are rapidly driven to the Executive Mansion, where the wife of the President joins him." Or "When the coming of the lunch hour has brought the morning reception to an end, and the President is again at his desk, one of the Cabinet officers appears by appointment, accompanied by a messenger with an armful or basketful of papers — chiefly made up of petitions and letters attached to appointments." Or "When there are rumors of important public transactions — and such rumors are perennial — some of the more prominent

of the newspaper men expect to have a few moments with the President
. . . Of course, confidential things are not disclosed; he does not give an
interview and is not quoted."

Tension, between fixity in the presidency and expansiveness in the
other branches and in the parties, was eased by the character of Harrison's predecessors. For all their legendary sorties, Grant was obliging,
Hayes distracted, Garfield equivocal, Arthur "chummy." Cleveland's
stubbornness and regard for the dignity of his office matched Harrison's,
but in the crucial matter of appointments Cleveland's party was at long
last indulged where Harrison had to contend with a spirit of reposses-
sion. To maintain a minimum of economy and self-respect, each president resisted the building of railroads and harbors and boodle; but only
Harrison was so blunt. Each administration allied itself with important
party machines; but only Harrison's alliance, with Blaine, was in the
manner of hostage-taking. Each president saw himself restricted in the
legislative realm to mainly negative methods; only Harrison developed
these into an art form.

Harrison's ancestry prepared him uniquely for the reign of "congressional government": father, Whig congressman from Ohio; grandfather,
general, Whig president, dead after one self-effacing month in office;
great-grandfather, general, Speaker of Virginia's House of Burgesses,
signer of the Declaration of Independence, member of the Continental
Congress; great-great-grandfather, general, member of the House of
Commons, one of the commissioners who ordered the execution of
Charles I. The effect was an overbreeding for punctiliousness and a feeling for tradition that, as Harrison aged, made it comfortable, even satisfying, to resist the currents of his own time.

His withholding nature appeared in its best constitutional clothes
at the opening of his inaugural speech: "There is no constitutional or legal requirement that the President shall take the oath of office in the
presence of the people. But there is so manifest an appropriateness in
the public induction to office of the chief executive officer of the nation
that from the beginning of the Government the people, to whose services the official oath consecrates the officer, have been called to witness
the solemn ceremony." He bore the same standoffish noblesse toward
Congress: the advice of Congress on appointments "is accepted. This is
a mere matter of custom." Recommendations are "followed as a rule,
unless something against the character or fitness of the applicant is
alleged," in which circumstance the president "exercises his prerogative

to make a selection of his own." (Here Harrison did vary custom. He notified senators of his choices only by public announcement.)

Appointments won Harrison his absolute least amount of affection. If a visitor brought up the subject of appointments, the president's back stiffened and his cheeks drained of their color. Resentments were famous. Mrs. Blaine never forgave him for places he wouldn't give son and son-in-law. Speaker Reed lamented he had only two enemies in Maine, and Harrison appointed one of them collector of customs. Harrison refused Hanna, who raised unprecedented funds for the 1888 campaign, the patronage of the Cleveland lighthouse. Boss Platt went to his grave claiming Harrison broke a promise to make him secretary of the treasury. Justice Brewer's appointment to the Supreme Court, already signed, was nearly withdrawn when Senator Plumb persisted to importune on Brewer's behalf.

Allegations on the other side, by new Civil Service Commissioner Theodore Roosevelt and historians since, that Harrison and Postmaster General Wanamaker acted in bad faith toward the merit service are unfounded. To agree with Roosevelt that these two Presbyterian elders — the one who held his five-coach train until Sunday midnight to depart on official journeys, who passed the plate in his Indianapolis church not only the week before he became president but the week after, the other who was not only America's First Merchant but guardian angel of the Y.M.C.A. — were, because of their party preferments, a "timid Psalm-singing" politician and a habitual liar, respectively, is to misunderstand the nineteenth-century Republican party and the last president for whom the party was a mighty reform movement all its own. Commissioner Roosevelt might — did — liken Cleveland and Harrison to the Walrus and the Carpenter; to Harrison, Blaine, Wanamaker, and most of their generation, for whom merit service was a mere program, the comparison was unthinkable.

A faith that no circumstance existed, in the party or elsewhere, that could not be bettered by an application of rectitude and hard work — if necessary his own — characterized all Harrison's endeavors. He named Blaine, the Plumed Knight, obvious choice by politics and experience, secretary of state, to disarm him; that he still considered it a magnanimous act is shown in his later praise of Lincoln's generosity in appointing the popular Seward to that office. He flanked him with an all-Presbyterian cabinet of lawyers, brigadier generals, and native Ohioans, as like himself and as unplumey as might be assembled. He could not predict Blaine's bouts of nervous exhaustion, to which full awareness of the conditions of his tenure and a prickly chief, as well as tragedy in the Blaine family, no doubt contributed. But both before and after 1891, when the secretary's incapacity became nearly total, the belligerent defense of treaty rights and equal trade that colored the administration's foreign relations was more consistently Harrison's than Blaine's. Mrs. Blaine, no admirer, remarked on the president's care and intelligence in affairs of her husband's department.

Harrison's foreign policy evidenced his constitutional scruples. If he strengthened the navy, if he set a tone for a future American empire, he himself demurred, delaying Hawaii's annexation by two months so as to pass the final decision to his successor. After leaving office he would speak against the arrogance of the United States as a world power with a subtlety and eloquence rivaling Lincoln's on slavery. Here, too, Harrison's lodestar was the past: "not the charge at San Juan Hill" but American guns in the Mediterranean, John Paul Jones, the *Alabama* off Cherbourg, their "relation solely to American rights and liberty and the freedom of the seas." In 1901, he defended the "anti-war party," which he said had been anticipated in the Gospel, against public invective. The previous year, he declined to campaign for the Republican ticket, which was pledged to foreign expansion.

As a lawmaker, Harrison demonstrated a fussiness that to associates (never mind enemies) was a hair short of sinful pride. As U.S. senator, he opposed Chinese exclusion, but only because it violated treaties; voted for the Pendleton Act, but only after a provision against election spending he believed was unconstitutional was removed; regarded temperance laws as a form of "fanaticism," though he himself was an abstainer. He did not break with family tradition. John Scott Harrison, a Whig in the House of Representatives in 1854, numbered himself a firm anti-Nebraska man. But he refused to partake in a last-ditch effort to kill the bill by the "revolutionary mode" of preventing a quorum.

Harrison's presidency is synonymous with the high McKinley tariff of 1891, a law the president championed; that, and the violence of the Homestead strike in 1892, caps his administration's reputation as staunchly probusiness. Harrison used threats of a veto to extract changes in the tariff and other important statutes — the Sherman Silver Purchase and the Sherman Antitrust acts of 1890, and the Forest Reserve Act of 1891; but in these, Congress, not the president, was the prime mover. It was on legislation where Harrison's backing was indispensable — where he resolved to offer it in full measure within constitutional bounds — that his tendency to defeat principle with principle operated at its most disheartening.

Harrison opposed slavery from his start in politics. At twenty, a reconsecrated Presbyterian, he broke with the Whigs over slavery. Supporting Frémont in 1856, he stood the next year as Republican candidate for Indianapolis city attorney; in 1860 he endorsed the "radical" Cassius Clay for the presidential nomination. Distinguished in battle, he was promoted to brigadier general in time to join Sherman's final push to Atlanta. As president, Harrison turned some of his finest oratory to connecting military glory with the war's high aims: the army desired to lay no "yoke on those who fought against us, other than . . . they shall yield to every other man his full rights under the law." And "Let those who would die for the flag on the field of battle give a better proof of their patriotism and a higher glory to their country by promoting fraternity and justice." He did not exempt business from his inspiration: "Mill fires were lighted at the funeral pyre of slavery. The emancipation proclamation was heard in the depths of the earth . . . men were made free and material things became our better servants."

Harrison's two acceptances as candidate and his inaugural address spoke to the disenfranchisement of black citizens. His four annual messages to Congress recommended education, civil rights, and criminal protection for "the colored people [who] . . . have, from a stand-point of ignorance and poverty, which was our shame, not theirs, made remarkable advances"; in 1890, 1891, and 1892 these endorsements were his closing appeals. In 1891, with the Lodge Bill to bring House elections under federal-court supervision already lost, Harrison suggested a commission to inquire "whether the opposition . . . is really vested in particular features supposed to be objectionable or includes any proposition to give . . . adequacy to the correction of grave and acknowledged evils."

The Lodge Bill and the Blair Bill were two measures Harrison promoted as Republican policy for the South. Antislavery sentiments aside, both matched his political style better than the strategies of manipulating men and jobs adopted by his predecessors. The Blair Bill proposed federal aid to public schools. It passed the Republican Senate in 1884, 1886, and 1888, failing the Democratic House; Harrison as senator sponsored an amendment against racial disparities in funding. In 1890 the Blair Bill became fatally tangled into midwestern Republicans' designs to attract Catholic voters distrustful of public schools. Senate defectors included John Sherman of Ohio, who before this transported black officeholders to national conventions in Pullman cars. The Lodge

Bill, by contrast, passed the House on a straight partisan vote despite opposition by the Mugwumps, who feared a "lesser federal morality," and by prominent African Americans, who wished for a stronger bill.

Harrison led. More than with school aid, he felt on solid constitutional ground with federal elections. The bill was conceived in 1889 at his Maryland retreat; his closest adviser, Louis Michener, consulted black leaders in Indiana; building on the president's message, his old comrade George Hoar drafted the bill in the Senate, followed in the House by Henry Cabot Lodge, also on friendly terms with Harrison. Before crucial votes, the president conferred with senators at the White House. Still, his most striking move was a refusal, an exquisite holdback, his unwillingness to issue the call for public support that Lodge urged as the only way to save his legislation when it ran aground on the Senate's packed calendar. In the event, the Lodge Bill was delayed until after the 1890 elections, when Republicans were dazed, business was alarmed, and silverites and Democrats dealed. The bill died on a procedural motion, by one vote.

The president, Harrison might have explained, of all public officers could not act the crusader, "an ignorant fellow who counts the empty sepulcher of our Lord of more value than His precepts." But now his rhetoric would be as dated as his sensibilities. The Blair Bill found the Grand Old Party's fabric of evangelism and antislavery worn thin; the Lodge Bill marked its decomposition. When the Republicans took control again in 1896, black rights were off the agenda. The Lodge Bill did not present the most spectacular of Harrison's foot-shooting exercises (that prize goes to liberalizing military pensions and punishments, then restricting the right of privates to reenlist). The Blair and Lodge bills were not "signature issues," in that they decided Harrison's subsequent fate. They were, however, the Republican party's, and the nation's.

— KAREN ORREN
University of California at Los Angeles

William McKinley

1897–1901

b. January 29, 1843
d. September 14, 1901

ETWEEN 1897 AND 1901, William McKinley opened the first pages of the history of the American Century, the era in which the United States replaced Great Britain as the world's greatest power. McKinley won the 1896 presidential election by promising to end a long depression that had not only been the worst in the nation's history economically, but had generated general strikes, bloody labor-management clashes, and mass marches on Washington by the angry unemployed, who threatened to paralyze the United States. By 1901, after winning a smashing reelection victory, he headed a country that was the world's economic leader, and, by a historic shift at the turn of the century, a global military and diplomatic power.

McKinley's accomplishments were little appreciated for many years. He was overshadowed by his colorful, ebullient vice president and successor, Theodore Roosevelt. Unlike Roosevelt, McKinley was self-effacing and worked quietly behind the scenes. He disdained Roosevelt's "strenuous life" and preferred quiet afternoon carriage rides with political advisers. Yet McKinley, not Roosevelt, was the first modern U.S. president, modern in the sense of being a pioneer in manipulating the new technology of telephones and movies to enhance his power (McKinley's 1897 inaugural was the first to be put on film); a trailblazer in exploiting the modern powers of the presidency, especially in foreign affairs; the initial chief executive when the nation turned from acting continentally to acting globally.

That this overly polite midwestern politician would transform a depression-plagued people into a global economic and diplomatic powerhouse did not seem likely when he ran in 1896. He had been born fifty-three years before in the small town of Niles, Ohio. After attending Allegheny College for only one term in 1860, he enlisted in the Twenty-third Ohio Volunteer Infantry as a private at the opening of the Civil War. McKinley emerged four years later as a major (the title intimates used until he became president). He was decorated for his bravery in delivering coffee and hot meals to troops in the field. With a helpful military background; self-education in the law; marriage to the belle of Canton, Ohio, Ida Saxton, in 1871; and a pleasing, modest personality, McKinley embarked on a career in the jungle known as Ohio politics. Nor did it hurt that his patron in the state was Rutherford B. Hayes, inaugurated president in 1877.

Under Hayes's sponsorship, McKinley won election to the U.S. House of Representatives and served from 1877 to 1882, and again from 1885 to 1891. By the late 1880s, he had become a political force in Washington because of his mastery of the tariff issue. Tariff debates were complex, highly statistical, tedious, and of consuming importance for every class of Americans. McKinley understood that the tariff was the key to both the creation of immense economic power by great corporations (as in the new steel industry) that could be protected from foreign competition, and also to the unlocking of corporate wealth for the support of ambitious politicians. Mastering the tariff, a fellow legislator said of McKinley, became "almost a religion" for him.

The historic 1890 "McKinley tariff" bill raised overall rates to the highest in the nation's history. But McKinley had come to realize (largely because of discussions with Secretary of State James G. Blaine) that many U.S. corporations were so productive that, more than protection, they needed help in selling their surplus goods abroad. Thus McKinley and Blaine inserted a reciprocity clause giving the U.S. government the power to obtain trading privileges in foreign markets in return for granting reciprocal privileges in the American market. This reciprocity principle became the centerpiece of twentieth-century U.S. foreign economic policy.

Despite (or because of) his growing national prominence, the Ohio legislature redrew the boundaries of McKinley's home district so that he lost reelection in 1890. He promptly won the governorship in 1892 and was overwhelmingly reelected in 1894, despite intensifying economic depression. He somehow kept labor unrest in Ohio contained (once by threatening to call out the state's national guard), yet succeeded in maintaining close relationships with labor leaders. Because of the depression, the Democrats, who ruled both the White House and Congress, suffered the worst electoral defeat in history to that point in the

1894 congressional races. The Republican presidential nomination was thus much worth pursuing in 1896, and McKinley won it with ease over divided opposition.

Several reasons for his victory stood out. Representing a Republican party known as anti-immigrant and intolerant of many ethnic groups, McKinley insisted instead on inclusive politics. He succeeded in winning support from the booming urban areas, home to many ethnic labor groups. Theodore Roosevelt later wrote that "more than almost any public man I have ever met, he has avoided exciting personal enmities. I have never heard him denounce . . . any man or any body of men." Meanwhile, McKinley's organization, led by the wealthy Cleveland industrialist Marcus Hanna, outspent, outorganized, and outsmarted the opposition.

In the presidential election, McKinley faced the charismatic "Boy Orator of the Platte," William Jennings Bryan of Nebraska. Bryan ran on a single issue: "free silver," the pledge to coin more silver (instead of coining mostly gold). Free silver, its supporters promised, would put more money in circulation, raise prices, enable the poor to pay off debts, and end the long depression. McKinley savagely countered by warning laborers that if Bryan won they would be paid in debased currency while having to pay higher prices for food. The Ohioan also cautioned that because many companies had to sell their surplus goods in overseas markets that used the gold standard (especially Great Britain), Bryan's silver policies would harm foreign trade and shut off the huge foreign investment that had built much of nineteenth-century America. Hanna raised millions of dollars from frightened capitalists so Republicans could spread this message and mobilize their supporters.

McKinley received 7 million votes to Bryan's 6.5 million, while triumphing in the electoral college 271 to 176. His victory proved so decisive, his appeal to all classes so effective, that it marked a historic realignment of American politics. McKinley's success in the rapidly growing industrial cities solidified a Republican grip on the north comparable to the Democratic lock on the south (where African Americans were systematically disenfranchised). After the 1896 excitement, therefore, politics became duller, many voters lost interest, and national politicians became freer to carry out policies without the public scrutiny generated by the fierce partisanship that had marked the years 1872 to 1896.

As president, McKinley quickly moved to control the national political agenda. Publicly he remained the kind, sensitive man who, as the nation knew, hovered lovingly over his invalid wife, insisted that she sit by his side at state dinners, and frequently had to cover her face with his handkerchief and whisk her out of the room when she had a seizure. The nation also knew that Mrs. McKinley's illness and the president's sensitivity had been shaped in part by the tragic deaths of two infant daughters. McKinley combined this kindness with a steel will and an insider's knowledge of Washington to dominate national politics after 1897. As a contemporary observer noted, the president surrounded himself with master manipulators, then manipulated them. Elihu Root, the leading corporate lawyer of the time, whom McKinley cajoled into becoming secretary of war in 1899, believed that "he had a way of handling men so that they thought his ideas were their own." Nevertheless, as a senator observed after the president's death, "I don't think that McKinley ever let anything stand in the way of his own advancement," for he "petted and flattered until he felt that all of the fruit on the tree was his."

Such instincts led McKinley to set up the first pressroom for White House reporters. Between 1890 and 1909, the daily circulation of U.S. newspapers jumped 300 percent. The president suggested that his aides and selected cabinet members regularly meet with the press. He even provided telegraph wires for the correspondents' use. McKinley understood (unlike his predecessors in the White House who had disdained reporters) that a leader who controlled the news could go far in controlling the politics. Thus he spoon-fed the press his version of events, and the newspapers (with a few notable exceptions, such as the sensationalistic New York yellow press) grew dependent on the handouts and even acquiesced when he later censored bad news from the war in the Philippines.

For all his Machiavellian talents, however, McKinley made several initial missteps in 1897. He quickly asked Congress to pass a new tariff bill that would both protect the home market for U.S. producers and include a reciprocity provision to open foreign markets. A strongly protectionist Republican leadership passed the Dingley tariff, which indeed jacked up tariff walls, but whose reciprocity provision proved too complex and, indeed, too contradictory to be useful. McKinley would have to find weapons other than the tariff if he hoped, as he told one senator, to make American producers supreme in world markets.

Certainly, McKinley's administration had no intention of trying to change the American industrial order. In his inaugural address, the president had condemned bad corporations "organized in trusts . . . to control arbitrarily . . . trade among our citizens." But Republicans were not about to break up the handiwork of their wealthiest supporters, especially amid the dark shadows and uncertainty cast by the depression.

Instead, they happily presided over the first great merger movement of industrial America. Railroads and such dominant companies as American Tobacco and Standard Oil ruthlessly took over competitors. The twelve major mergers of 1897 amounting to $1 billion became 305 mergers worth $7 billion by 1903. As McKinley winked at the formation of massive combinations, the resulting industrial corporations led by such new multinationals as Standard Oil, Singer Sewing Machine, and McCormick farm machinery conquered foreign markets without needing the reciprocity clause. Henry Adams, the shrewdest observer of the time, watched the executive mansion from his home across Lafayette Square, and concluded that the nation was ruled by "McKinleyism," that is, "the system of combinations, consolidations, trusts, realized at home, and realizable abroad."

The president tried again, more positively, to help this push for foreign markets by annexing Hawaii in 1897. As a congressman, he had shown great interest in Americanizing these islands and establishing a naval base. An 1893 coup in Honolulu by American sugar and fruit planters had nearly resulted in annexation. But four years later McKinley could not obtain the necessary two-thirds vote in the Senate. He was stopped by the fears of domestic sugar growers and also by strict constructionists who warned that the Constitution could not stretch so far across water in an attempt to control such a multiracial society without shattering.

The turn in McKinley's fortunes began with two overseas conflicts that flamed just as he entered office. The first had actually begun in 1868 when Cubans rebelled against Spanish domination. The United States managed to keep its distance from this war, and Spain restored order in 1878. But an 1894 U.S. tariff law triggered a new uprising when it discriminated against Cuban sugar imports and created massive unemployment on the island. By 1897, Spain's attempts to use brutal concentration camps, as well as military force, had only spurred on Cuban revolutionaries and created enormous U.S. sympathy (and financial help) for their cause. The Yellow Press and McKinley's own assistant secretary of the navy, young Theodore Roosevelt, led the cry for U.S. intervention.

At the same time, the most powerful European imperialists — Great Britain, Germany, Russia, and France — were maneuvering to carve up China. For a half century, U.S. policy had been well known: China was to remain whole (so as not to endure the fate of Africa, for example, which was being divided up by the Europeans). Only then could American merchants and missionaries sell to *all* of the gigantic China market and not be at the mercy of the colonial powers. Great Britain, whose efficient industry allowed it (like the United States) to compete successfully for Chinese customers, was the one nation that tended to agree with the American position. But when the British asked McKinley in early 1898 to help them protect China and keep it whole, the president had to decline. His first order of business, he knew, lay ninety miles south of Florida. He then came to realize that it might be possible to gain both prizes — Cuba and an undivided China market — with one war.

For Spain ruled over the Philippine Islands as well as over Cuba. The conquest of the port of Manila (and the annexation of Hawaii) would give the United States a newly powerful voice in Asian affairs. McKinley later said he could not locate the Philippines on a map, but that was a deception. During afternoon carriage rides in 1897 and 1898, he and Theodore Roosevelt discussed the islands, and the U.S. Navy already had detailed operational plans for seizing them. The real question was how and when the United States was to go to war with Spain.

McKinley did not want a war. He did, however, want the diplomatic fruits that only a war could produce: a peaceful Cuba open to U.S. development and a Philippines open for the establishment of U.S. bases. He could speak movingly of the horrors he had seen on the battlefields of the Civil War. He could, however, also speak movingly of the need to end Spanish brutalities and to find vast economic opportunities abroad that would put Americans back to work — and push Europeans behind Americans in the race for world markets.

By January 1898, McKinley was moving — at his own pace and for his own reasons — toward war. Neither Roosevelt's impatience nor the war cries of the Yellow Press (which wanted a war to boost newspaper circulation) moved him. He even banned the Yellow Press from the executive mansion. Three events especially pushed the president toward war. The first was the China crisis. The second was the late November 1897 warning from the U.S. consul in Cuba, Fitzhugh Lee, that the Cuban uprising was not just anticolonial, but was turning radical. "There may be a revolution within a revolution," Lee warned, a radicalization that threatened all foreign lives and property. The Americans, who had some $50 million of investments in Cuba, needed U.S. governmental action, in Lee's words, "for protection and the preservation of peace."

That observation led to the third event that pushed McKinley toward war: his realization that Spain could never pacify the island. Spanish weakness became glaring in late 1897 and early 1898 when officials could not prevent outbreaks in Havana itself. In January 1898, McKinley dispatched the new warship, *Maine*, ostensibly on a goodwill visit, actually on a mission to provide any necessary protection for U.S. citizens and their property around Havana. Marcus Hanna, now a senator, rightly said that sending the *Maine* resembled "waving a match in an oil well for fun." But given the radicalization of the revolutionaries and the weakness of Spain, McKinley believed he had no alternative. On February 15, 1898, a tremendous explosion sank the *Maine* and sent more than 250 American sailors to their deaths at the bottom of Havana harbor.

A heated American public opinion, fanned by the Yellow Press, demanded immediate war. Its cries became more fierce when a private letter written by Spain's minister to the United States, Dupuy de Lôme, was intercepted and published by Cuban sympathizers. In the letter, the minister called McKinley "weak" and a "would-be politician," among other unflattering labels. The president did not mind the insults. He did mind that the letter revealed the Spanish minister had not been negotiating with him in good faith.

McKinley partly quieted the cries for war with one of his favorite devices: appointing a commission to investigate the *Maine* disaster. He refused to go to war, moreover, until he was confident that the expenses of such a conflict would not push the nation back into economic depression, and until he believed the U.S. military, especially the navy, was fully prepared. By the end of March, both conditions were met. Rising exports and the discovery of gold in Alaska and South Africa had raised the amount of money in circulation and eased economic pressures. A $50 million military appropriation by Congress and a hyperactive Theodore Roosevelt had readied the military. (In late February, Roosevelt had, on his own, sent orders to the U.S. Navy units around the world to prepare for war. When he discovered this the next day, the president calmly canceled every order — except the telegram instructing Admiral George Dewey in the western Pacific to prepare to attack the Philippines.)

On March 28, the *Maine* investigation concluded that the ship had been destroyed by an outside agent, probably a submarine mine. (A more detached investigation seventy-eight years later concluded that an internal explosion, probably in the engine room, had doomed the ship.) McKinley, however, had already shaped his ultimatum to Spain. He demanded an armistice until October 1, an end to the concentration camp policy, and his own involvement in the effort to bring about permanent peace. The Spanish government accepted the first two demands, but not the third, fearing that it could lead to the overthrow of the Madrid government, as well as Spain's three-hundred-year empire, if McKinley insisted on Cuban independence.

On April 11, the president asked Congress for the authority to use military force to end the war in Cuba. The Senate, pushed by domestic sugar growers, passed the Teller amendment, pledging the United States not to annex Cuba. McKinley accepted this provision, but he refused to accept a House amendment that recognized the Cuban revolutionaries as the island's government. In a vicious political struggle, the president dug in his heels against recognizing a possibly radical regime and, on April 25, finally prevailed.

The United States fought the easiest and most profitable war in its history over the next three months. This "splendid little war," as Secretary of State John Hay called it, resulted in 2,446 Americans losing their lives. But only 385 were lost in battle. The remainder fell victim to diseases. The first encounter proved to be the most historic. On May 1, Dewey trapped and destroyed the Spanish fleet at Manila without any American loss of life. Even before he received official word of Dewey's easy triumph, McKinley ordered U.S. troops to embark from California to occupy the Philippine capital. It was the first time a president ordered the U.S. Army to fight outside the Western Hemisphere. In June, he again demanded that Congress annex Hawaii. This time he sidestepped the need to obtain the consent of two-thirds of the Senate and instead imaginatively used a joint resolution, requiring a mere majority of each house, to acquire the islands. After all, as McKinley told a close friend, "we need Hawaii just as much and a good deal more than we did California. It is manifest destiny."

The war in Cuba was marked by several decisive battles on land (in which Roosevelt and his Rough Riders won the fame that would later usher him into the presidency), while the U.S. fleet bottled up and destroyed the decrepit Spanish navy. McKinley's iron hand controlled American policy and news, especially through his new telephone and telegraph system, which for the first time gave a president access both to battlefield commanders and to reporters and editors in mere minutes. By late July, Spain was ready to discuss peace terms. The president knew he wanted to control (but not fully annex) Cuba and take Spain's longtime holdings of Puerto Rico in the Caribbean and Guam in the Pacific. All were important strategic points. The central question involved the Philippines and the nationalist movement led by Emilio Aguinaldo which had long fought for independence from Spain. McKinley believed that he needed at least Manila so he would have a base from which he could deal with the China crisis. If he had to take the remainder of the Philippines to protect Manila, however, he would incite Aguinaldo to attack the U.S. troops.

By October 1898, McKinley reluctantly decided that U.S. interests in Asia left him no choice. He had to take all the Philippines. Immediately, an "anti-imperialist" movement sprang up to stop him. It was based in New England and financed by Andrew Carnegie. As the Senate approached the decision on the treaty in early February 1899, McKinley appeared not to have the needed two-thirds vote. Then, on the night of February 4, the White House telegraph tapped out the news that Aguinaldo's troops had fired on U.S. soldiers (provoked, some claimed later, by the Americans trying to push out the Filipino contingents). The president hoped that this episode meant the Senate ratification of the peace treaty to show support for the U.S. troops under fire, and, in fact, the pact passed by a single vote. The United States was suddenly a Pacific power as well as dominant in the Caribbean.

But the supposed peace treaty produced a war in the Philippines that threatened to tear apart the United States much as the Vietnam conflict did seventy years later. As in Vietnam, the nationalists fought a guerrilla-style war of attrition that frustrated American commanders. The president believed he could contain the war with thirty thousand troops, only to discover that he had to send two and then three times that many.

Anti-imperialist protests that rocked the country threatened to defeat McKinley in 1900. He responded with intensified military pressure in the Philippines and more control, even censorship, of the news at home. Both sides used torture and committed brutalities against civilians. The death toll climbed to more than two thousand American lives and, according to one estimate, as many as two hundred thousand Filipinos.

As the presidential election approached, McKinley turned the tide of war by appointing a new U.S. commander, Arthur MacArthur (father of World War II General Douglas MacArthur), who employed new tactics and finally, in 1901, captured Aguinaldo. Perhaps the decisive turn, however, occurred in China. There the bitterly antiforeign Boxer movement had risen to kill missionaries and threaten the diplomatic compound in Peking with annihilation. Secretary of State Hay feared not only the Boxers, but the European powers who threatened to use the uprising as an excuse to colonize China.

In 1899, Hay tried to neutralize the imperialists with the first Open Door note. He asked each of them not to discriminate economically against other foreigners in their own spheres of interest. To Hay's surprise, the Europeans and Japanese reluctantly agreed. A year later, as foreign troops invaded to kill the Boxers, Hay sent a second Open Door note. This initiative asked the powers to preserve "Chinese territorial and administrative integrity." Again, to Hay and McKinley's surprise, the foreigners, who eyed each other warily, agreed. The Americans had pulled off a remarkable diplomatic victory by apparently keeping China whole. U.S. policy toward the Chinese was now fixed until the Communist victory of 1949.

McKinley backed up the Open Door notes in the summer of 1900 by sending five thousand troops to protect U.S. citizens and fight the Boxers. Thus, without consulting Congress, the president dispatched troops to the Asian mainland — an event unthinkable thirty-six months before — to fight the Chinese and closely watch the other invading foreigners. China responded by declaring war against the United States, but McKinley ignored it. By September, the foreign troops had brutally rolled back the Boxers. With MacArthur's help in the Philippines, McKinley and Hay had managed to save Republican political fortunes in 1900.

It was just in time. Bryan again led the Democrats. His early anti-imperialist broadsides deeply worried McKinley. But the turn in the wars in both the Philippines and China led Bryan to commit a fatal mistake: now he decided to downplay anti-imperialism and to emphasize the trust and money issues. "McKinley prosperity," however, had long since killed the voters' interest in breaking up the trusts or coining more silver. The president and Congress, moreover, had ended the silver debate by passing the Gold Standard Act of 1900. Only Bryan seemed not to understand that the money issue was dead. Even Andrew Carnegie reluctantly decided to vote for McKinley. The new Republican vice-presidential nominee (McKinley's first vice president had died in office), the war hero and New York Governor Theodore Roosevelt, finally lost his voice as he crisscrossed the country to trumpet the successes of American imperialism. The anti-imperialists, he proclaimed, were "simply unhung traitors, and . . . liars, slanderers and scandalmongers to boot."

The president won by a larger margin than he had in 1896: 7.2 million to Bryan's 6.3 million and a 392 to 155 electoral college margin. To Marcus Hanna's delight, the Republicans even took Bryan's home state of Nebraska. Hanna again ran a perfect campaign, so well oiled with money that afterward he returned $50,000 to surprised Standard Oil donors.

McKinley and the United States stood at the peak of their power. The silver issue was dead. Prosperity grew. Under the Platt amendment, which gave the United States the right to intervene in Cuba at Washington's discretion, U.S. forces used the new naval base at Guantánamo, but left the messy and racially charged day-to-day governing problems to a cooperative Cuban regime. The United States had proved to be a Pacific, even Asian, power, while no longer being challenged, even by the British, in the Caribbean. McKinley and Hay prepared a treaty giving the United States the sole right to build an isthmian canal in either Nicaragua or Panama. U.S. goods were so competitive that European producers complained of an "American invasion." McKinley meanwhile had developed new presidential powers, especially over information and as commander in chief of the military, that were to shape the "imperial presidency" of the later twentieth century. Hay caught the spirit: "The greatest destiny the world ever knew," he told McKinley in 1898, "is ours."

In early September 1901, the president traveled to the spectacular Buffalo, New York, Pan-American Exposition, where he was greeted with a fireworks display that culminated in the message written across the sky, WELCOME MCKINLEY, CHIEF OF OUR NATION AND EMPIRE. In his speech at the exposition, he declared that the remarkable productivity of American farms and factories had ended "the period of exclusiveness." He recommended the combining of trade reciprocity with a new merchant marine, a Pacific cable network, and an isthmian canal to solve "the problem" of finding "more markets." The next day, September 6, an anarchist, Leon Czolgosz, shot the president. Eight days later McKinley died. Theodore Roosevelt became chief of the nation, and also of the "empire" that his predecessor had built in less than four years to initiate the American Century.

— WALTER LaFEBER
Cornell University

Theodore Roosevelt

1901–1909

b. October 27, 1858
d. January 6, 1919

A LMOST A CENTURY after Theodore Roosevelt became an accidental president, he remains an arresting figure. In part, his continuing appeal is a result of his charisma: the energy still shines out from behind the thick lenses of his pince-nez glasses in the old photographs. In part, it is a result of his storybook life, which spanned Manhattan, the Badlands, Washington, San Juan Hill, Albany, and Africa. It is also a product of his stunning range of talents and interests. Author, naturalist, hunter, soldier, politician, bureaucrat, and journalist, Roosevelt was probably the last renaissance man to hold the nation's highest office.

More than this, Theodore Roosevelt commands our attention for his attempt to reconcile some of the most fundamental oppositions of modern American life. He stood at the divide between the nineteenth century and the twentieth, between the old presidency and the modern chief executive, between the old state and the new. Full of his own contradictions, this complicated man grappled with basic contradictions that define the twentieth-century United States. He was both an architect of modern governmental regulation and a defender of Victorian individualism. He was a conservative who promoted change. A fierce hunter, he was a pioneering conservationist. A war lover, he won the Nobel Peace Prize. An advocate of women's rights, he insisted women should do their duty as wives and mothers.

It is just this struggle to contain and reconcile oppositions that makes Roosevelt so persistently interesting. A man who can be claimed by both liberals and conservatives, he reminds us that the activist liberal state and conservative individualism have thrived side by side in the twentieth century. Roosevelt's political career illustrates how partisanship has accommodated independence. His social views show how an optimistic reform impulse and a profoundly pessimistic determinism have intertwined throughout modern American history. And his foreign policy reveals how much American diplomacy has been shaped by conflicting urges for engagement and isolation, for commerce, democracy, and power.

When he was sworn into office after McKinley's assassination in September 1901, Roosevelt had a fairly unusual opportunity to make a major impact on American life. There was no urgent national emergency, nothing like the Great Depression and Second World War that would allow his cousin, Franklin Roosevelt, to accomplish so much in the 1930s and 1940s. But at the beginning of the century, the consequences of large-scale industrialization were sweeping across the nation. Americans worried over the power of giant corporations, the friction of labor relations, the growth of cities, the decline of farms, and the persistence of poverty. Deeply split by ethnicity, race, class, and gender, people debated whether and how fundamental social problems could be solved. They also renewed an old debate, made more pressing by new colonial acquisitions and economic interests, about the proper role of the United States in the world. At the same time, the political system was in the midst of a basic transformation. The system of popular politics, built on passionate voter participation, powerful political parties, and relatively weak government, was giving way to a new politics, founded on limited participation, weak partisanship, and strong bureaucracies and executives. All these changes combined to make the office of the president — the head of the executive branch, the commander in chief of the armed forces — more important than it had been for decades. In 1901, a new president, even an accidental one, had an unusually good chance to shape the course of American history.

Theodore Roosevelt was particularly well suited to the demands of the presidency at the start of the century. To a remarkable degree, Roosevelt had emancipated himself from many of the conventional prejudices of his privileged background, and his unusually varied career in and out of government had prepared him for the expanding role. He was the epitome of the new breed of politician, able to master the conflicting demands of the changing style of American politics. Moreover, at a time when politics involved an especially wide range of issues, Roosevelt was a politician of unusual intellectual breadth and reflection.

Born to wealth in 1858, Theodore Roosevelt was nevertheless a self-

made man. The son of a prominent New York City merchant and philan-
thropist, he enjoyed the customary advantages of upper-class life — a
fine brownstone on East Twentieth Street, servants, tutors, a Harvard
degree. It was his father who first pushed him to remake himself. In
1870, Theodore, Sr., urged his "sickly, delicate" namesake, plagued by
asthma, to build himself up. *"I'll make my body,"* the boy vowed.
Through a rigorous program of hunting, hiking, and boxing, he lived up
to his promise.

That act of will set the tone for Roosevelt's career. As he remade his
body, so he remade his life. He was contemptuous of "men of wealth who
sacrifice everything to getting wealth"; he was even more contemptuous
of men of leisure with their "cultivated taste and easy life." Avoiding
business, leisure, law, and other vocations of the upper classes, Roosevelt went into politics. He did so, he recalled, even though the "men I
knew best . . . laughed at me." But Roosevelt, instinctively reaching for
power, went ahead and won election to the New York state assembly as a
Republican in 1882. His future lay in politics.

Roosevelt broke out of the pattern of the life he was born to in other
ways as well. When his first wife died suddenly after giving birth to their
daughter on Valentine's Day in 1884, Roosevelt buried his grief by going
west to the Dakota Territory, where he lived the life of a cattleman on
two ranches in the Badlands. When the Spanish-American War began in
1898, he brought together a cavalry regiment of cowboys, Indians, New
York policemen, and Ivy League athletes and led these "Rough Riders"
in the famous charge up Kettle Hill and San Juan Hill in Cuba. The
cowpuncher and warrior was also an unusually attentive father. While
Victorian convention encouraged wealthy men to devote themselves to
work rather than domestic life, Roosevelt roughhoused with his children
when he was home and wrote them long letters when they were apart.

Roosevelt's elite origins would always be clear from his self-confident
bearing, his manners, and his comfortable life. But in a period when the
sense of social class was probably more intense than at any other time in
American history, he managed to develop a critical distance from his
own set. Much more than most upper-class men and women, he had an
understanding of other Americans, an understanding acquired on the
range, the parade ground, and the hustings. That understanding did not
always make him more sympathetic to the less fortunate. And his contempt for the wealthy did not keep him from trying to act in their interest. But this unconventional man had a notably sophisticated, independent, and often controversial judgment of just what that interest was.

Roosevelt's political career trained him especially well for government

in the twentieth century. As he became president, power was shifting away from the legislative branch toward the executive, which commanded the armed forces and an ever larger complex of departments
and commissions. Unlike McKinley, Roosevelt had relatively little experience as a legislator — only his one term in the New York legislature. In
the eighties and nineties, he spent much more time as an administrator.
In 1889, President Benjamin Harrison appointed him to one of the pioneering agencies of the modern state, the Civil Service Commission. In
1895, he became president of New York City's Police Commission. In
1897, he became assistant secretary of the navy in the new McKinley administration. Then, as if to complete his training, Roosevelt won the
race for governor of New York in 1898. His term as chief executive of the
nation's most populous state, with its complex industrial and agricultural economy, was a preview of the larger challenges of the presidency.

Roosevelt was also well prepared for the emerging political style of the
twentieth century. Starting his public career during the attack by liberal
independents on parties and partisanship, he had learned the tricky
balancing act that would be required of all ambitious politicians in the
future. In 1884, he declined to join the Mugwumps in bolting the Republican ticket headed by James G. Blaine and supporting the Democratic nominee, Grover Cleveland. Although Roosevelt stuck with his
party, he criticized old-style party machines and appealed to indepen-
dent-minded voters. As a member of the Civil Service Commission, he
was part of the reformers' attempt to replace the parties' control of patronage with the merit system of appointments to government jobs. But
he still knew how to work with the party bosses, and they needed his
reputation as a reformer. That reputation helped earn him the Republican nominations for governor of New York and then, two years later, for
vice president.

Roosevelt was perfectly suited to another major feature of the new
politics of the twentieth century, the increasing focus on the personal
qualities of candidates and leaders rather than on parties and voters.
Colorful, outgoing, and egocentric, Roosevelt was ready for the attention. Running for vice president in 1900, he sailed through the grueling
long-distance speaking tour that Americans were just beginning to expect of national candidates. Roosevelt knew, too, how to cultivate the
press in order to build his public image. He was adept at using pictures
and stories about himself and his family to maximum effect. Only William Jennings Bryan rivaled Roosevelt in establishing the model for the
twentieth-century American presidential candidate and leader.

Roosevelt was well suited to the demands of the presidency, finally, because of his unusual intellectual breadth. The first decade of the twentieth century was a time when a strikingly broad range of basic issues, involving intimate private life as well as public business, was openly contested. Americans debated the morality of divorce, the control of prostitution, and the relationship of the sexes as well as race relations, corporations, tariffs, and overseas expansion. More than any other political leader of the day, Theodore Roosevelt was able — and eager — to participate in those debates.

Roosevelt's easy command of this broad range of issues was partly a reflection of his varied career. In addition to his diverse roles in politics, government, cattle ranching, and the military, he was an avid hunter and an accomplished naturalist. At the young age of nineteen, he published *The Summer Birds of the Adirondacks.* That was the first of many books. Even as his political career developed, he built a successful career as a writer. His major works were biographies and histories: *Gouverneur Morris, Thomas Hart Benton, The Naval War of 1812,* and the sweeping, multivolume *Winning of the West.* But Roosevelt wrote about a stunning array of topics: politics, warfare, overseas expansion, labor, hunting, conservation, "The Ancient Irish Sagas," "The Mongols," "The American Boy," "Women and Science," "Christian Citizenship," and "Character and Success." Writer, soldier, politician, naturalist, rancher, Roosevelt was as close to Thomas Jefferson as any modern man, living in a more specialized age, could come.

The range of Roosevelt's topics suggests the breadth of his intellect. He had a striking capacity to think in fundamental terms about the whole range of human experience. Throughout his career, Roosevelt focused his writings and speeches on the building blocks of human life: man, woman, family, race, nation, work, war. Seldom original in dealing with any single aspect of life, Roosevelt was truly creative in the way he related the parts of experience and thought about the interplay of private and public, family and politics, war and peace, ideas and action.

Despite his qualities as a thinker, politician, executive, and self-made man, despite all his training for the presidency, the forty-two-year-old Roosevelt faced an uncertain future in 1901. The gilded obscurity of the vice presidency might well have marked the end of his political career. But then McKinley was shot. That tragedy allowed Roosevelt, the youngest president ever sworn into office, to harness his many talents in the White House. He succeeded well enough that he won election to the presidency in his own right in 1904, easily defeating the rather staid Democratic challenger, Alton B. Parker.

Roosevelt's popularity reflected several factors: the enduring Republican electoral majority created by McKinley's victory over Bryan in 1896; the persisting internal divisions that plagued the Democratic party; and the ongoing economic revival after the hard times of the 1890s. The president succeeded, too, because of his ability to fascinate Americans and personify some of the central aspects of American society and culture. Trailed eagerly by the press, Roosevelt made the presidency into a constant whirl of activity, the center of national attention. He was the first president to ride in an automobile, submerge in a submarine, and travel overseas. Even his misadventures were arresting. He lost the sight in one eye in a boxing match, flew through the glass window of a lighthouse tender in a collision in the Gulf of Mexico, and suffered a bad leg injury when a trolley car hit his carriage and killed the secret service man next to him. When the president spared a helpless bear during a hunting trip in Mississippi in 1902, he inspired a national craze for the stuffed toy teddy bear. But Roosevelt's popularity reflected more than his ability to captivate a national audience. For many people, this youthful, dynamic president perfectly embodied the optimism and energy of a still young, increasingly powerful nation.

The president also flourished because of his approach to the central issues of the day: the control of business, the pursuit of social reform, and the clarification of America's international role. Where he could, Roosevelt labored to reconcile the conflicts over these matters. Sometimes he succeeded. Sometimes he managed only to reflect the contradictions of American society. And occasionally he avoided confronting conflict altogether.

The problem of business was unquestionably the central issue of Theodore Roosevelt's presidency. It could not have been otherwise. The first billion-dollar corporation, United States Steel, emerged the year he took office. By then, Americans were at once awed, thrilled, and appalled by the vast corporations or "trusts" that increasingly controlled the economy. An intense debate continued over what to do about the consequences of large-scale industrialization, among them the economic might of big business, the vulnerability of consumers, the damage to the environment, and the weak status of labor.

In essence, there were five approaches to this issue by the early twentieth century. One was laissez-faire, the preferred, do-nothing strategy of much of the upper class. The second was socialism, the government ownership of business, which was probably a more popular notion than at any other time in American history. In between these extremes were three moderate solutions favored especially by the mostly middle-class reformers known as progressives. The antitrust idea would limit the power of business and preserve competition by using the authority of the federal government, under the Sherman Act of 1890, to break up big corporations. Compensation would tax businesses and use the proceeds to mitigate their harmful effects on society. Regulation would use government oversight to make sure that businesses did not harm society in the first place.

On this issue, probably more than any other, Roosevelt broke with most of his own class and aligned himself with the progressives. The president believed that society could no longer allow powerful businessmen to act on their own, free of restraint. He accepted the fundamental premise of progressivism — and of twentieth-century liberalism — that the capitalist economy required governmental intrusion in order to preserve it. Roosevelt *did* want to preserve capitalism, from capitalists themselves and from the socialism he abhorred. He even wanted to preserve the giant corporations, so long as they did not misbehave. Throughout his presidency, Roosevelt worked to build up the power of the federal government to confine corporations and the rest of business to appropriate channels.

Although Roosevelt did not want to eliminate large corporations, he used antitrust prosecutions to enhance the authority of the executive branch. In 1901, a pitched battle between E. H. Harriman and James J. Hill for control of northwestern railroads led to the formation of a single giant railroad holding company, the Northern Securities Company. Amid public fears about the power of this "great railway trust," the federal government shocked Wall Street by filing suit to dissolve the company early in 1902. The news brought the great financier J. P. Morgan, who had helped create the Northern Securities Company and U.S. Steel, to Washington. "If we have done anything wrong," he told the president, "send your man to my man and they can fix it up." "That can't be done," Roosevelt snapped back. The financier asked whether the president was "going to attack my other interests, the Steel Trust, and the others?" Roosevelt said no, "unless we find out that in any case they have done something that we regard as wrong."

Roosevelt's account of his conversation with Morgan, however embroidered, pointed to the future of business-government relations. Roosevelt could accept, and even welcome, the formation of U.S. Steel; he could not accept the way Morgan treated the president of the United States like a mere "rival operator." The government's suit went ahead. In 1904, the Supreme Court ruled 5 to 4 that the Northern Securities Company had indeed violated the Sherman Act. The case, the first successful federal prosecution of a single, tightly integrated interstate corporation, was a signal victory for Theodore Roosevelt. So was the administration's successful suit in 1902 to prevent the unpopular meatpacking companies of the "beef trust" from conspiring to fix prices and restrain competition.

With these and other successful suits, Roosevelt won acclaim as the great "trustbuster." But he was not interested in a full-scale crusade to break up big business. Neither was he particularly interested in using income or corporate taxes to compensate the rest of society. Instead, having sent corporations a warning about his power, he devoted his administration to establishing the federal government's capacity to regulate business formally and informally. In 1903, Congress responded to Roosevelt by creating the Bureau of Corporations, housed in the Department of Commerce, to investigate and publicize the behavior of giant companies. Two years later, after considerable struggle, Roosevelt helped obtain passage of the Hepburn Act, a precedent-setting measure granting the Interstate Commerce Commission, an arm of the executive branch, power to set the rates that railroads charged to shippers. The Hepburn Act helped protect the railroads from more stringent regulation at the state level. But it also forced them to surrender one of the capitalist's cherished privileges — the unilateral right to set prices.

The Northern Securities case and the Hepburn Act demonstrated the basic political reality that shaped Roosevelt's attempts to control the economy. The president won authority for the executive branch when he could offer himself as the representative of a public frightened and angry over a new threat from business. On those occasions, people would overlook their traditional fear of a powerful, activist government in Washington. That was what happened when threats to consumers and the environment appeared during the Roosevelt administration.

In the early 1900s, Americans were shocked by revelations about sickening conditions in the meatpacking industry and fraudulent advertising in the patent medicine business. Those revelations stimulated the most visceral fears of a consumer society, fears of being cheated and poisoned. As progressives and even some businessmen pressed for legislation, Roosevelt shrewdly exploited those fears. On the same day in 1906, he signed two laws that significantly enhanced the regulatory power of the federal government. The Meat Inspection Act empowered inspectors from the Department of Agriculture to go into packinghouses to prevent bad meat from coming to market. The Pure Food and Drug Act empowered the department to fine and imprison producers caught selling adulterated or misbranded goods in the marketplace. These two measures were not strong enough to please many progressives. They contained benefits for corporations, especially the biggest meatpackers. Nevertheless, businessmen had once again surrendered some of their freedom to the government.

Roosevelt also profited from public fears about the environment. By the early 1900s, Americans were realizing that economic growth endangered the supply of natural resources, the survival of birds and animals, and the beauty of the landscape. The growing conservation movement, quintessentially progressive, demanded that individuals give up some of their freedom to consume resources and damage the environment for the sake of the public good. Roosevelt embodied the contradictions of the movement. The man who loved birds and loved the west wanted to protect nature. But the man who hunted animals wanted to preserve natural resources so that they could ultimately be consumed for development.

Roosevelt vigorously pursued the conservationist agenda. By executive order, he created the Pelican Island wildlife refuge in Florida, the Crater Lake National Park in Oregon, and more than fifty bird reserves in twenty states and territories. He established national monuments and national forests and withdrew millions of acres of federally owned land from commercial exploitation, mostly in the west. In 1902, he signed the Newlands Act, a land reclamation measure that allowed federal support for irrigation projects. As with other economic issues, Roosevelt's environmental record disappointed the most ardent progressives. But the president outraged many businessmen, conservatives, and westerners by establishing the government's capacity to regulate resources and promote conservation.

Labor relations was another arena in which Roosevelt expanded his power by claiming to act on behalf of the public. He took office against a backdrop of intense labor conflict. In 1901, there were more than three thousand strikes across America as labor and capital battled to control work and its rewards.

Believing in individual initiative, Roosevelt had never been an advocate of unions. But in 1902, a strike by the United Mine Workers in the anthracite coal region of northeastern Pennsylvania led the president to support organized labor and, once again, to increase the power of the federal government. Whatever Roosevelt and most Americans thought of the union movement, they were appalled by the arrogant behavior of the railroad bosses who ran the coal mines. The owners refused to recognize the union or worry about the northeastern businesses and consumers who faced a winter without coal. When Roosevelt summoned the owners to a conference with the UMW leadership in Washington, he was affronted by their "condition of wooden-headed obstinacy and stupidity . . . utterly unable to see the black storm impending."

As the public's anger mounted, the Roosevelt administration brokered a settlement. A presidential review commission examined the facts of the situation and offered a binding solution. It was only a partial victory for the UMW, which did not win official recognition or much of its demands. It was a much greater victory for Roosevelt and the power of the state. Drawing "the permanent lesson of the strike," Roosevelt declared that it was "essential that organized capital and organized labor should thoroughly understand that the third party, the great public, had vital interests and overshadowing rights in such a crisis." "I wish that capitalists would see," Roosevelt said plaintively, "that what I am advocating . . . is really in the interest of property, for it will save it from the danger of revolution."

Roosevelt never intervened so dramatically in labor-capital disputes again. He denounced labor radicals such as Big Bill Haywood of the International Workers of the World. In 1907, he sent federal troops into a strike in Goldfield, Nevada, where they helped intimidate miners and the IWW. Over the course of his presidency, Roosevelt did not do much more to help or hurt most American workers. What he did offer in the settlement of the anthracite strike was a design for the liberal future. The federal government, representing the public interest, would supervise relations between workers and employers. Organized labor would have a place at the conference table. Business, in return for these unpleasant concessions, would avoid something much worse, the "danger of revolution."

In all, Roosevelt's response to industrialization amounted to a considerable — and controversial — expansion of presidential authority. Many Americans, not just businessmen, feared or resented the growth of federal power. In the last years of his administration, Congress rebuffed the president's demands for more power over corporations and the environment. Nevertheless, the Roosevelt administration had altered the relationship between government and business.

It was ironic that Roosevelt did so much to build up institutional power and constrain individual freedom. In an age of economic concentration and developing organization, Roosevelt still focused on the individual. "[I] have always," he confessed in the 1890s, "been more interested in the men themselves than in the institutions through and under which they worked." Despite the complex interdependencies of industrial society, Roosevelt continued to believe in the preeminent importance of the individual man or woman — in his or her power to affect the world. The central social problem, the one that animated so much of his speeches and writings, was how an individual governed him or herself; next to that issue, the creation of legal and organizational restraints was important but secondary. "Much can be done by wise legislation and by resolute enforcement of the law," he declared in 1906. "But still more must be done by steady training of the individual citizen, in conscience and character."

One senses that Roosevelt saw no contradiction in what he had done to build the coercive power of the federal government. From his point of view, a single individual, the president, was ultimately responsible for all the growing regulatory apparatus of the executive branch. Roosevelt's activist state was no impersonal monolith: it had an individual, human face.

Roosevelt's individualistic convictions were more clearly and simply translated in his approach to social reform, the second major issue of his administration. In the 1900s, there were many crusades, led mostly by progressives, to ameliorate the impact of industrialization and to improve the quality of private and public life. These crusades reflected a deep split in social thought. On the one hand, there was an optimistic belief that people changed when their environment changed. Better housing, compulsory education, neighborhood playgrounds, and prohibition of alcohol would, it was hoped, remake Americans in the image of middle-class reformers. On the other hand, there was a profoundly pessimistic belief that hereditary differences — race above all — were ineradicable and potentially troublesome. The best that could be done was to keep different groups apart: the age of reform was also the age of segregation.

Roosevelt, like many Americans, managed to hold both views. Well versed in the scientific racism of the day, he placed great weight on racial differences. And he certainly favored various environmentalist reforms. But Roosevelt did surprisingly little to advance the agenda of social reform during his presidency. That was partly because Roosevelt, the careful politician, did not want to get caught in the crossfire over such divisive issues as Prohibition. It was also because Roosevelt's commitment to individualism held him back.

Believing that human beings should succeed or fail on their own, the president was reluctant to allow an activist government to dictate their fate. In this respect, Roosevelt, the confident self-made man, was not a progressive. Most middle-class reformers, raised on Victorian individualism, believed that their parents' values had become a justification for the power of big businessmen and the weakness of the rest of society. So the progressives generally downplayed personal freedom and looked for institutional means of controlling and reshaping human behavior. Roosevelt, however, wanted to reform, rather than abandon, the individualist values of the nineteenth century. He demanded a rehabilitated individualism, grounded in self-control and social responsibility rather than selfishness.

That was why the president mostly confined his involvement with social problems to sermonizing. His prominent office — the "bully pulpit" — gave him an ideal forum from which to urge men and women to build their "conscience and character." The White House also enabled him to publicize pressing social issues. During Roosevelt's administration, conferences explored the decline of rural life and the threats to dependent children. The Department of Commerce and Labor studied the rise of divorce.

Roosevelt's conflicted views on some issues limited his advocacy of social reform. The president was, for instance, quite sympathetic to many of the aims of the women's movement. He admired educated women, favored equal rights, deplored the double standard of sexual behavior, and condemned "male sexual viciousness" and "the flagrant man-swine." But Roosevelt expected women, like men, to do their duty. And that duty meant marriage and childbirth. So the president lamented the rising divorce rate. He also excoriated middle- and upper-class "race suicide," the decreasing birth rate that jeopardized, he thought, the survival of his class, his race, and his nation. In all, Roosevelt did little to undermine women's traditional role. He certainly did not make a major attempt to legislate his attitudes.

Roosevelt's ideas also led him to embrace a pessimistic view of race. He became president when race relations were under enormous strain. African Americans faced lynching in the south. Native Americans, beset by federal armies and bureaucrats for generations, seemed to be wasting away. Asian Americans faced virulent hostility, especially on the West Coast. Even sympathetic whites did not believe that people of color could attain equality in America. Segregation, unjust as it now appears, seemed to be the humane solution: the races would be separated in order to protect the supposedly inferior.

Roosevelt accepted and even promoted segregation. Believing in African American inferiority himself, he did little to change white Americans' attitude and behavior toward blacks. In 1901, Roosevelt angered southern whites by inviting the African American leader Booker T. Washington to dine at the White House. After the furor over that occasion, Roosevelt offered few other affronts to the racial status quo. One night in 1906, black federal troops apparently shot up the inhospitable town of Brownsville, Texas, killing a bartender and wounding a police lieutenant. Roosevelt, infuriated that the soldiers would not tell what had happened, dishonorably discharged 170 of them, including 6 winners of the Medal of Honor. Confronted with black anger, Roosevelt was unrepentant. "If the colored men elect to stand by criminals of their own race because they are of their own race," the president warned, "they assuredly lay up for themselves the most dreadful day of reckoning." It was a sorry performance, one that Roosevelt omitted from his memoirs.

Although Roosevelt admired the fighting spirit of Native Americans, he did not think they were capable of much progress. In fact, the president feared that their culture would disappear if they had too much contact with whites. He criticized the federal policy of "assimilation," the attempt to use education and land reallotment to make Native Americans into independent farmers with white cultural values. Accordingly, the Roosevelt administration slowed the assimilation program and tried to keep Native Americans more separate from the larger society.

Roosevelt also furthered white Americans' attempts to separate themselves from Asians. With an eye toward the electoral vote of California, he successfully supported the continued exclusion of Chinese immigrants from American shores. The president was angered by the anti-Japanese sentiment in California that threatened diplomatic relations with Japan. But he nevertheless entered into the so-called "Gentleman's Agreement" with Japan in 1908 to restrict migration to the United States.

Overall, Roosevelt compiled a prudent, contradictory, and ultimately disappointing record on social issues. Practicing the politics of the possible, he avoided controversy and tolerated injustice.

The nation, the focus of patriotism, was fundamental for Roosevelt. He believed that the nation required the same virtues as the individual; it, too, was defined by duty. To Roosevelt, the United States, virtually a chosen people, had a special mission in the world. Before, during, and after his presidency, he worried that Americans would shirk their mission. As president, Roosevelt did what he could to build up the nation's power to carry it out. But he had difficulty clarifying just what the mission should be.

The nation's international role was, in fact, rather confused at the turn of the century. Americans' longtime desire to avoid foreign entanglements had been increasingly compromised by the expansion of the economy and national power. In this age of imperialism, the United States, despite its revolutionary heritage, had joined the rush of Western powers to grab colonies around the world. With its victory in the Spanish-American War of 1898, the nation had taken Cuba, Guam, Puerto Rico, and the Philippines. It was now engaged in a brutal war to subdue Filipino guerrillas. As Roosevelt took office, the United States no longer seemed like a unique experiment in democracy and freedom in the New World. Rather, it looked like any other imperial power of the Old World.

That did not particularly bother Roosevelt. After all, the former assistant secretary of the navy and the hero of Kettle Hill had been deeply implicated in the imperial expansion of the United States. Yet as president, he accepted the tacit national understanding that the United States should not continue to acquire colonies. Responding to Americans' uneasiness about their imperial role, he brought the war in the Philippines to an end in 1902 and promised to prepare the islands for eventual self-government. But he also rejected ostentatious idealism. Unlike Woodrow Wilson, Franklin Roosevelt, and other liberal presidents, the Republican Roosevelt did not talk of making the world safe for democracy or the four freedoms.

Roosevelt did not do much to put commerce at the center of American foreign policy either. In the early 1900s, many businessmen wanted Washington to help them expand their ventures abroad. The Roosevelt administration made sure that the consular service helped support overseas trade. But Roosevelt, disdainful of business values, would not let the immediate interests of businessmen dictate American diplomacy.

Theodore Roosevelt worked to build up American strength and influence within the limits of the international balance of power. As with social reform, Roosevelt knew how to accept reality in foreign affairs. "Speak softly and carry a big stick," he liked to repeat, "you will go far." The president spoke softly when necessary. He did not contest Russian control of Manchuria or Japanese control of Korea. He largely left Europe to the European powers.

Roosevelt was determined to assert American power, in the form of the Monroe Doctrine, in the Western Hemisphere. In 1902, he staged a show of naval force in the Caribbean to urge the Germans and the British to end a dispute over Venezuelan loans. The same year, he sent troops to Alaska to push the British to settle a dispute over the boundary line between the United States and Canada. By 1904, the president was offering the "Roosevelt Corollary" to the Monroe Doctrine: the United States would do its "duty" and intervene to stop any "general loosening of the ties of civilized society" in the Western Hemisphere.

To back up such talk, Roosevelt had to have that "big stick." As president, he worked to increase America's capacity to project power around the globe. He launched a vigorous and fairly successful campaign to reform the army. He also won a substantial increase in the size of the navy. And, in the "great bit of work of my administration," he pursued the construction of an isthmian canal through Central America that would make it easier for American ships to circle the world. Roosevelt's immense enthusiasm for this project led him to dubious measures. To get the rights to a ten-mile swath of Panama, the president covertly supported a Panamanian revolt against Colombia in 1903. Roosevelt denied any impropriety at the time, yet it was colonialism in all but name. "I took the Isthmus," he bragged after leaving office. By then, the difficult work had begun to build the Panama Canal, a critical channel for American power.

Roosevelt knew how to increase American influence by more subtle means as well. In meetings at Portsmouth, New Hampshire, in 1905, the president artfully mediated the negotiations to end the Russo-Japanese War, the feat that won him the Nobel Peace Prize. In 1906, Roosevelt successfully played the role of mediator again in a dispute between France and Germany over Morocco in 1906.

Through deft diplomacy, military reform, and the Panamanian landgrab, Roosevelt built up American power for the twentieth-century world. But he had not done much to clarify the uses of that power beyond the defense of a sphere of influence in the West. Roosevelt's United States was not an aggressive agent of overseas commerce, an impassioned champion of international democracy, or, for that matter, a militaristic apologist of imperialism. On the whole, Americans did not mind. They could live with an uncertain international mission in the last decade before the world wars.

Roosevelt's approach to diplomacy, social reform, and big business ensured his continuing popularity. He could no doubt have won reelection in 1908. But in the afterglow of the 1904 campaign, he had vowed not to run again. Giving way to his handpicked successor, William Howard Taft, Roosevelt left for a long hunting trip to Africa in 1909. Private life frustrated him, however, and Taft disappointed him. After unsuccessfully challenging Taft for the Republican presidential nomination, Roosevelt launched a stirring third-party candidacy at the head of the Progressive ticket in 1912. He managed only to break up the Republican electoral coalition and elect the Democratic nominee, Woodrow Wilson.

From 1901 to 1909, Roosevelt's passionate nature and complex talents had found a perfect balance and a perfect outlet in the White House. Out of power, the former president lost that balance, with sometimes sad and embarrassing results. For a time, he served as a contributing editor of *Outlook,* a magazine of political and cultural criticism. Eager to bring the United States into World War I, he made a bellicose, troubling campaign for military "preparedness." When America did join the Allies in 1917, the hero of Kettle Hill tried to recapture his old martial glory by volunteering to lead a division to France. Wilson turned him down. Now even war, his old friend, failed him. His youngest son, Quentin, was shot down over France in 1918. "I am not what I was," Theodore Roosevelt admitted. He died less than a year later, at the age of sixty, in January 1919.

Roosevelt's memory has lived on — because of his personality, his talents, his adventures, his accomplishments, and his significance. Few presidents before or since were as colorful or as capable. Some certainly accomplished more. Still, Roosevelt helped to recast the presidency, accumulate power for the executive and the nation, and thrust the state more deeply than ever into the workings of the economy.

In all this, his administration served as a model for the liberal future. Roosevelt did, indeed, seem quite modern and forward-looking. But his presidency was also evidence of how much the past persists and complicates American life. In his celebration of individualism, Roosevelt reminds us that eighteenth- and nineteenth-century values have survived and even flourished in an increasingly organized, centralized, and bureaucratized society. This contradictory man reminds us, too, that the modern United States is built on contradictions — between liberalism and conservatism, isolation and intervention, partisanship and independence, exploitation and conservation, feminism and sexism, racism and democracy, convention and reform, pessimism and optimism. Those contradictions energized Theodore Roosevelt and defined his presidency.

— MICHAEL McGERR
Indiana University at Bloomington

The Meteoric Rise of Teddy Roosevelt

Teddy Roosevelt blazed into national prominence in the late 1890s while still in his thirties. In 1901, at age forty-two, he became the youngest man to become president. It was a remarkably short, steep path to power.

Until 1897 the only national office Roosevelt had held was a stint on the U.S. Civil Service Commission, where his scrupulously nonpartisan administration had upset leading powers in his Republican party. From there he carried his reform politics back to New York, taking on Tammany Hall as head of the New York City Board of Police Commissioners. He was making something of a name for himself, but hardly building a political base. In fact, he was sowing deep distrust in the Republican establishment.

His appointment as assistant secretary of the navy in 1897, however, placed Roosevelt in the right place at exactly the right time. When the USS *Maine* exploded in Havana harbor in January 1898, Secretary of the Navy John D. Long was incapacitated. Roosevelt acted decisively, cabling orders to Admiral George Dewey to prepare to take the Philippines should the United States and Spain go to war. Soon after war broke out, Roosevelt himself entered the fray, organizing and training a volunteer cavalry regiment known as the "Rough Riders."

Just months later, Roosevelt rode his military exploits in Cuba to victory in the New York gubernatorial election. His momentum carried into the Republican National Convention in 1900, where, despite President McKinley's reservations, Roosevelt was nominated to succeed the deceased vice president, Garret A. Hobart. Like most other Republican regulars, McKinley's campaign manager, Ohio Senator Mark Hanna, was deeply suspicious of the new running mate. "Your *duty* to the country," Hanna warned McKinley, "is to *live* for *four* years from next March." A little over a year later, McKinley was dead and Roosevelt was president.

William Howard Taft

1909–1913

b. September 15, 1857
d. March 8, 1930

WILLIAM HOWARD TAFT entered the presidency as the handpicked successor to the most popular and dynamic public figure of his age, and he never managed to escape from Theodore Roosevelt's shadow. Although he claimed to be trying to continue his predecessor's policies and cement his legacy, Roosevelt and his progressive supporters considered Taft's timorous, legalistic approach to his office first a disappointment, and then a betrayal. Their repudiation of him proved fatal to his presidency. He left office the most thoroughly disavowed incumbent in American history. But Taft's unhappy term as president was a painful aberration from a successful, even brilliant, career that both preceded and followed his single term in the White House.

Taft was born on September 15, 1857, in Cincinnati, Ohio, to a prosperous family of considerable local distinction. After a comfortable middle-class childhood in the Midwest, he attended Yale and then returned to Cincinnati to earn a law degree. A few years later he married a socially prominent Cincinnati woman, Helen Herron, with whom he had three children and whose strong political ambitions helped buttress his own much less intense ones. In the years after his marriage in 1886, Taft became ever more deeply involved in local and eventually national Republican politics. He was an assistant county prosecutor, the Ohio collector of internal revenue, a state superior court judge, and from 1890 to 1892 the solicitor general of the United States. Upon his departure from the Justice Department, he spent eight years as a federal judge on the Sixth Circuit, based in Cincinnati, where he was also the dean of the University of Cincinnati Law School. He liked serving on the bench, and he soon set his sights on a seat on the Supreme Court.

In 1900, his fellow Ohioan William McKinley named him the first civil (that is, nonmilitary) governor of the Philippines, which had become a United States possession in 1898. He acquitted himself well in that visible and demanding position, in which he sought to create a system of laws to replace the harsh military rule that the United States had imposed on the island before he arrived. He left his post to serve as Theodore Roosevelt's secretary of war from 1904 to 1908. Taft's service in Roosevelt's cabinet changed his life. It made him a prominent national figure and brought him into closer contact with Republican party leaders. More important, it led to a relationship with the president that would become one of the most important, and ultimately most complicated, of his life.

In retrospect, it seems difficult to understand how two such different men could form so close an alliance, and how the dynamic Roosevelt could ever have imagined the cautious, somewhat lethargic Taft as a suitable successor. But at the time, their differences were less evident, for Taft had many qualities that Roosevelt greatly admired. Taft could be relied upon for sound political and legal advice. He was an able negotiator and a surprisingly adept bureaucratic politician, capable of smoothing over disputes within the executive branch or between Roosevelt and Congress. Perhaps most of all, he was unflinchingly loyal and looked upon Roosevelt much as he had earlier looked upon his own parents and his wife — as a tower of strength on whom he could rely and from whom he could draw his own power. Roosevelt quickly came to trust Taft completely.

But some of the same qualities that made Taft so appealing and effective as a subordinate contributed to his failings as president. He was a natural follower who operated best within clearly defined structures created by others. He was a man of cautious, legalistic temperament, who instinctively recoiled from doing what Roosevelt reveled in doing — bending the rules, and at times the law, to achieve his objectives. Taft was a born procrastinator who agonized over decisions, complained constantly that he did not have enough time to make up his mind, and always looked for some existing regulation, precedent, or law that would tell him what to do. Although he was often very skilled at the internal politics of the government, he had little talent and even less liking for electoral politics. He considered campaigning for the presidency "a nightmare." Taft was only one year older than Theodore Roosevelt (who was and remains the youngest man ever to have become president), but he seemed like a figure from a different generation. He dressed in the style of the conservative Midwestern bourgeoisie from which he had come and had an already old-fashioned upturned mustache. And unlike the athletic Roosevelt, Taft was sedentary and obese. He weighed more than three hundred pounds when he was elected president, and he attracted popular derision for having an oversized bathtub installed in the White House to accommodate him.

The unflattering contrasts with his predecessor mostly came later, however. For during the 1908 campaign and its aftermath, Taft seemed to be that rare figure among politicians: a leader acceptable, even appealing, to virtually everyone. Because he had been Roosevelt's most trusted lieutenant and anointed heir, progressives believed him to be one of their own. Because he had been one of the first viceroys of the American empire and had traveled widely over the preceding eight years, imperialists and internationalists trusted him to maintain America's active role in world affairs. And because he had been a restrained and moderate jurist with a punctilious regard for legal procedure, conservatives expected him to abandon Roosevelt's aggressive and, as they saw it, reckless use of presidential powers.

He won election in 1908 almost ridiculously easily. With the simultaneous support of the president, Roosevelt's progressive allies, and the Republican Old Guard, he received his party's nomination virtually uncontested. In November, he received not only the votes of Roosevelt's enormous constituency, but also the votes of business leaders and others who had come to despise the president's policies. John D. Rockefeller wired his congratulations after the Republican convention. J. P. Morgan, learning of Taft's nomination, commented, "Good! Good!" Andrew Carnegie donated $20,000 to Taft's campaign. Taft won by a smaller popular margin than Roosevelt had in 1904, but his victory was nevertheless decisive. He received 51 percent of the popular vote, to 43 percent for the Democratic candidate, William Jennings Bryan, running forlornly for the third and last time. He won 321 electoral votes to Bryan's 162. Delighted progressives expected a presidency that would consolidate their achievements. "Roosevelt has cut enough hay," they proclaimed. "Taft is the man to put it into the barn." Conservatives rejoiced that they were rid of the "mad messiah" and that a man of judicial temperament had replaced him. Taft entered office on a wave of good feeling.

But the new president himself was anxious and uncertain as he moved into the position that most Americans now strongly identified with his predecessor. Shortly after his Inauguration, as Roosevelt prepared to depart for an extended safari in Africa, Taft wrote him a letter that hinted at the insecurities that were already plaguing him. "When I am addressed as 'Mr. President,'" Taft said, "I turn to see whether you are not at my elbow. . . . I have not the facility for educating the public as you had . . . , and so I fear that a large part of the public will feel as if I had fallen away from your ideals; but you know me better and will understand that I am still working away on the same old plan." At the same time, however, Taft was already suggesting to others that his predecessor was not in all ways a model to whom he would aspire. Roosevelt, he told his colleagues, in the bland, understated language that characterized his public and private discourse, "ought more often to have admitted the legal way of reaching the same ends." It was not long, of course, before Taft discovered that he could not please both the progressives and the conservatives in his party. Increasingly, without really intending it, he found himself pleasing the conservatives and alienating the progressives.

Within days of taking office, Taft called Congress into special session to consider one of the most controversial and divisive issues of the time: the tariff. The protective tariff on imports into the United States was the principal source of revenue for the federal government. It was also a vehicle for protecting domestic industries from foreign competition, and for generations it had been the subject of great battles between agrarian regions, for which the tariff raised the price of manufactured goods, and urban, industrialized areas, which benefited from the protection. By 1909, tariff reform had been a consistent demand of progressives for more than a decade, less because of their commitment to free trade than because they believed foreign competition would weaken the power of the great trusts and thus help redistribute wealth and lower prices to the consumer. Theodore Roosevelt had made a number of tentative gestures toward tariff reduction but had always pulled away from the issue in the end. Taft was determined to confront it in a battle he hoped would establish his presidency as a strong and important one. "I believe the people are with me," he wrote in January 1909, shortly before taking office, "and before I get through I think I will have downed Cannon and Aldrich too."

Joseph Cannon was the conservative and, many claimed, tyrannical Republican Speaker of the House. Nelson Aldrich was an almost equally conservative senator from Rhode Island. Both were part of a new breed of powerful Republican leaders in Congress, and both were closely allied with large corporate and financial interests. They were committed to preserving as much of the existing tariff system as the politics of the moment would allow, and thus determined to weaken what Cannon called "this babble for reform" even if they could not kill it. Taft's private pledge to "down" them both was, in fact, little more than a rhetorical wish, for he was unwilling to intervene in the deliberations of Congress as the tariff bill took shape. If he was sent a bad bill, he said, he would veto it.

Cannon could not block the power of progressive insurgents in the House, who — under the direction of Sereno Payne, chairman of the House Ways and Means Committee — produced a bill substantially lowering rates on more than eight hundred items (and imposing the first federal inheritance tax, on estates of more than $10,000). Some of the most advanced progressives condemned Payne's bill for not going far enough, but Taft praised it as coming as close to the administration's goals "as we can hope." In the Senate, however, Aldrich stepped in where Cannon had failed and managed to win approval for amendments that raised rates on more than six hundred of the items listed in the House bill, dramatically weakening it. Senate progressives fought tirelessly against the conservative assault, with help from the reform press, which published devastating editorials and savage cartoons ridiculing the conservative position. But Taft refused to intervene, even rhetorically, convinced that it was improper for the president to meddle in the legislative process. By the time the Senate was done, the House bill was in tatters — with rates on most items far higher than the House had set them and with some rates higher than they had been before the legislation was drafted.

The conference committee that reconciled the two versions, most of its members appointed by Cannon and Aldrich, produced a final bill very close to the one the Senate had passed. Republican progressives in both houses voted against it. Taft — having remained silent during the prolonged and bitter debate in Congress, and having backed away from an insurgent effort to limit the power of Speaker Cannon — now expressed unhappiness with some of its provisions but decided to sign it nevertheless. He then unwisely defended it as an important progressive victory. "This is the best tariff bill the Republican party has ever passed," he told an audience in Winona, Minnesota, "and therefore the best tariff bill that has been passed at all." Republican progressives, already disillusioned with Taft for his passivity during the battle, responded to the Winona speech with incredulity and fury — greatly deepening the wedge that had already emerged between them and the president.

The legislative battle over what became known as the Payne-Aldrich tariff was a critical moment in the history of the Republican party. In the course of the debate, the tensions that had been building throughout Theodore Roosevelt's presidency between progressive insurgents and the conservative Old Guard burst into plain view and destroyed the party's unity. Taft's passivity, and his ill-chosen words in Winona, contributed to the damage. But there may have been nothing he could have done in the end to prevent the rupture that occurred and nothing that would have protected him from being weakened by it.

In his conduct of America's relations with the world, Taft continued Roosevelt's efforts to extend the nation's influence abroad but rejected many of the methods that had allowed his predecessor to succeed in those efforts.

Taft himself exhibited a progressive sensibility in international affairs that was far less bellicose than that of his predecessor. In 1911, he reached an agreement with Great Britain to submit all future Anglo-American disputes to arbitration. It would, he said, be "the crowning jewel of my administration," but he could never persuade the Senate to ratify it. Similarly, he negotiated an agreement with Canada to lower trade barriers between the two nations, only to have the Canadian government back away from the measure in the end. In other cases, he was more successful in settling disputes through peaceful, legal channels. A fishing dispute with Newfoundland, a disagreement over the U.S.-Canadian boundary, and other relatively small but troubling controversies came to tranquil ends through arbitration.

But Taft's belief in arbitration and international law was a small footnote in a foreign policy that was mostly devoted to extending American economic influence. His unfortunate choice for secretary of state was Philander C. Knox, a corporate lawyer with little talent in international relations and a strong commitment to protecting American business interests abroad. The Roosevelt administration had, of course, been concerned with American business interests too, but Knox seemed to regard the interests of the corporate community as his principal concern. He aggressively encouraged American investment in less-developed regions, and those efforts were so visibly at the center of his performance in office that critics soon labeled his policies "Dollar Diplomacy."

Taft's greatest diplomatic failure was in the Far East. He and Knox allowed themselves to be persuaded by Willard Straight, a former diplomat now serving as an agent of American bankers, to step up the American presence in China, where European powers and Japan were already busy carving the nation into "concessions" within which they could monopolize trade. Taft and Knox tried to persuade the British, the French, the Germans, and the Japanese to allow the United States to participate in their Chinese ventures — which included mining and railroad building. All such efforts were rebuffed.

American efforts in the Caribbean were not very much more satisfying to the advocates of Dollar Diplomacy. Taft considered American investment in the region to be a policy that would simultaneously serve American business, limit European influence in Latin America, and create stability in the target nations. He would, he said, "substitute dollars for bullets." But the Taft-Knox efforts made almost no visible contribution to bringing peace to Latin America. American policies tried and failed to resolve an economic crisis in Honduras and a bitter civil war in Nicaragua. Nor did American dollars always substitute effectively for bullets. Taft sent American troops into Nicaragua in 1910 to prop up a new government that the United States had helped create, and those troops remained there for most of the next twenty years. American diplomacy — both economic and military — created deep resentments in much of Latin America, and it was not until Franklin Roosevelt introduced his Good Neighbor Policy in 1933 that relations between the United States and Latin America significantly improved.

With Taft's standing among progressive Republicans already declining in the aftermath of the Payne-Aldrich tariff controversy, and with the party growing more and more deeply divided in general, a sensational controversy broke out late in 1909 that shattered the president's uneasy relationship with the allies of Theodore Roosevelt. The controversy emerged in connection with an issue that had become especially important to reformers: conservation.

Many progressives had been unhappy when Taft had replaced Roosevelt's secretary of the interior, James R. Garfield, an aggressive conservationist (and the son of the martyred president), with Richard Ballinger, a corporate lawyer with a much weaker commitment to preserving or managing the natural environment. When Ballinger invalidated Theodore Roosevelt's removal of a million acres of forests and mineral reserves from public land available for development (claiming that the former president had acted illegally), suspicion of the new secretary grew stronger. No one was more unhappy with Ballinger than Gifford Pinchot, the head of the National Forest Service, a devoted ally of Roosevelt and a doggedly partisan bureaucratic politician. Taft considered Pinchot a dangerous radical. Pinchot in turn soon came to view the president and Ballinger as enemies of conservation and reform.

In the meantime, Louis Glavis, an Interior Department investigator, produced an inflammatory report charging Ballinger with having conspired to turn over valuable public coal lands in Alaska to a businessman with whom he had once worked. Glavis had helped block the land transfer during Garfield's term as secretary. When Ballinger took over the Interior Department, he reopened the matter and approved transferring the lands to his former associate. Unwilling to trust the Interior Department with a report critical of its head, Glavis took his charges to Pinchot, and Pinchot took them to Taft. Ballinger quickly submitted a brief in his own defense. The president — after laboriously studying the conflicting documents and consulting with many others — accepted Ballinger's claims and fired Glavis for "disloyalty to his superior officers." Taft hoped not to antagonize Pinchot with his decision, but Pinchot was by now so committed to defeating Ballinger that he was unwilling to accept Taft's assurances that the secretary had done nothing wrong. In November, Glavis took his charges public and demanded a congressional investigation of Ballinger. A few weeks later, Pinchot publicly attacked the president for his part in the affair — in effect daring Taft to fire him. When the president reluctantly obliged, Pinchot was jubilant.

A congressional investigation, dominated by Old Guard Republicans, exonerated the president and the secretary of the interior but revealed behavior by both men that dismayed progressives and aroused the press. Progressives throughout the country rallied to the support of Pinchot, whom they considered a martyr to anticonservation reactionaries, and condemned Taft, unfairly, as a conservative determined to undo Roosevelt's legacy. In fact, Taft was in many ways as sympathetic to the idea of conservation as Roosevelt was; but he was also, as always, uncomfortable with Roosevelt's (and Pinchot's) aggressive methods, as he made clear in his tepid, procedural defense. "It is a very dangerous method of upholding reform," he told a protesting progressive congressman, "to violate the law in so doing, even on the ground of high moral principle, or saving the public." The Pinchot-Ballinger controversy, as it became known, all but destroyed Taft's presidency. It confirmed the division of the Republican party and earned the president the enduring enmity of its reformers. And it helped draw Theodore Roosevelt back into politics to rescue progressivism from those whom he believed were betraying it.

Theodore Roosevelt spent the first year of Taft's presidency on an extended safari in Africa and a tour of Europe. But his shadow loomed large over his unhappy successor even then. American newspapers gave extensive coverage to his exploits in the wilds and even more to his triumphal visits to European capitals, where he was greeted by heads of state as if he were still the president. His return to New York in the spring of 1910 was a major public event; reporters quickly noted that while he declined an invitation from Taft to visit the White House, he met at once with Pinchot, who had already traveled to England to see him several months before.

Publicly, Roosevelt claimed to have no plans to reenter politics. Privately, he was furious with Taft, who had, he believed, "completely twisted around the policies I advocated and acted upon." On September 1, 1910, he delivered a remarkable speech in Osawatomie, Kansas, in which he spelled out his own, greatly enlarged vision of social justice and reform and called, even more emphatically than he had as president, for a strong federal government that acted as the "steward of the public welfare." He made no direct reference to Taft, but his challenge to the administration's legalistic conservatism was clear.

Taft suffered another political blow in the 1910 congressional elections, in which conservative Republicans — who were by now his only allies in the party — suffered massive primary defeats at the hands of insurgent challengers, while Republican progressives won renomination without exception. In the general election, moreover, the Democrats — who were touting progressive candidates of their own — won control of the House of Representatives for the first time in sixteen years and greatly expanded their strength in the Senate.

Increasingly dispirited, Taft continued on the path he had chosen and, in October 1911, announced a decision that finally drove Roosevelt into open opposition. Taft had from the beginning of his presidency been much more active both in increasing regulation of corporations and in enforcing the antitrust laws than Roosevelt (his reputation as a "trustbuster" notwithstanding) had been. He proposed, but was unable to win passage of, an act that would have answered a long-standing desire of antitrust partisans: a measure to require corporations engaged in interstate trade to incorporate under the jurisdiction of the federal government, not with state governments unable or unwilling to regulate them. He supported the 1910 Mann-Elkins Act, which gave the Interstate Commerce Commission authority to regulate rates in the telephone, telegraph, cable, and radio industries. And Taft's justice department launched ninety suits (compared to Roosevelt's forty-four) against corporate combinations that it believed to be operating in "restraint of trade": including highly publicized cases against Standard Oil and American Tobacco, in which the government's position was upheld in landmark Supreme Court decisions. To Roosevelt, such actions were troubling in themselves, for he believed that regulating the trusts was a better course than trying to break them up. But what outraged him was the announcement on October 27, 1911, that the administration was filing an antitrust suit against the United States Steel Corporation.

United States Steel — a vast conglomerate created in 1901 by J. P. Morgan, which combined Andrew Carnegie's great steel empire with a number of other companies — was the largest corporation in the world, controlling over two thirds of the nation's steelmaking capacity. In 1907, during a financial panic, Morgan had arranged for U.S. Steel to purchase the Tennessee Coal and Iron Company; it was, he claimed, the key to an elaborate plan to shore up shaky banks. Roosevelt had assured him at the time that the administration would not interfere by filing an antitrust suit. But in its 1911 announcement, the Taft administration cited the Tennessee Coal and Iron acquisition in its suit — clearly implying, many believed, that Roosevelt had acted improperly. The episode led to Roosevelt launching open and direct attacks on Taft for the first time.

By now, a year before the next presidential election, Taft himself was virtually resigned to defeat. As early as mid-1910, he was acknowledging his failure as a political leader. "I have had a hard time," he wrote Roosevelt in one of their last cordial exchanges. "I have been conscientiously trying to carry out your policies but my method of doing so has not worked smoothly." By September 1911, he was writing privately to his family that "I am not very happy with this renomination and reelection business." And by the time of his party's convention in 1912, he was confiding in friends that "I have long been making plans for my [postpresidential] future." In fact, the 1912 election was as cruel an experience as any incumbent president has ever encountered. Roosevelt handily defeated Taft in every primary election. But the majority of the delegates were still in the control of party leaders, who brushed aside the Roosevelt challenges and renominated the president. Roosevelt and his followers walked out of the Republican Convention, created their own organization (the Progressive party), and nominated Roosevelt for president. At about the same time, the Democrats nominated one of their own most progressive leaders, Governor Woodrow Wilson of New Jersey.

Taft recognized that he had no hope of reelection and hardly participated in the campaign at all. "There are so many people who don't like me," this most amiable of public men explained miserably to a friend. He finished third, winning only 23 percent of the popular vote, the lowest percentage of any incumbent — and indeed any major-party candidate — in American history. He won only 8 electoral votes, from Vermont and Utah. In March 1913, he accompanied Woodrow Wilson to his Inauguration and then moved to New Haven, where he became a professor of constitutional law at Yale.

But Taft's retirement from public life was never complete and, in the end, short-lived. During World War I, he served as one of the chairs of the War Labor Board. And in 1921, the newly elected Republican president, Warren G. Harding, appointed him chief justice of the Supreme Court, a position he held until a few weeks before his death in 1930 and that provided him at last with the contentment and satisfaction that he had never achieved in the White House. Taft was an effective administrator of the nation's court system and did much to streamline the processing of litigation. He also usually helped sustain the Court's traditional conservatism on economic matters and continued the movement within the judiciary — symbolized by the Lochner decision of 1905 — of blocking any public policies that infringed on what the justices considered the basic right of freedom of contract, which included the right of individual employers to make agreements with individual employees without interference from the law or from unions. And yet even on the Court, Taft revealed on occasion the remnants of the progressive commitments that had first brought him to national political prominence. In 1918, for example, he wrote a dissent in a notable decision — *Adkins v. Children's Hospital* — in which the Court struck down a law mandating a minimum wage for female workers in the District of Columbia. Taft argued that the very low pay most women received was dangerous to their own and their children's health and was, therefore, a proper issue for the law to address.

"Politics makes me sick," Taft wrote frequently to friends and relatives during his troubled years in the White House. And a distaste for politics goes far to explain his failure as president. He was a man of considerable intelligence, talent, and integrity. But he was temperamentally unsuited for the intensely political character of the presidency; and his uneasiness with the demands of the office seemed to evoke all his worst qualities — his tendency to procrastinate, his excessive legalism, even a kind of physical and intellectual laziness.

But Taft's failure was not solely a result of his own shortcomings. He was elected to office in the shadow of one of the most remarkable and beloved figures of his age. It would have been difficult for any leader to avoid unflattering comparisons with Theodore Roosevelt. And he was elected at a time when the schisms within the Republican party, and within national politics at large, were bursting to the surface, creating an unusually adversarial and ideologically charged political climate. Taft was a man who prized stability and wished nothing more than to preside over a tranquil, prosperous age. Instead, he found himself caught up in an impassioned struggle over the future of government and the character of American economic life — a struggle he inadvertently intensified by trying to remain true to the principles that had guided him throughout his career, and one that finally rendered him, even as president, virtually irrelevant to the politics of his time.

— ALAN BRINKLEY
Columbia University

Woodrow Wilson

1913–1921

b. December 28, 1856
d. February 3, 1924

W OODROW WILSON is neither very fondly remembered nor well understood by most Americans. Yet he occupies a secure position within the exclusive pantheon of great presidents. The domestic legislation that he signed into law and the new directions that he charted in foreign policy during the First World War shaped the politics and diplomacy of the United States throughout the twentieth century. Among all presidents, only Franklin Roosevelt and Lyndon Johnson have matched Wilson's record in enacting a significant legislative program. (In appraising Johnson's Great Society in 1965, the political commentator Tom Wicker suggested that the early New Deal was an insufficient measure; rather, Wicker observed, one had "to go all the way back to Woodrow Wilson's first year to find a congressional session of equal importance.") As for the realm in which he carved out his most monumental legacy, no chief executive has ever communicated more effectively to the people of the world the ideals of democracy, or set in motion a more original idea for the prevention of war than the twenty-eighth president. Writing of the Covenant of the League of Nations, General Jan Smuts credited Wilson with having actuated "one of the great creative documents in human history." Nevertheless, few presidents, after accomplishing so much, experienced a reversal of fortunes as tragic as the one he met near the end of his term. As a consequence, he remains one of the most controversial presidents.

Thomas Woodrow Wilson was born to Jessie Woodrow Wilson and Joseph Ruggles Wilson on the night of December 28, 1856, in Staunton, Virginia. Both parents were the children of Presbyterian ministers (the path that Joseph also chose) and transplanted Northerners sympathetic to the Southern cause. In 1858 they moved to Augusta, Georgia. Their son's earliest memory was of hearing, at the age of four, that Lincoln had been elected president and that there was to be war. Indeed, the Civil War and Reconstruction overshadowed Wilson's childhood. As a seven-year-old, he witnessed the solemn march of Confederate troops on their way to engage Sherman's army. He watched wounded soldiers die inside his father's church, which had become a makeshift hospital. Later he would see Jefferson Davis paraded under Union guard through the city streets.

Even so, Tommy, as he was then called, had a fairly normal upbringing in a religious household. As a youngster he played baseball. Football became a lifelong devotion. (In adulthood he played golf and, along with William Howard Taft, helped to establish it as the favored recreation of presidents.) Although dyslexia kept him from learning his letters until his ninth year, he was an unusually bright boy; the Reverend Wilson attended to his education until he went away to school. Wilson spent one year at Davidson College in nearby North Carolina and then transferred to the College of New Jersey (renamed Princeton University in 1896), where he excelled in debate. After graduating in 1879, he began pursuing a law degree at the University of Virginia, but soon found the legal profession unsatisfying. His father's vocation appealed to him even less. Instead, Wilson was drawn to a different sort of pulpit — politics. Since adolescence he had longed to be "a leader of men." He once inscribed calling cards with the words "Thomas Woodrow Wilson, Senator From Virginia." "I should be complete," he wrote, "if I could read the experiences of the past into the practical life of the men of to-day and so communicate the thought to the minds of the great mass of the people as to impel them to great political achievement." Believing that to be a leader was to be an educator, he enrolled in the Ph.D. program at the Johns Hopkins University — to study history and political science and thence to become a scholar and teacher.

Wilson's main intellectual interest was the perfection of democratic government. His first book, *Congressional Government*, argued that the American system was woefully inefficient compared to its British parliamentary cousin. Published in 1885, the treatise launched him on a distinguished academic career. He taught at Bryn Mawr College from 1885 to 1888 and then for two years at Wesleyan University until Princeton offered him a professorship in 1890. A steady flow of articles and books, including *The State* (1889) and the five-volume *History of the American People* (1902), as well as a reputation as a brilliant lecturer, led to the presidency of his beloved alma mater in 1902.

"I feel like a new prime minister," Wilson said as he readied for the assignment that would make him the country's most famous university president. Quickly he set about recruiting renowned scholars, revising the curriculum, and expanding the physical plant. He also introduced the "preceptorial" method of instruction (small discussion groups held in conjunction with lectures). But bitter disappointments accompanied the brilliant successes. In 1907, he was resoundingly defeated in an attempt to abolish Princeton's exclusive undergraduate eating clubs and in another highly publicized struggle, in 1909 and 1910, over the location and control of the new Graduate College. Though he now contemplated resigning, Wilson's innovations nonetheless had transformed the genteel college for wealthy young men into a place of higher learning rivaled by few others in the world.

Meanwhile, his prestige as a commentator on national affairs was flourishing. On the public platform his style never failed to impress. One journalist, despite overlooking the trademark pince-nez eyeglasses, nicely captured Wilson's mien: "There is a look of seasoned scholarliness about him which accents his affability and charm. He is of medium height, inclined to be spare of frame, and to show in his long, lean face the features of his Virginia Presbyterian antecedents." According to another news report, "his diction was beautiful in its simplicity. He spoke clearly and distinctly, in a rather loud tone, and with perfect enunciation." Given his background, it was not surprising that New Jersey's Democratic political bosses thought he would make a fine gubernatorial candidate. When, in 1910, they proffered the nomination, he accepted with alacrity.

In all of this Wilson was singularly blessed in his personal life. In the spring of 1883 he had met Ellen Louise Axson of Rome, Georgia. Ellen was an intelligent young woman who, like her future husband, was reared in Presbyterian manses. Well versed in philosophy and literature, she was a talented artist, having studied at New York's famed Art Students' League. (Landscapes, painted in the American impressionist style, were her specialty.) At least for Woodrow, it was love at first sight. Three months later they met again, by chance, in a hotel lobby, whereupon he took her in his arms, kissed her, and confessed his love. They were married in 1885. By 1889 they had three daughters — Margaret, Jessie, and Eleanor.

The story of Ellen and Woodrow is one of the most romantic in presidential history, and it is chronicled in the 2,500 letters they exchanged during their twenty-nine-year union. "My love for you released my real personality, and I can never express it perfectly either in act or word away from you," he wrote in a typical letter after fifteen years of marriage. "I want you so much," she wrote while they resided in the White House. "Oh how I *adore* you! I am perfectly sure that you are the greatest, most wonderful, most lovable man who *ever* lived!"

Yet their marriage was not one long, uninterrupted idyll. In 1907, on vacation in Bermuda, Wilson encountered Mary Allen Hulbert Peck, a sophisticated, worldly woman noted for her diverting dinner parties. They soon fell in love. (In subsequent years, Mrs. Peck would contemptuously rebuff Wilson's political enemies who offered to buy the letters he had written to her.) Whether the two were ever physically intimate remains conjectural; but the relationship caused great pain to everyone involved, especially Ellen, whom Wilson still deeply loved. Filled with remorse, he eventually ended the affair with his "Dearest Friend" and Ellen forgave him. Until her death in 1914, in good times and bad, she remained the most important influence on her husband's life and career. As ever, he depended on her not only for love and understanding, but also for counsel on all of his endeavors, including politics. When the New Jersey governorship beckoned, Ellen was as excited about the prospects as he.

Wilson launched out onto the sea of politics during an era of growing discontent with the status quo of industrial America. In the final two decades of the nineteenth century, seven hundred thousand workers were killed, for lack of safety standards, in on-the-job accidents. Child labor, from the sweatshop to the coal mine, was pervasive. Farmers were in revolt over plummeting commodity prices and ruinous transportation costs. Corruption plagued municipal and state government. Massive immigration strained social services to the breaking point. At the same time, so-called captains of industry arrogated to themselves enormous economic power. The efforts of many progressive reformers, including former president Theodore Roosevelt, had ameliorated some of these problems, but there was still much to be done. Because the two major parties had been slow to respond, others rose to challenge them. By the 1910s, for example, the Socialist party and its quadrennial standard-bearer, Eugene Debs, had attained respectability. Hundreds of socialists held public office across the nation and as many as three million Americans read socialist newspapers on a weekly basis. As the historian Frederick Jackson Turner observed, the age of reform was "also the age of socialistic inquiry."

Into this welter entered the "Princeton Schoolmaster." Although he once had leaned toward conservatism, the burgeoning progressive movement had caused him to rethink his views. Modern industrial organization had so distorted competition that the rich and the strong virtually tyrannized the poor and the weak, he had concluded. "In the face of such circumstance," he asked, "must not government lay aside all timid scruple and boldly make itself an agency for social reform and political control?" Upon winning the governorship, Wilson turned against the machine that had promoted him and pushed through the legislature a package of timely reforms — a workmen's compensation act, laws to regulate public utilities and railroads, the direct primary, and corrupt practices legislation. Suddenly, the New Jersey governor had become the new hope of national progressivism. With the Republicans split wide open between the incumbent Taft and a resurgent Roosevelt, the Democrats scented victory in 1912. At their convention in June, on the forty-sixth ballot, they nominated Wilson for president.

The field of candidates that season — Wilson, Roosevelt, Taft, and Debs — was arguably the most impressive of the twentieth century. Roosevelt and his newly formed Progressive party set the terms of the debate. Their platform, the New Nationalism, pledged to make the country's vast corporate structure accountable to the public by using federal power both to regulate monopolies (instead of doing away with them) and to safeguard the rights of working people. Wilson responded with the New Freedom, a program to rehabilitate democracy not by regulating, but by restoring competition. This task could be accomplished, he argued, through new antitrust legislation (a remedy that Roosevelt had abandoned) as well as by lowering tariffs and creating a more elastic currency system. Exactly how Wilson felt about federal legislation to protect workers, however, was not altogether clear. But that had little effect on the outcome of the election. He carried forty states, with 42 percent of the popular vote to Roosevelt's 27 percent, Taft's 23 percent, and Debs's 6 percent.

Wilson's inauguration opened a new chapter in the modern presidency. He became the first chief executive since John Adams to appear before Congress in person, and did so on more occasions — five times in 1914 alone — than any president to this day. He was also the first to hold regular press conferences and the last to write all of his own speeches. Wilson revitalized the notion of party government as well, as if he were prime minister, by sitting in on committee meetings on Capitol Hill and driving the Democrats, who now enjoyed majorities in both houses of Congress, toward their legislative goals.

His methods resulted in a historic breakthrough in October 1913, when Congress enacted the Underwood-Simmons bill, the first downward revision of tariffs since the Civil War. This, the New Freedom's initial assault on the special interests, was followed in December by a complete restructuring of the currency system. The Federal Reserve Act helped to curb Wall Street's domination over the nation's finances, balanced the needs of small and large banks by establishing twelve regional Federal Reserve banks, and, through the creation of the Federal Reserve Board, imposed federal supervision over the whole complex. The act is generally regarded as Wilson's single greatest legislative accomplishment. In grappling with the problem of the trusts, however, Wilson effected a merger of the New Freedom and the New Nationalism. In the fall of 1914, he signed the Clayton Act, an antitrust law of modest scope; then, upon the urging of progressives, he endorsed the proposition of a regulatory agency empowered to exercise continuous governmental supervisory authority over big business. The results, the Federal Trade Commission Act of September 1914, amounted to the Rooseveltian solution.

Wilson's leadership in the enactment of these measures elicited handsome praise from every corner. In two years he had achieved more than most presidents do in two terms. Yet the New Freedom had some serious limitations. For example, the president allowed the introduction of racial segregation in the Treasury and Post Office departments, "one of the worst blots on the administration's record," his most distinguished biographer, Arthur S. Link, has written. Only under extreme pressure from African American leaders and the northern liberal press did Wilson countermand his subordinates in late 1914. Nor had he thrown his support behind social justice legislation to shield men, women, and children from the harsher consequences of the nation's industrial achievements — though in 1916 he would far exceed the expectations of progressives devoted to that cause.

Wilson occupied the White House not only during an age of reform at home, but also at the dawning of a new epoch in world history characterized by profound revolutionary movements in China, Mexico, and Russia. These revolutions were informed by the socialist critique of industrial capitalism and imperialism. Wilson tended to see them as the repercussion of political and economic exploitation of great masses of people by reactionary governments determined to thwart orderly change at any cost. At once an activist and idealist in foreign affairs, he believed that America should serve as nothing less than "the light which will shine unto all generations and guide the feet of mankind to the goal of justice and liberty and peace."

To that immodest end, in 1913, he extended diplomatic recognition to the new republic of China (the first world leader to do so) and then pulled the United States out of a multinational banking consortium that was impairing China's sovereignty. In confronting the Mexican Revolution, the president undertook a more hazardous mission, with mixed results at best. When the leader of the new liberal government was executed by a military dictator, Wilson could have easily recognized the counterrevolutionary regime, as the European powers had done, thereby safeguarding American business interests. Instead, he pursued a measured military interventionist policy that succeeded, in 1914, in bringing down the reactionary junta. As the revolution resumed course, Max Eastman, a prominent American leftist, commented that Wilson's purpose was "to let the Mexican people govern, or not govern, themselves." Yet, while he had resisted constant demands (mainly from Republicans) to impose a protectorate, the progressive forces in Mexico deeply resented Wilson's meddling, and it nearly led to war in 1916.

Wilson's incontrovertibly Herculean labor, of course, awaited him in Europe. In August 1914, no person living anywhere in the world could imagine a spectacle as violent and complicated as the one into which humanity was about to be plunged. The magnitude of the opening phases of the conflict strained human comprehension. In September, during the first Battle of the Marne, the Allies and the Central Powers together sustained more than a million casualties. By the end of 1914, France alone counted 900,000 dead, wounded, or missing. In 1915, the figure for Germany was 850,000 and for Great Britain 313,000. By the time the czar was overthrown in March 1917, Russia had suffered some 3.6 million dead or otherwise incapacitated. In all, at least 10 million people would go to their deaths as a result of "the Great War."

To Americans during the otherwise quiet summer of 1914, the titanic struggle came "as lightning out of a clear sky," as one editorialist wrote. In the final days of July, however, the cause of Wilson's most intense anxiety was his wife's failing condition. (In March the First Lady had been diagnosed with tuberculosis of the kidneys.) Meanwhile, on August 1, Germany declared war on Russia, and on France two days later. At midnight on August 4, Great Britain declared war on Germany. Wilson thereupon issued a proclamation of neutrality. He spent the following day at Ellen's bedside. She died on August 6. "Oh my God," he whispered, "what am I to do?"

The beginning of the First World War marked the beginning of the historically crucial period of Wilson's life. In the immediate circumstances, the administration found itself beset by innumerable conundrums for which there existed few guiding precedents. The earliest problems centered on Great Britain's naval blockade of northern Europe and its intermittent seizures of American merchant ships suspected of carrying contraband to the Central Powers. By the summer of 1916, Anglo-American relations had fallen to their lowest ebb since the British burned Washington, D.C., in 1814.

Allied economic warfare, though severe, was soon eclipsed by Germany's novel method of retribution — submarine warfare. On May 7, 1915, the Germans virtually forfeited the contest for public opinion when a submarine sank, without warning, the British liner *Lusitania*, drowning 1,198 men, women, and children — civilians all, among them 128 Americans. Even so, the vast majority of Americans expected their president to keep his head and save them from Europe's awful mess. For their part, neither the Allies nor the Central Powers wanted to provoke the world's most powerful neutral to armed retaliation. Each side made well-calculated concessions at critical moments. As crises came and went, Wilson alternately protested and sought ways to accommodate the conduct of both sides, while striving to preserve American neutral rights and public sensibilities.

American neutrality was always a fragile thing. The best way to keep the country out of armed conflict, Wilson reasoned, was to bring about a negotiated settlement between the warring alliances. Twice, in 1915 and 1916, he sent his emissary, Colonel Edward M. House, to Europe for direct (albeit futile) parlays with the belligerent governments. In the meantime, he had begun to sketch out a plan for a postwar peacekeeping organization. It included provisions for settling disputes through the process of arbitration, for eliminating the production of armaments by private enterprise, and for the imposition of collective military and economic sanctions against any nation that attacked another. Wilson was by no means the sole author of the concept of a league of nations. He drew many of his ideas from a new internationalist movement that had come into being in the United States at a fairly early stage in the war. Two divergent, broadly based groups of activists — "progressive internationalists" and "conservative internationalists" — composed this movement.

Feminists, liberal reformers, and socialists filled the ranks of the progressive internationalists. Their organizations — for instance, the Woman's Peace Party, led by Jane Addams, and the American Union Against Militarism — were the impassioned proponents of both the so-called New Diplomacy and an Americanized version of social democracy. For them, peace was indispensable to the survival of the labor movement, to their campaigns on behalf of women's rights and the abolition of child labor, and to social justice in general. If the war in Europe raged on indefinitely, they believed, then the United States could not help but get sucked into it and reform would die. Their manifestos called for an immediate armistice, disarmament, an end to colonialism, and a "Concert of Nations" to supersede the old balance-of-power system.

The program of the conservative internationalists was developed mainly by Republicans of the League to Enforce Peace (LEP), founded in 1915 and led by former president Taft. Conservative internationalists, too, advocated a world parliament and, in principle, endorsed arbitration and collective security; but they believed as well that the United States should expand its army and navy, resist any diminution of sovereignty, and reserve the right to exercise force independently. They did not seek a negotiated end to the war; nor were disarmament and self-determination among their concerns. Wilson was cordial with the LEP, but his sympathies lay decidedly with its progressive counterparts. The differences between progressive and conservative internationalists, in domestic and foreign policy alike, would grow more conspicuous as the next electoral cycle drew near.

By that time, after long months of unrelieved depression, Wilson's private life had taken a happy turn. In the spring of 1915 he had met and fallen in love with a stylish, Virginia-born, forty-three-year-old widow, Edith Bolling Galt. Around the White House the president could be heard singing "Oh, You Beautiful Doll!" Perhaps inevitably their courtship became the subject of gossip and a minor political issue (in part because of a revived whispering campaign about Mrs. Peck), as well as the object of cheerful tidings. From their wedding day in December 1915, Edith would remain her husband's closest adviser, passionately dedicated to what she considered his best interests.

However much remarriage boosted his confidence and made him whole again, Wilson would not have been continued in office — nor could he have made a truly plausible case for the sort of peace he envisioned — if he had not been willing to move to the left of center in American politics. For a generation, the Democrats had been the minority party; they had prevailed in 1912 only because of Republican feuding. By 1916, Theodore Roosevelt had reconciled with the G.O.P.'s conservative chieftains, and it appeared that Wilson was destined to be a one-term president. Then, in January 1916, he appointed Louis D. Brandeis, the "People's Lawyer," to the Supreme Court. That summer he secured passage of the Keating-Owen bill, which imposed restrictions on child labor; and the Adamson Act, which established the eight-hour day for railroad workers. Finally, he signed a federal workmen's compensation measure and a revenue act heavily weighted against corporations and the wealthy in order to pay for a military preparedness program. Progressive internationalists delighted in Wilson's second impressive burst of reform and worked hard for his reelection, while conservative internationalists lined up as his harshest detractors.

But there was more. As the epiphenomenon of his advanced progressivism, the president called for American membership in a future league of nations, a theme that complemented the campaign slogan "He Kept Us Out of War!" As things turned out, he had deprived his opponent, Charles Evans Hughes, of any completely serviceable issue. From child labor to the European war, Wilson had made the causes of reform, peace, and internationalism his own. On election day, large numbers of former Bull Moose Progressives and Socialists swelled the normal Democratic vote for president. By a narrow margin Wilson managed a stunning upset over Hughes. Yet, therein, the league had already begun to take on a partisan complexion.

Reconfirmed by the electorate, Wilson now decided on a bold stratagem for ending the war, based on progressive internationalist principles. On January 22, 1917, he went before the Senate and called for "peace without victory." In this manifesto, the president set forth a penetrating critique of European imperialism, militarism, and balance-of-power politics — the root causes of the war, he said. In their stead, he held out the promise of a new world order sustained by arbitration, a dramatic reduction of armaments, freedom of the seas, self-determination, and security against aggression. The chief instrumentality of this sweeping program was to be the League. Thus the concept of Wilsonianism itself was born.

One week later, Germany announced the resumption of unrestricted submarine warfare against all flags. After three American ships were sunk without warning, and after the "Zimmermann Telegram" exposed a German effort to persuade Mexico to invade Texas, public opinion shifted markedly. By the end of March, Wilson, too, had reluctantly concluded that belligerency had been "thrust upon" the United States. In his address to Congress, on April 2, 1917, he declared, "The world must be made safe for democracy. Its peace must be planted upon the tested foundations of political liberty." Americans would be fighting, then, not for conquest or even in defense of neutral rights, but "for a universal dominion of right by such a concert of free peoples as shall bring peace and safety to all nations and the world itself at last free" — a program attainable, now, apparently only through the crucible of war.

The United States mounted a mobilization effort that bordered on the miraculous. New federal agencies (such as the War Industries Board, the Railroad Administration, the National War Labor Board, and the Food Administration) coordinated virtually every sector of the economy to put America's industrial and agricultural might in harness for military purposes and in order to ship supplies by the convoy to the Allies. To sing the virtues of the cause and to discredit all things German, the government empowered the Committee on Public Information to inaugurate a propaganda campaign of unprecedented proportions. Under the Selective Service Act, the size of the armed forces grew from one hundred thousand to five million. By the summer of 1918 some two million "doughboys," under the command of General John J. Pershing, had arrived in France in time to tip the balance in favor of the Allies.

On October 6, 1918, the German government appealed to Wilson to take steps for the restoration of peace based on his Fourteen Points address of the previous January. This celebrated restatement of progressive war aims (with the League of Nations as its capstone) was the ideological cement that had held the Allied coalition together after the Bolsheviks had seized power in Russia in November 1917, pulled their ravaged nation out of the war with Germany, and then challenged all the belligerents to repudiate plans for conquest. An armistice was signed on November 11. Preparing to take part personally in the Paris Peace Conference, the president's supreme ambition now seemed on the threshold of accomplishment.

Yet the American political calendar — specifically, the congressional elections of November 1918 — greatly compounded his task. Against the Democrats and the Wilsonian peace plan the Republicans raised a fiercely partisan, ultraconservative campaign, even as diplomats hammered out the armistice. Wilson, most historians maintain, committed the worst blunder of his presidency by responding with an appeal to the public to return a Democratic Congress. When the Republicans captured majorities in both houses, they claimed that he had been repudiated. They also gained control over important committees, including the Senate Foreign Relations Committee, which would be chaired by Wilson's arch antagonist, Henry Cabot Lodge.

The forces of reaction suffused the impending League controversy in other ways. During the war Wilson had sorely neglected to nurture the left-of-center coalition that had elected him to a second term; if it had remained intact, he might have prevailed in the midterm elections and thus been able to secure American leadership in a peacekeeping organization intended to serve progressive purposes. But he had begun to lose his grip on his former base of support as a tidal wave of anti-German hysteria and the superpatriotism known as "One Hundred Percent Americanism" swept the country. Acts of political repression and violence, sanctioned by federal legislation, were committed practically everywhere against not only German Americans, but also pacifists and radicals. (To cite but one example, Eugene Debs, aged sixty-two and in failing health, was sentenced to ten years' imprisonment for making a speech against the war.) Because he acquiesced in the suppression of civil liberties and the radical press, Wilson himself had contributed to the gradual unraveling of his once ascendant coalition. Many progressive internationalists would never forgive him.

In contrast to his troubles at home, Wilson's arrival in France was triumphal. To the war-weary peoples of Europe, the Fourteen Points had acquired the status of sacred text, and "Wilson" was becoming something more than the name of a president. Into the streets of Paris, London, and Rome, millions turned out to hail the "Moses from across the Atlantic." These unprecedented demonstrations strengthened the hand of the "Savior of Humanity" in the early phases of the peace conference and helped to ensure the inclusion of the Covenant of the League as an integral part of the peace treaty.

But Wilson still paid a heavy price. His fellow peacemakers — David Lloyd George, Georges Clemenceau, and Vittorio Orlando — held grave doubts about the New Diplomacy. ("God gave us the Ten Commandments, and we broke them," Clemenceau quipped. "Wilson gives us the Fourteen Points. We shall see.") Fully aware of the arithmetic in the Senate, the statesmen of Europe leveraged their acceptance of the Covenant to gain concessions on other contentious issues. During six months of acrimonious negotiations, the president was able to moderate some of the Allies' more extreme territorial demands against Germany; but, showing the strain of his heavy work and finding himself in a minority of one, he was just as often compelled to compromise his principles. Perhaps most egregiously, he permitted the Allies to impose upon Germany a huge reparations burden and, on top of everything else, a "war guilt" clause — saddling it with the moral responsibility for allegedly having started the war. Because he alone had promised so much, Wilson would bear the main burden of criticism for the punitive qualities of the Treaty of Versailles. His one hope was that eventually the League would be able to rectify the injustices contained in the treaty itself.

Had the question been put to a national referendum upon his return in July 1919, the United States almost certainly would have joined the League. Why it failed to do so is still debated by historians. To begin, Wilson had already lost the active support of most left-wing progressives and socialists. Many mainstream liberals, too, turned away after reading the treaty. They felt that, regardless of his motives, he had forsaken the Fourteen Points and conceded too much in the territorial compromises — in short, that his precious League would be bound to uphold an unjust peace.

As for the Senate, sheer partisanship motivated much of the opposition. (Senator Lodge's chief goal, frankly, was to deny Wilson his crowning glory.) At the same time, many of the objections were grounded in ideological conviction. Like Lodge, the majority of Republicans were conservative internationalists; as such, they were convinced that Wilson had consigned too many vital national interests to the will of an international authority — even though he had agreed to provide for withdrawal from the League and to exempt the Monroe Doctrine from its jurisdiction.

Then there were the "Irreconcilables," a small but sturdy knot of fifteen senators who flat-out opposed the League in any form. Not all of the Irreconcilables were partisans or reactionaries; some of them based their opposition on grounds similar to those of progressive and socialist critics. Indeed, only a few of Wilson's adversaries were isolationists, strictly speaking. "Internationalism has come," one Democratic senator aptly declared. "And we must choose what form the internationalism is to take." That, fundamentally, was what the great debate was all about.

The Foreign Relations Committee, under Senator Lodge's skillful leadership and packed with Wilson's enemies, honed the grievances down to a total of (curiously) fourteen. The most significant reservation contradicted the controversial Article 10: "The United States assumes no obligation to preserve the territorial integrity or political independence of any country . . . unless in any particular case Congress . . . [shall] so provide." But the Republicans' worries did not end there. They were concerned about membership in the International Labor Organization; about the League's provisions for disarmament; and about the important corollary to collective security — that is, the restrictions that Article 10, in tandem with arbitration, imposed against the unilateral exercise of force, thus possibly preventing the United States from exerting its power without consulting the League first. (The United States, Wilson earlier had asserted during an exchange with congressional leaders, would "willingly relinquish some of its sovereignty . . . for the good of the world.")

After a series of unproductive meetings with senators called "Mild Reservationists," Wilson came to a momentous decision. Against his doctor's advice and Edith's pleading, he embarked upon a strenuous speaking tour to take his case directly to the American people. For three weeks in September 1919, the president, his physical reserves nearly spent, traveled ten thousand miles by train throughout the Middle and Far West, making some forty speeches to hundreds of thousands of people. In expounding Article 10, he explained his view that military sanctions probably would not come into play very often — in part because arbitration and, especially, disarmament would help to eliminate potential problems from the start. He also spoke to the question of sovereignty as it related to arbitration and the implied hindrance to unilateral action. "The only way in which you can have impartial determinations in this world is by consenting to something you do not want to do" and to refrain, as well, from doing something that you *want* to do. There might be times, he said, "when we lose in court [and] we will take our medicine." The Lodge reservations would "change the entire meaning of the Treaty," Wilson insisted. The United States could not join the League grudgingly, on conditions of its own choosing. If it did not go in as a full-fledged member, he would feel obliged to stand "in mortification and shame" before the boys who went across the seas to fight and say to them, "'You are betrayed. You fought for something that you did not get.'" And there would come, "sometime, in the vengeful providence of God, another struggle in which, not a few hundred thousand fine men from America will have to die, but as many millions as are necessary to accomplish the final freedom of the peoples of the world."

As the throngs grew larger and the cheers louder, Wilson looked more haggard and worn out at the end of each day. At last, his doctor called a halt to the tour and rushed him back to Washington. On October 2, he suffered a stroke that nearly killed him and paralyzed his whole left side. From that point on Wilson was but a frail husk of his former self, a tragic recluse in the White House shielded by his wife and doctor. Mrs. Wilson would serve as the arbiter of what and whom the president saw. Contrary to a popular invention, however, she did not run the executive branch of the government; that task was performed, as always, by the various departmental heads.

Between November 1919 and March 1920, the Senate voted on the treaty three times. But, whether on a motion to approve it unconditionally or with the fourteen reservations attached, the tally always fell short of a two-thirds majority. In November 1920, Warren G. Harding, the Republican presidential candidate, won a landslide victory over the Democrat, James M. Cox. The Republicans were pleased to interpret the returns as the "great and solemn referendum" that Wilson had said he had wanted for his Covenant. "So far as the United States is concerned,"

Lodge now affirmed, "that League is dead." A few weeks later, Wilson was awarded the Nobel Peace Prize.

In surveying the ruins, many historians have cited the president's stroke as the primary ingredient of the debacle. They argue that a healthy Wilson would have grasped the situation and found some middle ground on the question of reservations. (A recent study of John Milton Cooper, Jr., has established beyond a reasonable doubt that Wilson's affliction had seriously impaired his political judgment.) Other historians have contended that his refusal to compromise was consistent with his personality and behavior throughout his life, that his psychological make-up would never have permitted him to yield to the Republicans (especially to Lodge), regardless of the state of his health. Still another school of thought maintains that these factors alone did not cast the die, for they do not take into account the untoward domestic political conditions that had taken shape long before, or the ideological gulf that had always separated progressive and conservative internationalism. Thus, although Wilson's collapse surely worsened the gridlock, the Senate's rejection had roots in events going back to the birth of the American internationalist movement and the forging of Wilson's victory coalition in 1916. The dissolution of progressive internationalism — a result of wartime repression and Wilson's failure to rekindle the coalition as the parliamentary battle was getting under way — sealed the fate of a *Wilsonian* League.

That the president should have resigned there can be no doubt. (Had the Twenty-fifth Amendment, which pertains to presidential disability and succession, been in effect, he would have had no choice.) Yet if the stroke made him less amenable to compromise, he was no less alert to the reality that the reservations were designed to establish a strict construction of the Covenant wholly different from what he had intended. "The imperialist wants no League of Nations," he had written just before the Senate rendered its final verdict, "but if . . . there is to be one, he is interested to secure one suited for his own purposes." International security involved the acceptance of both constraints and obligations, or "a renunciation of wrong-doing on the part of powerful nations," including the United States. For Wilson, Article 10 constituted "the only bulwark against the forces of imperialism and reaction." In the end, he preferred that the United States forgo membership rather than join a Lodgian league, a conservative league that would "venture to take part in reviving the old order." To the very end, the "stern covenanter" possessed the soul of a progressive internationalist.

Whatever the central cause of his historic failure, Wilson's conservative and partisan adversaries believed that his was a dangerously radical vision, a new world order alien to their own understanding of how the world worked. His severest critics among progressives believed he had not done enough to resist the forces of reaction — either in America or at the peace conference. "What more could I have done?" he asked the historian William E. Dodd in a cry of anguish shortly before he left the White House to live out his remaining three years in quiet retirement. As Dodd observed, it had all been "one long wilderness of despair and betrayal, even by good men." Perhaps more perceptively than anyone, the journalist Ray Stannard Baker commented on Wilson's fate: "He can escape no responsibility & must go to his punishment not only for his own mistakes and weaknesses of temperament but for the greed and selfishness of the world."

— THOMAS J. KNOCK
Southern Methodist University

Wilsonianism

"It would be the irony of fate if my administration had to deal chiefly with foreign affairs," Wilson remarked after winning the presidency, following a campaign that focused exclusively on domestic issues. Yet he went on to craft a foreign policy that, in many ways, shaped the course of the twentieth century.

In the 1920s and 1930s assessments of Wilson's legacy were unfavorable. Many Americans came to believe that intervention in World War I had been engineered by profit-hungry munitions manufacturers and British propagandists, and that the world had hardly been "made safe for democracy." Criticisms of the Treaty of Versailles multiplied as another European conflict loomed.

After 1941, however, Wilson's reputation soared to new heights. A new wisdom held that World War II might have been averted if America had joined the League of Nations. Scholarly works now reinterpreted his quest as a struggle with prophetic overtones. Even Hollywood, in Darryl Zanuck's Oscar-winning epic, *Wilson* (1944), helped prepare the way for an American commitment to postwar internationalism. In 1945, President Truman characterized the United Nations Charter as Wilson's vindication.

Since then, perspectives on "Wilsonianism" have varied widely. In the 1950s, foreign policy practitioners and historians of the "Realist" school condemned Wilson's ideas as unsound for a bipolar world in the grip of the cold war; whereas, so-called New Left historians of the 1960s contended that "Wilsonian values" actually had "their complete triumph in the bipartisan cold war consensus." In the 1990s, one historian argued that neither the Realists nor the New Left fully understood Wilson — that cold war "containment" and "globalism" did not necessarily constitute internationalism, and that America's hostility toward the United Nations was fundamentally anti-Wilsonian.

As the cold war ended, world leaders, including all presidents since the first President Bush, strove to envision "a new world order." Yet, as the twenty-first century dawned, the main tenets of authentic Wilsonianism — adherence to international law and the concept of multilateralism, disarmament, the notion of a community of nations, and self-determination — still awaited sorting out by the makers of American foreign policy.

Warren G. Harding

1921–1923

b. November 2, 1865
d. August 2, 1923

T HE PRESIDENCY OF WARREN G. HARDING began in medi-
ocrity and ended in corruption. But despite Harding's own
brief and largely undistinguished tenure, his era signaled the
emergence of a new American century. These years saw the
beginning of new social policies (in immigration especially) and the rise
of a modern culture for which the Harding administration and its politics were as often the foil as they were the illustration. The years from
1921 to 1923 were also an economic and diplomatic entr'acte, a time
when serious economic instability and depression gave way to the material prosperity for which the 1920s is best remembered, and America's
uneasy posture as an international power was first displayed.

Almost any politician who succeeded Woodrow Wilson in 1921 would
seem small to the historian, in the shadow of the man who had helped
to define progressivism at home and internationalism on the world
stage. But it was precisely in this Wilsonian context, and maybe because
of his limited experience in politics, that Warren Harding had distinct
political advantages at the time. Wilson's activism and strenuous virtue
had become overwrought and had exhausted the optimism of the prewar world, an exhaustion visible in both the debacle of the postwar
peace treaty and in Wilson's own physical incapacity. Unlike Wilson's
crumbling physique and ghostlike absence at the end of his second term,
Harding had a robust figure (which he dressed with great care), a dark
handsome face, and a resonant voice, all of which made him seem presidential. And he had an agreeable, friendly personality, which contrasted
with Wilson's stern frostiness. Genial, warm, and easygoing, Harding
provided an excellent antidote to the later years of Wilson's era, which

included the war and the failed campaign for the League of Nations. Harding's personal and physical characteristics played well with the public at large, and also within the Republican party, which had its own recent experiences with activism and schism surrounding the charismatic figure of Theodore Roosevelt. Harry Daugherty, Harding's manager and the political wheeler-dealer who "arranged" his nomination after the two front-runners deadlocked at the 1920 Republican convention in Chicago, told him: "You're the best balanced man I know in public life today." As a result, the convention, which Daugherty, in anticipation of the deadlock, had carefully primed to turn to the obscure Harding, finally did so enthusiastically early on the sweltering morning of June 12, 1920. And in November, so did the electorate.

Harding won an unprecedented 60.4 percent of the votes in the first election in which women, newly enfranchised by the Nineteenth Amendment (to which Harding had given his early support), participated; and he received 404 electoral votes against the Democratic ticket of fellow Ohioan James M. Cox, and Cox's running mate, Franklin D. Roosevelt. In fact, the election turned out to be a broad Republican landslide. It not only brought the little-known governor of Massachusetts, Calvin Coolidge, to within the proverbial heartbeat of the presidency, but it gave Republicans an enormous margin of 303 to 131 in the House of Representatives and a 24-person majority in the Senate. The campaign had been guided by the theme of a return to "normalcy" (a word Harding has been credited with inventing, but which he actually brought from the margins of the language and made current), which Harding introduced in a speech he gave in Boston in May 1920. "America's present need is not heroics, but healing; not nostrums but normalcy; not revolution but restoration . . . not surgery but serenity." And during the campaign, he explained what he meant by the unusual word: "By 'normalcy' I don't mean the old order, but a regular steady order of things. I mean normal procedure, the natural way, without excess." Whatever its long-term rhetorical value, the word and what it suggested hit the right chord with the voters.

The man who struck that note had spent four years in the Senate (1915–1919) as part of the standpat segment of the Republican majority, but had made no special mark. More importantly, he had made no real enemies. In fact, Harding's natural milieu was not Washington, but the small-town world of Ohio which the writer Sherwood Anderson was in the process of surgically dissecting in his stories and against which the sophisticated world of literature was to turn its venom in the 1920s. It was a world of middlebrow ambitions and ambiguous respectability.

Harding had been born into that world on November 2, 1865, in Blooming Grove, and he spent most of the rest of his life in it. Warren, whose biblical middle name, Gamaliel, had been carefully chosen by his deeply religious mother, was the oldest of the six surviving children of Dr. George Tryon Harding, a veterinarian who later practiced homeopathy, and Phoebe E. Dickerson. Raised along strictly Protestant lines, Harding graduated from Ohio Central College and started his working life, before he was eighteen, as a schoolteacher in a town just north of Marion. His most indelible experience was as a successful small-town editor of the *Marion Star*. Harding bought the paper in 1884 and transformed it from an insignificant, small circulation sheet into a major organ of the Republican party. In 1891, Harding married Florence (Flossie) Kling DeWolfe who, despite being five years older and divorced, had set her sights on the popular young editor. Flossie too became a fixture at the newspaper. Together they created an extremely successful business venture as she managed the books and he managed the boys and the news. The world of the newspaper was Harding's most beloved site. When William Allen White, the legendary Republican journalist and political reformer, visited Harding at the White House years later, the president seemed more interested in discussing the minutiae of the day-to-day operations of a newspaper than in the growing crises of his own administration. Even then, Harding wanted to know how an editor could meet the needs of distant subscribers who could add nothing to the prosperity of the local businessmen whose advertising was the paper's strength. Harding's concerns were those of a big man in a small town whose enterprise and influence helped to keep the pot boiling and his friends happy. It was the Ohio small town and its genial good fellowship that both made Harding an attractive (and malleable) candidate and set severe limits on his performance as president when, after the conclusion of World War I, the United States emerged as a great industrial nation perched at the beginning of an age of awesome world power.

Harding brought that peculiar mixture into his administration when he appointed an odd assortment of former political cronies and men of genuine talent and vision commensurate with America's new stature. Most prominently, Harding appointed Herbert Hoover, the recent administrator of European relief, to the Commerce Department; Charles Evans Hughes, a brilliant lawyer and internationalist, as secretary of state; and the millionaire financier Andrew W. Mellon, who understood the worldwide dimensions of America's economic potential, to the Treasury Department. In addition, Harding selected as his secretary of agriculture Henry C. Wallace, a man of substantial talent and a strong advocate for farmers' business interests. This choice demonstrated considerable insight into the special difficulties and needs that farmers were experiencing in a new era. But at the other end of Harding's cabinet were the political insiders whose sole distinction lay in their friendship with the president or their contribution to his unexpected rise to power. Among them were men whose behavior would help to bring ignominy and scandal to the administration, like Albert Fall as secretary of interior, Harry Daugherty as attorney general, and Will Hays as postmaster general. And they, in turn, brought with them to the capital a nest of subordinates and hangers-on whose approach to Washington was not unlike their vision of themselves as the managers of small towns throughout America. These were all arenas for the spoils and pleasures of office.

Recent historians of the 1920s, like Ellis Hawley, have argued that Harding's instincts in his selection of these men suggested more than political acumen; that he was, in fact, participating in a recreation of the state in the context of the economically centralizing experience of the war; that behind his notion of "normalcy" lay not a vision of the local life of nineteenth-century communities, but the drive to organize the twentieth-century economy. And certainly men like Hoover, Mellon, and Wallace understood very well the benefits of large organizations and the potential role of government policy in the coordination of economic activity.

But there is very little evidence that Harding himself was either engaged by these issues or framed his actions and choices accordingly. And while the dismissive stance toward Harding and his administration adopted by an earlier generation of historians such as Arthur Schlesinger, Jr., and William Leuchtenburg that long organized historical memories of the period is probably unwarranted, Harding as president during the early years of the 1920s was more like a juggler than the impresario of a new vision of government. Throughout most of his administration Harding demonstrated his rejection of Wilsonian activism by declining to interfere with the agenda laid down by Congress. In the early 1920s, deliberation in the legislative branch was dominated by repeated disputes among the remnants of the progressive faction, whose chief aim seemed to be to keep America isolated from world affairs, and the often vociferous spokesmen for farmers deeply hurt by the postwar economy. In that context, Harding's natural inclinations to mediation and policy restraint may well have been the most reasonable course.

Despite congressional bickering and the president's tendency to adopt a low political profile, Harding's Washington became the site for a major

American foreign policy initiative, and the source of two important pieces of legislation that emerged early in the twenties to anticipate crucial later laws, the Fordney-McCumber emergency tariff and the Johnson immigration bill. Both laws were approved by Harding and belie the image of this period as the beginning of a Republican ascendancy whose power resulted from a unified commitment exclusively to business interests. Indeed, the legislation revealed how uneven and ambivalent was America's emergence into its new role as international hegemon, and how much the politics of the decade were as often rooted in cultural as in economic concerns.

Americans had rejected Wilson and the League of Nations, but they could not and did not turn their backs on the world. One of the highlights of the Harding years was the successful conclusion of the Naval Disarmament Conference, which convened on November 12, 1922, in Washington. At its opening events Charles Evans Hughes dazzled the conferees by offering to scrap American warships in the Pacific in order to lead the way to ending the dangerous buildup in progress among the great naval powers, especially Japan and Britain. In the end, the Five Power Treaty kept the capital ship tonnage of the United States, Great Britain, Japan, France, and Italy in a rigid ratio that not only calmed the tensions in the Pacific, but signaled America's willingness to assume real leadership on the world stage.

At almost the same time, Washington sent out very different signals about America's new position in the world. Like its more notorious successor in 1930, Smoot-Hawley, the Fordney-McCumber Tariff of 1922 imposed high protectionist rates on a large variety of foreign agricultural and manufactured goods perceived to be in competition with American products, from shelled almonds to wallpaper. It represented, in the view of the *New York Commercial*, "the composite selfishness of the country." The tendency of the bill to reflect the jockeying among economic interests, many of which coexisted within the Republican party, was clear at the time. Less clear was the fact that, in the light of America's new position as the premier creditor nation, such stiff walls to trade were a severe economic miscalculation, even with the bill's apparent flexibility (giving the president the power to adjust rates up and down by as much as 50 percent). The U.S. in lending money to the world needed to become a market for international goods, not, as it now positioned itself, a protector of a host of narrowly defined interests.

The United States was also erecting walls of another kind. While American immigration policy had begun to be selective in the late nineteenth century, excluding first Asian immigrants and then a range of those deemed undesirable on medical, moral, or political grounds, European immigrants had been free to enter the United States unimpeded for almost the entire 150 years of the nation's history. Since about 1890, more and more of those immigrants had come from eastern and southern Europe, and included Russian and Polish Jews, Italians, Greeks, and various Balkan and Slavic peoples. Patriotic societies and labor groups had been pushing for restricting this "new" immigration for decades. The confluence of world war with its superheated patriotism, the succeeding labor unrest and government repression of radicals, and the increased popularity of racialist theories which spilled from the infatuation with the Darwinism of the late nineteenth century finally created the conditions for what the historian John Higham has called the "Tribal Twenties." Congress had several times passed legislation restricting immigration, only to be defeated by a combination of opposition from business leaders (who feared the loss of cheap labor) and presidential vetoes. But in 1921, under the tutelage of its chairman, Albert Johnson — whose mentors included the Nordic supremacist Madison Grant, and Kenneth Roberts, who was writing incendiary articles about the dangers of the new immigration for the *Saturday Evening Post* — the House Committee on Immigration passed a bill that largely suspended immigration from eastern and southern Europe. In the Senate, the bill was altered to establish quotas among different groups of immigrants according to their proportion of the foreign born in the population as of the 1910 census. The final bill included the quota provision and established a maximum of 350,000 immigrants per year (most of whom were to come from northwestern Europe). Harding signed the first bill in American history restricting European immigration based on racialist principles. It passed the Senate by 78 to 1 and served as the model for the more famous and more permanent immigration bill of 1924, which enforced the principles established in 1921 with even greater rigor and bias.

In fact, the immigration legislation was only one piece of a wider movement in the early 1920s toward a rigid commitment to a certain vision of Americanism. The resurgence of the Ku Klux Klan and a series of race riots all pointed to a society still very uneasy with its urban centers and their cosmopolitan populations. The 1920 census had confirmed the shift from a rural to an urban society, but the culture of Americans had long been farm- and small-town-oriented as well as deeply Protestant. Those cultural roots had been symbolically enshrined in the 1919 Prohibition Amendment, which first took effect during Harding's term in office. The Volstead Act, which Congress passed to enforce Prohibition, entrusted various ill-paid officials and revenue agents to administer its provisions. Prohibition cut off urban immigrants from the beer and wine that had long been part of their social lives, while turning much illegal drinking toward harder liquor (sometimes home-stilled but increasingly smuggled in from Canada and the Caribbean). That change helped to give the period one of its symbols, as bathtub gin sold illegally in speakeasies became a form of sophisticated indulgence.

As laws like immigration restriction and the Volstead Act tried to enforce the values and ways of life of an older and more uniform culture, some Americans, often taking their cues from college students, began to identify with the very urban and culturally complex society whose development the laws were poised to impede. Jazz music and jazz dancing, all the rage on college campuses and drawn from the rich African American history of music in the United States, became not only another symbol of the twenties, but an indicator of a very different tempo — urban-centered and based on the newly mixed population that racists abhorred. That tempo also resulted from new technologies, like the radio, and a new orientation to leisure. Radio not only broadcast music and other entertainments, but became a source for the rapid dissemination of news and information. Indeed, the Harding-Cox campaigns had served as the occasion for the public debut of radio broadcasting.

The increasingly complex American population was visible in another innovative form of leisure, among the growing audiences at movie houses, where the silent screen could speak to all regardless of language. It was even more conspicuous among actors like Rudolph Valentino, Pola Negri, and Greta Garbo, whose glamour came from their exotic habits and foreign extraction. Many Americans were scandalized when respectable women began to imitate the mannerisms and dress of screen stars and to adopt the more liberated attitudes that lay behind them. But the period ushered in a decade that brought many changes to women's lives, as they voted, and worked in more conspicuously public places, as well as consumed new entertainments and products. Well aware of the potential voting power of this new constituency, Harding threw his support behind the Sheppard-Towner Act of 1921, which provided for well-baby clinics and milk stations. Harding had been the first president elected with the help of women, and while the Sheppard-Towner legislation had progressive roots, Harding clearly recognized that, contrary to some contemporary opinion, his good looks alone would not assure women's future allegiance.

While younger people often flouted their parents' old-fashioned mores and beliefs through a moderate adoption of the products of modern culture, it was a prominent class of literati and intellectuals who articulated this liberated perspective most flamboyantly. Notable among these was the talented F. Scott Fitzgerald, a self-proclaimed naughty boy who had written *This Side of Paradise* in 1920 to provide Americans with a peephole on their children's misbehavior. In 1922, he confirmed his position as member of the liberated younger generation and helped provide other symbols of its culture in *The Beautiful and the Damned* and *Flappers and Philosophers*. While Fitzgerald and his wife, Zelda, became spokespersons for a new modern attitude, other writers were explicitly attacking what they saw as the old ways and narrow bigotry of small-town America. Especially influential was Sinclair Lewis, who reached millions through popular novels like *Main Street,* in 1920, and *Babbitt,* in 1922. With these books, Lewis gave an indelible form to the characteristics of the American businessman, his ethics, his limited desires, his artificial family life, and his hypocrisies. Although Lewis wrote no overtly political novels during this period, all the social novels of the early twenties had a political edge, because it was the business culture and the politics it supported that seemed basically to define American life. That perspective was made explicit in Harold Stearns's 1922 collection *Civilization in the United States,* in which an array of young intellectuals condemned one American institution after another as hopelessly subverted by business standards and destructive of artistic sensibilities and humane values. This volume became a manifesto of expatriation as Stearns and many of his coauthors left for Europe, which was both more hospitable and cheaper. While the writers rarely mentioned Harding and his administration, the genial president with his business club associations, who advocated "the least possible . . . government interference with business," including large corporations, no doubt symbolized for them everything that was wrong with postwar America.

While the increasingly vocal voices of writers defined the dominance of business as the source of America's inhospitality to matters of the spirit, the success of business in the Harding years was not quite so clear. Not only did farming interests continue to exert considerable influence within the Congress, and legislation like immigration restriction and the Sheppard-Towner Act suggest the multiple sources of political influence, but the general environment of the early twenties presented a very mixed economic picture. The possibilities of American economic might and business leadership had been visible during the war, but the immediate postwar period had produced an uncertain cycle of extreme inflation in 1920 followed by depression and unemployment, and farmers suffered both from the collapse of prices for agricultural products after the war and renewed international competition. That uncertainty followed Harding into office. It was not until well into his short tenure that the coordination between business and government policy finally cleared a path toward the prosperity that would subsequently define the decade. By 1922 real wages in manufacturing began to rise strongly, representing the new industrial vision and technological achievements for which Henry Ford was the most conspicuous spokesman, and their payoff for working Americans.

When early news of various scandals involving the leasing of federal government oil reserves in California and Wyoming to favored and well-paying businessmen began to surface in 1923, the emerging culture of the time, as much as the particular presidency, seemed to be in question. No one initially blamed Harding (though a treacherous Daugherty would later hint at his complicity), but Albert Fall and other members of the cabinet close to the president had certainly profited from these betrayals of the public trust. The most damaging disclosures about Fall's role in leasing the immense Teapot Dome reserve in Wyoming and the Elk Hills reserve in California to Henry Sinclair's Mammoth Oil Company and to Edward Doheny did not come out until after a stern and virtuous Calvin Coolidge had become president. The country would be stunned and riveted by revelations at Senate committee hearings, but the trials did not take place until many years later. When Harding, who collapsed of a heart attack during a tour of western states, died of a thrombosis on August 2, 1923, in a hotel room in San Francisco, Americans sincerely mourned a popular and well-liked president. Indeed, Harding had been not only likable, but very human. He even advised Americans to relax and enjoy life more. When his funeral train carried him back to Ohio, enormous crowds greeted the train along the way. Many people kneeled in prayer, sang hymns, and wept as they waited all through the night to pay their respects. But his reputation was already under a cloud. That cloud would grow darker (and with it rumors of possible suicide or foul play in the president's death) as what became known as the Teapot Dome scandal erupted on front pages that fall. As Harding's successor quietly took his oath of office amid widening news about political misconduct, the role of business in American life was about to become much clearer.

— PAULA S. FASS
University of California at Berkeley

Calvin Coolidge

1923–1929

b. July 4, 1872
d. January 5, 1933

THE PRESIDENCY came to Calvin Coolidge on August 2, 1923. Coolidge was visiting his father, Colonel John Coolidge, at the family farm in Plymouth, Vermont, when President Warren G. Harding unexpectedly died in San Francisco. The vice president's boyhood home had no phone (or electricity), so a Western Union night messenger from nearby Bridgewater had to drive over rutted roads to hand-deliver the shocking news. At 2:47 A.M., in the sparse old farmhouse, Coolidge's taciturn father, a notary public, read out a presidential oath of office before eight hastily gathered witnesses. The slightly built, sandy-haired, taciturn son, who had taken the time to change into a black suit for the unphotographed, pompless occasion, with the family Bible at hand and his beloved wife at his side, repeated the oath and became the thirtieth president of the United States. Thirteen minutes later, at 3:00 A.M., President Coolidge blew out the kerosene lamps, changed back into his night clothes, and went to sleep.

Such phlegmatic indifference in the face of a momentous event reveals much about the character and presidential conduct of Calvin Coolidge. President Coolidge would make his mark in American history by doing his best to make no mark at all.

His presidential reticence has been noted by all of Coolidge's biographers, friend and foe alike. Donald McCoy, the most balanced of Coolidge scholars, subtitled his life of the flinty New Englander *The Quiet President.* William Allen White, the sage newspaperman of Emporia, Kansas, called his still entertaining 1938 Coolidge biography *A Puritan in Babylon.* Arthur M. Schlesinger, Jr., appalled by Coolidge's presidential torpor, was yet more biting in his tellingly titled account of the pre-

New Deal years, *The Crisis of the Old Order:* "His frugality sanctified an age of waste, his simplicity an age of luxury, his taciturnity an age of ballyhoo. He was the moral symbol the times seemed to demand." While no fully developed revisionist account of his presidency has appeared, Coolidge has been championed by the conservative British historian Paul Johnson. At a 1995 Library of Congress symposium on Coolidge, the keynoter Johnson delivered a glowing tribute: "No one in the twentieth century defined more elegantly the limitations of government and the need for individual endeavor, which necessarily involves inequalities to advance human happiness." For Johnson, as for Coolidge, when it comes to government, less is more. And because Coolidge did so little, Johnson instructs us, "his true place in the pantheon of American presidents . . . [is] a high one."

Most historians, though, aware that the United States suffered a stock market crash, a financial panic, and then a long steady slide into the Great Depression just seven months after Coolidge turned over the presidency to Herbert Hoover, rate President Coolidge much lower.

But the voting public loved Calvin Coolidge, even if most historians have not. Before becoming president, he had been city councilman, state representative, mayor, state senator, lieutenant governor, governor, and vice president. The man perhaps best known to posterity for proclaiming that "the chief business of the American people is business," believed that public service was among men's (he would have chosen the masculine noun) chief duties, and that political office was the best means for so serving that public. From 1898, when he became a city councilman, to 1929, when he left the White House, Calvin Coolidge spent most of his adult life in elected political office.

Coolidge served not in quest of reform or social change. His presidential tenure entailed saying less (hence his famous nickname, "Silent Cal"), doing less, and sleeping more than any other twentieth-century president. Still, Coolidge's retreat from decision making, bully-pulpit gesticulating, and public-policy leadership indicated neither indifference nor ineptitude. When stirred, he could and did act. But Coolidge believed that government's roles were few and far between. He believed that the free market economy, not the federal government, should provide Americans with their opportunities and their rewards. In Coolidge's America, property rights guaranteed freedom and liberty. To infringe on those rights through government regulation, legislation, or taxation was, with rare exception, to attack the foundation upon which the nation was built. Calvin Coolidge believed that he served the public by protecting the rights of property.

Coolidge's unpredictable rise to the presidency was due, in large part, to his predictable set of beliefs. His high regard for property rights and low regard for government intervention came at the right time — an era of rapid economic expansion and low unemployment. From 1922 to 1928 industrial production rose 70 percent, the gross national product 40 percent, and per capita income about 30 percent. In the year 1923, when Coolidge became president, more cars were built than had been made in the previous fifteen years. Half the cars sold were Henry Ford's Model Ts, which came in "any color you want as long as it's black." That same year, to great fanfare, U.S. Steel had cut its workday from twelve hours to eight while maintaining the daily pay rate. Over at General Motors, Alfred P. Sloan was inventing modern corporate management, marketing, and distribution, paving the way for GM's rise to a global colossus. Business was booming, and no one boosted business more than Calvin Coolidge. "The man who builds a factory builds a temple, the man who works there worships there," he orated in his oddly stilted, epigrammatic way. For Coolidge and a majority of the voting public, the nation's businessmen had earned the nation's respect, as well as a free hand. In his 1924 inaugural address, Coolidge proclaimed that the United States had reached "a state of contentment seldom before seen." He concluded that his duty as president was to do nothing that would interfere with that blessed contentment.

Coolidge's rise to the presidency came on the wings of one of his few positive actions. As governor of Massachusetts he'd made national news with his firm handling of the Boston police strike of September 1919, which came at the height of America's post–World War I Red Scare (precipitated by the success of the Bolshevik Revolution in Russia). The Boston police, underpaid and demoralized by abysmal working conditions, had spent months seeking redress from the municipal authorities. Achieving little, they gained a union charter from the American Federation of Labor, attempted to negotiate, and, when rebuffed, declared a strike. On September 9, 1,117 of Boston's 1,544 policemen walked off their jobs. Gleefully, hoodlums took to the wide open streets. As Donald McCoy writes, "The Boston Common, which had become a gathering place for night-owl toughs, became one big dice parlor." Small riots broke out, several respectable citizens were robbed (a rarity at the time), windows were smashed, and hooligans, as the saying went, roamed the avenues. Many Bostonians feared that the Red Revolution had reached their shore. Newspapers across the land headlined what several described as anarchy.

Through the tense weeks leading up to the strike, Coolidge followed one of his cardinal rules: "If you see ten troubles coming down the road, you can be sure that nine will run into the ditch before they reach you." He had done his best to do nothing. But with chaos threatening his middle-class and Brahmin constituents, Coolidge was forced to act. Deploying a rhetorical style that some fifty years later would be called the art of the sound bite, Coolidge stated: "[There is] no right to strike against the public safety by anybody, anywhere, any time." Coolidge busted the police union and brought order back to Boston. The mainstream press hailed him as a hero. Suddenly, the unexceptional Massachusetts officeholder was a national figure. The stage was set.

The next move, like so much else about this passive man, came through no great effort of his own. At the 1920 Republican Convention, Ohio Senator Warren G. Harding had been selected for presidential nominee because better qualified candidates — like Illinois Governor Frank Lowden, California Senator Hiram Johnson, and General Leonard Wood — had so divided the convention delegates that the party needed a lesser candidate acceptable to nearly all, if generally admired by few. On June 12, Harding was chosen as a compromise candidate on the tenth ballot. Coolidge, in turn, was not Harding's choice. But the convention delegates, having skipped over the party's lions to make the amiable Harding their champion, paid little heed to Harding's wishes and made Coolidge, the restorer of law and order, their surprise vice-presidential nominee. The Democrats, for their vice-presidential nominee, chose the nationally unknown assistant secretary of the navy, the then strapping Franklin D. Roosevelt.

From June through November, Coolidge campaigned in a dozen eastern states. He spoke around the great issues of the day, quoting Benjamin Franklin on the need for thrift and industry. Harding remained moored to his front porch in Marion, Ohio. Rather than campaign against his actual Democratic opponent and fellow Ohioan, Governor James M. Cox, Harding ran against the stricken, lame-duck President Woodrow Wilson, champion of "the war to end all wars," which had instead ended in cynicism and a failed effort to bring the United States into the League of Nations. With a smile, Harding urged voters to turn away from unenumerated Wilsonian "nostrums" and join him in bringing "normalcy" back to America. "Normalcy" beat "nostrums" in the electoral college by 404 votes to 127; Cox lost every state but the Democratic Solid South. (Memories of Abraham Lincoln and the Union Army occupation made white Southerners loath to support any Republican.)

From Harding's inaugural in March 1921 until his sudden demise on August 2, 1923, Vice President Coolidge did little. A man of limited means, Coolidge found himself in the awkward position of not being able to afford a respectable house in the District of Columbia (the vice president would not be provided with an official residence until 1977). Always careful with his coin, Coolidge rented a suite of four rooms at the New Willard Hotel, noting in his autobiography with some bitterness, "It is difficult to conceive a person finding himself in a situation which calls on him to maintain a position he cannot pay for." In Washington, Coolidge was best known for sitting silently through the dinner parties that his office required him to attend. While few would have guessed it, Silent Cal enjoyed the dinners and listened intently to the conversations swirling around him, learning the ways of the nation's capital.

When the presidency came to Coolidge upon Harding's death, he was, in his own mind, ready. His core political belief — less government — was firmly in place. The dinner party conversations, his presiding over the Senate, his attendance at cabinet meetings (then a relatively rare privilege for a vice president) left him confident that he understood the processes of national political power, which he knew full well he would use in only the most limited fashion. And while Coolidge had no real agenda of his own, he understood that an agenda was awaiting him. Harding's death had left a nation in mourning even as the scandals of that administration, summed up, but by no means limited to "Teapot Dome," were boiling. Harding had left a monumental mess and Coolidge had to clean it up.

Perhaps no president (with the possible exception of Gerald Ford) was better equipped to take over a scandal-ridden administration and restore the public's faith in their government. Whatever Coolidge's failings, his integrity was unimpeachable and his stoic steadiness was soothing. Throughout the early months of 1924, Coolidge swept up after one corrupt Harding administration official after another. Secretary of the Interior Albert Fall was publicly charged with accepting bribes in exchange for leases to the Elk Hills, California, and the Teapot Dome, Wyoming, naval oil reserves. Harding's attorney general, Harry M. Daugherty, had allowed a slew of nefarious and illegal activities to take place in the Justice Department. The Veterans' Bureau, led by a Harding crony, was a hotbed of fraud and kickbacks. Coolidge forced resignations, appointed special prosecutors, insisted that justice be done, and when indictments resulted in jail sentences for Secretary of the Interior Fall and others, he breached no interference and issued no pardons. What could have been a political disaster for the Republican party and a severe blow to the body politic was so capably and so honestly handled by President Coolidge that the scandals had disappeared by the time the 1924 election season began.

As politician and public trustee, President Coolidge had earned his party's trust and its nomination to run for president on his own. What should have been a year of grand triumph for the president was instead the time every father fears most. On July 7, 1924, Calvin Coolidge, Jr., age sixteen, died. His father wrote: "If I had not been president he would not have raised a blister on his toe, which resulted in blood poisoning, playing lawn tennis in the South Grounds. In his suffering he was asking me to make him well. I could not. When he went the power and the glory of the presidency went with him. . . . I do not know why such a price was exacted for occupying the White House." Calvin Coolidge was never a lighthearted man, never one who believed the world was made to bring ease and contentment into every heart. After his son died, everything, ever after, looked even more gray.

Coolidge soldiered on. In 1924, he ran against the Democrat John W. Davis, a conservative corporate lawyer (he had represented J. P. Morgan and Co.) who had served in the Wilson administration, and Wisconsin Senator Robert La Follette, running for a brand-new and short-lived Progressive party. The Democrats had taken 103 ballots to select Davis at their convention. In the 1920s they were in the throes of a culture war, deeply divided between a rural, Protestant, pro-Prohibition, pro-Ku Klux Klan, "100 percent American" faction and an urban, largely Catholic, anti-Prohibition, anti–Ku Klux Klan, pro-immigrant and pro-immigration faction. Neither side of the party was particularly enthusiastic about Davis — nor, for that matter, was any significant faction of the electorate. La Follette represented a different sort of threat to Coolidge. The Progressive party meant to be an American version of England's Labour party. They advocated the nationalization of key industries such as the railroads and utilities, labor reforms, public works programs during times of high unemployment, and aid to farmers. La Follette hoped to take away enough votes from each party to deny Coolidge an electoral college victory and throw the election into the House of Representatives, where he believed — a long shot — a progressive coalition would vote him into the presidency.

The Republicans focused their efforts on destroying the La Follette challenge. Vice-presidential nominee Charles G. Dawes, a Chicago banker, traveled the country accusing La Follette of being a "Red." Coolidge, while rarely leaving the White House (in part because he was in mourning for his son), did, in his few public utterances, declare that the people must decide "whether America will allow itself to be degraded into a communistic or socialistic state or whether it will remain American." Such attacks, combined with the Progressive party's lack of campaign funds, worked to limit La Follette's support. Overwhelmingly, those Americans who voted decided to "Keep Cool with Coolidge." Coolidge pulled in 15,718,211 votes to 8,385,283 for Davis and 4,831,289 for La Follette. A record low turnout of eligible voters — a mere 48.9 percent — indicates just how many Americans felt that national politics played little role in their lives.

As president in his own right, Coolidge worked to shrink the federal government and diminish the role of the executive. He offered, in the words of one of his more surprising supporters, the pundit H. L. Mencken, "government stripped to the buff." His main effort in 1924 had been to veto bills that would have given World War I veterans a bonus and post office employees a salary raise. In 1926, he signed the most important piece of legislation passed during his presidential tenure: the Revenue Act of 1926. Driven through Congress by Secretary of the Treasury Andrew Mellon, one of the richest men in the country, the act cut estate taxes in half, cut the surtax on great wealth by 50 percent, repealed the gift tax, and greatly lowered income tax rates across the board. Almost all of the savings went to upper-income Americans. All told, the act cut federal revenues by around 10 percent in just one year. Another tax cut in 1928 further reduced corporate taxes and, thus, federal revenues. Mellon added to the tax reduction of America's wealthiest citizens and corporations by personally refunding, crediting, or abating some $3.5 billion during the eight years he headed Treasury. Coolidge approved the actions of the worldly Mellon, whom he held in some awe. Less federal revenues and more money in the hands of the wealthy, Coolidge believed, made for a more productive nation.

Just as Coolidge worked to do little with the power of government, so too did he endeavor to offer little leadership on the great social issues of the day. In 1925 the Ku Klux Klan, claiming some five million members, marched through the streets of Washington, D.C. In their robes and with their burning crosses, Klan members across the nation — with more members in Connecticut than in Mississippi and more in Oregon than in Louisiana — attacked not only African Americans, but Jews, Catholics, and women who refused to accept subordination to men. During the Klan march through the capital, Coolidge was silent. His own views on women were, in fact, quite similar to those of the Klan. While Coolidge clearly adored his wife, Grace, he also commanded that she not drive, ride horseback, bob her hair, wear slacks, or state any political view. Coolidge also noted, in one of his rare public statements on racial issues, that "Nordics deteriorate when mixed with other races." Coolidge, while no leader on the matter, went along with the national Republican party's deliberate attempt throughout the 1920s to shun the legacy of Lincoln and Reconstruction by turning its back on African Americans and courting the votes of white supremacists in the South. Coolidge's views on women, African Americans, and the dangers of race mixing were perfectly in accord with the great majority of his constituents. Still, no profile in courage for President Coolidge.

Mostly, Coolidge just watched in his silent way. In 1925, while reporters wired some two million words a day on the Scopes evolution trial in Dayton, Tennessee, to a fascinated nation, Coolidge said nothing. While Europe came apart over World War I loan payments to the United States, Coolidge (a firm member of the "Neither a borrower, nor a lender be" school of high finance) was heard only to mutter, "they hired the money, didn't they?" And when Congress moved to aid farmers, who had been suffering hard times since agricultural prices began falling in the early 1920s, Coolidge, with little comment, after no real involvement in the legislative wrangling, simply vetoed their effort — twice. Coolidge made no attempt to lead, or even affect congressional policy making on virtually any issue beyond vetoing measures he saw as too costly (fifty bills met his veto as compared to six under Harding). Similarly, he let his cabinet officers, most of them holdovers from Harding (except for those who had been caught with their hand in the cookie jar), run their departments as they saw fit. Secretary of Commerce Herbert Hoover, a ball of fire who struggled mightily to turn the Commerce Department into a clearinghouse of economic information and a facilitator of business efficiency, was allowed to pursue his aggressive plans even though Coolidge found Hoover's incessant activism a foolish waste of time. Even when Coolidge met visitors in the White House, he often just sat mute as they entreated his involvement in some issue of the day. He later wrote: "Nine tenths of a president's callers at the White House want something they ought not to have. If you keep dead still they will run down in three or four minutes."

Coolidge himself ran down in the summer of 1927. On August 2, exactly five years after he took over the presidency, Coolidge summoned reporters, told them to line up, and offered each of them a slip of paper: "I do not choose to run for president in 1928." He allowed no questions, gave no comments. He had had enough. He would not run for reelection. The job, he believed, had cost him his son and had damaged his health and that of his wife. He felt that little good would come from another term in office.

In 1929, after Coolidge handed over the presidency to Herbert Hoover (whom he pejoratively called "wonder boy"), the ex-president retired to Northampton, Massachusetts, where he had first been elected to public office in 1898. Thanks to his frugality while in the White House — he had saved most of his presidential salary — and thanks also to the House of Morgan, which had managed a stock portfolio for him, Calvin and Grace Coolidge had a comfortable nest egg. To keep busy, Coolidge wrote a popular autobiography, and for a short while, a syndicated newspaper column, "Thinking Things over with Calvin Coolidge." He watched the ravages of the Great Depression with dread and offered neither his successor nor the public any thoughts on ending or even simply enduring it. On New Year's Day, 1933, he stated: "In other periods of depression it has always been possible to see some things which were solid and upon which you could base hope, but as I look about, I now see nothing to give ground for hope — nothing of man." Four days later, the gray man from Vermont died of coronary thrombosis.

The historian Barry Karl explains Coolidge's popularity in the 1920s, as well as his faltering reputation ever after, as "one image of leadership: simple virtue in rebellion against the growing complexities of industrial life. If, afterward, that image seemed unreal, it was the change in the public's sense of leadership that made it so, not its failures to meet the requirements of its own times." For several decades after his death, Coolidge's reputation was that of a do-nothing president who had exacerbated the wounds of racism and class conflict, turned a blind eye to Americans in need, and contributed through his inaction to the speculative fervor, global economic instability, and domestic economic inequity that gave rise to the Great Depression. A more recent president, however, rejected that assessment. Shortly after Ronald Reagan was inaugurated in 1981, his staff had a portrait of Harry Truman taken down from a place of prominence in the Cabinet Room of the White House. In its stead, they hung an oil of that first champion of tax cuts and reduced federal domestic expenditures. Calvin Coolidge, after decades of ignominy, was back in the White House.

— DAVID FARBER
University of New Mexico

Herbert Hoover

1929–1933

b. August 10, 1874
d. October 20, 1964

T HE COLD, gray morning of March 4, 1929, sent shivers through the crowd huddled in front of the Capitol. Herbert Clark Hoover had just been sworn in as the thirty-first president of the United States. His voice, always a monotone, came booming over the loudspeakers: "I have no fears for the future of our country. It is bright with hope." Engineer, businessman, humanitarian, and commerce secretary for eight years, Herbert Hoover personified executive competence. "We were in a mood for magic," wrote the journalist Anne O'Hare McCormick. "We summoned a great engineer to solve our problems for us; now we sat back comfortably and confidently to watch the problems being solved."

Within seven months of his Inauguration, a "depression" — a term coined by Hoover to minimize the crisis — struck. Try as he might, he could not beat it. The magic failed, and the nation turned against him. "People were starving because of Herbert Hoover," one angry mother told her son in 1932. "Men were killing themselves because of Herbert Hoover, and their fatherless children were being packed away to orphanages . . . because of Herbert Hoover."

The charge was unfair, but it stuck. Hoover's presidency, begun with such bright promise, became the worst ordeal of his life and a personal tragedy that was nearly Greek in proportion. Contemporaries vilified him. History judged his administration a failure. Herbert Hoover confronted what Theodore Roosevelt called the "great moment" and fell woefully short.

Yet the historical significance of Hoover's presidency has only grown.

He embodied notable firsts in the history of the modern executive: the first president with a degree in engineering, the first to come from the Far West, and the first to bring with him a wife already established as a public figure. He was the first president to take responsibility for managing the ups and downs of the economy, to abandon the do-nothing policies of predecessors who faced its collapse, and, finally, the first to use the power of presidential publicity in a systematic way.

But Hoover also carried with him an older ethos, sometimes (and mistakenly) called "rugged individualism," as well as a limited conception of the presidency and an obsessive fear of big government. In the end, he served as a transitional figure. In failing to meet his moment, Herbert Hoover nonetheless heralded the rise of a more dynamic executive and greater federal activism, even as he formulated prescient critiques of both.

In the early years of the Great Depression, as belts tightened across the nation, Herbert Hoover dined at the White House in regal splendor. Glittering trumpets announced his arrival. Liveried butlers stood ready to serve. Night after night, the president entered in black tie, seated himself stiffly, and consumed seven full courses, even when the only guest was his wife. He had thought about economizing but decided against it. Any variation in his routine might be taken for lost confidence.

It was not that Herbert Hoover was insensitive — far from it. The cruel rumors — that dogs instinctively disliked him, that roses wilted in his hands, that he had masterminded the kidnapping and murder of Charles Lindbergh's infant son in 1932 — were as painful as they were absurd. Hoover was doing all he could to promote recovery, more than any president in hard times past. Still he was scorned. His natural sullenness turned to self-pity. "You can't expect to see calves running in the field the day after you put the bull to the cows," Calvin Coolidge reassured him. "No," said an exasperated Hoover, "but I would expect to see contented cows."

Hoover's frustration was understandable. He had never failed before. He was born in 1874 in West Branch, Iowa, a tiny hamlet of subsistence farmers. A youth spent on the edge of poverty was leavened by the simple joys of the hinterland: hunting rabbits in snow-white winters, fishing crystalline streams in summers, meeting friends and neighbors on the street or at worship. Religion lay at the center of life. Year-round, young Bertie learned the ways of the Society of Friends, commonly known as Quakers, with their commitments to conscience, community, and service. Such an upbringing only reinforced the values of the rural frontier — a dread of waste, a dedication to thrift, a belief in individual exertion, and a deep regard for neighborliness and cooperation.

Family life as Hoover knew it ended in the early 1880s, when his father and mother died just over three years apart. His natural shyness deepened, his reticence grew, and he threw up an impenetrable stoicism around himself. Relatives in Oregon took him in, and soon his capacious intelligence earned him a spot at Leland Stanford's new university in California. After graduating in 1895, he turned his degree in engineering into one of the most successful mining firms in the world. By the age of forty, Herbert Hoover was a millionaire with offices in New York, London, and Melbourne.

With the impulses of a good Quaker, Hoover balanced private gain and public service. In the autumn of 1914, just after the outbreak of the First World War, he organized the Commission for Relief in Belgium to aid starving refugees. He spent the next two years working fourteen hours a day from London to distribute over two and a half million tons of foodstuffs to nine million war victims. He took no salary and soon became known as the greatest humanitarian of his generation. Finns added the word *hoover* to their language. It meant "to help."

In 1917, as the United States entered the war, Hoover returned home to join the "big game" of government service. Within weeks, he accepted an appointment as food administrator in Woodrow Wilson's burgeoning war bureaucracy. Understanding the necessity of organization and management, Hoover sought "absolute power" over the nation's food supplies, then strictly limited the use of that power. He worried that big government, however essential in war, risked "the total eclipse of individual initiative." In Hoover's mind, such an eclipse would be catastrophic, for the greatness of the United States lay in the "free and rightful play of individual effort."

Hoover insisted on a "cooperative individualism," one not so rugged as to "run riot" nor so solitary as to ignore the needs of society. So he demanded that the food agency be powerful but temporary, to exist no longer than the war. And he mobilized what he called "the spirit of self-sacrifice" through "engines of indirection." Huge publicity campaigns promoted voluntary cooperation — "wheatless" and "meatless" days, backyard Victory gardens, exhortations to children to "finish every last bite." Hoover also endorsed agricultural price supports and encouraged cultivation of marginal lands. Production soared, the real income of farmers jumped 25 percent, and America became the breadbasket of the Allies.

The end of the war brought Hoover to the Paris Peace Conference, where he counseled President Woodrow Wilson as director general of relief for Europe and as a member of the President's Committee of Economic Advisers. So formidable was Hoover's intelligence, so impressive his command of the issues, so mute his personal ambition, that John Maynard Keynes, the British economist and himself a delegate, concluded that Hoover was "the only man who emerged from the ordeal of Paris with an enhanced reputation."

Returning home in 1919, Hoover confronted a world of political possibilities. At one point, Democratic party bosses looked on him as a potential candidate for the presidency. "There could not be a finer one," chimed in a rising young star from New York named Franklin Roosevelt. Hoover rejected the call of Democrats, confessing that he could not run for a party whose only member in his boyhood home had been the town drunk. Instead he accepted an appointment from Republican President-elect Warren Harding to head the lowly Commerce Department, but only after receiving assurances that he would have a say in every commercial aspect of public policy.

Hoover transformed the Commerce Department into one of the most powerful agencies in Washington. Among insiders he became known as "secretary of commerce and under secretary of everything else." Unlike Treasury Secretary Andrew Mellon, the aluminum magnate, Hoover was no narrow conservative. Dedicated to efficiency, distribution, cooperation, and service, he espoused a progressive capitalism called "associationalism." It sought to bring stability to the private sector without sacrificing Hoover's cherished individualism or his faith in social responsibility. Industrywide organizations (known as trade associations) would plan and regulate business, with government as a handmaiden providing advice, statistics, and forums to businesses; setting standards; and developing markets. The aim of such cooperation both within industries and between business and government was to keep the economy stable and prosperous and the marketplace safe from over-regulation. Only when business grossly overstepped its bounds would government act as umpire.

Equally important to Hoover's scheme was "welfare capitalism." An enlightened form of personnel management, the philosophy called on owners to sponsor company unions, institute profit sharing and pension plans, pay decent wages, and protect workers from factory hazards

and unemployment. Less than 5 percent of American companies instituted such policies, but in Hoover's mind, at least, progressive capitalism could not succeed without them.

Under Hoover's leadership, government and business extended their
wartime partnership and dropped all pretense of a laissez-faire economy. "Never before, here or elsewhere, has a government been so completely fused with business," boasted the *Wall Street Journal*. And never
before had one man been associated so thoroughly with success in both
or in bringing the two together. It came as no surprise, then, that when
delegates assembled at the Republican National Convention in Kansas
City in June 1928, they nominated Herbert Hoover for the presidency.
The legendary campaign with his Democratic rival Al Smith, the first
Catholic to be nominated for president by a major party, left Hoover in
command of nearly 60 percent of the popular vote and solidified the already strong hold of Republicans on Congress.

Hoover brought to the White House a penetrating mind, a passion
for order, and a breathtaking capacity for hard work, all qualities that
had helped to make him a master organizer. With him, too, came liabilities of temperament and outlook: a numbing stiffness (he wore high,
starched collars and a tie even when fishing); a lack of political magne-
tism (listening to Hoover, said a colleague, was like taking a bath in a tub
of ink); almost no experience with electoral politics (before the presidency, the only office for which he had campaigned was class treasurer
at Stanford); and an altogether too rigid philosophy of government and
economics (in which businesses would voluntarily cooperate with government and one another to keep the economy stable and prosperous).
In the absence of such cooperation, he had virtually no vision of public
policy at all.

To the White House, Hoover also brought a wife, Lou Henry Hoover,
who, unlike most of her predecessors, had already carved out a reputation of her own. A graduate of Stanford, she was the only woman in her
class with a degree in geology. Although she had never practiced her
profession formally, she remained very much a new woman of the 1890s
— intelligent, robust, and possessed of a sense of female possibilities.
She camped and fished; rode bicycles and horses; played tennis, baseball, and basketball. She spoke fluent Chinese, translated a little-known
mining manual from Latin to English, and researched the gold mining
techniques of the ancient Egyptians as she accompanied her husband on
his globetrotting career. And she shared his devotion to volunteerism,
proving to be as public-spirited and as gifted an organizer as he. She was
a member of several women's clubs and a sponsor of the Girl Scouts. In

1922 she served as the only woman vice president of the new National Amateur Athletic Federation and organized its Women's Division. In the ill-defined office of First Lady, Lou Hoover would use her energies to begin slowly to reshape it.

Despite the suddenness of the stock market crash in 1929, neither the breadth nor depth of the Great Depression was immediately apparent. From a level of 1.5 million, or 3 percent of the labor force, unemployment rose to 4.3 million by 1930, then jumped to 7.9 million by 1931. Not until 1932, the last full year of Hoover's presidency, did the picture turn catastrophic, with nearly 12 million workers — over one in four — unemployed.

By then, the story was the same everywhere: too little money, too many destitute. Private charities, the principal source of aid, had long since run dry, while the institutional solutions of an earlier day, like Jane Addams's beloved Hull House, stood flooded in misery. The simple neighborliness and volunteerism that had seen Americans through other crises were no longer enough. In New York, city employees had been donating 1 percent of their salaries to feed the needy since 1930; yet New Yorkers were starving to death anyway, forty-two in 1932 alone. "Compared to the size of the problem," wrote one historian, "it was like using a peashooter to stop a rhinoceros."

City resources were quickly depleted by an estimated thirty million destitute people. And after a decade of extravagant spending and sloppy bookkeeping, many states were already in financial trouble. Until New York established its Temporary Emergency Relief Administration in 1931, no state even had an agency to handle the problem. While the citizens of seven Latin American countries overthrew their governments, most people in the United States blamed themselves. And contrary to myth, most were reluctant to accept aid. An older conception of poverty as a sign of personal failure still ruled and with it, an abhorrence of public assistance. The "right people" never took "something for nothing," explained the wife of an unemployed railroad worker. When Americans did, they sometimes risked their rights as citizens. In 1932 residents of Lewiston, Maine, voted to bar all welfare recipients from the polls. In Herbert Hoover's America, the destitute were being disfranchised.

From the fall of 1930 onward, Hoover accepted responsibility for ending the crisis and as humanely as possible, two more firsts for a sitting president. When Treasury Secretary Andrew Mellon urged him to "liquidate labor, liquidate stocks, liquidate the farmers," Hoover liquidated Mellon by easing him out of the cabinet. When Democratic leaders in

Congress demanded all federal employment and salaries be cut by 10 percent, Hoover called the idea "heartless and medieval." Instead he cut his own salary and developed the first presidential program ever to fight a depression. He was no "economic fatalist," Hoover declared, dedicated to the common belief "that these crises are inevitable and bound to be recurrent. . . . The same thing was once said of typhoid, cholera, and smallpox."

Measured against past depression presidents — Martin Van Buren in the 1830s, Ulysses. S. Grant in 1873, Grover Cleveland in 1893, Theodore Roosevelt in 1907, Warren Harding in 1921 — Hoover was a whirlwind of activity. He did not hesitate, in the first instance, to use publicity. He invited business leaders to the White House and secured public pledges to maintain employment, wages, and prices. At every chance, he reassured Americans that "conditions are fundamentally sound" (so often that the journalist Edward Angley published the president's failed forecasts in 1931 and called the book *Oh Yeah!*). Meanwhile he appointed two commissions, the President's Emergency Committee on Employment and the President's Organization on Unemployment Relief, to stimulate local efforts.

Not content with publicity alone, Hoover tried to invigorate the economy directly. He understood the vicious cycle of rising unemployment and falling demand and knew the necessity for investment. Just after the crash he demanded a tax cut to put more money into people's hands. In 1930 Congress happily complied. The Federal Farm Board, an innovative Hoover agency created in 1929 to stabilize agriculture, promoted the sale of farm commodities through cooperatives. Despite reservations, Hoover endorsed the Smoot-Hawley Tariff in 1930 to wall out cheap foreign goods. In 1931 he prompted Congress to expand the lending powers of the Federal Land Banks by $125 million and later urged the establishment of home-loan banks to discount mortgages. Finally he increased spending on public works to prime the economic pump. Before he was done, his public building projects overshadowed those of all of his predecessors combined.

The program was in keeping with Hoover's associational philosophy of using government to promote private and local action. Yet the strategy was ultimately grounded in the past. Hoover's commitment to voluntarism, his faith in the ability of capitalism to revive itself, his conviction that too much government action would threaten public freedoms and private initiative — all derived from an earlier code of individualism anxious over the forcible power of government and naive about the efficacy of volunteerism and neighborliness to cope with the problems of a modern industrial economy. Not that Hoover eschewed government activism; rather he approached government with the traditional skepticism of an American worried over its coercive power, aggravated by the regional distrust of a westerner who had seen the steady spread of federal regulation in his own lifetime. "If government were to solve every difficult problem," Hoover warned in 1928, "it would destroy not only our American system but with it our progress and freedom as well."

Yet nothing Hoover did worked. Employers tried to keep their pledges, but within months, the downward spiral of the economy forced them to reduce wages and lay off workers. Tax cuts were followed by tax increases when reduced revenues threw the federal budget out of balance. It was widely believed that recovery could not proceed without a balanced budget, and in 1932 Congress enacted the largest peacetime tax increase to date to make up the difference. The resulting Revenue Act of 1932 stymied recovery by undermining investment and consumption. The Smoot-Hawley Tariff was perhaps the greatest failure of all. It left American loans as the only prop for European exports, brought on a wave of retaliation, and choked world trade. The Federal Farm Board was indeed innovative in 1929, but it had never been designed to handle the much more sweeping collapse of agriculture in the 1930s. Even public works did not induce recovery. Spending, which ran to more than $1 billion by the end of Hoover's administration, scarcely approached what was needed.

Between 1930 and 1932, some 5,100 banks failed as panicky depositors withdrew their funds. Losses amounted to over $3.2 billion. Yielding to pressure from congressmen and some bankers, Hoover called for a revival of the War Finance Corporation of 1918. Early in 1932, Congress created the Reconstruction Finance Corporation to save the credit structure of the nation. Capitalized at $500 million (with the power to borrow four times that amount), the RFC could lend money to banks and their chief corporate debtors, insurance companies and railroads. Within three months, bank failures dropped, temporarily, from 140 to 1 every two weeks. The Glass-Steagall Banking Act (1932) eased credit by allowing the Federal Reserve to use government bonds as a basis for $2 billion in new currency.

From the start Hoover rejected the idea of federal relief for the unemployed. He feared that a dole or giveaway program of the kind being used in Britain would damage the character of recipients. "We are dealing with the intangibles of life and ideals," he reminded Americans. Federal relief experiments could also have unhealthy results for the whole nation, perhaps creating a permanently dependent underclass. Administration would be expensive, with a new bureaucracy to police the program. Not only would the state be meddling in the private lives of citizens, but direct federal relief "would bring an inevitable train of corruption and waste." Hoover assumed exhortation, volunteerism, and neighborly cooperation would be enough.

As if to prove the president's point, the First Lady exhorted her forces to service. She pressed the more than 250,000 Girl Scouts nationwide to join in relief work and helped to promulgate the Rapidian Plan in 1931 to achieve that end. As the first First Lady to use the radio, she rallied support for volunteerism, encouraging groups such as the 4-H Clubs to devote themselves to local relief. Behind the scenes, she mobilized informal networks of friends and women's organizations and ensured that appeals to the White House found their way to local sources of aid. When none were available, she furnished assistance anonymously.

Early in 1931, as unemployment topped four million, a new Congress came to Washington with a slim Democratic majority in the House. After rejecting Democratic proposals for federal public works and a federal employment service, Hoover softened his stand on federal relief. Through the 1932 Emergency Relief and Construction Act, he authorized the RFC to lend up to $1.5 billion for "reproductive" public works such as toll bridges. The Treasury could lend an additional $350 million for public buildings such as post offices. Another $300 million in loans went to states for direct relief of the unemployed. They were wholly inadequate. When the governor of Pennsylvania requested funds to give 13 cents a day to the needy, the RFC sent him enough for 3 cents a day.

Hoover had given ground on relief. But like the rest of his Depression program, by 1932 it was too little and too late. Bands of desperate people began taking matters into their own hands. In Wisconsin the Farm Holiday Association dumped thousands of gallons of milk on highways in a vain attempt to raise prices. Ten thousand striking miners formed a forty-eight-mile motorcar Coal Caravan that snaked its way in protest across southern Illinois. In March, a demonstration turned ugly when three thousand protesters surged toward the gates at Henry Ford's Rouge Assembly Plant in Dearborn, Michigan. Ford police drenched them with hoses, then opened fire at point-blank range. Four marchers were killed and more than twenty wounded.

Hoover was compassionate but only to a point, as a ragtag army of World War I veterans soon learned. In June 1932, more than seventeen thousand veterans arrived in Washington to petition for immediate payment of their Adjusted Compensation Certificates. In 1924 Congress had voted these so-called "bonuses" for service in the First World War. They were due to mature in 1945. Penniless and hungry, the veterans wanted their money now. Their leaders met with congressional representatives, but the president refused to see them. Hoover's unresponsiveness, however, did not mean heartlessness. Quietly, the First Lady distributed food and blankets.

Worried about unbalancing the budget, Hoover dismissed the veterans as a special-interest lobby, the payment of whose bonuses would have nearly doubled the deficit. The House nonetheless enacted a bonus bill, but the Senate spared Hoover the trouble of vetoing it by blocking the legislation. About two thousand veterans stayed in the capital to dramatize their plight. When two were killed late in July as Washington police tried to evict them from federal buildings, Hoover called in the army. He wanted only unarmed military support for the police. What he got was a bloody rout. Army Chief of Staff General Douglas MacArthur, leading four troops of cavalry, six tanks, and a column of infantry, cleared the Federal Triangle in the center of the city, then razed the Bonus Army encampment on the flats across the Anacostia River, despite Hoover's orders to halt. In Albany, New York, Governor Franklin Roosevelt was outraged at the president: "There is nothing inside the man but jelly!"

A year earlier, similar charges echoed in some circles, when Japan invaded Manchuria and Hoover did nothing. Not until January 1932 did Secretary of State Henry Stimson formally respond with notes to Japan and China refusing to acknowledge any territory acquired contrary to the Kellogg-Briand Pact of 1928 outlawing war. To advocates of a foreign policy more muscular than moral sanctions, the "Stimson doctrine" of nonrecognition seemed as paralytic as Hoover's Depression program.

Here, too, Hoover was doing more than met the eye. Although he never abandoned his commitment to self-interested nationalism, he was proceeding cautiously toward internationalism. The United States, he once noted, could no longer afford "the pretense of an insularity we do not possess." He favored American participation in the League of Nations and the World Court, strongly endorsed the Kellogg-Briand Pact, and supported continuing efforts at disarmament. In Latin America, he began speaking about the United States as a "good neighbor" and acted with restraint when violence erupted in Nicaragua and Cuba in the early 1930s.

As the Depression deepened, Hoover recognized the need for international cooperation, believing the crisis was bred overseas. In 1931, he called for a one-year moratorium on all war debts and reparations. By July, it was in place, and by September a Hoover-promoted standstill agreement, preventing further withdrawal of short-term credit from Germany, was functioning as well. And in 1932 he committed the United States to participating in the London Economic Conference to stabilize world currencies and trade.

Hoover's internationalism had strict limits. Like his domestic policies, it relied on moral suasion and eschewed government intervention. Abroad, as at home, voluntary cooperation, not government coercion, was Hoover's rule. His internationalism recognized the growing interdependence of nations but insisted always on American freedom of action.

As the election of 1932 approached, Republicans refused to abandon Hoover. When their national convention opened in June, delegates endorsed his Depression program to the last detail. But they could muster little enthusiasm for the man. Hoover was renominated by acclamation in a hall one-third empty.

The campaign was over before it began. "I had little hope of reelection," Hoover later explained, "but it was incumbent on me to fight it out to the end." In his mind, it was an epic battle between two philosophies of government: the dangerous "State-ism" of Democrats against the voluntarism and prudent leadership of Republicans. Franklin Roosevelt would increase federal spending, inflate the currency, reduce the tariff, and "build a bureaucracy such as we have never seen in our history." Hoover called instead for balanced budgets, a cautious program of public works and government loans, and the mobilization of private resources.

Election day brought Hoover a thundering rebuke. Roosevelt received nearly 58 percent of the popular vote. Except for 1912, when the party was divided, no Republican presidential candidate had ever lost so badly. For the next thirty years, Democrats would campaign against the memory of Herbert Hoover and his pitiless incompetence, no matter whom Republicans ran.

Like the hero of a classical tragedy, Herbert Hoover came tumbling down. Never again would he hold elective office. For the rest of his long life, and well beyond, his presidency would be condemned as a failure. Until his death in 1964, he advised presidents from Truman to Kennedy. From time to time, he reentered public service, most notably as head of two important commissions on the organization of the executive branch of government in the 1940s and 1950s. But mainly he spent his postpresidential years as a symbol of conservative condemnation of New Deal liberalism.

The real tragedy, of course, was that one of the most incisive thinkers of his generation and its greatest humanitarian would be remembered not for his many good works or innovations but for the perceived hardness of his heart and head. Partly it was his own fault. Hoover's humanitarianism could never overcome the rigidities of his intellect or the shortcomings of his temperament. But partly, too, it was his moment. Hoover was a transitional figure, a nineteenth-century man from the west confronting the puzzles of twentieth-century America: the role of government in the new industrial age, the place of individual and community in mass society, the survival of democratic capitalism in an era of crisis and change, the position of the nation-state in an interdependent world. By the end of the twentieth century, Hoover's critique of big government, his concern over the loss of individualism, and his stress on voluntary cooperation and nonintervention would seem more attractive to a generation with no living memory of the Great Depression but still wrestling with the problems he tried to solve.

— MICHAEL B. STOFF
University of Texas at Austin

Franklin D. Roosevelt

1933–1945

b. January 30, 1882
d. April 12, 1945

T O MANY AMERICANS, the modern presidency *is* Franklin Delano Roosevelt. But to scholars, at least, his historical preeminence has often been a source of frustration. Like Roosevelt's closest associates, historians have found the political and personal masks of FDR virtually impenetrable. How could a president be so revered yet so vilified, so familiar yet so opaque? Conventional in his beliefs in God, country, and capitalism, Roosevelt proved almost eerily flexible in considering economic responses to the Depression. He was inclined to view sound balanced budgets as moral imperatives, even while countenancing unprecedented deficits that in World War II became the largest ever in relation to national production in U.S. history. A labor paternalist with little commitment to unions per se, he presided over a breakthrough in the numbers and influence of organized labor. Justly renowned as the "savior of capitalism," he cultivated an image as a class warrior, the champion of the "forgotten man at the bottom of the economic pyramid." A frequent vacationer on Vincent Astor's yacht, he was loved for the wealthy enemies he had made.

The puzzling contradictions are visible from the very beginning of Roosevelt's life: his birth on January 30, 1882, to James and Sara Delano Roosevelt on their country estate overlooking the Hudson River near the village of Hyde Park in Dutchess County, New York. The Delanos and the Roosevelts had long English-Dutch patrician pedigrees and the business associations — shipping, financial interests in coal and railroads — that often accompanied them. Sara could link Franklin to a dozen *Mayflower* passengers; the tacky opulence of the parvenu Vanderbilts up the road earned the Roosevelts' aristocratic condescension. Franklin's was a world elegantly insulated from the turbulence of the late nineteenth century — a world of genealogy and geniality, of a security and confidence in his own worth as an adored only child, of private tutors, of European travels, of the pony he received at age four. He was to the manor bred.

Yet this idyllic picture is incomplete. Franklin's winsome manner served as emotional armor, sometimes to shield his ailing older father from unpleasantness (James died of a heart condition at age seventy-two when his son was eighteen) and often to circumvent the devotion of his beloved but smothering and strong-willed mother. Even though Franklin followed the predictable educational path of the Social Register elite — Groton and then Harvard — it was not an entirely smooth one. Parental protectiveness prevented him from entering Groton until age fourteen, after others in his class had attended for two years. He never quite overcame his outsider status as the new boy — all the more so because he was too spindly and inexperienced to excel at what mattered most in Groton student culture: team sports, especially football. But the influence on Franklin of the school's celebrated headmaster, the Reverend Endicott Peabody, was profound; unlike all but a few Groton graduates, he took to heart Peabody's call to manly Christian social responsibility through public service.

At Harvard, Franklin joined seventeen other Grotonians in the class of 1904 and lived in a privately owned residence hall on the so-called Gold Coast. No longer an outsider, he achieved more social than academic distinction, most prominently election as president of the Harvard daily newspaper, *The Crimson*. Revealingly, what his wife later claimed was his life's "greatest disappointment" was his failure to be selected to Harvard's most prestigious "final club," Porcellian. But also, amid his gentleman's Cs and his social striving, Franklin made a notably fortunate choice. He courted his distant cousin, Anna Eleanor Roosevelt, who was as socially conscious as he was socially adept, as sincere as he was artful, and as insecure (the product of a wrenching, orphaned childhood) as he was ebullient and engaging. Eleanor also possessed an intelligence, compassion, and idealism that reveal something of Franklin's aspirations. They married in 1905.

For the next several years, the couple hardly seemed as if they were on their way to a rendezvous with destiny. Eleanor had one child after another; four of the six children she ultimately bore arrived in the first five years of marriage. She often felt intimidated and entrapped by a domestic life of servants, nurses, and, most of all, the ubiquitous presence of her mother-in-law.

Franklin, characteristically, was undaunted. An undistinguished performance at Columbia University Law School prepared him to pass his bar exams in 1907. He then left Columbia without his degree and worked several years for a Wall Street firm. But this job bored him. He soon moved into something that did not: politics. He discovered that he was uncommonly good at it. Handsome and self-assured, he had a quick sense of humor, a remarkable ability to persuade, and an intuitive grasp of public opinion. When leaders of Duchess County's Democratic party (with whom Franklin's father had had a gentlemanly association) suggested that he consider running for office, he jumped at the chance, even though the state senate seat for which he was nominated had been a Republican stronghold. His hostility to political bossism and corruption played well in 1910, seeming "progressive" rather than merely aristocratic. The Roosevelt aura, which might have been clouded by the inconvenient fact that he now stood across party lines from his idolized cousin Teddy, worked very much to his advantage, as a searing division within the Republican party encouraged many progressive Republican voters to embrace this new Roosevelt over his conservative opponent. Benefiting from a national Democratic tide and from his own energetic and effective campaign (in politics, FDR was a quick learner), he moved into the state senate at age twenty-eight.

With an instinct for publicity and drama (the Roosevelt name again did not hurt), he immediately gained something of a national reputation as a progressive who had defied the party directives of New York City's Tammany Hall political bosses. Timely prenomination support for Woodrow Wilson culminated in Roosevelt's appointment in 1913 to the assistant secretaryship of the navy, an office he held for nearly eight years. FDR was thrilled. Not only did it appeal to his lifelong love of ships and the sea, but it was the very position that had propelled TR to national attention fifteen years earlier. FDR used it to build his influence in New York politics (eventually learning the lesson — especially after a crushing defeat in the 1914 New York primary for senate as the anti-Tammany candidate — that an unwillingness to work with the New York City machine was political suicide in a statewide campaign). He also pursued his political education by cultivating ties with both Wilsonian and Teddy Roosevelt progressives, projected himself as an adept administrator able to cut through red tape, and parlayed his responsibility for civilian labor relations in navy yards into a reputation as a friend of labor.

But marital difficulties almost annulled these political gains. In 1918, Eleanor was devastated to discover love letters to Franklin from her social secretary Lucy Mercer, a woman with qualities that Franklin valued and shared, and that Eleanor had always felt she lacked: good looks, easy humor, and social grace. Faced with an ultimatum from Eleanor, a threat from his mother to sever his financial lifeline, and the prospect that Mercer, a devout Catholic, might refuse to marry a divorced man, Franklin agreed never to see Mercer again (though he surreptitiously violated this pledge as president). At one level, their marriage was a shell, salvaged for the sake of the children and Franklin's political future. What remained of it? Not love, although in better times regard, affection, and even an emotional bond survived. Not trust; Eleanor would never forget or forgive. Instead, the marriage that emerged from this crisis paired a more tempered husband and a more self-directed wife in the most prominent and consequential political partnership of the century.

Franklin's infidelity therefore did not slow his meteoric political ascent. In 1920, at age thirty-eight, he was nominated as the Democratic party's candidate for the vice presidency. He has positioned himself well. The magic Roosevelt name, his ably burnished reputation as a Wilsonian progressive, his captivating zest, and the exigencies of geographic ticket-balancing all recommended him to the presidential nominee emerging from the divided and brokered Democratic Convention, Governor James Cox of Ohio. In a doomed campaign that would end in a Republican landslide, with the Democrats' support for progressive change and the League of Nations falling on deaf ears, Roosevelt's barnstorming gained him national exposure and an invaluable network of political contacts.

Then disaster struck. In the summer of 1921, Roosevelt fell victim to a devastating polio epidemic. Despite assiduous efforts, he would never walk again — unless one counts the excruciating charade by which he made his way to a podium or another short distance with the aid of a cane on one side and a fiercely gripped strong arm on the other, swinging his hips from one heavily braced lifeless leg to the other, all the while beaming and bantering with onlookers to distract attention from his ordeal.

How did Roosevelt's paralysis shape his political career? By all odds, it should have ended it. Yet to most voters, his disability remained either unknown or an abstraction. In this pretelevision age, a gentleman's agreement assured that Americans saw no pictures of Roosevelt being carried up stairs or out of his car; political cartoons, in fact, might show him running, jumping, or in otherwise robust engagement. Americans heard his resonant, reassuring radio voice or saw photos or newsreels of a vigorous, confident, powerful man (swimming and other rigorous ex-

ercises had transformed his slender torso) speaking from behind a podium or a desk, or waving from a car. His props — the encompassing
grin, the cigarette holder, the upthrust chin, the easily removable cape
— heightened the effect.

FDR's illness also removed him from electoral politics at a time when
the Democratic party was weak and bitterly divided. With the vital assistance of a superb campaign strategist and political confidante (the
gnomelike Louis Howe, who guided FDR's career for almost a quarter of
a century beginning in 1912), he played a remarkably prescient and
crafty political game. He maintained a thriving correspondence with
Democratic leaders, soothing conflicts between warring camps, all the
while anticipating that an economic downturn might ultimately provide
him and his party an opportunity for national leadership. He continued
to attract public notice with well-received nominating speeches at the
Democratic National Conventions of 1924 and 1928 for presidential
candidate and New York Governor Al Smith.

Finally, FDR's incapacitation helped to extend both Eleanor's independence and her political importance to her husband. She literally
became his political stand-in and a key conduit for social reform ideas,
gaining new confidence and a critical place in his counsels that she
never relinquished. Speeches and articles, goodwill tours, political
strategizing, Democratic party publicity and platform writing, work
with women's and social welfare organizations — all drew her well beyond the traditional role of politician's wife as helpmate.

FDR returned to electoral politics sooner than he would have liked. In
1928, he suspended his convalescence in favor of an uphill race for governor of New York at the insistent urging of Democratic leaders who
hoped that his candidacy would help Smith carry his home state in the
presidential election. Though Herbert Hoover defeated Smith even in
New York itself, FDR waged a vigorous, spirited, and — by a slim margin — victorious campaign.

FDR's tenure as governor marked him as a progressive activist, a
capable administrator, an eloquent master of the new medium of radio,
a magnet for talented subordinates and advisers, and the Democratic
party's best hope. Even before the bottom fell out of the stock market in
October 1929, he supported assistance to upstate farmers, cheaper electricity through public water power development, and old-age insurance.
By 1931, he pioneered efforts to aid the unemployed and to recognize —
in stinging refutation of President Hoover's early optimistic prophecies
— their swelling ranks and needs. He was the first governor to advocate
unemployment insurance and the first to set up a state agency to admin-

ister a program of state welfare payments to the jobless. As he took pains to demonstrate, he was no radical; he would carry his criticism of Hoover's large budget deficits right into the presidency. But he had established a mood: an assertion of government responsibility, an invocation of social duty in what he called an "age of social consciousness," a sense of forward movement and engagement with public needs. With his landslide reelection to the governorship in 1930, he had blazed a path to the presidency.

In positioning himself for the nomination, he readily exploited the scapegoat status of Herbert Hoover, attacking the forlorn president from left, right, and mushy center. He became the nation's most prominent and lucid voice for economic change, adroitly sidestepping specific programmatic commitments with sometimes pointed rhetoric. FDR targeted "the faults in our economic system" and advocated "a wiser, more equitable distribution of the national income." When combined with his progressive but by no means radical record as governor of New York, such language heightened both his broader appeal and the intensity of his opposition, opening him to charges of demagogic class warfare from his former ally and main competitor at the Democratic Convention, Al Smith. But with timely support from key leaders who feared that another deadlocked convention, for which Democrats had recently become infamous, could again jeopardize victory (no Democrat, after all, had won the presidency with a majority of the popular vote since 1852), FDR captured the nomination with the necessary two-thirds of delegate votes on a tense fourth ballot. Roosevelt dramatically boarded a plane the next day to become the first nominee to deliver his acceptance speech at the convention. In his speech to the delegates and the nation, he pledged himself "to a new deal for the American people." This symbolism of departure, change, and repudiation of "Republican leadership" was the key to the ensuing campaign.

Raymond Moley, Rexford Tugwell, and Adolf Berle, Jr. — the "Brain Trust" of advisers from Columbia University, assembled by FDR in 1932 and effectively dissolved in 1933 — also helped to provide a framework for some of FDR's speeches and later policies with a bold set of recommendations for economic planning to rectify the nation's maldistribution of income. But Roosevelt's principal strategy was to leave options open and to maintain his broad appeal. Less flexible, more systematic thinkers, such as Herbert Hoover, might dismiss him as a "chameleon on plaid." But FDR's campaign achieved its ends. He demonstrated a sense of command, compassion, and understanding of the gravity and extent of Depression conditions (allowing Hoover to seem all the more out of touch or unfeeling when he downplayed the avalanching crisis, partly out of defensiveness).

Most important, he won the psychological battle. FDR was buoyant, quick-witted, infectiously optimistic — so much in contrast to the tortured, dour Hoover, of whom it was said that any rose placed in his hand would have wilted. Alternative candidates with more forthright recovery and relief programs to support the working class, Socialist Norman Thomas and Communist William Z. Foster, made a poor showing, garnering less than a million votes out of almost forty million cast. Roosevelt won in a landslide, receiving over 57 percent of the popular vote and carrying forty-two of the forty-eight states. He was joined by a heavily Democratic House and Senate.

Defeating the Depression would prove more difficult than defeating Herbert Hoover, for Roosevelt inherited an economic crisis far greater than any in American history. Between its 1929 high and its lows in 1932 and early 1933, the national income halved. Foreign trade collapsed. And 83 percent of the dollar value of the stock market vanished. A quarter of the workforce had no jobs at all, and many of the rest could find only part-time work. Crop prices had fallen so far that some farmers took to burning their corn for warmth, and thousands lost their land to defaulted mortgages and unpaid taxes. In cities, desperate, dispirited unemployed people stood in breadlines, scavenged in garbage piles, or congregated in shantytowns called Hoovervilles. There were few places to turn. Charities and city welfare offices were so overextended that relief payments, when they were available at all, were a pittance, even compared to their earlier low levels. Many company welfare programs or neighborhood and ethnic networks buckled under the strain.

Just days before FDR's Inauguration, the banking system all but collapsed. One out of five banks had already folded since the onset of the Depression, wiping out nine million savings accounts. By the first few days of March 1933, runs on the remaining banks brought withdrawals from panicked depositors at the rate of 10 percent a week, which no bank, no matter how sound, could long endure. Governors across the country closed the banks, declaring what were euphemistically called "bank holidays."

A cataclysm of this magnitude was an invitation to strong leadership. Roosevelt entered office amid calls — even from former and future opponents — for emergency dictatorial action. But for all his qualities of charismatic leadership, Roosevelt moved in a different direction, one that set the tone for his entire presidency. Reassurance and serenity were his preferred approach. Even a failed assassination attempt less than three weeks before his first Inauguration underlined this message, thanks to the unharmed Roosevelt's evident calm, instant control, and solicitude for others who were shot.

FDR's inaugural address, delivered in his determined, sonorous voice, exhibited his major approaches to rhetorical leadership in combating the Depression. The most famous line — "The only thing we have to fear is fear itself" — was at one level absurd. Americans had much to fear from the Depression. But paired as it was with FDR's recognition that "this nation asks for action, and action now," it confirmed his message of confidence in change. Most revealing was Roosevelt's reference to the banking crisis. An indictment of the banking community, it set the stage for a whirlwind performance during his first days in office: a national bank holiday to keep banks closed while the Treasury Department worked on an Emergency Banking Act, a special session of Congress to pass the law virtually sight unseen, his first "fireside chat" radio address to explain his resolution of the crisis, and then the successful reopening of the banking system.

FDR had identified himself with public fury toward bankers and their failures. But the banking reform he signed, which would soon be expanded to include various restrictions on speculative bank practices and greater power for the Federal Reserve Board, propped up bankers instead of displacing them. It provided financial ballast for the system by allowing a creation of the Hoover years, the Reconstruction Finance Corporation, to purchase nonvoting stock in banks. The Emergency Banking Act, in fact, was worked out in consultation with bankers themselves and within a framework established by officials from the Hoover administration's Treasury Department. Even so, the wand of confidence that Roosevelt waved over the banking system performed its magic. When banks holding 90 percent of the nation's deposits were allowed to reopen, many with the most minimal showing of solvency, Americans actually began returning the savings they had squirreled away instead of resuming panic withdrawals.

From the start, FDR's prescription for the Depression and for the damaged public psyche was what he trumpeted in May 1932 as "bold, persistent experimentation." He explained, "It is common sense to take a method and try it; if it fails, admit it frankly and try another. But above all, try something."

That "something" has become known as the First Hundred Days, the most concentrated period of legislative reform in U.S. history. It began with FDR's message to Congress on banking. Fourteen presidential messages, fourteen historic laws, and a hundred days later, the New Deal had arrived. The National Industrial Recovery Act, a bundle of energetic contradictions all by itself, set the tone. As with most New Deal legislation, Congress played a key role in inspiring it. Partly to counter a Senate-passed "radical" bill to spread available work to more of the unemployed by mandating a maximum thirty-hour workweek, the NIRA combined a $3.3 billion authorization for job-creating construction projects under a new Public Works Administration. The National Recovery Administration called for firms in various industries to reach code agreements mandating production standards along with labor protections (including unionization rights under Section 7a, minimum wages, and maximum workweeks).

Despite the efforts of the flamboyant head of the NRA, Hugh Johnson, to harness public enthusiasm to bring noncooperators into line, the program soon reeled out of control. Each industry — and usually the most powerful forces in the industry at that — exploited its dominance of code enforcement in an attempt to stabilize itself and to raise its own prices, sometimes at the expense of the law's stated goals of bolstering mass purchasing power or supporting union rights. Given limited bureaucratic expertise and FDR's own stated commitment in 1933 to a "concert of interests" that would bring recovery by working through powerful economic actors already in place, this result may have been inevitable. But it should not overshadow the psychological and political impact of the NRA's dramatic injection of government into economic management, which helped to dispel the bewildered fear of economic disintegration that FDR had faced as he moved into the presidency.

This drama was heightened by the accompanying drumbeat of legislation in the spring of 1933. The Agricultural Adjustment Act did for farmers what the NIRA did for industry, and in much the same way. A permissive omnibus piece of legislation, the AAA responded to congressional inflationary pressures by including the Thomas Amendment, which FDR immediately used — despite scandalized cries from traditionalists like his budget director — to slash the gold value of the dollar, thereby expanding the money supply and freeing the United States from the international gold standard.

As with the NIRA, the AAA's implementation was guided more by power holders within the regulated sector than by a transcendent public interest. Its plan to boost farm prices by subsidizing farmers to take part of their land out of production placed impoverished consumers in the peculiar position of paying farmers to grow less food. It was enforced through "county committees" dominated by influential landowners who used the law to their advantage, sometimes at the tragic expense of marginalized sharecroppers left without land to plant. Yet it helped to stabilize farm prices, markedly increased total farm income, and worked in conjunction with other New Deal legislation to improve soil conservation, to offer limited aid to poorer farmers, and to save thousands of farms by refinancing mortgages. As a result, the AAA yielded substantial public — and especially farmer — support.

The federal government quickly moved to the center of attention in these crowded early years of the New Deal, with the United Press wire service, for example, releasing four times as much news from Washington, D.C., in 1934 as it had in 1930. As Washington became the place to take public economic problems, the New Deal bureaucracy served as a magnet for socially committed young people.

Activity, reassurance, and innovation were only part of FDR's impact in lifting the Depression psychology of despair. The Tennessee Valley Authority, passed in May 1933, fell short, like all other New Deal "planning" programs, of its ambitious visions of imaginative social reconstruction. But its combination of flood control and government-generated hydroelectric power joined later New Deal initiatives like the Rural Electrification Administration (1935) in revolutionizing — literally, electrifying — rural America. In 1930, 90 percent of America's farms lacked electricity — and with it electrically pumped running water, refrigerators, electric lights, and access to the outside world through radio; by 1941, with thousands of miles of government power lines and lower rates, that figure fell to 60 percent.

In 1933, the New Deal also began to build a set of protections against economic calamity which has since come to be seen as a social right to a "safety net." Congressional pressure, for example, resulted in a Federal Deposit Insurance Corporation to fend off future financial panics by guaranteeing deposits if a bank failed. Federal welfare and work programs, a number of them administered by the colorful and dedicated Harry Hopkins, offered jobs and other assistance to the unemployed and the needy (although never to all, or even most, of them).

None of these emergency programs survived Roosevelt's presidency, but their impact was substantial nevertheless. Their contributions to the nation's infrastructure — schools, playgrounds, airports, parks, bridges, and municipal buildings — were immense. The emotional and physical sustenance they provided to millions of Americans was at the core of the image that tied Americans to the New Deal and that forged their future expectations for a responsive government. And they helped lay the groundwork for one of the most important of the New Deal's achievements, the Social Security Act of 1935, which became the cornerstone of the American welfare state for the rest of the century. It provided pensions for the elderly, insurance for the unemployed, and direct assistance to the disabled, the elderly poor, and single mothers. For all the criticisms of regressive financing, incomplete coverage, surrender to the medical lobby, and the creation of dependency which have been directed against the Social Security Act in the years since its passage, its protections against the privations of old age and unemployment are one of the New Deal's most enduring monuments.

Franklin Roosevelt had a far less coherent "New Dealer" vision than many of the idealistic young lawyers and policy intellectuals whom he brought into his administration. But he was the embodiment of these changes. His compelling gift for conveying energy, empathy, and confidence operated at a personal and a public level, in his free-wheeling twice-weekly press conferences or in his twice-yearly (a number vastly inflated in popular memory) "fireside chats" on the radio.

The millions of letters that workers, farmers, and clerks sent FDR confirm the caring connection that he forged with the public. The contrast to Hoover, who received a small fraction of this volume (even including hate mail) from ordinary Americans, is instructive. Workers' homes commonly included photos of FDR, often torn from newspapers. Paternal, religious (FDR as savior or Moses, for example), and personal references ("He saved my home"; "He gave me a job") to FDR underlined the intimacy of this linkage. It overwhelmed any uncomfortable realities or limitations of New Deal programs. "I am sure the president, if he only knew, would order that something be done, God bless him," one supporter wrote. "I know he means to do everything he can for us; but they make it hard for him; they won't let him," declared another.

But the veneration accorded FDR, far from dissolving challenges to his New Deal, ironically contributed to them. By giving people hope, by lifting fears of disintegration and chaos, FDR and the New Deal helped to thaw an undercurrent of dissent. Spurred by FDR's (and the NRA's) recognition of union legitimacy, workers turned 1934 into a banner year for labor militancy. A citywide general strike that temporarily closed down much of San Francisco, a massive national textile worker walkout, and other violent and brutally suppressed worker uprisings ushered in a spirit of class conflict that would contribute, in 1935, to the formation of a Committee for Industrial Organization (CIO) to unionize mass-production workers.

Others successfully exploited the gap between the New Deal's crusading image and its concessions to those in power. From 1934 until his assassination in September 1935, Louisiana Senator Huey Long gained millions of followers with flamboyant denunciations of plutocrats and an "Every Man a King" program to redistribute their wealth. The Reverend Charles Coughlin's fulminations against international bankers drew the nation's largest weekly radio audience. Millions signed petitions supporting the Townsend Plan for lavish old-age pensions for retirees, and insurgent political forces such as Upton Sinclair's "End Poverty in California" movement or Minnesota's Farmer-Labor party made dramatic advances. Whether Roosevelt's dissident critics could have coalesced to dent his electoral prospects, and whether the most popular of those critics represented a challenge to capitalist premises that was any more fundamental or wide-ranging than that posed by FDR himself, is doubtful. But however one assesses these pressures for change, they transformed the political equation in 1935. The 1934 congressional election had already confirmed that. The conventional wisdom had held that Democrats were in for the customary midterm drubbing. Instead, they actually picked up seven seats in the House and nine in the Senate, leaving the Republican party outnumbered by three to one in Congress. Many new Democrats had been elected by calling for public ownership or income redistribution, and third parties now held ten House and two Senate chairs.

This election was in part a vote of confidence in FDR and in part a repudiation of conservative alternatives to reform. The business-banking-government partnership for economic recovery sought by FDR became political poison. More to the point, it became unsustainable. Once economic collapse was averted, businessmen's traditional suspicions of expanded, activist government reemerged. By May 1935, the Chamber of Commerce condemned FDR and his programs at its national convention. Later that month, the conservative Supreme Court declared unconstitutional the leading mechanism for cooperation, the NRA.

In the summer of 1935, FDR stemmed the threatened disintegration of his political coalition. He combined a reform surge, sometimes known as the Second Hundred Days, with sharp, polarizing rhetoric. By ostracizing the "special interests" that had already abandoned him, he cut the ground out from under protest leaders. In June, he denounced the nation's "unjust concentration of wealth and economic power," calling for "very high taxes" on "vast fortunes" and "inherited economic power." The resulting Revenue Act of 1935 institutionalized this rhetoric, creating a politically sensational top income tax bracket (applying only, it turned out, to John D. Rockefeller, Jr.). Yet its $250 million revenue yield was too small to allow real income redistribution. The New Deal tax system continued to be dominated by regressive taxes, such as those accompanying Prohibition repeal, the AAA, and Social Security.

Also indicative of FDR's changed political calculus was his belated support for the Wagner Labor Relations Act, the most important piece of pro-union legislation in U.S. history. In contrast to Section 7a of the NRA, this law had a real enforcement mechanism: a National Labor Relations Board. Under its supervision, employers were prohibited from specified "unfair labor practices," such as firing union activists, and were required to bargain with whatever union received a majority vote as the sole bargaining unit in an NLRB election. When combined with the Fair Labor Standards Act of 1938, which (with significant exemptions) banned child labor, phased in a minimum wage, and mandated overtime pay, the system of New Deal labor relations simultaneously channeled worker militancy into more controllable bureaucratic channels and bolstered unionization campaigns. Between 1933 and 1941, union membership almost tripled (from under three million to more than eight million), and unions became one of the strongest pillars of Democratic party support.

Roosevelt took to the 1936 presidential campaign this image as a happy warrior willing to appeal to class interests. In campaign addresses, he ceremoniously denounced the "economic tyranny" of hazily defined "economic royalists" who "are unanimous in their hate for me — and I welcome their hatred." The lackluster Republican moderate Alf Landon was unable to escape his party's blame for the "Hoover Depression." His pointed attacks on the popular new Social Security Act only underlined the Democrats' themes of legislative accomplishment and economic recovery. FDR's landslide not only buried Landon (who received a humiliating 36.5 percent of the vote and carried only two small states) and the remnants of the Long-Coughlin-Townsend challenge (only 2 percent of voters backed Union party candidate William Lemke), but the president carried with him an unprecedentedly lopsided Democratic Congress.

At this point, it seemed, the sky was the limit. But although the early years of his presidency may have suggested that Roosevelt possessed near perfect pitch in politics, stalemate ruled FDR's second term. Explanations are various. FDR's campaign pyrotechnics had failed to lay the groundwork for specific new reforms. The magnitude of his 1936 landslide, combined with the menacing international spread of fascism, heightened popular and congressional sensitivities to anything that smacked of a "dictatorial" power grab. And some of the damage was self-inflicted.

Buoyed by an illusory sense of political invulnerability, FDR moved in February 1937 to eradicate the threat to New Deal programs posed by a conservative judicial system. The Supreme Court, which had already nullified the NRA in 1935 and the AAA in 1936, faced crucial decisions in a docket that included constitutional challenges to the Social Security program and the Wagner Act. Roosevelt clearly saw the 1936 elections as liberating him from previous political constraints rather than as imposing lame-duck liabilities.

This assurance seems to have emboldened him to launch a frontal assault on the court's composition, and a devious one at that, for Roosevelt coyly refused to place policy at center stage. Instead, he disingenuously framed the issue as one of assisting judicial efficiency by adding a member to the Supreme Court for each justice over seventy — a proposal that would have dramatically reoriented the court with six additional appointments. The president quickly found that he had vastly overestimated the amenability of the huge Democratic majorities in Congress. What soon became known as his "court-packing" plan repelled not only his opposition, but also many moderate supporters of his New Deal, with its blatant challenge to the separation of powers.

The mounting popular distaste for his proposal, much augmented by the unexpected court decision in April 1937 to uphold the constitutionality of the National Labor Relations Act, gave Democratic congressional power brokers, their loyalties already frayed by years of reform initiatives, the excuse they needed to let the measure die. Because the Court had apparently learned the lesson of the recent electoral landslide, and because FDR gained the opportunity to name no fewer than five justices over the next two years, the president would win his "Roosevelt court" anyway, but at a political price, and with a loss of political capital, that only underscored the fissures in his nominal congressional majority.

On the heels of that debacle came the 1937 recession. The unanticipated economic downturn caught Roosevelt off-guard. Signs of economic improvement had led him back to the fiscal orthodoxy that appears to have been a sincere component of his "Dutch house-holder" ethos. Accordingly, he put an abrupt stop to the expansionary federal spending of 1936 by seeking deep cutbacks in New Deal jobs programs in the pursuit of a balanced budget. The politically hostile dubbed this the "Roosevelt recession," an economic setback that piled further political damage on a presidency already tarnished by the court fight. Before the economy picked up in the late spring of 1938, industrial production fell by a third, stock prices plunged, and unemployment shot up by nearly four million.

Desperately groping for a plausible path to recovery, FDR agreed to prime the economic pump, the first clear addition of Keynesian ideas to the president's arsenal of economic management tools. His renewed spending programs did not represent a wholehearted conversion to countercyclical spending. But policy momentum was moving clearly toward the use of expansionary fiscal tools to fight recession.

Despite, and also because of, the setbacks Roosevelt faced since his reelection in 1936, he approached the 1938 Democratic primaries determined to recapture the initiative, and reconstitute his congressional power base, by using his continuing personal popularity to defeat particularly troubling enemies in Congress. This "purge," as it was branded, failed, an indication of the power of local, conservative forces that had hemmed in the New Deal from the start. The midterm elections themselves brought even worse news. Not a single Republican congressman met defeat, as Republicans picked up eighty-one seats in the House, eight in the Senate, and thirteen governorships. Most importantly, the elections consolidated an enduring coalition of Republicans and southern Democrats. This "conservative coalition" entrenched itself in Congress, its influence augmented by a congressional seniority system that ensconced conservative southern Democrats in strategic committee chairmanships in the House and Senate. After 1938, it was clearly poised to block or attenuate subsequent Rooseveltian reform initiatives, placing Roosevelt, and his Democratic presidential successors, on the defensive on a number of key policy issues.

What then had the New Deal accomplished? It had hardly resolved the economic crisis. Nor had it offered a coherent new blueprint of state-society relations. Roosevelt was a master of symbolic politics, improvisation, and experimentation, but hardly the architect of a redistributive renovation of the political, economic, and social order. The fundamental forces of social marginalization — segregation of African Americans, discrimination against women, exploitation of sharecroppers, exclusion of most Jewish refugees fleeing Nazi horrors in Europe — were still in place. Roosevelt's New Deal served less as a radical departure than as a stabilizer.

FDR's unique contribution, perhaps, was to restore public confidence in the American system by orchestrating an enlargement of the government's role and by fostering a broader sense of inclusion. The New Deal "broker state" incorporated and mediated core interests: most notably capital and organized labor. This shift can be measured in dollars and cents — by 1939, a near doubling of the federal budget and a 50 percent expansion of government employment over the Hoover presidency — but this yardstick falls short of capturing the dimensions of the change. FDR had helped to expand popular expectations of what an activist federal government could do to enhance the security of workers, of bank depositors, of retirees, even of the economy as a whole. In the process, he had assembled a durable "New Deal electoral coalition" centered on urban, lower-income ethnics, African Americans, and unionized workers, millions of whom had found little to draw them into national politics in the 1920s.

The onset of World War II reconfigured the priorities of Roosevelt's presidency. Although imbued with the internationalist ethos of his service in the Wilson administration, Roosevelt's primary concern during the 1930s had been economic recovery at home rather than conflict mediation abroad. Even the Good Neighbor Policy, designed in part to refurbish the image of the United States in Latin America and fortify hemispheric defense, was also in part a device to stimulate the domestic economy through increased trade with Latin America.

Ultimately, however, international crisis thrust itself onto the agenda. When war erupted in Europe with the German invasion of Poland in 1939, Roosevelt's sympathies unequivocally lay with Hitler's victims, but the political reality of American politics — the strength of isolationist organization and sentiment at home — tested the skill and resolve of this canny politician. As Hitler's victories mounted in 1940, FDR maneuvered cautiously but consistently to provide what support he could to the beleaguered and isolated British. He sought and won the repeal of the Neutrality Acts that would have prevented subsequent sale of weapons and supplies to Britain, and even resorted to the ingenious expedient of bartering American destroyers for British naval bases in the Caribbean. The inauguration of his famous Lend-Lease system, which would eventually provide approximately $48 billion of aid to at least thirty-eight countries in the course of the war, capped the year's improvisations in support of Adolf Hitler's foes.

It was in this context of international emergency as well that a wavering Roosevelt, previously torn between a further defense of his New Deal and the allure of retirement to his presidential library, finally committed to run for a third term in office. The catastrophic international situation ultimately resolved the dilemma and undermined opposition to this tradition-breaking initiative as well. The fact that he faced, in

Wendell Willkie, an advocate of collective security forced to adopt isolationist campaign rhetoric by the same political dictates that constrained Roosevelt, only underscored the relevance of the international context to his successful, precedent-shattering third term election in 1940.

As the international situation grew grimmer with Hitler's victories, Roosevelt's deviation from the spirit of neutrality grew bolder. In deploying U.S. naval vessels to patrol the Atlantic and escort merchant convoys, Roosevelt was clearly flirting with the possibility of direct embroilment in the conflict, a fact that did not go unnoticed by his critics. Indeed, so strong was the suspicion of his motives and cunning that when the United States was finally drawn fully into the war by the Japanese assault on Pearl Harbor in December 1941, there were, and continue to be, those willing to believe that he had deliberately ignored the impending attack — and thereby sanctioned the death of 2,403 men and the crippling of the Pacific fleet — on that "date which will live in infamy," in the interests of forcing America to go to war. Such suspicions are unfounded. But there is no doubt that Pearl Harbor resolved in a stroke the dilemma of neutrality, assisted in no small part by the fact that Hitler himself obligingly linked the two theaters of war within days after Pearl Harbor by declaring war on the United States.

The entry of the United States into the war produced an often tense alliance between the West and the Soviet Union, only recently a pariah. In the consultations among the Allies, the Soviet Union would participate on equal footing with the British and Americans. The Big Three — Stalin, Churchill, and Roosevelt — would become the focus of Allied war decision making.

As the war continued, FDR worked toward the construction of a postwar organization. He first outlined the idea of a United Nations during the Atlantic Charter discussions of 1941. At the Teheran Conference of 1943, he sketched for Stalin a concept of a United Nations that would mediate conflict, spearheaded by an executive council of the United States, Soviet Union, Britain, and China. These Four Policemen would be the enforcers of international peace in crises.

But before those long-term issues could be resolved, it was necessary to come to terms with the complexities of coalition warfare and the conflicting war priorities of the Big Three. Stalin was well aware of the sizable body of opinion in the West which would not have regretted the prospect of the two totalitarian powers — Germany and the U.S.S.R. — fighting each other to the death. He was adamant in holding the Allies to the objective of opening a second front in Europe. Until that happened, the Soviet Union was essentially alone in engaging the full weight of the

German army on the continent. The continual postponements and re-orientation of forces, first in North Africa, and then in Italy, tried Stalin's patience and the credibility of the second-front pledge until the long-awaited D day invasion was launched in 1944. A second source of tension was the nature of German defeat. The lessons of World War I seemed to dictate a policy of total surrender, lest militarist forces in Germany once more revive the stab-in-the-back myth of 1918. Stalin was particularly vehement in pursuit of this goal, fearing that a separate peace between Germany and the West might leave the Soviet Union alone in confrontation with Nazi power.

Much has been written in criticism of Roosevelt's handling of this delicate relationship with the Soviet ally. The bulk of this criticism has been directed toward Roosevelt's concessions at Yalta in 1945, his final wartime conference. The ailing president, critics have charged, sacrificed Eastern Europe on the altar of other postwar goals. In reality, Roosevelt had little latitude at Yalta. The Soviet Union already occupied most of the territory that the West supposedly ceded to it. But there is little doubt that Roosevelt, unduly optimistic, did indeed hope to "handle" Stalin.

The tangled threads of the Roosevelt war legacy came together in 1945. That was the year in which Germany surrendered to the combined Allied forces, the year in which American nuclear capability — developed in secrecy in a crash insurance program authorized by Roosevelt even before the outbreak of war — triggered Japanese surrender and launched the nuclear threat that defined the limits of the subsequent cold war, the year in which FDR's United Nations project came to fruition in San Francisco.

World War II would thus reconfigure the global balance of power. But it would also reconfigure home-front politics, its economic context, and the New Deal itself. Rising wartime demand proved the solution to an economic crisis that had eluded Roosevelt's vigorous efforts throughout his first two terms. War mobilization stimulated demand for underutilized industry and spurred national production to a 75 percent increase between 1939 and 1944, reducing unemployment to a minuscule level of 1 percent. Renewed prosperity created the greatest anomaly of all; in a devastating global struggle that claimed tens of millions of lives of soldiers and noncombatants, including 405,000 American lives, overall life expectancy of Americans actually rose.

The challenge to the Roosevelt presidency, then, became the mobilization of this revitalized economy toward the goal of military victory. Massive defense conversion efforts (which ultimately allowed the war to claim half of U.S. production) converted assembly lines in such core industries as automobile manufacturing — responsible for 20 percent of war production — to turn out aircraft engines, tanks, and other motorized vehicles. The resulting output supplied a significant portion of the entire anti-Axis military effort; by 1944, the United States was producing more than 50 percent more war material than Japan, Germany, and Great Britain combined.

This "production miracle" was not the product of free enterprise, except in the perverse sense that it at times was almost risk-free. Cost-plus contracts assured profitability. A host of additional tax write-offs and investment incentives spearheaded the production effort. In some cases, government bore the costs of new plants, subsequently leasing them to the business in question; by war's end, the government owned roughly a sixth of the total industrial capacity of the nation, almost all of which was later sold to the private sector, usually at bargain-basement prices. Moreover, the interests of business could expect a sympathetic ear at the main war mobilization agency, the War Production Board, which was dominated by dollar-a-year men from the business community, often still on private salaries. In the process, the war effort, with a push from advertisers who assumed a central role in the government's propaganda campaigns, indissolubly linked the ideas of freedom, democracy, and free enterprise as the American way of life that overcame all enemies.

World War II put the final touches on the derailment of the New Deal, but not so much because, in Roosevelt's words, "Dr. New Deal" had given way to "Dr. Win the War." Conservatives used the war to cashier "unnecessary" New Deal programs. More fundamentally, as the historian Alan Brinkley points out, the war narrowed New Deal liberalism itself in certain crucial respects: jettisoning even a vision of restructuring capitalist institutions or challenging concentrated economic power. New Deal liberalism had reconciled itself to the structure of the economy, casting government not as a planner but as a Keynesian balance wheel to modulate the business cycle, facilitate economic growth by fostering mass consumption, and provide a safety net for those left behind.

The limits of wartime liberalism were also evident in perhaps the most serious violation of civil liberties of the twentieth century: the internment of Japanese Americans. By an executive order in February 1942 that was subsequently upheld by the wartime courts, Roosevelt authorized the relocation of more than 112,000 Japanese Americans on the West Coast.

But if some of the broadest aspirations of the New Deal were circumscribed, the war did not expunge its permanent achievements or reverse its direction. Wartime exigencies produced lasting effects on the role of federal government. The war set the seal on a permanently expanded federal government, quadrupling civilian employment and establishing a baseline for government spending that surpassed that of the Depression years. In support of this expansion came increased revenue. Wartime income taxation policy dwarfed the New Deal's crusade against wealth by converting this class tax into a mass income tax — underpinned for the first time with a payroll deduction system — that outstripped all other taxes in revenue-generating potential.

Both depression and war had thus left a legacy of expanded government responsibility and institutional support. In 1944, FDR spoke of an "economic bill of rights." Americans, he said, had a right to the security of adequate income, jobs, social insurance, and education. But Roosevelt did not live to test the political feasibility of this vision. His health faltered visibly as the war approached its end, and he won election to his fourth term in November 1944 already close to death. Franklin Delano Roosevelt died of a massive stroke on April 12, 1945. The New Deal welfare state survived to shape American politics and society for the next half century.

— MARK H. LEFF
University of Illinois at Urbana-Champaign

FDR, Commander in Chief

FDR's leadership through the Great Depression dominates his image in history, but he played an equally formative role in crafting a national response to the second great crisis of his presidency, World War II. His wartime leadership is often recalled through memorable lines in his speeches — "a date which will live in infamy," he said of December 7, 1941, when the Japanese attacked Pearl Harbor — as well as through his dealings with other Allied leaders in shaping general military strategy, especially in Europe. Yet as commander in chief, FDR was closely engaged in the day-to-day progress of the war. Inspired by Churchill, he had a Map Room set up in the White House and visited it several times a day for updates and conferences with his top military advisers. He followed naval developments especially closely, drawing on his experience as assistant secretary of the navy during World War I.

Early in the war Roosevelt made a series of important military decisions. He encouraged Doolittle's air raid on Tokyo, designated North Africa as the focus of the first Allied offensive in the Atlantic theater, and selected Eisenhower to lead the invasion of Europe. Only as he acquired trust in his military commanders did Roosevelt's direct involvement taper off.

Roosevelt exerted his influence both overtly and indirectly, but he made sure that the chain of command culminated in the White House. In 1939, before the war broke out, he directed the Joint Army-Navy Board to report to the executive office, and shortly after the U.S. entered the war, he reorganized the board as the Joint Chiefs of Staff. The resulting mechanism, with its overlapping jurisdictions, struck the British as irregular and unorganized; "the whole organization belongs to the days of George Washington," one British emissary sputtered. But it suited FDR's executive style, maximizing flexibility and leaving the president firmly in place as the final arbiter of military decisions.

Harry S. Truman

1945–1953

b. May 8, 1884
d. December 26, 1972

O N APRIL 12, 1945, Vice President Harry S. Truman was having a late-afternoon bourbon at the Capitol when a message arrived from the White House. "Please come right over," it said, "and come in through the main Pennsylvania Avenue entrance." Truman grabbed his hat and raced out the building to his unguarded limousine. At the White House, Eleanor Roosevelt calmly took him aside. With remarkable tenderness, Truman recalled, she put her arm on his shoulder and said, "Harry, the president is dead."

The nation's grief was overwhelming. Franklin Roosevelt had been president forever, it seemed, leading his flock through the Great Depression and World War II. Following in his footsteps would not be easy, as Truman discovered at the White House funeral service. When he entered the room, the shaken mourners neglected to rise.

The need for respect and recognition dominated Harry Truman's life. Born in the farm village of Lamar, Missouri, on May 8, 1884, Truman grew up in Independence, a railroad center near Kansas City, where his family moved in 1890 to take advantage of the town's fine public schools. His father, John, a distant, hot-tempered man, worked at numerous jobs without much success. His mother, Martha, raised the three children — including John Vivian, born in 1886, and Mary Jane, born in 1889 — with extraordinary care. It was Martha Truman who encouraged Harry's love of reading, music, and history. A devoted Democrat, she dispensed political advice to her son — often unsolicited — until her death in 1947, at age ninety-four.

A product of rural Missouri, Harry Truman reflected both the certainties and prejudices of small-town, nineteenth-century America. His values were clear, fixed, traditional; his friendships lasted a lifetime. He possessed egalitarian instincts, remarkable courage, and a pioneer's faith that big things could come from small stakes.

Throughout his life, Truman struggled — not always successfully — to overcome provincial stereotypes and suspicions. After visiting New York City for the first time during World War I, he complained: "If only I could have stayed these two days in Kansas City, instead of this . . . Kike Town, I'd have felt much better." He disliked Paris even more. The Folies-Bergère performance was "disgusting," he wrote, as was most of French life. He dismissed modern art as "the vaporings of half-baked, lazy people."

Nothing came easy to Truman. As a child, his small stature and thick eyeglasses set him apart from the crowd. Before and after school, he worked at a drugstore and practiced the piano for hours. "Why no, I was never popular," he once admitted. "The popular boys were the ones who were good at games and had big, tight fists. I was never like that." A fine high school student, Truman expected to attend college. But poor eyesight prevented him from applying to West Point, his childhood fantasy, and financial problems kept him from going anywhere else. Though remarkably well read, and eager to keep learning, he remains the only twentieth-century president without a college degree.

Truman moved from job to job after high school — railroad clerk, bank teller, mine owner, oil speculator, farmer — in search of a permanent career. In 1905, he joined the Missouri National Guard, a natural decision for a young man who loved military history and hoped to expand his contacts in the local business community. Three years later, he helped charter a Masonic Lodge and became a lifelong member. Yet the most consuming passion of Truman's early adult years was his courtship of Elizabeth Virginia "Bess" Wallace. His devotion to her would span a lifetime. As president, he wrote: "You are still on the pedestal where I placed you that day in Sunday School in 1890" — when Bess was five years old.

For Harry Truman, as for millions of Americans, World War I stirred deep feelings of patriotism and duty. Well past draft age, at thirty-three, he volunteered for military service, joined an artillery regiment, and rose quickly through the ranks. Commanding a tough, working-class unit known as Battery D, Captain Truman took part in the Allied offensives at Meuse-Argonne and at Verdun. The war brought out his leadership qualities, offered him masculine satisfaction, and made him a host of new friends. It was, in many ways, the defining experience of his life.

After returning from Europe, he married Bess and opened a men's clothing store in Kansas City with a fellow veteran named Eddie Jacobson. The business collapsed during the recession of 1921, which hit the nation's Farm Belt especially hard. But fortune came knocking when Tom Pendergast, the political boss of Kansas City and neighboring Jackson County, asked Truman to run for public office in 1922. Pendergast's brother had served with Truman in the army, a powerful bond in postwar America, strengthened by Truman's active membership in veterans' groups like the American Legion. Pendergast, an Irish Catholic, saw Truman, a Baptist and a Mason, as an asset to his Democratic political machine. Truman saw Pendergast as the key to a successful new career.

With Pendergast behind him, Truman was elected judge for the eastern district of Jackson County, part of a three-member administrative court that handled county affairs. In that office, he walked a fine line between efficient service to his constituents and partisan loyalty to a corrupt political machine. Fair and honest himself, Truman went about the business of building better roads and improving public services while ignoring the squalor and thievery of those who put him in office. "Looks like everybody got rich in Jackson County but me," Truman wrote to Bess after Pendergast went to federal prison for income tax evasion. "I'm glad I can still sleep well even if it is a hardship on you . . . for me to be so damn poor."

Working with the Pendergast machine was both a blessing and a curse. It sensitized Truman to the needs of diverse groups, fueled his belief in the welfare system, and got him elected to the U.S. Senate in 1934. On the other hand, the label of "machine politician" would plague him for years. It was hard to earn respect as a legislator, despite his substantial accomplishments, when the press kept referring to him as "the senator from Pendergast."

Truman labored hard in Washington and kept a low profile. Preferring committee work to floor speeches, he played a significant role in formulating important but unspectacular legislation in the fields of transportation and interstate commerce. A loyal New Dealer, he supported the Wagner Act, Social Security, the Fair Labor Standards Act, and even the "court-packing" bill, which went down to disastrous defeat. Reelected on his own in 1940 — the Pendergast machine lost momentum after "Big Tom" went to jail in 1939 — Truman gained national exposure by heading a special committee to investigate the country's defense program. Established in 1941, the Truman Committee looked into the costs of military contracts, the fairness with which they were distributed, and their impact on small businesses and working people. By all accounts, Senator Truman did a splendid job. Press reports credited him with saving billions of tax dollars, making the production process more equitable, and bolstering public confidence in the war effort.

In 1944, President Roosevelt decided to run for an unprecedented fourth term. Democratic party leaders respected his enormous vote-getting ability, but worried about his failing health. Furthermore, they did not believe that Henry Wallace, the enigmatic vice president, was qualified to run the nation if Roosevelt did not finish out his term. After intense squabbling, FDR agreed to replace Wallace with Harry Truman, the choice of most party leaders. But getting him to agree was no easy matter. Bess strongly opposed the move, and Roosevelt made no public endorsement — a silence that wounded Truman's considerable pride. At the Democratic National Convention in Chicago, the party chairman, Robert Hannegan, summoned the Missouri senator to his hotel suite at the Blackstone, where Truman listened in on a telephone conversation between Hannegan and the president.

"Bob, have you got that fellow lined up?"

"No, he is the contrariest Missouri mule I've ever dealt with."

"Well, you tell him if he wants to break up the Democratic party in the middle of a war, that's his responsibility."

Truman paced the room in silence for several minutes. "If that's the situation, I'll have to say yes," he replied, "but why the hell didn't he tell me in the first place?"

The two men met at the White House a few weeks later. Truman was shocked by the president's gaunt appearance and trembling hands — the way his mouth dropped open as he gasped for breath. Afterward, he told reporters that Roosevelt "looked fine" and remained "the leader he's always been." Privately, he wondered how much longer the president could carry on.

After a relatively easy Democratic victory in 1944, Truman settled into his mundane vice-presidential chores. Contact with the White House was minimal. The only controversial moment occurred when Truman flew to Kansas City to attend the funeral of his disgraced political mentor, Tom Pendergast. Loyal to a fault, the new vice president brushed aside criticism with the simple statement: "He was always my friend and I have always been his."

On April 12, 1945 — less than two months into his fourth term — Roosevelt died of a massive stroke at his vacation retreat in Warm Springs, Georgia. The man who had been president for twelve years was gone, replaced by a relative unknown. Truman not only looked ordinary, he publicly portrayed himself — perhaps unwisely — as a simple man overwhelmed by the prospect of filling a giant's shoes. After taking the presidential oath, he turned to reporters, and said, "Boys, if you ever pray, pray for me now."

In fact, Truman had great confidence in his abilities. "It won't be long," he wrote to Bess a few days later, "until I can sit back and study the whole picture and . . . there'll be no more to this job than there was to running Jackson County and not any more worry." Of course, it wasn't that simple. In his first year in office, Truman ordered atomic weapons dropped on Japan, fenced with Stalin over the future of Eastern Europe, faced severe economic pressures generated by the switch from wartime to peacetime production, and fought to keep the spirit of New Deal liberalism alive.

Few issues proved more sensitive than the use of the atomic bomb. Truman knew nothing about the weapon (or the Manhattan Project that created it) until after he became president. Acting quickly, he set up a blue-ribbon advisory committee, chaired by Secretary of War Henry Stimson, which recommended the bomb's deployment against Japan (the only remaining enemy) as soon as it became available. In July 1945, following a successful A-bomb test in New Mexico, Truman declared that the Japanese must surrender unconditionally or face "prompt and utter destruction."

A number of top scientists from the Manhattan Project opposed the committee's recommendation on humanitarian grounds. Having pleaded with President Roosevelt to build an atomic bomb to counter Nazi Germany, they now found themselves in the odd position of urging President Truman not to use it against Japan. Joining these scientists were a handful of ranking civilian and military officials who contended that Japan was already close to collapse, that America's tight naval blockade and fierce air campaign were working well, and that the use of an atomic weapon would trigger a dangerous arms race with the Russians.

But Truman held firm, believing that the bomb would save American and Japanese lives by ending the war quickly. After learning about the intensity of Japanese resistance on Iwo Jima and Okinawa, he feared that a bloodbath would result from a full-scale invasion of the Japanese home islands. On August 6, 1945, planes from the 509th Composite Group dropped the first uranium bomb (known as "Little Boy") on the city of Hiroshima, killing an estimated 140,000 people. When Japan did not immediately surrender, a plutonium bomb (nicknamed "Fat Man") was dropped on Nagasaki three days later, killing 70,000 more. Japan capitulated on August 12.

Though Americans overwhelmingly approved of these bombings, the decision remains controversial to this day. Some insist that Truman used the bomb to bully the Russians in the emerging cold war. Others claim that his decision was based on racism against Asians as well as revenge for the bombing of Pearl Harbor and prisoner-of-war atrocities like Bataan. Still others contend that Truman should have provided a "demonstration" of the bomb's enormous destructive power for the Japanese or made it clear that "unconditional surrender" did not automatically include the emperor's removal. Yet, for all the debate over Truman's decision, one brutal certainty remains: Japanese leaders did not bring themselves to surrender until two atomic bombs had been dropped on their cities.

The United States faced two major challenges following World War II — domestic reconversion and Soviet expansion. Could Truman manage the difficult economic transition that lay ahead? Could he maintain Roosevelt's Grand Alliance and keep Soviet-American relations on an even keel? The early results were discouraging, to say the least.

Following World War II, a surge of consumer demand held out the promise of prosperity based on peace. For the past five years, Americans had worked overtime in offices and factories, banking their paychecks, buying savings bonds, and dreaming of the day when autos, appliances, prime beef, and nylon stockings would reappear in the nation's stores and showrooms. Between Pearl Harbor and the Japanese surrender, the public had accumulated an astonishing $140 billion in savings and liquid securities, while the average weekly wage had almost doubled, from $24.20 to $44.20. "I'm tired of ration books and empty shelves," said one factory worker. "I'm ready to spend."

But factories could not change from fighter planes to station wagons overnight. Reconversion took time. With the demand for consumer goods far outracing the supply, President Truman hoped to keep inflation in line by extending wartime price controls. His plan met strong opposition from the business community and from Republican leaders, who lobbied to "strike the shackles from American free enterprise." In June 1946, Truman vetoed a compromise bill that extended the life of the Office of Price Administration, but effectively limited its power. As controls ended, prices shot up. This fueled the demand for higher wages, which brought a flurry of labor unrest. In 1946 alone, five million workers went out on strike, totaling 120 million days of lost labor. When two railroad brotherhoods threatened a national strike to shut down the country's rail service, President Truman — a strong union supporter — forced them to back down. "If you think I'm going to sit here and let you tie up this whole country," he told union leaders, "you're crazy as hell."

A few weeks later, the United Mine Workers went on strike, leading power stations and factories to close for lack of fuel. Truman responded with an angry radio address, demanding that the miners return to work. They did, coaxed along by a federal court injunction that imposed millions of dollars in damages against the UMW. For Truman, these victories came at a heavy cost. Not only did he offend large parts of the labor movement, but he also appeared incapable of governing a nation wracked by consumer shortages, labor strife, soaring inflation, and an approaching cold war.

In November 1946, the Democratic party suffered a crushing defeat at the polls. Campaigning against the ills of reconversion with the slogan "Had Enough?" the Republicans captured both houses of Congress for the first time since 1928. When the Truman family returned to Washington from a campaign trip on election eve, no one showed up to greet them. The train station was deserted. "Don't worry about me," the president told his daughter, Margaret. "I know how things will turn out and they'll be all right."

In foreign affairs, meanwhile, relations between the United States and the Soviet Union moved swiftly downhill. The wartime alliance had been just that — a *wartime* alliance. It began to unravel, in American eyes, at the first sign of Soviet aggression in Europe. The Soviet takeover in Poland, mass arrests in Hungary, the tightening Russian grip on East Germany, Bulgaria, and Rumania — all added fuel to the fire. The American press began to play up the astounding — and generally accurate — reports of Soviet brutality in Russia and the occupied lands. Public opinion polls showed most Americans agreeing that Russia was aggressive, imperialistic, and determined to rule the world.

In February 1946, the Soviet dictator Joseph Stalin delivered a major address predicting the collapse of capitalism. The following month, with Truman at his side, the former British prime minister Winston Churchill warned that Russia had drawn an "iron curtain" across Europe "from Stettin in the Baltic to Trieste in the Adriatic." From Moscow, meanwhile, a U.S. foreign service officer laid out the doctrine of "containment" in an eight-thousand-word telegram designed to stiffen America's resistance to Communist expansion. America must be patient, wrote George F. Kennan. It must carefully define its vital interests and then be prepared to defend them through "the adroit and vigilant application of counterforce at a series of constantly shifting geographical and political points."

Kennan's telegram was perfectly timed. Rather than focusing on Eastern and Central Europe, which were firmly under Russian control, it shifted attention to other areas, far more vital to American security, such as Western Europe, the Mediterranean, and the Middle East. Here the line could be drawn — and Communism contained.

The first trouble spot appeared to be the Mediterranean, where Stalin was seeking concessions from Iran and Turkey, and where Communist-led rebels were battling the Greek government in a bloody civil war. Early in 1947, Great Britain, the traditional power in that region, informed the United States that it could no longer provide military and economic assistance to Greece and Turkey. Exhausted by World War II, England urged its American ally to maintain that assistance in order to prevent further Soviet expansion.

Led by Dean Acheson — then an assistant secretary of state — the Truman administration won support for its Mediterranean policy by portraying Soviet intentions in excessively alarmist terms. If Greece fell to the Communist rebels, Acheson insisted, other nations would follow "like apples in a barrel infected by one rotten one." On March 12, 1947, the president declared — in the so-called Truman Doctrine — that the United States would aid the democratic struggle against totalitarianism by supporting "free peoples who are resisting the subjugation by armed minorities or by outside pressures." Although Congress allocated $400 million in military aid to Greece and Turkey, it did so with reluctance and unease. Neither recipient practiced democracy, after all; and the Truman Doctrine, with its apocalyptic vision of Communist expansion, appeared to open the door for a costly, ever escalating military crusade against left-wing forces around the world.

Three months later, the Truman administration unveiled a more ambitious — and palatable — proposal known as the European Recovery (or Marshall) Plan, to revive the war-torn nations of that continent with massive economic aid. "Our policy is not directed against any country or doctrine," said Secretary of State George C. Marshall, "but against hunger, poverty, desperation, and fear."

Though all European countries were invited to participate, the Russians and their satellites — unwilling to divulge critical information about their economies and fearful that capitalist aid would undermine the Communist system — soon refused the offer. President Truman was relieved. He had extended a blanket invitation to avoid the appearance of worsening the cold war, yet he knew that Congress would reject any package that offered aid to the Stalin regime. After months of debate, Congress authorized more than $10 billion in loans and subsidies for European recovery, with the largest sums going to England, West Germany, and France.

The Marshall Plan proved a tremendous success on both sides of the Atlantic. By creating jobs and raising living standards, it restored economic confidence throughout Western Europe while curbing the influence of Communist parties in Italy and France. At the same time, the Marshall Plan increased American trade and investment in Europe, opening new markets for U.S. goods. As President Truman noted, "Peace, freedom, and world trade are indivisible . . . We must not go through the 1930s again."

The iron curtain that descended on Europe had a powerful impact inside the United States. Americans were frustrated by the turn of global events. Soviet aggression had robbed them of the fruits of victory. One form of totalitarianism had been replaced by another. The result was an erosion of public tolerance for left-wing activity, encouraged by officials like FBI Director J. Edgar Hoover, who claimed that Communists were now a major force in labor unions, newspapers, radio, movies, churches, schools, fraternal orders, and the government itself.

President Truman scoffed at such descriptions. He considered domestic Communists and their supporters to be wrongheaded and obnoxious, but not a menace to national security. Furthermore, he believed that reactionary Republicans were using the "Communist issue" to smear their liberal opponents as dangerous radicals and disloyal subversives.

Initially, Truman helped the Red Scare along. In mobilizing support for containment, he sounded at times like the Red-baiting alarmists he thoroughly despised. And in attempting to defuse the issue of domestic subversion, he established a Federal Loyalty Program in 1947 that greatly alarmed civil libertarians. The program called for extensive background checks of all civilian workers in the executive branch — not simply those in sensitive positions — and provided minimal safeguards for the accused. Even worse, the program stirred up congressional Red-hunters by conceding the possibility that a serious problem existed. The assault was led by the House Committee on Un-American Activities (HUAC), composed of conservative Republicans and southern Democrats who were united mainly by their hatred of the New Deal.

In the summer of 1948, a conservative journalist named Whittaker

Chambers testified before HUAC about his Communist past. Chambers not only described the formation of a "Communist cell" in Washington in the 1930s, he also named the members, one of whom, Alger Hiss, had been a prominent New Deal official. Hiss demanded — and received — an HUAC invitation to respond. "I am not and have never been a member of the Communist party," he declared. "To the best of my knowledge, I [have] never heard of Whittaker Chambers."

The headlines were spectacular. In September 1948, Hiss filed a $75,000 slander suit against Chambers for calling him a Communist on the national radio show "Meet the Press." Chambers responded by producing numerous documents showing that Hiss had almost certainly engaged in espionage for the Soviet Union during the 1930s. These documents — known as the Pumpkin Papers because Chambers had hidden them in a pumpkin patch on his Maryland farm — helped convince a federal grand jury to indict Hiss for lying under oath before HUAC. (The ten-year statute of limitations on espionage had just run out.)

In Alger Hiss, American conservatives found the perfect symbol for all that they distrusted and despised. Hiss was the quintessential New Dealer — young, intelligent, sophisticated, Ivy League. His liberal admirers included Dean Acheson, Eleanor Roosevelt, Adlai Stevenson, and Felix Frankfurter, to name a few. Even President Truman called the Hiss case a "red herring," cooked up by right-wing elements to discredit progressive ideals. The president's feelings were understandable, yet his remark was unwise. Although conservatives routinely exaggerated the "Communist issue" for political purposes, Truman's visceral response to Hiss — who was convicted and sent to federal prison in 1950 — served to bolster the false notion that his administration was oblivious to problems of national security.

The Republican landslide of 1946 appeared to signal the demise of American liberalism. In the following months, the Republican Congress brushed aside Truman's proposals for national health insurance, full employment, and federal aid to education, while passing a controversial bill, known as Taft-Hartley, to curb the power of organized labor. Republican leaders viewed the union movement as both an appendage of the Democratic party and a threat to the employer's authority in the workplace. The Taft-Hartley bill had strong public support, due to the crippling national strikes of 1946. Its most important provisions outlawed the "closed shop," which required workers to join a union at the time they were hired, and encouraged individual states to pass "right to work" laws, which made union organizing more difficult. Although Truman opposed the bill, calling it an "affront to working people," Congress easily overrode his veto.

The president also confronted the issue of racial discrimination, long ignored by the White House and the Congress, by forming a special task force on civil rights. Its final report included a series of bold recommendations, such as desegregation of the armed forces and formation of a civil rights division within the Justice Department. Although Truman courageously implemented these recommendations by executive order, his personal feelings were mixed. As a political leader, he had to balance the interests of two distinct voting blocs within the Democratic party: southern whites and northern blacks. As an individual, he believed that all citizens deserved equal treatment, but he felt uncomfortable with the notion of *social* equality for African Americans, as did most white people of that era.

In many ways, Truman viewed America as an extension of Jackson County and the politics he knew best. His job was to provide services to loyal constituents and legitimate interest groups. Sometimes, this meant swallowing old prejudices; mostly, it meant doing what he thought was right. Truman struggled hard with his conscience, as his private letters and diaries attest. On the eve of integrating the armed forces, he complained to his sister that Eleanor Roosevelt "has spent her public life stirring up trouble between whites and blacks — and I'm caught in the middle." As he moved to recognize the new state of Israel, against the wishes of State Department advisers, he fumed at the intense Jewish pressure on him, claiming, "Jesus Christ couldn't please them when He was here on earth, so how could anyone expect that I would have any luck?"

As president, Truman vented most of his considerable anger in private. His diary entries spoke of hanging "traitors" like John L. Lewis, who had led the miners' strike in 1946, and even bombing the Russians. He fumed at every imagined slight and displayed a hatred of the press that easily rivaled Richard Nixon's. Aides regularly checked his personal correspondence to make sure that inflammatory letters, dashed off in a rage, did not reach the mail sack. Sometimes they missed. In one instance, Truman sent a furious note to a *Washington Post* music critic who had panned a vocal concert by his daughter, Margaret. "Some day I hope to meet you," the president wrote on White House stationery. "When that happens you'll need a new nose, a lot of beefsteak for black eyes and perhaps a supporter below!" The critic, no doubt thrilled, sold the note to a collector.

*   *   *

Almost no one expected Truman to win the 1948 presidential election. A host of blunders and problems had sapped his political strength. His relations with Congress were poor, especially in domestic affairs. The media, recalling the elegant, ubiquitous FDR, portrayed Truman as too small for the job. Pundits joked about his failings. "Would you like a Truman beer? You know, the one with no head."

His party was badly divided. In December 1947, a group of left-wing Democrats formed the Progressive Citizens of America, with an eye toward the coming election. Led by Henry Wallace, the Progressives blamed Truman for starting the cold war at home and abroad. They opposed the Truman Doctrine, the Marshall Plan — and virtually all criticism of the Soviet Union. Though Wallace had no chance of winning the presidential election in 1948, his independent candidacy appeared certain to split the Democratic vote.

Some urged Truman not to run. Other possible candidates, including General Dwight D. Eisenhower and Supreme Court Justice William O. Douglas, were suggested by party leaders. *The New Republic,* a favorite of liberals, ran the headline: "HARRY TRUMAN SHOULD QUIT." When word reached the sweltering Democratic National Convention in Philadelphia that Eisenhower was unavailable, Jersey City political boss Frank Hague crushed out his cigar. "Truman," he mumbled. "Harry Truman, oh my God!"

The news quickly got worse. To bolster Truman's meager chances, the convention nominated Alben Barkley, the popular Senate majority leader from Kentucky, for vice president. Yet even Barkley could not prevent most southern delegates from bolting the convention after northern liberals attached a strong civil rights plank to the Democratic platform. Assembling later that week in Birmingham, Alabama, amid rebel yells and Confederate symbols, these southerners formed the States' Rights (or "Dixiecrat") party. Demanding "complete segregation of the races," they selected Governors Strom Thurmond of South Carolina and Fielding Wright of Mississippi to be their presidential and vice-presidential candidates.

The Republicans, meanwhile, were confident and united. Their national ticket — headed by Governors Thomas Dewey of New York and Earl Warren of California — was the strongest in years. Yet Truman plunged ahead, portraying himself as the champion of common people, determined to protect America from a heartless Republican assault. To stress this image, he demanded passage of an eight-point program that included civil rights, public housing, a higher minimum wage, and greater storage facilities for farmers. Calling Congress back into special session, he blasted the "do-nothing" Republicans, claiming they were interested only in the rich.

World events in 1948 also strengthened Truman's hand. In June, Soviet troops blockaded the roads and rail lines to West Berlin, which lay deep inside Communist-controlled territory, to protest the merging of French, British, and American occupation zones into the unified nation of West Germany. Ruling out military force to break the blockade, Truman decided to supply West Berlin from the air. Western pilots made close to three hundred thousand flights into the city, delivering food, fuel, and medical supplies. The Soviets eventually backed down; West Berlin survived.

Truman could see his prospects rising as the 1948 campaign progressed. Traveling the nation by train, he hammered out the same message to large, friendly crowds. "They don't like the New Deal. They want to get rid of it," he repeated. "This is a crusade of the people against the special interests, and if you back me we're going to win." His momentum went unnoticed; few experts gave him a chance. A survey of fifty political analysts in *Newsweek* showed every one of them predicting a Republican victory. On election eve, the staunchly conservative *Chicago Tribune* went to press early with the now classic mistaken headline: "DEWEY DEFEATS TRUMAN."

When the ballots were counted, Truman had won the tightest presidential election since 1916, with 24.1 million votes to Dewey's 22 million, and 303 electoral votes to Dewey's 189. Governor Strom Thurmond garnered 1.1 million votes and four southern states, while Henry Wallace won no states and barely a million votes. Many observers contended that the Democratic split probably aided Truman's centrist message by removing both the left-wing tinge of the Wallace camp and the racist cries of the Dixiecrats. In the end, the voter's chose the president's common touch and mainstream liberalism over Dewey's stiff and evasive demeanor.

The triumph was short-lived. Within months, a rush of world events and political assaults overwhelmed his presidency. The trouble began in 1949, when an American spy plane picked up evidence that Russia had successfully tested a nuclear device. Combined with alarming developments in Asia, the loss of America's atomic monopoly served to heighten global tensions and dramatically increase the fear of Communist subversion at home.

For most of Truman's presidency a civil war had raged in China between the Nationalist government of Chiang Kai-shek and the Communist forces of Mao Tse-tung. In 1949, Secretary of State Dean Acheson issued a 1,054-page "White Paper on China," conceding that the world's most populated country was about to go Communist, despite billions of dollars in American aid. "The unfortunate, but inescapable fact," said Acheson, "is that . . . nothing [the United States] did or could have done within reasonable limits . . . could have changed that result." True or not, these words read more like an excuse than an explanation. How well did "containment" really work when more than 600 million people were "lost" to Communism?

The China debacle all but ended bipartisanship in foreign affairs. By 1950, Republican attacks on the president combined the Hiss case, the Russian A-bomb, and the fall of China into a single theme, packaged most effectively by a reckless young Wisconsin senator named Joseph R. McCarthy, who claimed that America was losing the cold war because the Truman administration was filled with "traitors" who wanted the Communists to win. The real enemy was not in Moscow, thundered McCarthy, but rather in Washington, D.C.

The senator not only provided a simple explanation for America's "decline" in the world, he also generated enormous support and publicity for his cause. Prominent Republicans, sensing the political benefits of the "Communist issue," rallied to his side. Senator Robert Taft of Ohio, known as "Mr. Republican," privately dismissed McCarthy's charges as "nonsense." Yet he told McCarthy to keep punching — "if one case doesn't work, try another."

President Truman viewed McCarthy as a shameless publicity hound who would say anything to make headlines. In a letter to Vice President Barkley, Truman recounted an old fable to describe the senator's behavior. Long ago, there had been a mad dog that went around biting people. To deter him, the master had fastened a clog to the dog's neck. Even though the clog was a badge of dishonor, the dog foolishly viewed it as a symbol of power. The moral, Truman concluded, was that some men "often mistake notoriety for fame, and would rather be remarked for their vices and follies than not be noticed at all."

Truman was right about McCarthy, but helpless to stop him. The Red Scare kept expanding, aided ominously by another Asian war. On June 27, 1950, Communist troops from North Korea crossed the 38th parallel to invade anti-Communist South Korea. The attack put great pressure on the Truman administration. To ignore it, the president believed, was to encourage aggression elsewhere. What exactly would containment mean if the Communists were allowed to overpower a neighbor without

5555

fear of retaliation? Furthermore, a weak response by Truman would serve to reinforce McCarthy's charge that he was "soft on Communism."

The president moved quickly, dispatching ground troops to defend South Korea. He did so through the United Nations, without consulting Congress, a move that would hurt him as the war turned sour. In September 1950, troops led by General Douglas MacArthur routed the enemy after a brilliant amphibious landing at Inchon, on South Korea's west coast. In October, U.N. troops crossed the 38th parallel in pursuit of the routed North Korean army. A month later, Chinese Communist troops attacked en masse, trapping MacArthur's army near the Yalu River that divided North Korea and Manchuria. In desperate winter fighting, U.N. troops — composed mainly of American Army and Marine units — retreated through mountain blizzards and a wall of Chinese infantry to form a defense line just south of the 38th parallel. Though disaster had been averted, the nation was stunned by the quick military reversal.

When the Communist offensive stalled in March 1951, General MacArthur demanded an escalation of the war, including a naval blockade of China, a massive bombing of its war factories, and an invasion of the Chinese mainland by the forces of Chiang Kai-shek. The president disagreed. Expanding the war would frighten other U.N. participants, he realized, and severely weaken America's troop strength in Europe, which Truman saw as the prime cold war battleground. Moreover, escalation might bring Russia into the conflict, with the threat of nuclear attack. As General Omar Bradley put it, MacArthur's strategy would "involve us in the wrong war, at the wrong place, at the wrong time, and with the wrong enemy."

MacArthur refused to keep silent. Despite repeated warnings from the president, he kept up his drumfire for an expanded Asian war. Encouraged by right-wing Republicans, the general sent a letter to Joseph Martin, the House minority leader, that criticized Truman's refusal "to meet force with maximum counterforce," and ended with the oft-quoted line: "There is no substitute for victory." Furious at such insubordination, the president relieved MacArthur of his command.

Though Truman had made the proper choice, the reaction was severe. For millions, MacArthur symbolized traditional military values in a world complicated by the horrors of atomic war. Letters and wires flooded the White House, running twenty to one against the firing. On Capitol Hill, angry legislators placed some of their telegrams into the *Congressional Record:* "IMPEACH THE IMBECILE"; "IMPEACH THE B

WHO CALLS HIMSELF PRESIDENT"; "IMPEACH THE RED HERRING
FROM THE PRESIDENTIAL CHAIR." Truman's popularity never recovered. The president seemed to be paralyzed by the bloody stalemate —
afraid to win the war on the battlefield, unable to end it at the bargaining table.

Korea and McCarthy served to bury Truman's second term. The president had hoped to further the New Deal tradition — to implement liberal reforms in an era of economic growth and widespread prosperity.
He wanted to expand social security and unemployment insurance, increase the minimum wage, advance the government's role in public
housing, health insurance, farm supports, and civil rights. But the "Fair
Deal" he proposed in 1949 was sidetracked by the explosive issues of
Communist aggression and domestic subversion. By 1951, public opinion polls put Truman's approval rating at 23 percent — the lowest ever
recorded for an American president. When he decided not to run again
in 1952, the nation heaved a collective sigh of relief.

As an ex-president, Truman expressed himself freely. In an age of television sound bites and political deception, his plain speaking seemed
candid and refreshing. By the time of his death in 1972, Truman had become an American hero, admired for his honesty, combativeness, and
courage under fire. His decisions to use nuclear weapons, confront the
Soviet challenge, revive the battered economies of Europe and Japan, resist Communist aggression in Korea, further civil rights, strengthen the
welfare state, and fire General MacArthur — all looked better as the
years rolled by. Historians ranked him among the "near-great" presidents, a notch below Washington, Lincoln, and FDR. At his death, one
of the many eulogists, Senator Adlai Stevenson III of Illinois, noted that
the triumph of Harry S. Truman was the triumph of America itself: "the
ability of this society to yield up, from the most unremarkable origins,
the most remarkable men."

— DAVID M. OSHINSKY
University of Texas at Austin

Dwight D. Eisenhower

1953–1961

b. October 14, 1890
d. March 28, 1969

T HE EISENHOWER PRESIDENCY was one of the most unusual in modern American history. Both Eisenhower himself
and many of his top aides had no previous experience in public office. Even more atypical, he and they had spent most of
their adult lives rising to the top in other fields of endeavor, most notably
the military, business, law, and education. In no other twentieth-century
presidential administration did the professional politician enjoy less
prestige and influence. Despite, or perhaps because of, those circumstances, Eisenhower's presidency was highly competent, effective, and
successful, the most so of any presidency since World War II.

Dwight David Eisenhower was born on October 14, 1890, in Denison,
Texas, the third of six children, all sons, of David and Ida Stover Eisenhower. Eisenhower's parents, although very well educated by the standards of their time (they met in college), suffered a sharp drop in social
status when David Eisenhower's general store failed, and his partner
absconded with what little money was left. Traumatized by these developments, Eisenhower's father eventually found a low-paying job at a
creamery in Abilene, Kansas, where Dwight spent most of his youth.
Sustained by their own hard work, frugality, and deep religious convictions as members of a religious sect called the River Brethren, Eisenhower's parents succeeded in providing modestly for their large family.

Dwight Eisenhower thus grew up in a household informed by both
formal education and economic adversity. Like many Americans raised
in those circumstances, he and his brothers soon demonstrated considerable ambition. Even so, most people who knew him liked Ike, as he
was called, precisely because his desire to rise did not manifest itself in

intense or unpleasant ways. Eisenhower was unpretentious, easygoing, and a good but not exceptional student and athlete. He did have a hot temper, but he learned to control it. In later life, it was seldom visible in public. Although Eisenhower's passion for reading history led to a prediction in his high school yearbook that he would one day be a history professor at Yale, his family's lack of money led him instead to seek higher education at one of the nation's military service academies. At first Eisenhower applied to the Naval Academy, but upon learning that he was too old to be admitted he turned next to the U.S. Military Academy at West Point. On June 14, 1911, he enrolled there and began an army career that lasted (with one brief interruption) until 1952.

Nothing in Eisenhower's record at West Point indicated future greatness. Away from home and from his parents' moral influence for the first time, Eisenhower got into trouble for such things as mocking his commander's order to report to his quarters in a full dress coat (Eisenhower wore the coat and nothing else), dancing too informally with a young woman (for which he was demoted from sergeant to private), and smoking (a habit acquired at West Point and kept for most of his adult life). Much more interested in athletics than schoolwork, he ultimately excelled at neither. A serious knee injury acquired in a varsity football game put an end to Eisenhower's athletic ambitions during his second year as a cadet. Thereafter his exercise was limited to simple gymnastics, walking, calisthenics, and a little rowing. Depressed by this turn of events, Eisenhower gave even less attention to his studies than before and compiled a mediocre academic record. He graduated sixty-first out of a class of 164 in 1915 and was commissioned as a second lieutenant in the infantry, his knee injury having precluded service in the more glamorous cavalry.

The army that Eisenhower joined as an active-duty officer in the spring of 1915 was tiny by the standards of the major industrial societies, and unpromising as a career choice. Total troop strength was 120,000 men, including officers, of whom he was one of the most junior. Prospects for rapid promotion seemed slight, as they had been since the end of the Civil War half a century earlier. Since that time, the army had mostly been used to "pacify" American Indians, and patrol the nation's southern border. Only the Spanish-American War (and the revolution in the Philippines which it helped encourage) had disturbed that pattern in 1898, and then only briefly. Over the thirty-five years after Eisenhower entered, however, the U.S. Army underwent a transformation in size and mission, making military service an avenue of extraordinary opportunity for its most able and ambitious members. In that sense, the timing of Eisenhower's entry could hardly have been more fortuitous.

His first posting was to Fort Sam Houston in San Antonio, Texas, where he met Mamie Geneva Doud of Denver, who was spending the winter in the city with her parents. The two soon began a serious courtship and became engaged in February 1916. Five months later, they married. He was twenty-five, and she, nineteen. Perpetually short of money, as were so many married junior officers, Eisenhower supplemented his modest army salary by playing poker with his fellow officers and coaching high school football. In September 1917, the Eisenhowers' first child, a son they named Doud Dwight, was born.

With American entry into the First World War that same year came a big change in Eisenhower's activities. He spent the duration of the conflict training officers for combat in France, an assignment he performed so effectively at Fort Oglethorpe in Georgia, Camp Meade in Maryland, and Camp Colt in Pennsylvania that he himself never saw action. While some might have welcomed this safer existence during wartime, Eisenhower was deeply disappointed and frustrated by his failure to get to France. Despite this apparent setback, his effective organizational skills during the mobilization effort earned him promotion to the rank of major in 1920, placing him ahead of many of his West Point classmates.

The American military experience in World War I had another lasting consequence for Eisenhower. It made him a tank enthusiast. Convinced that this new machine had the potential to transform ground warfare, Eisenhower became a serious student of tank doctrine and a strenuous advocate of the new weapon, over the objections of some of his more cautious Army superiors. His interest in tanks led him into a friendship with a like-minded fellow officer, Colonel George S. Patton. Both angered superiors by urging the use of tanks en masse in battle, rather than simply as an aid to front-line infantry. The chief of the infantry, who feared these ideas might diminish support for the traditional infantry, threatened Eisenhower with court-martial should he publish his recommendations. Although deeply angry at this heavy-handed move to silence him, Eisenhower complied; but he did not change his mind about the tank's future.

This episode proved unexpectedly important when Patton introduced Eisenhower to Brigadier General Fox Conner, who had served on General John J. Pershing's staff in France as operations officer. Conner expressed interest in Eisenhower's and Patton's ideas about tanks, and the result was a long, detailed discussion during which Conner directed

THE AMERICAN PRESIDENCY ★ 384

most of his questions at Eisenhower. Evidently impressed by this young major, Conner invited him a few months later to join his staff as executive officer. The assignment meant relocating to the Panama Canal Zone, where Conner commanded an infantry brigade, but Eisenhower welcomed it as a chance to learn from a highly respected army thinker and as a change of scene for him and Mamie after the traumatic death of their first son in 1921 of scarlet fever. In Panama, Conner, without Eisenhower being fully aware of it at the time, groomed him for much higher things. Conner intensively tutored his young executive officer in military history and doctrine, philosophy, and literature. Conner also gave Eisenhower a daily written assignment, composing field orders that familiarized him with preparing plans and orders for operations, including their logistics.

All of this study helped make Eisenhower ready for a major breakthrough in his career, admission in 1924 to the Command and General Staff School at Fort Leavenworth, Kansas, which Conner arranged. There Eisenhower fully applied himself, and finished first in his class. In his spare time, he helped Mamie raise their second child, John, and learned to play golf.

Success at Leavenworth finally set Eisenhower's military career on the fast track. Six months after finishing his coursework, he was ordered to Washington to serve in the office of General Pershing and attend the War College. From there he went to the assistant secretary of war's office, to research and prepare a mobilization plan for the U.S. economy in the event of another major war. Seemingly an empty exercise in 1929, Eisenhower nonetheless applied himself thoroughly. Bernard Baruch, who had headed the War Industries Board, which oversaw mobilization of the U.S. economy during World War I, instructed him in the complexities of the task, thereby giving Eisenhower his first serious lessons in political economy. Eisenhower's conclusions impressed Army Chief of Staff Douglas MacArthur, who in January 1933 asked Eisenhower to join his staff as senior writer. The two men worked together for the remainder of the decade, in Washington and then the Philippines, where MacArthur served as U.S. military adviser to the Philippine government. Eisenhower learned a great deal from working under the temperamental and egocentric MacArthur, who delegated many sensitive political tasks to his discreet, persuasive, and thoroughly agreeable aide. Eisenhower became close to Philippine President Manuel Quezon, who preferred Eisenhower's company to that of his commanding general. Quezon and Eisenhower had long conversations about the wide range of issues confronting Quezon as head of government, such as taxes, education, and defense.

Late in 1939, Eisenhower, by then a colonel and tired of working for MacArthur, returned to the U.S., where he soon began applying the lessons learned from his earlier postings. Highly regarded as a staff officer, he handled several assignments effectively, most notably the planning connected with the biggest peacetime maneuvers in the army's history, held in the summer of 1941. Shortly after they ended, Eisenhower was promoted to brigadier general. When Pearl Harbor was attacked, Eisenhower went to work for the army chief of staff, General George C. Marshall. The two had met a few times earlier, and Eisenhower had done his best on those occasions to make a good impression, acting on Fox Conner's advice that he should try to serve under Marshall if he ever had the chance. For the next several months, he and Marshall worked side by side on a daily basis, preparing war plans. Marshall soon began referring to Eisenhower in official correspondence as his "subordinate commander," and saw that he was rapidly promoted to major general and then lieutenant general.

In June 1942, Marshall appointed Eisenhower commander of U.S. forces in Europe, the job that ultimately made him famous. It was a very difficult task, requiring great political as well as military skills because he needed to work closely with senior British government and military officials. Eisenhower, who once defined leadership as "the ability to decide what is to be done, and then to get others to want to do it," handled the more political and diplomatic aspects of his new role superbly. He adroitly persuaded such difficult personalities as Churchill, Montgomery, de Gaulle, and Patton to work together effectively, an extraordinary and militarily invaluable achievement. American and British military success in North Africa, Italy, and, above all, in France owed much to Eisenhower's skill at building a smoothly functioning Allied command.

Eisenhower's leadership abilities were tested to the utmost during preparations for the Allied invasion of France across the English Channel. The invasion involved enormous logistical and political challenges, which Eisenhower handled effectively. The successful Normandy invasion on June 6, 1944, D day, and subsequent Allied defeat of Nazi Germany made Eisenhower known throughout the world, and a hero to many Americans.

After the war, General of the Army Eisenhower worked briefly at four different tasks. First he served as army chief of staff, overseeing the army's rapid demobilization and pleading before congressional committees without much success for a less drastic shrinkage in the nation's armed forces. After leaving that job early in 1948, Eisenhower composed a wartime memoir, which he called *Crusade in Europe*. Writing at a pace of sixteen hours a day for many weeks, Eisenhower managed to finish the 550-page book and see it published by the fall of that same year. Scholarly and well written, the book proved a great success (it sold over a million copies) and made Eisenhower a wealthy man. Next he moved on to the presidency of Columbia University, recruited for that seemingly unlikely post by Columbia trustee and IBM chief executive officer Thomas J. Watson, Sr. Almost alone among the nation's leading business executives in perceiving that President Harry Truman might win the 1948 election, Watson wanted to keep Eisenhower, whose abilities he greatly admired, in the public eye and prepare him for a presidential race four years later. In keeping with these objectives, Watson began introducing him to leading businessmen once he arrived at Columbia that fall. Some of them became Eisenhower's closest friends for the rest of his life. When war in Korea broke out, Eisenhower took leave from his post at Columbia to accept Truman's request that he become commander of the newly formed North Atlantic Treaty Organization.

While serving at NATO headquarters in Paris during the fall of 1951, Eisenhower began receiving a series of moderate and liberal Republican visitors, who urged him to seek the GOP's presidential nomination in 1952. Anxious to end the party's twenty-year string of failure in presidential races, and fearful that its leading candidate, Senator Robert Taft of Ohio, was too conservative and partisan to be elected, these Republicans turned to Eisenhower as their only real hope. Although receptive to their requests, he at first ruled out running actively, in view of his current position at NATO. To encourage a draft-Eisenhower movement, he wrote a letter on October 14, 1951, his sixty-first birthday, to Senator James Duff of Pennsylvania, a leader of the GOP's moderate-to-liberal wing. In the letter, Eisenhower declared that he was a "liberal Republican" and willing to accept the party's presidential nomination.

When the draft-Eisenhower forces managed to bring about a victory for his candidacy in the New Hampshire Republican presidential primary held on March 11, 1952, Eisenhower decided to return home and campaign actively. He and his supporters eventually prevailed by a narrow margin over the Taft forces at a tense convention that summer. Wanting and needing to placate the so-called Old Guard Republicans, Eisenhower agreed to accept their candidate for the vice-presidency, California Senator Richard Nixon. Having succeeded in uniting his party behind him, Eisenhower went on to defeat soundly the Democratic candidate, Governor Adlai Stevenson of Illinois, in an election marked by the highest voter turnout since 1908. So popular was Eisenhower's candidacy and so wide his margin of victory that the Republicans also managed to gain control of both houses of Congress.

Somewhat less clear was what an Eisenhower presidency portended. Throughout his campaign, Eisenhower had emphasized the need to continue the struggle against communism as effectively as possible, to root out corruption and waste in the federal government, and to end the war in Korea on favorable terms. Beyond that, he had offered few specifics as to what he would do. Characteristically, he was much clearer about what he would not. His presidency, Eisenhower explained in various campaign addresses, would not undertake to dismantle the most important public policy achievements of the previous twenty years. Nor would it seek any large-scale expansion of, or change in, the federal government's role in society. Rather, the Eisenhower administration would make such incremental changes in existing policies and programs as became necessary — and no more. Eisenhower called this approach "Modern Republicanism."

Eisenhower's selections for his cabinet and White House staff were consistent with that philosophy. He chose men drawn mostly from the worlds of business, corporate law, the military, and education. They, like him, shared the view that the most durable New Deal programs and policies, such as Social Security, the system of fairly high and progressive income tax rates, and acceptance of labor unions could not, and should not, be undone. The same went for the large military and national security establishment and system of foreign alliances aimed at containing the spread of communism that had emerged under Roosevelt and Truman. Indeed, some of the new Eisenhower appointees had earlier contributed to the existence of those policies. The new secretary of defense, Charles E. Wilson of General Motors, had endorsed a stable and enduring relationship with the auto workers' union while head of that company. The new secretary of state, John Foster Dulles, had helped shape the nation's foreign policy commitments during Truman's presidency by acting as an unofficial adviser to and spokesman for congressional Republicans. Although Eisenhower's cabinet selections were dismissed by liberal Democrats as "eight millionaires and a plumber" (the latter a reference to the new labor secretary, plumbers' union head, Martin Durkin), their outlook differed significantly from that of their counterparts in the Republican administrations of the 1920s.

No one understood that better than the leader of the GOP's Old Guard, Robert Taft. His influence on staffing the new administration proved slight. The new treasury secretary, Cleveland banker George Humphrey, sympathized with the Old Guard view. But Humphrey was a practical man who distrusted theories of all kinds, and he went along with leaving the government's economic policies pretty much undisturbed. Similarly, the new agriculture secretary, Ezra Taft Benson of Utah, sounded quite conservative, but he acquiesced to Eisenhower's decision that the practice of government subsidies for farmers be continued for the time being. The only other high administration official who enjoyed a cozy relationship with the Old Guard, the new vice president, Richard Nixon, mattered even less to the making of public policy. Disliked by Eisenhower and many others in and out of the government, Nixon found himself and his ideas usually ignored and sometimes even belittled. All of this became very clear to Taft when the first Eisenhower budget proposal emerged and differed only marginally from Truman's. Determined to be a loyal Republican, Taft went along, but privately he felt defeated. And when he died suddenly in the summer of 1953, the Old Guard lost its most effective and influential leader.

But if GOP conservatives felt alienated from the policies and personnel of the new Eisenhower administration, many others did not. In its first term especially, Eisenhower's presidency enjoyed considerable success by taking the Modern Republican path. Despite vigorous resistance from some congressional Republicans, his first budget plan (for fiscal year 1954) passed. Old Guard efforts during 1953 and 1954 to rewrite the nation's labor laws in ways opposed by the major industrial unions also came to naught. And when the economy turned down in the fall of 1953, Eisenhower succeeded the following year in pushing through an economic stimulus package only slightly different from the one Truman had used to fight a recession in 1949 and 1950.

Rather than dismantle Roosevelt's and Truman's domestic policy legacies, Eisenhower's presidency in its first term tended to consolidate and slightly expand them, especially so once the Democrats regained control of Congress in 1955. In 1954, Eisenhower recommended and Congress approved an expansion of the Social Security system that extended coverage to ten and a half million new workers, and an increase in monthly benefits. Eisenhower supported a rise in the minimum hourly wage from 75 cents to a dollar, which passed in 1956. His administration also pushed through the Federal Aid Highway Act that same year, which provided 90 percent of the funds needed to construct a comprehensive interstate highway system (localities were expected to contribute the remaining 10 percent). Congress, with Eisenhower's blessing, appropriated $25 billion to launch this initiative in 1956, the largest sum for any domestic program in the country's history to that point. Eisenhower also successfully supported legislation in 1955 to encourage the creation of a seaway linking the Atlantic Ocean and the Great Lakes through the St. Lawrence River. Four years later, he helped dedicate the completed project. Congress also went along with the Eisenhower administration's proposals to liberalize and extend unemployment insurance and workmen's compensation acts.

As revealing of the nature of Eisenhower's domestic policy during its first term was the administration's unsuccessful quest for even more such policy changes. Eisenhower proposed legislation to extend health insurance coverage through federal underwriting of the policies issued by private carriers. Congress blocked the move, however, which had antagonized both the most extreme New Deal liberals, who wanted a program of public health insurance, and the most conservative Republicans and their Dixiecrat allies, who wanted the health care system to remain an entirely private one. And in the area of public housing, Eisenhower pushed for the creation of significantly more new construction than Congress ultimately agreed to finance.

Also indicative of Eisenhower's domestic priorities during his first term were his three appointments to the U.S. Supreme Court: California Governor Earl Warren, named chief justice in the fall of 1953, John M. Harlan, appointed in 1955, and William J. Brennan, named to the Court one year later. The first was one of the most liberal leading Republicans in the country, and the third a moderately liberal Democrat. Even John Harlan, although more moderate than either Warren or Brennan, shared no enthusiasm for conservative activism on the Court of the kind that had provoked a political crisis during the 1930s. An apostle of judicial restraint in most areas, Harlan favored allowing the popularly elected branches of government to take the lead in making public policy at a time when many Old Guard Republicans still considered the Court's primary role to be the protection of private enterprise, power, and "rights" from government intrusion.

The outlines of Eisenhower's foreign policy during these years proved similarly consistent with that of his two immediate predecessors. He and his key foreign policy adviser, Secretary of State John Foster Dulles, supported the huge military and national security establishment created under Roosevelt and Truman, as well as the Marshall Plan and NATO alliance that Truman had pushed for successfully. And although Eisenhower considered Truman's policy of endeavoring to contain communism everywhere in the world no matter what the cost in American lives and dollars to be excessive, he and Dulles made no radical retrenchments in the nation's overseas commitments. Rather, Eisenhower and Dulles continued to speak the same hawkish language when dealing with the Soviets and their allies that Truman had, and with even greater credibility. For example, in seeking to end the war in Korea, Eisenhower and Dulles indicated privately to the Chinese government (through the Indian prime minister) that the U.S. might use atomic weapons in Korea and China to break the Korean military stalemate. Although it is unclear whether such belligerence persuaded the Chinese and their North Korean allies to agree to an armistice, one materialized soon thereafter, on July 27, 1953.

The new administration's policy toward the emerging nations of the Third World was similarly informed by the containment policy that Truman and his advisers had earlier developed. Although sympathetic to the former European colonies' quest for national independence, Eisenhower and Dulles feared that the termination of colonial systems of government might lead to the emergence in many places of weak, multiparty governments that would be susceptible to communist infiltration and overthrow. Rather than run such a risk, the Eisenhower administration intervened directly when it believed those circumstances had begun to develop in a country vital to U.S. national security. For those reasons, Eisenhower and Dulles used the CIA to topple the government of Mohammed Mossadegh in Iran during the summer of 1953, and that of Jacobo Arbenz Guzmán in Guatemala one year later.

In Southeast Asia, Eisenhower's presidency similarly adhered to Truman's commitments, and even expanded them, most notably in Vietnam. Eisenhower and Dulles increased the military assistance program to pro-Western elements in Vietnam which Truman's administration had begun. Eisenhower did refrain from sending in U.S. troops when French forces there were decisively defeated in the battle of Dien Bien Phu and subsequently withdrawn. However, his administration continued and increased its support for the new government of South Vietnam after the nation was partitioned in 1955 at a conference in Geneva, Switzerland. Eisenhower and Dulles also encouraged South Vietnam's new leader, Ngo Dinh Diem, not to allow the election throughout all of Vietnam called for in the Geneva accords, for fear that it would bring the North Vietnamese leader, Ho Chi Minh, a communist, to power there. And to contain the spread of communism in the region more generally, Eisenhower and Dulles established the Southeast Asia Treaty Organization in September 1954. Modeled on the NATO agreement (but much weaker), SEATO committed the United States to consulting with the governments of France, Great Britain, the Philippines, Pakistan, Australia, and New Zealand should they be menaced by a "common danger" in that region. A separate protocol extended this commitment to Laos, Cambodia, and South Vietnam.

A similar pattern of basic continuity emerged with respect to such issues as Eastern Europe and arms control. Eisenhower and Dulles verbally encouraged freedom-loving people in Eastern Europe to resist Soviet-dominated governments there, but when a rebellion broke out in Hungary during 1956, the U.S. government offered no military or economic assistance to the rebels, who were soon violently repressed by Soviet military might. This refusal to intervene was consistent with the policy Truman's administration had earlier established. With respect to arms control, Eisenhower and Dulles expressed cautious interest in the idea, but during the administration's first term took no bold steps in that direction. The most they would do was agree to meet with Soviet leaders to discuss this and other matters at a conference held in Geneva in July 1955, the first such "summit" meeting in ten years. The only concrete agreement between the two sides that emerged was one aimed at encouraging more cultural exchanges between East and West, as a way of fostering greater mutual understanding. A useful first step, this accord did not lead to any immediate progress toward an arms control treaty.

The arms control issue had become in some ways more important once Eisenhower became president precisely because he favored a greater reliance on nuclear weapons to protect national security, and reductions in conventional U.S. military forces. Eisenhower's great knowledge of military affairs had made him acutely aware of how the emergence of weapons of mass destruction such as the hydrogen bomb had made conventional armed forces less useful in conflicts involving the Soviet Union. The huge size of the communist Chinese army, Eisenhower also understood, undermined the feasibility of using conventional military force in Asia to halt the spread of communism there. He feared, too, that attempting to match the Chinese and Soviets in conventional forces would prove so costly to the U.S. economy and traditions of limited government as to endanger that which the country's defenders most wished to save. Eisenhower also knew that heavy military spending had reduced the amount of resources available for such worthwhile domestic initiatives as the construction of new schools and hospitals. For all of these reasons, he and Dulles pushed for what they called a "New Look" in national security policy, which would lead to a smaller, less expensive military establishment. The inevitable consequences were to place greater reliance on nuclear weapons and the development of ever more sophisticated and accurate means to deliver them. Eisenhower's New Look also implied greater reliance on covert operations by the CIA, and other kinds of counterinsurgency efforts.

Even this change in foreign policy, more significant than in other areas, was really only one of degree. When the Korean War ended, the size of U.S. conventional forces would surely have been reduced, even if a Democrat had been president. And most revealing, perhaps, was that military spending throughout Eisenhower's first term in office remained fairly stable at about one tenth of U.S. gross national product. That was extraordinarily high for a major industrial society ostensibly at peace. As the term "New Look" unwittingly implied, the change had more to do with appearances than underlying realities.

Such spending did contribute to an extraordinary economic boom in 1955 and 1956, without much immediate inflation, and a federal budget surplus in the second of those two years. This impressive achievement helped carry Eisenhower to a landslide reelection victory over Adlai Stevenson, despite concerns about Eisenhower's health (he had been hospitalized in September 1955 for a massive heart attack and again in June 1956 to undergo surgery for ileitis, a digestive disorder). Also crucial to Eisenhower's electoral success was his administration's acceptance of the most durable and important New Deal programs and policies. He himself made that clear when speaking to his loudly cheering supporters on election night in 1956. "I think," he said then, "that Modern Republicanism has now proved itself. And America has approved of Modern Republicanism."

Despite his broad popularity with voters, Eisenhower encountered sharp criticism, even during the early and middle years of his presidency. The most common complaints about Eisenhower's performance during that time concerned his handling of McCarthyism and the Supreme Court school desegregation decision. Eisenhower's cautious approach to both situations persuaded many liberals that he had acted timidly when boldness had been required.

Eisenhower, to be sure, never directly confronted Wisconsin Senator Joseph R. McCarthy, and, in fact, refused during the 1952 campaign to rebuke him publicly when the two men shared a speakers' platform in Wisconsin, for McCarthy's earlier description of General George Marshall as a communist dupe. Eisenhower came to regret that decision, but not his unwillingness to confront McCarthy more generally. Eisenhower found direct public confrontations distasteful. Even more important, he believed that Truman's direct attacks on McCarthy had backfired, by drawing attention to the undocumented allegations of a hitherto obscure U.S. senator. And Eisenhower was determined to preserve the dignity of his office, lest the power of his presidency more generally be eroded, to the nation's detriment. Instead, Eisenhower and his aides tried to ignore McCarthy and to undermine him indirectly. By taking this course, Eisenhower in effect compelled the Senate to act once McCarthy's own recklessness had turned public opinion against his actions, which the Senate eventually did. Eisenhower indicated privately that he strongly disagreed with McCarthy's allegations and tactics, unlike many of the most conservative Republicans.

With respect to the Supreme Court's 1954 decision in *Brown v. Board of Education of Topeka, Kansas,* which struck down racially segregated public schools as unconstitutional, Eisenhower remained silent in public about the ruling, while telling one senior aide (speechwriter Arthur Larson) that he privately disagreed with the Court's reasoning in the case. Eisenhower's equivocal response to the Brown decision did not reflect a reactionary animus toward blacks. Eisenhower privately deplored racial prejudice, although some of his remarks while president indicate insensitivity at least in this area. He remained silent about the Brown ruling because he disagreed with the Court's conclusion that separating whites from blacks during schooling was necessarily injurious to the latter, and because he believed that government attempts to compel racially integrated schools would backfire. Until attitudes among prejudiced whites changed, he argued privately, efforts to compel racially integrated schools would lead only to efforts on the part of such whites to avoid them, one way or another.

Eisenhower eventually made clear that he as president would enforce the Court's decision. When resistance to school desegregation in Little Rock, Arkansas, turned violent in September 1957 and the state's governor, Orval Faubus, sided with the protesters, Eisenhower federalized the state's national guard, ordered units deployed in Little Rock, and forced the protesters (and Faubus) to back down. Had he made stronger statements in support of the Brown decision earlier, this confrontation might never have occurred. Eisenhower believed, however, that such statements would have had a polarizing and even inflammatory effect, on many white southerners especially, and might have produced exactly the result they were intended to forestall. Having spent much of his life in southern and border states, Eisenhower had acquired an acute familiarity with the difficulties involved in trying to dismantle Jim Crow and a pessimistic view of the likely consequences of trying to use the federal government's power to do so. To his liberal critics, Eisenhower's tepid response to the Brown ruling, whatever the reason, was the single greatest failure of his presidency.

A few of Eisenhower's harshest detractors went farther. They suggested that his handling of McCarthy and the Brown decision revealed a basic orientation toward Hooverite conservatism. Some critics thought him inarticulate and even stupid. They were wrong, for although Eisenhower did sometimes speak of the relationship between government and business in ways that sounded reminiscent of Herbert Hoover, Eisenhower's presidency endorsed and perpetuated an increase in federal government authority that Hoover had staunchly opposed. The idea that Eisenhower was a poor public speaker or even unintelligent seems to have developed mostly in the aftermath of his stroke on November 25, 1957, when Eisenhower's speech was seriously affected, and by his penchant for military usage, which was clear and precise to those in the armed forces, but awkward to many civilian ears. His occasional willingness to sound confused when he wished to leave reporters uncertain about where he stood on some matter of public policy also bolstered that impression. Eisenhower's lack of familiarity with high culture seems also to have misled some highly educated people into underestimating his intelligence. Clearly not a fully rounded man, Eisenhower was just as obviously not a stupid one.

The more negative assessments of Eisenhower and his presidency also grew out of its difficulties during his last three years in office. Eisenhower's Modern Republican approach had worked well during his first five years as president precisely because social conditions then had not demanded any major shifts in public policy. These were the golden years of the 1950s, when the fighting in Korea ended, suburbia mushroomed, the postwar baby boom peaked, and the middle class grew bigger and ever more prosperous and secure. During the late 1950s, however, conditions began to change in ways that demanded more than the incremental adjustments Eisenhower's presidency had thus far made. Required increasingly to think and act in new ways, Eisenhower and his aides found doing so very difficult.

Many of the country's new problems by then were unexpected byproducts of earlier public policy successes. Strong labor unions had raised workers' wages so high as to make whole industries vulnerable to increasing foreign competition. Rapid economic growth and middle-class affluence led to ever more pollution of the nation's environment, from cars especially. Higher wages, rapid economic growth, and government spending also combined to create a rising wave of inflation that menaced the long-term prosperity of all Americans. Increased levels of education among women and nonwhites encouraged them to demand greater equality, in employment especially. Ever larger stockpiles of U.S.

armaments bred a feeling of national security, but also increased the risk of a nuclear Armageddon should war with the Soviets come. The end of colonial rule in many Third World countries made revolutionary change more likely there.

Faced with these growing challenges, Eisenhower's presidency moved haltingly to meet them. Deeply concerned about the related problems of growing foreign competition and inflation, Eisenhower and his aides pushed successfully for curbs on union power, cuts in federal spending, and higher interest rates during the last few years of his presidency. Administration officials strongly opposed to these moves either left voluntarily or were forced out. These changes in policy and personnel did bring inflation down, but also caused the economy to stall in 1958, recover, and then stall again in 1960, an unpopular result that cost the Republicans heavily in the elections held in those years. Any major environmental initiatives would likely have reduced economic growth (and increased unemployment) even further, and none materialized. Eisenhower's growing concern about black equality before the law did lead his administration to push successfully for a new federal civil rights statute in 1957, the first such measure since Reconstruction. Fierce resistance from southern Democrats in Congress, however, succeeded in amending Eisenhower's proposal so thoroughly as to make it almost totally ineffectual. Efforts to meet the growing concerns of women in the paid labor force produced little more than administration studies.

Eisenhower's foreign policy record in these years was even more unsatisfactory. Efforts to keep cutting military spending stalled in response to the public outcry over the Soviets' successful launch of the *Sputnik* satellite in September 1957. Eisenhower and his aides tried hard to negotiate an arms control agreement with the Soviets, only to see talks collapse in May 1960 when an American U-2 surveillance plane was spotted and shot down over the U.S.S.R. That same year the new Cuban leader, Fidel Castro, began to move toward a pro-Soviet policy, thereby giving the U.S. another major national security concern in Latin America. Around the same time, a Marxist-inspired rebellion in Laos began to destabilize that country.

All of these problems contributed to the defeat of Eisenhower's would-be successor, Vice President Richard Nixon, when he ran against Massachusetts Senator John F. Kennedy for the presidency in 1960. Hampered by very deep divisions within the GOP over how best to address the nation's growing problems, Nixon ran a campaign lacking in clarity and confidence and narrowly lost, a result Eisenhower interpreted as a public rejection of his administration's eight-year record. In retrospect the outcome seems to have reflected doubts about what a Nixon presidency would have done, rather than disapproval of what Eisenhower's had accomplished.

Eisenhower spent the last eight years of his life in retirement with Mamie at their farm in Gettysburg, Pennsylvania. Although one of the most popular living Americans during the 1960s, he quickly became a very dated figure, unable to exert much influence over the direction of his party or the country. His hawkishness on the Vietnam War issue also diminished his luster with many moderates and liberals. When he died on March 28, 1969, the event signaled to many the passing not just of a man, but also of an era.

The Eisenhower presidency remains historically significant for at least three reasons. Its success rehabilitated the Republican party in the eyes of many Americans old enough to remember the Hoover administration's ineffectual and seemingly callous response to the Great Depression. Even more important, Eisenhower's presidency reflected a shift in the GOP away from the Old Guard position. A moderately liberal Republican president, he was the first to fit that description since Theodore Roosevelt. Eisenhower's presidency is important, too, because it demonstrated both the strengths and weaknesses of an incremental approach to politics and public policy. Effective in the quiet years of the mid-1950s, it seemed much less so in the turbulent decade and a half that followed.

— DAVID L. STEBENNE
Ohio State University

John F. Kennedy
1961–1963

b. May 29, 1917
d. November 22, 1963

J OHN F. KENNEDY was more important dead than alive. Of no other president in the twentieth century could that blunt judgment be made. Certainly Franklin D. Roosevelt has had a significant posthumous presence in American life. But FDR spent a dozen years in office. He made decisions, created institutions, and nurtured ideas aplenty for a host of admirers and critics. In contrast, Kennedy served in the White House fewer than three years (the fabled "thousand days"). His was an administration short on practical accomplishments. At the time of his assassination, the president had signed no major piece of domestic legislation and had failed to achieve his main foreign policy objective — turning back the tide of Communist-led revolutions in the developing world. He was still favored to win reelection, but racial tensions that would soon rend the Democratic party were already beginning to cut into his approval rating.

Yet, since his death, John Kennedy has had a profound and remarkably durable impact on American politics and culture. His identity as the first Catholic president — and the scion of a ferociously ambitious Irish-American family — helped speed the ethnic diversification of the public sphere. A reluctant reformer, he nevertheless proposed — near the end of his life — far-reaching bills for civil rights, education funding, and health insurance for the elderly which became cornerstones of a new liberalism after his successor, Lyndon Johnson, pushed them through Congress. JFK also proclaimed the bold goal of a manned voyage to the moon by the end of the decade; it was a costly pledge that both Presidents Johnson and Nixon felt bound to honor. And Kennedy's vigorous efforts to oppose anti-American radicals in what was coming to be known as the Third World helped bring on the debacle of the Vietnam War and may even have motivated his killer, a devotee of Fidel Castro named Lee Harvey Oswald.

But one cannot explain Kennedy's posthumous aura through demography and policy alone, particularly given their often painful legacy. Why, a generation after the assassination, did millions rush to view a film, directed by Oliver Stone, that alleged a powerful conspiracy had caused the president's death? Why, in the 1990s, did a Washington, D.C., radio station that airs both rock and roll and the antic Howard Stern dub itself WJFK? Why do portraits of the president and his wife still adorn the walls of many working-class homes? Why did Americans, in opinion polls conducted from the 1970s to the 1990s, routinely rate John Kennedy the greatest of all presidents?

For millions in the U.S. and abroad, JFK, in memory, came to symbolize the lost promise of the 1960s. He embodied the visions of sensual, flamboyant change that flowed from an era of unparalleled economic abundance. The president and his young wife, Jacqueline, were remembered as democratic monarchs, tragically dethroned just as they were beginning to accomplish great things. As it was, the First Couple, with the aid of ubiquitous television cameras, established a dazzling style of athletic beauty and an elegant, witty manner. Subsequent legions of male politicians and their mates sought to mimic the Kennedys, rarely with success.

What made Kennedy so alluring as a man also contributed to his limitations as a president. Leader of the dominant nation on earth, he believed that concentrated brilliance and energy — beaming from a tight, self-referential squad of White House advisers — could vanquish "any one to assure the survival and the success of liberty," as he put it in his inaugural address. Having captured the presidency at the age of forty-three, less than a decade after first being elected to the Senate, Kennedy thought he could beguile his critics into acquiescing in his plans.

However, a growing collection of rebels — in the Caribbean, Southeast Asia, and the American south — were immune to his charms. He spent much of his presidency responding to their challenges, even while he kept burnishing an image of regal audacity. In a self-reflective moment, JFK might have agreed with a forerunner who was the first chief executive to be murdered in office. "I claim not to have controlled events," wrote Abraham Lincoln in the midst of the Civil War, "but confess plainly that events have controlled me."

*　　*　　*

As much as any man of his generation, John F. Kennedy was groomed from his youth to run for the White House. His father practically insisted upon it. Joseph P. Kennedy, the son of a prosperous saloon owner and grandson of a refugee from the great Irish famine, was determined to best the Protestant elite in both business and politics.

The first part of the task he handled himself. After graduating from Harvard in 1912, Joseph Kennedy quickly learned how to make millions. By the mid-1920s, an age that certainly "roared" for that small minority of Americans able to master the roguish ways of Wall Street, he had become one of country's richest men. Subsequent investments in the movies made him even wealthier. So did a shrewd decision to sell off most of his stock just months before the Crash of 1929 and to jump back into the market in the early 1930s when share prices were at their nadir.

Kennedy handled his nine children as aggressively as he did his portfolio. With the expert assistance of his wife, Rose, the self-disciplined daughter of a leading Massachusetts politician, Joseph trained his boys for political office and his girls for spirited refinement. Pride of place fell to John (nicknamed Jack) after the patriarch's eldest son, Joseph Jr., a navy pilot, crashed and died during a risky mission in 1944. John suffered from a variety of physical ailments that frequently confined him to bed and plagued him throughout his life. But his father still demanded that the boy, while at their vacation homes on Cape Cod and in Palm Beach, join his siblings for rigorous calisthenics and grueling sailing races and touch football games. Rose pinned notes on his pillowcase urging him to memorize the presidents, and both parents drilled John and his brothers in current events and history. Competition was the family creed. "Don't play unless you are captain," Joe Sr. advised his brood. "Second place is failure."

The tycoon was no chilly martinet; he adored his children, wrote them long letters when he was away, and spent lavishly to guarantee their happiness and further their careers. But he wanted them to prove that no position was too lofty for a Kennedy. On occasion, the multimillionaire Harvard graduate revealed a touch of plebeian resentment: in the 1930s, a reporter blithely called him an Irish American. "I was born here," sputtered Joseph. "My children were born here. What the hell do I have to do to be an American?"

His son's ascension to the White House would supply the answer. The patriarch blazed the path by serving the New Deal in two significant posts, as the first chairman of the Securities and Exchange Commission ("it takes a thief to catch a thief," quipped FDR) and then as ambassador to Great Britain, beginning in 1938. Loose talk about the need to "coop-erate" with Adolf Hitler forced him to resign the latter in 1940, but his money and ethnicity (Irish-Catholics anchored the Democrats in the north) kept Joseph Sr. in the thick of party affairs.

The diplomatic job had at least one advantage: it gave son John his first chance to get noticed for something other than his handsome face and his talent for seducing women (the latter a trait he shared with his father). In 1939, the young man interrupted his studies at Harvard to serve as one of the ambassador's secretaries. John admired the aristo-crats he met in London and had ample time to travel on the continent. That fall, after war broke out, he returned to Cambridge with a fine idea for a senior thesis: the British government's failure to prepare itself for armed conflict. The paper was sloppily written and lacked rigorous anal-ysis. But his father persuaded his friend Arthur Krock of the *New York Times* to revise it thoroughly for publication. The resulting book, *Why England Slept,* sold quite well, in part because Joseph Sr. purchased thirty thousand copies and relegated most of them to his spacious attic. At age twenty-three, the Kennedy with his name on the cover was sud-denly a celebrated author.

John's next big, if unplanned, career move was to become a war hero. He enlisted in the navy a few months before the bombing of Pearl Har-bor and soon volunteered to command a PT (patrol torpedo) boat in the South Pacific. The small crafts, with their plywood hulls and bulky weapons, were ineffective machines; their skippers tended to be rich and well-educated men who prized courage over results. One August night in 1943, near the Solomon Islands, a Japanese destroyer rammed Kennedy's *PT-109,* slicing it in half. The young skipper rallied his men and, the next morning, took a leather belt in his teeth and pulled an in-jured sailor to an islet several miles away.

The extraordinary rescue was an invaluable political asset. It made Jack Kennedy the epitome of what he later called "a new generation of Americans, tempered by war"; it also helped him escape the image of a privileged and pampered son. Of course, his father helped him make the most of the ordeal. Joseph Sr. induced the *Reader's Digest,* the magazine with the highest circulation in the country, to run excerpts from John Hersey's laudatory article in *The New Yorker* about the rescue. And, dur-ing the 1946 campaign, thousands of reprints appeared on subway and bus seats throughout the congressional district adjoining Boston which Jack Kennedy sought to represent. He won the traditionally Democratic seat by a landslide. Six years later, he defeated the incumbent Republi-can Senator Henry Cabot Lodge, Jr., by a much narrower margin. That victory was particularly gratifying. Back in 1916, Lodge's grandfather had beaten John's grandfather for the very same seat.

Kennedy's fourteen years in Congress, from 1947 to 1961, were notable mainly for political adroitness. He supported broadly popular bills to raise the minimum wage and extend Social Security. He curried favor with organized labor, a vital liberal constituency, while being careful not to alienate the millions of Catholics who liked Senator Joseph McCarthy (a family friend for whom his brother Robert once worked) or the southern whites who feared that "a federal dictatorship" would mandate racial equality. In 1954, when hospitalized for back surgery, Kennedy was the only senator who avoided casting a vote on McCarthy's censure.

On civil rights, not yet a major issue in Congress, Kennedy carved out a moderate position. In 1957, he backed a civil rights law that segregationists deplored and then helped to undercut it by voting for a provision to require a jury trial for violators. Lily-white southern panels, its supporters knew, would always acquit white defendants in such cases.

That same year, Kennedy's second book, *Profiles in Courage,* won a Pulitzer Prize — thanks, in part, to the backroom lobbying of Arthur Krock. The brisk study of eight American politicians who had stood up for principle, despite the cost to themselves, was almost entirely a ghostwritten affair (this time the spectral wordsmiths were Theodore Sorensen, Kennedy's Senate aide, and the historian Jules Davids). Its bestseller status helped propel JFK into a contest for the vice-presidential nomination in 1956, which he narrowly lost. During the battle over the civil rights bill, a congressional critic asked of Kennedy, "Why not show a little less profile and a little more courage?"

But, by then, the Massachusetts senator was already running for president. Since 1960, many admirers have portrayed JFK as the virtually inevitable victor, a figure whose looks, demeanor, and campaign strategy simply overwhelmed his opponent, the clumsy and anxious Richard Nixon. This interpretation ignores the extreme closeness of the contest, both in polls conducted during the race (neither candidate ever led by more than seven points) and in the final results. The shift of a mere 8,900 votes in Illinois and 46,000 in Texas — two states in which there were allegations of widespread voting irregularities — would have landed the Republican vice president in the Oval Office.

Certainly Kennedy ran an artful campaign. Choosing Lyndon Johnson for the second spot on his ticket undoubtedly helped him win Texas and may have diminished opposition elsewhere in the south. His endorsement of an "absolute" separation of church and state put Protestant bigots on the defensive. His brother (and campaign manager)

Robert's successful plea that a Georgia judge release Martin Luther King, Jr., from a maximum-security prison where he was being held for a minor traffic charge, rallied black voters to JFK's side.

As he stumped the country, vaguely promising a New Frontier, Kennedy whipped up the kind of adulation formerly associated with such entertainers as Rudolph Valentino and Frank Sinatra. After witnessing Kennedy speak that summer, the novelist Norman Mailer marveled that "the Democrats were going to nominate a man who, no matter how serious his political dedication might be, was indisputably and willy-nilly going to be seen as a great box-office actor, and the consequences of that were staggering and not at all easy to calculate." Political stardom helped JFK draw eight million more votes than had Adlai Stevenson in 1956.

Yet, in the end, a majority of Americans in 1960 cast their ballots along a dividing line as old as the Reformation. Protestants favored Nixon, 62 to 38 percent; Catholics backed one of their own by a margin of 78 to 22. With solid support from Jews and African Americans, John Kennedy could have claimed to be the champion of an embryonic coalition spawned by the great twentieth-century migrations to urban America from Europe, Mexico, and the states of the former Confederacy. He was not just the first president born in that century, but also the first chief executive who owed his victory to a congeries of religious, racial, and ethnic minorities.

But Joseph P. Kennedy's boy had no intention of flaunting his or anyone else's ethnic heritage. He would strive to be president for *all* Americans, gradually winning over domestic opponents with his vigorous style and aggressive policies — which made a sharp, and purposeful, contrast with Dwight Eisenhower, the man he replaced. The preeminent concern of the new administration — as well as the greatest force for national unity — was the cold war.

JFK and his key advisers (Attorney General Robert Kennedy, Secretary of Defense Robert McNamara, and National Security Adviser McGeorge Bundy) beheld two fronts in the conflict with Communism. First, they sought to contain the Soviet Union, led by the mercurial premier Nikita Khrushchev, inside its client bloc in Eastern Europe and make sure that the U.S.S.R. didn't gain an advantage in the nuclear arms race. Second, they wanted to reverse gains made by Communists and other anti-American radicals in the Third World, where extreme social and economic inequalities often gave revolutionaries the upper hand. In his inaugural address, Kennedy talked of nothing else but foreign policy, proclaiming the nation's duty to be "defending freedom in its hour of maximum danger." Now he had to deliver.

His initial undertaking — an adventure planned in the final days of the Eisenhower administration and eagerly embraced by the new president — was an utter disaster. On April 17, 1961, a brigade of 1,400 Cuban exiles, trained and armed by the United States, began an invasion of their homeland. The purpose of the landing at the Bay of Pigs — on a beach the Cubans call Playa Girón — was to facilitate the overthrow of the government of Fidel Castro. Since taking power in 1959, Castro had incurred American hostility by confiscating foreign-owned land, making a trade agreement with the U.S.S.R., and becoming an increasingly open Communist. But he was a popular leader and had advance warning of the attack. It took Castro's planes and troops less than twenty-four hours to force the outmanned enemy to surrender and to take most of the *brigadistas* prisoner. Just before the invasion, President Kennedy called off a planned air strike on Playa Girón for fear it would reveal U.S. sponsorship of the whole affair. Such an attack would only have prolonged the exiles' agony, but its withdrawal created the suspicion that Kennedy had "sold out" the anti-Castro cause.

In fact, the embarrassing failure at the Bay of Pigs — for which Kennedy publicly assumed the blame — only increased his popularity. It also strengthened his obsession with a regime he viewed as nothing more than a Soviet beachhead ninety miles off the coast of Florida. Within six months, the administration had initiated Operation Mongoose, a secret program that did everything short of war to destroy Castro's regime. Under Mongoose, 2,400 Americans and Cubans, with CIA funds, gathered intelligence, sabotaged the Cuban economy, and several times tried to assassinate Castro. One botched plot involved a poisoned cigar. The largest covert operation of JFK's term, Mongoose convinced the Cuban leader to seek ironclad protection against the Yankee behemoth.

His search helped set in motion events that nearly ended in nuclear war. Castro pleaded with his Soviet patrons to ward off a future U.S. invasion. In the summer of 1962, Khrushchev decided to begin construction of missile bases on the island with warheads capable of destroying a major city in the eastern U.S. He feared losing Moscow's sole ally in the Western Hemisphere and hoped to boost the confidence of the Soviet military. A year earlier, Kennedy and Khrushchev had bristled rhetorically over the latter's threat to restrict Western access to Berlin, which had long been a pawn in the cold war. The Communists ended that crisis in brutal fashion when they erected a high wall separating the two sectors of the city and began shooting East Germans who tried to scale it. Kennedy condemned the building of the wall but did nothing to stop it. He was content that the territorial status quo in Europe would last. But a superpower conflict in the Caribbean over genocidal weapons was much harder to resolve.

For nearly two weeks in October 1962, the president and the premier, abetted by teams of feverish advisers, engaged in what one historian has called "the most frightening military crisis in world history." Essentially, each side gambled that the other would recoil from incinerating tens of millions of human beings. After considering a full-scale assault on Cuba, Kennedy opted for a "quarantine" on Soviet shipping to the island to prevent any warheads from getting through. Khrushchev relented, knowing, as did few citizens of either nation, that the U.S. nuclear arsenal was much larger than his. Red navy vessels transporting materiel cargo returned to port. But the Soviet leader demanded, via consecutive diplomatic notes, that the U.S. promise not to invade Cuba and to remove missiles long stationed in Turkey which threatened the Soviet homeland. Robert Kennedy persuaded his brother to publicly accept the first demand and to ignore the second. In secret, the attorney general also pledged to withdraw the missiles from Turkey at a later date (a decision the U.S. had already made, as long-range U.S. bombers had made the missiles obsolete). Khrushchev agreed, and the crisis was over.

Resolution of the missile crisis ended the most dangerous phase of the cold war. JFK had come to understand the lunacy of nuclear confrontation. The expensive arms race did continue, as the Soviets worked mightily to catch up. But after the fall of 1963 each side avoided making gestures that would trip the ultimate fear of the other. A kind of confidence slowly grew between two powers whose leaders came to accept the unthinkability of a third world war. Months after the crisis, the U.S. quietly took its missiles out of Turkey. And negotiations began on a treaty to stop nuclear testing in the atmosphere. It was ratified in the fall of 1963.

Eight thousand miles away in South Vietnam, however, Kennedy was not so cautious. The National Liberation Front (or Vietcong), a guerrilla army led by Communists, was fighting to topple a regime the U.S. had helped set up in 1956, during the Eisenhower administration, and had been supplying with military equipment and American advisers (almost seventeen thousand by JFK's death). South Vietnam's president was the aloof, autocratic Ngo Dinh Diem, a Catholic and a large absentee landowner in a society composed overwhelmingly of Buddhist peasants. Diem was unable to counter the appeal of the Vietcong, heirs to the powerful nationalist movement that had broken the French hold over the country. For self-protection, he instructed his generals to avoid combat and unleashed the secret police, commanded by his sadistic brother-in-law, to quash all civilian opponents. In the late spring of 1963, a Buddhist monk burned himself to death in protest. The next day, a grisly photo of the immolation appeared on front pages around the world.

Privately, JFK sometimes voiced doubts about the wisdom of U.S. policy. Yet he had sworn to stop more "communist takeovers" and feared, as he told a close aide, "If I tried to pull out . . . from Vietnam we would have another Joe McCarthy red scare on our hands." In 1962, the U.S. Air Force began spraying herbicides on rural areas of South Vietnam to strip away jungle cover for the Vietcong and to poison food supplies. This tactic did not prevent the guerrillas, armed with captured weapons, from besting Diem's unmotivated troops. On November 1, 1963, with the tacit support of the U.S. embassy, dissident South Vietnamese officers staged a coup d'etat. While under arrest, Diem and his brother-in-law were murdered. As soon as Kennedy heard the news, he rushed out of the room, "a look of shock and dismay" on his face.

In recent years, some writers have maintained that JFK at the time of his death was bent on withdrawing from Vietnam. If he had lived, goes the argument, the long ordeal in Indochina would have been avoided. The bulk of the evidence refutes that contention. After the overthrow of Diem, Kennedy, despite his forebodings, was planning to augment U.S. troops. To accept defeat would have reversed, without warning, the entire thrust of his foreign policy. Years earlier, the Vietnamese Communist leader Ho Chi Minh had predicted, "the white man is finished in Asia." America's youthful president never understood the power of that sentiment in an anticolonial age.

Back home, John Kennedy moved cautiously to remedy the problems of a society that did not seem ripe for an updated New Deal. In the wake of the recession of 1958 to 1960, his first and highest priority was to spur economic growth. Without more secure jobs and higher corporate profits, it would be impossible to sell even a modest reform agenda. Ever the practical politician, JFK knew that, at best, only a narrow majority in Congress was willing to back favorite liberal measures such as health insurance for the aged and expanded federal aid to education. Conservative southern Democrats chaired many key committees in the

House and Senate, and the Republican minority saw no need to placate a man who had eked out one of the narrowest electoral victories in history.

Thus Kennedy, following the counsel of the economist Walter Heller, pursued a strategy to boost overall growth and consumer confidence without disturbing class inequalities. The president proposed a big tax cut for all levels of income and pressured both unions and corporations to accept only small increases in wages and prices. The economy seemed to respond, although because JFK took office during an upswing in the business cycle, it is difficult to gauge the precise impact of his policies. Nevertheless, each year from 1961 to 1965, the real gross national product shot up by over 5 percent.

This kind of performance convinced the president and his men that the government could afford to be more aggressive at home as well as abroad. The administration worked with its congressional allies to fund construction at colleges and to draft the ambitious bills that eventually established Medicare and the War on Poverty. None of these initiatives passed while JFK was alive, but his efforts did signal a desire to extend the welfare state that earlier Democratic presidents from Wilson to Truman had been so instrumental in building.

The one domestic issue that grabbed the public's attention during Kennedy's term was civil rights — a profound moral issue that previous occupants of the White House had largely shirked. During the early 1960s, the massive nonviolent actions of a growing black freedom movement gradually forced officeholders to make a choice — to help advance equality under the law or to resist it.

Until the last six months of his life, JFK tried to avoid the inevitable. He delayed, until November 1962, fulfilling a campaign promise to order the end of discrimination in federal housing and then did so with a minimum of publicity. The Justice Department urged grassroots activists in the south to register voters instead of confronting segregation laws directly through sit-ins and freedom rides. Then, when black citizens attempting to exercise their constitutional rights were beaten and shot, the government was slow to come to their aid. JFK often seemed more concerned about the harm that racist violence did to America's image abroad than about the people being harmed or the justice of their cause. He was not a callous man, just a prudent politician. Absent the white south, how could the Democrats continue as the majority party?

The spring of 1963, however, was a time for choosing. That April, the Southern Christian Leadership Conference, headed by Martin Luther King, Jr., began a well-planned series of street protests against the rigidly segregated order in Birmingham, Alabama. Local police attacked the demonstrators, who included thousands of children, with billy clubs and attack dogs, while city firemen blasted them to the pavement with high-pressure bursts of water. Television cameras recorded the violence, and most viewers were horrified. Then, in June, Alabama Governor George C. Wallace symbolically defied a federal order to allow two black students to attend the state university in Tuscaloosa. And Medgar Evers, leader of the NAACP in the neighboring state of Mississippi, was shot in the back and murdered.

Finally, JFK took a stand. On June 11, he told the nation that the time had come "to treat our fellow Americans as we want to be treated." He then outlined the broadest civil rights law in the nation's history. In late August, the president welcomed a throng of 250,000 black and white demonstrators to Washington, D.C. — after first warning organizers that the march "for jobs and freedom" might alienate Congress and persuading them to temper their goals and their rhetoric. By fall, the costs of his new political profile were indeed apparent. According to a Gallup poll, 50 percent of Americans believed that the administration was "pushing integration too fast" (only 11 percent answered "not fast enough"). A majority of southerners were ready to vote for Senator Barry Goldwater, a conservative Republican, instead of JFK, in the 1963 presidential election.

It seemed an excellent idea for the president to travel down to Texas, where state Democrats were at war over civil rights and other issues. As he rode through downtown Dallas in an open limousine, two rifle bullets instantly ended his life.

The dramatic public murder of the president on November 22, 1963, appeared to throw the nation into a painful political decline from which it has never truly recovered. The event, reported and debated more widely than any other killing in world history, soon became a marker symbolizing the end of U.S. confidence and supremacy — and the beginning of mass suspicion of the plans and character of America's top leaders.

For over three decades, the question of who assassinated JFK and why has seldom lost its power to haunt and fascinate. Some five hundred books and countless articles are devoted to the matter — far more than to any aspect of Kennedy's life. The alleged conspirators belong to almost every conceivable locus of national and even foreign power in the 1960s: the CIA, the FBI, the military, the Mafia, leaders of the Teamsters Union, Fidel Castro, Cuban exiles, oil magnates, and Lyndon Johnson himself. Each of the conspiracy theories neglects or minimizes details that could refute it. But most Americans have been persuaded. As of the mid-1990s, more than 80 percent believed there was a plot of some kind to kill the president.

At the same time, JFK remains the model of how a leader should act — a harmonious mix of grace, wit, idealism, and toughness. As Garry Wills writes, Kennedy "did not so much elevate the office [of president] as cripple those who held it after him. His legend has haunted them; his light has cast them in shadow." Skeptical historians have barely scratched the heroic reputation. Disclosures that the president was a routine, flagrant adulterer — including sexual liaisons with Marilyn Monroe, the girlfriend of a leading mobster, and young White House staffers — may have actually enhanced it. One can't expect a superman to play by the ordinary rules.

The memory of JFK thus contains a double meaning. His death was a tragedy — for his family, his society, and for his rather belated commitment to liberal reform. Yet the image created during his lifetime continues to fuel many Americans' yearning for a good and vigorous ruler, one who can make the old dreams live again, if only in words and pictures. There he is, our JFK, looking back at us from book jackets, movie posters, and Web sites: the upswept hair, the decisive gesture, the buoyant grin. He will always be glancing toward a future that never arrives.

— MICHAEL KAZIN
Georgetown University

Lyndon B. Johnson

1963–1969

b. August 27, 1908
d. January 22, 1973

MEASURING PRESIDENTIAL LEADERSHIP is a singular challenge in assessing Lyndon Baines Johnson. The man and the period of his five-year presidency were a case study in an outsized man coinciding with uncommonly volatile times.

LBJ was a larger-than-life character. Born in Stonewall, Texas, in 1908 and reared in Johnson City, a tiny hamlet in the state's south central Hill Country, fifty miles west of Austin, Johnson became an amazing American success story. He came from prominent Texas families, the Johnsons and the Baineses, who enjoyed standing in the state's political, religious, and educational institutions, but he grew up relatively poor. Nevertheless, inheritance and environment — nature and nurture — combined to make him a driven, grandiose, insatiable character who insistently used politics to become as powerful, influential, consequential, and wealthy a public figure as possible.

During his college years at Southwest Texas State Teacher's College in San Marcos, as a congressional secretary from 1931 to 1935, as the Texas director of FDR's National Youth Administration from 1935 to 1937, a congressman from 1937 to 1949, a senator from 1949 to 1961, and vice president from 1961 to 1963, no one who met him ever forgot LBJ.

He was an imposing figure who had to hold stage center, needed to dominate and control. One observer in the Senate remembers him as someone who appeared to be moving even when seated. "Wherever he'd sit down there was a cloud of dust," Bryce Harlow says. "Something was happening all the time, even if he were seated in the committee room. . . . He was getting up; he was sitting down. His alarm clock watch, which he just loved to harass people with. . . . It was a stunt . . . he always set it off at some propitious time to attract attention to himself. He couldn't stand not being the cynosure of all eyes. He had to be at the head of the table. . . . And people had to do what he thought they should do."

The journalists Rowland Evans and Robert Novak described Johnson's dealings with other senators as "the treatment." It included "supplication, accusation, cajolery, exuberance, scorn, tears, complaint, the hint of threat. It was all these together. It ran the gamut of human emotions. Its velocity was breathtaking, and it was all in one direction. . . . He moved in close, his face a scant millimeter from his target, his eyes widening and narrowing, his eyebrows rising and falling. . . . Mimicry, humor, and the genius of analogy made The Treatment an almost hypnotic experience and rendered the target stunned and helpless."

Benjamin Bradlee of the *Washington Post* compared meeting the six-foot three-and-a-half inch Johnson with going to the zoo. "You really felt as if a St. Bernard had licked your face for an hour, had pawed you all over. . . . He never just shook hands with you. One hand was shaking your hand; the other was always someplace else, exploring you, examining you. And of course he was a great actor . . . the greatest." Bradlee remembers an LBJ encounter as "a miraculous performance." Vice President Hubert Humphrey compared his boss to a tidal wave. Like Napoleon, Johnson was a tornado in pants.

And Johnson was a predictably self-serving character. As with so many other self-made men in the country's history, he was ambitious for fame and fortune. But what set Johnson apart from other Horatio Algers was his commitment to reach beyond himself and help others. As he told his biographer Doris Kearns Goodwin, "Some men want power simply to strut around the world and to hear the tune of 'Hail to the Chief.' . . . Well, I wanted power to give things to people — all sorts of things to all sorts of people, especially the poor and the blacks."

The psychiatrist Robert Coles describes Johnson as "a restless, extravagantly self-centered, brutishly expansive, manipulative, teasing and sly man, but he was also genuinely passionately interested in making life easier and more honorable for millions of hard-pressed working-class men and women. His almost manic vitality was purposely, intelligently, compassionately used. He could turn mean and sour, but he . . . had more than himself and his place in history on his mind."

In the fifty-two years before he became vice president, Johnson had converted his ambition into a series of political and legislative triumphs. As secretary to a wealthy, self-indulgent south Texas congressman, he had taken over the duties of the office, making it responsive to Depression-stricken farmers, businessmen, and army veterans. In the mid-thirties, he became the youngest and best state NYA director, a Texas wunderkind, who at age twenty-eight designed means to help impoverished youngsters and then beat several better-known opponents for south central Texas's Tenth District congressional seat. After eleven years in the House, where he established himself as a supremely effective advocate for middle-class and poor constituents, he defeated, by fair means and foul, Coke Stevenson, one of the most popular governors in Texas history, in a U.S. Senate race. Twelve years in the Senate, where he became the most effective majority leader in American history, made him a viable but unsuccessful candidate for president.

Pained by childhood poverty in rural Texas, educated by national deprivation during the Depression, and inspired by FDR, the greatest liberal reformer in the nation's history, Johnson had produced many "good works" during his congressional service. His commitment to New Deal and Fair Deal programs, the liberal nationalism of the 1930s and 1940s, helped transform America, particularly his native south and west. LBJ's public-service agenda included aid to education, dams providing flood control, conservation, cheap rural electrification, public works modernizing the nation's infrastructure, low-cost public housing for millions of poor Americans, Social Security benefits, minimum wages, subsidies serving the elderly, unskilled workers, and farmers, black civil rights, and space exploration through a National Aeronautics and Space Administration.

His thousand days as vice president was a painful time of lost influence. Although John Kennedy sent him on several foreign trips to help represent the United States in the competition with communism, although he made him head of a committee to oversee the development of the Apollo space program, the manned mission to the moon, and although he appointed him chairman of a committee on equal employment opportunity to help advance equal rights for blacks, Johnson saw himself as a minor influence in an administration seeking to conquer the "New Frontier." He later described his vice presidency as "filled with trips around the world, chauffeurs, men saluting, people clapping, chairmanships of councils, but in the end it is nothing. I detested every minute of it."

Johnson's political activism and rise to the presidency was facilitated by the Great Depression, the transformation of the role of the federal government, the rise of the national security state in response to World War II and the cold war, and, of course, Kennedy's assassination. If

Johnson had come of political age in the late nineteenth century or the 1920s, he never would have had the chance to support the many domestic programs enacted to revive the American economy and to forestall future downturns. The massive defense spending of the 1940s and 1950s allowed him to provide federal largesse to the south and the west, helping turn the Sunbelt into the region of greatest population growth and prosperity in the post–World War II era.

Likewise, the Supreme Court's desegregation decisions and the emergence of the civil rights movement of the 1950s led by Martin Luther King, Jr.'s, Southern Christian Leadership Conference (SCLC) opened the way to bold reforms, which helped make Johnson a national figure and vice-presidential candidate. Without the struggle for black rights, with which Johnson identified himself through the 1957 and 1960 civil rights laws, he never could have emerged as an acceptable running mate for John Kennedy and as the man who inherited the presidency following JFK's death.

Johnson's presidency from November 1963 to January 1969 was the product of his political imagination, of the personal stamp he put on everything his administration did in domestic and foreign affairs. At the same time, however, domestic and international developments beyond his control went far to make his presidency one of the most important in the country's history, a story of great success and terrible failure.

Kennedy's assassination traumatized the nation, challenging LBJ to overcome the demoralization and lost confidence in national institutions produced by JFK's death. Johnson's response was masterly. A reassuring address to the country on November 27, 1963, set the tone for the next twelve months. Johnson promised to carry forward Kennedy's program — an $11 billion tax cut, the civil rights bill, and an antipoverty package that aimed to make the 23 percent of underprivileged Americans into what LBJ described as "tax payers rather than tax eaters."

In his January 1964 State of the Union message, Johnson demonstrated the sense of moral surety for the work of reform which would characterize his more than five years in the White House. He was an evangelist declaring an "unconditional war on poverty," which would "cure" current want and "prevent" future neediness. In his speech, he asked the Congress to do "more for civil rights than the last hundred sessions combined" and "to build more homes, more schools, more libraries, and more hospitals than any single session of Congress in the history of the Republic."

In February, Johnson summoned aides Richard Goodwin and Bill Moyers to join him in the White House swimming pool to discuss plans for a reform campaign. "We entered the pool area," Goodwin recalls, "to see the massive presidential flesh, a sun bleached atoll breaching the placid sea, passing gently, sidestroke, the deep-cleft buttocks moving slowly past our unstartled gaze. Moby Dick, I thought. . . . 'It's like going swimming with a polar bear,' Moyers whispered. . . . Without turning his body, Johnson called across the pool: "Come on in, boys. It'll do you good." Stripping on the spot, they joined the president in the pool, where he "began to talk as if he were addressing some larger, imagined audience of the mind." He declared his intention to go beyond Kennedy and create a "Johnson program." Goodwin "felt Johnson's immense vitality," as he outlined a program that could move the nation "toward some distant vision — vaguely defined, inchoate, but rooted in an ideal as old as the country."

In May, Johnson used a commencement address at the University of Michigan in Ann Arbor to stimulate excitement for his Great Society. "Will you join the battle to give every citizen the full equality which God enjoins and the law requires, whatever his belief, or race, or the color of his skin?" he asked the audience. "Will you join in the battle to give every citizen an escape from the crushing weight of poverty? Will you join in the battle to make it possible for all nations to live in enduring peace — as neighbors and not as mortal enemies? Will you join in the battle to build the Great Society, to prove that our material progress is only the foundation on which we will build a richer life of mind and spirit?"

The result of Johnson's antipoverty crusade and reach for a Great Society was an explosion of groundbreaking and far-reaching laws passed by Congress between 1964 and 1968. In addition to passing JFK's proposed $11 billion tax cut to stimulate a sluggish economy, the U.S. Congress, under President Johnson, enacted or created the Civil Rights Act, forbidding racial segregation in places of public accommodation — bus terminals, hotels, restaurants, movie theaters, buses, parks, and swimming pools, for example (1964); the Voting Rights Act, eliminating devices for barring African Americans from exercising the suffrage (1965); a fair housing law, forbidding discrimination in the rental and sale of private homes (1968); antipoverty agencies, such as the Office of Economic Opportunity, and programs administering food stamps, Head Start, job training, and community action (1964); measures to protect the environment (1965); conservation of natural resources (1965–1968); consumer protections for children, roads, traffic, tires, mines, as well as natural gas line safety and truth in lending laws (1965–1968); urban renewal through a Department of Housing and Urban Development (1965) and Model Cities (1966); national management of air, rail, road, and water transportation through a Department of Transportation (1966); Medicare to insure the elderly against the costs of illness and hospitalization and Medicaid to provide health care to the country's least affluent citizens (1965); federal aid to elementary and secondary schools, as well as institutions of higher learning to improve the quality of schooling and expand educational opportunity (1965); immigration reform to replace the 1924 national origins system of ethnic quotas with rules favoring family cohesion and nationally needed skills (1965); and cultural supports in the creation of a public broadcasting system (1967), national endowments for the arts and the humanities (1965), freedom of information laws, allowing citizens to sue executive agencies (1966), and the Kennedy Center for the Performing Arts in Washington, D.C. (1967).

And the list of reforms does not end there. At the close of his presidency, LBJ's cabinet presented him with a catalog of domestic achievements during the five years and two months of his administration. The front inside cover of Johnson's memoirs, *The Vantage Point,* lists two pages of "landmark laws" with which he and the Congress "wrote a record of hope and opportunity for America."

There is no denying Johnson's role in making this record. In the days immediately after JFK's death, LBJ urged Kennedy's aides to forgive the fact that "he did not have the education, culture, and understanding that President Kennedy had . . . but he would do his best. . . . I want you to draw the threads together on the domestic program," he told Ted Sorensen. "But don't expect me to absorb things as fast as you're used to."

It was false modesty. Johnson was as great an executive legislator as ever sat in the White House. He had largely learned his craft as Senate majority leader, where he had cultivated the art of exchanging favors for votes. Leading a bill through the Congress, as he later told Doris Kearns Goodwin, meant attending to legislative business "continuously, incessantly, and without interruption. If it's really going to work," he said, "the relationship between the President and the Congress has got to be almost incestuous. He's got to know them even better than they know themselves. And then, on the basis of this knowledge, he's got to build a system that stretches from the cradle to the grave, from the moment a bill is introduced to the moment it is officially enrolled as the law of the land."

One aide recalls: "The best liaison we had with the Congress was Lyndon Johnson. He spent an enormous amount of time persuading congressmen to vote for particular issues." The columnist Drew Pearson, in a memo he wrote to himself, marveled at all the personal attention

Johnson was giving to congressmen. "He phones them, writes them notes, draws them aside at receptions to ask their opinions, seeks them out to thank them for political favors. Almost every Tuesday and Thursday he invites a group of Congressmen over for an evening at the White House. While Lady Bird shows the wives around the upstairs living quarters, the President talks policies and politics with the men." Johnson also had Larry O'Brien, his principal congressional liaison, arrange an evening for legislative aides, who did so much of the work on the Hill.

Texas Congressman Jake Pickle remembers how Johnson, "contrary to almost any president I've ever heard of, was on the phone constantly, talking to members of the House and the Senate. It would be nothing for him to talk to fifteen, twenty, or thirty different congressmen or senators during a day about some matter." And he was "awfully persuasive. He's about the best close-in, eyeball salesman that you'll run across."

Massachusetts Republican Silvio Conte remembered how LBJ called to thank him "on behalf of the nation for your vote." Conte "damn near collapsed right on the spot. . . . It's the only time since I have been in Congress that a President called me. I will never forget it," Conte said. Similarly, Johnson took pains to assure that his Great Society proposals would not antagonize the most conservative members of Congress. He instructed Larry O'Brien, for example, to send an aide to the Capitol to tell Mississippi Senators Jim Eastland and John Stennis that the president was ready to see them "any time a serious problem arises."

Not everyone succumbed to Johnson's wooing. Missouri Republican Thomas Curtis, who had served in the House since 1950 and was a member of the Ways and Means Committee, had little patience with Johnson's methods of winning congressional support. He remembers being called to the White House to receive the treatment. When "Lyndon sat down right next to me and started out, 'Now, Tom, your president wants to go over some matters with you,' I interrupted rudely and said, 'Come off it, Lyndon, my president your ass, what do you want to talk about?'" Johnson made no substantive gain with Curtis.

Defiant legislators paid a price for their independence. When Senator Frank Church, the Idaho Democrat who went against the President on a major bill, defended his vote by saying that the noted columnist Walter Lippmann shared his views, Johnson replied: "I'll tell you what, Frank, next time you want a dam in Idaho, you call Walter Lippmann and let him put it through for you." If Democratic congressmen persisted in defying the president's wishes, the House Speaker John McCormack arranged to strip them of committee seniority, a substantial blow to the influence of long-term members.

Johnson also understood that mastering Congress meant having a first-rate staff. As president, he could not interact with senators as he had as majority leader, though this limitation was difficult for him to accept. O'Brien remembers holding the president back from involving himself too much in the hour-to-hour management of pending bills. "There was no time day or night he wasn't prepared to charge in," O'Brien says. Early one morning, after a losing an all-night struggle in the House, O'Brien called Johnson to report the defeat. "When did this happen?" Johnson asked. In the middle of the night, O'Brien replied. "Why didn't you call me?" Johnson said. "You should have called me and told me about it. You know, when you're up there bleeding, I want to bleed with you. We have to share these things."

Johnson also relied on his cabinet to make legislative gains: "I have encouraged each one of you, and your deputies and key assistants, to meet frequently and informally with members of Congress," he told them. "It is vital to everything we do. Make an effort to get to know the men who sit on the committees that oversee your operations. You will find that this will make your job easier; you will get better results in your legislative programs. . . . The job of day-to-day contact with Congress is the most important we have. Many battles have been won, and many cases settled 'out of court' by the *right* liaison man being there at the *right* time, with the *right* approach."

The labor leader George Meany believed that Johnson "had a greater knowledge . . . of what was going on over on Capitol Hill than any other president ever had." Director of the Budget Bureau Charles Schultze remembers Johnson's unrelenting absorption with getting bills passed. "Never let up" was LBJ's motto: "don't assume anything; make sure every possible weapon is brought to bear . . . ; keep everybody involved; don't let them slacken."

Johnson's legislative agenda was so ambitious that he needed a "two-shift day" to achieve it. Rising at 6:30 or 7:00 each morning, he worked in bed until 8:00 or 8:30. Arriving at the Oval Office at about 9:00 A.M., he worked until 2:00 P.M., when he exercised by vigorously walking around the White House grounds or taking a swim. The second half of his "day" started at 4:00 P.M. after a nap in his pajamas, a shower, and fresh clothes. It lasted until at least midnight and often until 1:00 or 2:00 in the morning. Sometimes the two shifts turned into an uninterrupted fourteen or sixteen hours.

For all Johnson's energy and savvy about getting laws passed, his legislative record also relied heavily on public developments over which he had little or no control. First, Kennedy's assassination created a national mood of which Johnson took advantage. Reaching for a Great Society was an effective antidote to feelings that JFK's murder showed America as "a sick society." A war on poverty, a crusade to right national wrongs, to advance social justice and reduce economic inequality, struck resonant chords with a nation wondering if Kennedy's death reflected the country's affinity for violence and mayhem.

At a more practical level, LBJ's landslide victory and that of liberal Democrats in the 1964 elections opened the way to a reform agenda that dated back to the early years of the century. As Johnson himself put it shortly after becoming president, "Everything on my desk today was here when I first came to Congress" twenty-six years ago. Though Johnson was able to win passage of three Kennedy proposals in 1964 — the tax cut, civil rights, and antipoverty laws — the great burst of change came in 1965 after the election of the Eighty-ninth Congress.

This Congress of "accomplished hopes," of "realized dreams," as House Speaker John McCormack called it, was as much the product of Republican blundering in the 1964 campaign as anything Johnson and the Democrats did. After Vice President Richard Nixon's narrow loss to JFK in 1960, right-wing Republicans mounted a four-year campaign that gave them control of the presidential nomination and the party.

Arizona Senator Barry Goldwater, the Republican nominee, was an uncompromising anticommunist and New Deal antagonist. He preached victory over Soviet Russia and Red China and the restoration of individual freedom through the dismantling of federal programs and controls. His rhetoric frightened the country into believing that he was an extremist who would provoke a nuclear war or deprive middle-class Americans of an economic safety net.

Johnson brilliantly exploited Goldwater's vulnerabilities in the 1964 campaign. He encouraged the media, which was decisively on Johnson's side in the election, to alarm the public with stories of Goldwater's extremism. Johnson also initiated negative television advertising as an effective means of scaring the electorate into giving him one of the greatest victories — 61 percent of the popular vote, 486 to 52 in the electoral college — in presidential history. His "Daisy Field" ad was a model of how to attack an electoral opponent: It portrayed a beautiful child in a field of daisies as she pulled the petals off a single flower. A voice-over counted down to a nuclear explosion and a message urged LBJ's election.

Johnson had more trouble with his own party in the campaign than with Goldwater and the Republicans. Eager to win election to the presidency in his own right and get Attorney General Robert F. Kennedy,

John's younger brother, out of his administration, Johnson had to overcome popular sentiment favoring Kennedy's selection as vice president.
Believing that his victory would be attributed to RFK's presence on the
ticket, Johnson was determined not to have him as his running mate. He
declared his intention to bar any member of his cabinet from the ticket
as a way to forestall Kennedy's selection. At the same time, Johnson had
to head off a rebellion of the Mississippi Freedom Democratic party to
prevent disruption of the Democratic Convention. LBJ worked out a
compromise with competing black and white delegates to assure the
party harmony he believed essential to his success in the fall campaign.

Johnson's effective use of Goldwater's hyperbole against him and his
ability to mute Democratic party divisions allowed him not only to win a
landslide, but also to give the Democrats two-thirds majorities in both
houses of Congress. The sea change in Congress provided Johnson with
an opportunity that no liberal president had had before or since FDR.
And Johnson, of course, used it to good effect. But it seems fair to say
that his greatest allies in putting across the Great Society in 1965 were
Goldwater and conservative Republicans.

In foreign affairs as well, especially in policy making toward Vietnam,
peripheral developments played a significant part in setting Johnson's
course. Johnson inherited a set of commitments that he could not easily
ignore. Eisenhower had committed the United States to membership
in the anticommunist Southeast Asia Treaty Organization (SEATO) and
more specifically to the independence of South Vietnam. At the height
of the cold war, with Communist China describing wars of national
liberation, including the struggle in Indochina to replace French imperial rule with socialism, Eisenhower and Secretary of State John Foster
Dulles provided more than $1 billion in financial and military help to
Ngo Dinh Diem's pro-Western government in Saigon. By 1960, more
than fifteen hundred Americans were assisting the Vietnamese, and the
U.S. mission in Saigon was our largest in the world. Eisenhower also
pronounced a domino theory for the region — if Laos or South Vietnam
fell to Communism, the rest of Southeast Asia would fall inevitably into
Communist China's orbit.

John Kennedy subscribed to Eisenhower's concern. His public and
private actions made clear that he saw the independence of Laos and
South Vietnam from Communist control as vital to America's national
security. He also feared another round of McCarthyism in the United
States if his administration "lost" Indochina. Remembering the debate
in the 1950s over "Who lost China?" JFK was determined to insulate

himself and the Democrats from renewed right-wing attacks on liberal failure to meet the Communist challenge.

In response to a Communist insurgency in South Vietnam, JFK expanded the number of military advisers to 16,700 and substantially hiked the amount of aid to keep the Diem government going. In 1963, when it became clear that Diem had lost whatever popular support he had, Kennedy accepted the need for a military coup, which toppled Diem and then, to JFK's dismay, took his life. American endorsement or at least acceptance of the coup created a U.S. involvement with South Vietnam which no president could now easily abandon.

Nevertheless, Kennedy was leery of America's growing commitment to Vietnam. He worried that we would be drawn into an open-ended guerrilla war from which we could not escape. He told an aide that once we put in combat troops, it would be "like taking a drink. The effect wears off, and you have to take another." There is some evidence that JFK intended to remove U.S. advisers after he won reelection in 1964. But most of what he said and did indicated a determination to preserve South Vietnamese freedom from Communism.

Lyndon Johnson had no doubt that JFK intended to defend South Vietnam, and from the first day of his presidency he pledged himself to honor that "commitment." In dealing with Vietnam, prior events joined with the judgments of a powerful president to set America on a difficult and ultimately failed course that echoes in American foreign policy twenty-five years after our defeat.

During the first sixteen months of his term, LBJ struggled to avoid greatly expanding U.S. involvement in Saigon's military struggle against Vietcong guerrillas and North Vietnam's Communist regime under Ho Chi Minh. While he worried about the dangers of a wider war, fearful of the potential costs in American lives, dollars, and domestic tranquillity, he believed that he had no choice but to continue what Truman, Eisenhower, and Kennedy had been doing to meet the Communist threat around the globe.

Johnson feared that if he allowed a Communist takeover in Vietnam, it would mean the loss of all of Southeast Asia. He also expected it to embolden Moscow and Peking and heighten the chances of a third world war. In shaping policy toward Vietnam, LBJ never forgot the Munich experience of 1938. He also worried about the domestic consequences of failing to block Communism in Vietnam. Like Kennedy, he saw it provoking a right-wing outcry that would destroy his administration and defeat the Democratic party for as far into the future as he could see. In Johnson's estimate, Vietnam's independence was not only essential to American national security, but also to American domestic peace and the political survival of a Democratic party committed to national progressive reforms.

Johnson's first major expression of these concerns came in August 1964 during the presidential campaign. Determined to assure that Goldwater won no advantage over him in the competition to convince voters that he was prepared to meet the Communist threat, LBJ responded to North Vietnamese attacks on U.S. destroyers in the Tonkin Gulf with an air raid and a request for a congressional resolution endorsing U.S. determination to resist Communist aggression in Vietnam. Winning almost unanimous consent in Congress, the resolution, Johnson said, was like grandma's nightshirt, "It covered everything." As the next four years demonstrated, Johnson was largely mistaken.

In 1965, as South Vietnamese political instability and Communist military gains indicated that Saigon could not survive for long without direct U.S. military intervention, Johnson agreed to use American power to block a Communist victory. In March, he launched Rolling Thunder, a sustained bombing campaign against North Vietnam, which promised to force Hanoi into peace talks and an end to aggression against its southern neighbor. In July, after it had become clear that bombing alone would not deter the Communists, Johnson committed U.S. ground troops to the fighting.

The next three and a half years were a nightmare of confusion and disappointment. Every prediction about the "light at the end of the tunnel," about North Vietnamese incapacity to sustain a war against U.S. firepower, proved to be wrong. By 1968, Johnson had committed 525,000 American troops to the fighting and unleashed more tonnage on Vietnam than had been used by American air forces during all of World War II. But it did not bring the Communists to their knees nor even to the peace table. And just as bad, the war forced Johnson to put the Great Society on hold. Though repeatedly saying publicly that the United States could afford both guns and butter, he privately recognized that domestic spending would produce an intolerable amount of inflation and a possible recession. Consequently, he did all he could to stint on Great Society programs.

Johnson's inability to master a Third World country or what he described as a "pissant" power, unhinged him. Instead of recognizing that it would take the U.S. many years, if ever, to win a limited guerrilla war ten thousand miles from home, he convinced himself that the source of his troubles was American domestic opposition, which he believed was Communist-controlled. Even when the North Vietnamese demonstrated their resiliency by launching the Tet offensive in January-February 1968, Johnson could not accept the proposition that Hanoi saw the conflict not as primarily an episode in the international cold war, but as a civil war that they would not abandon short of national self-determination. To his mind, without Chinese and Russian help and Communist-inspired domestic dissent, North Vietnam and the Vietcong would not be able to hold out against American might.

Events in Vietnam beyond Johnson's control now combined with an affinity for unrealistic political judgments to destroy his presidency. In escalating the war, Johnson had refused to make his intentions clear to the Congress, the press, and the public. In 1965, he wouldn't reveal the extent of American military commitments for fear it would provoke a national debate and sidetrack the Congress from enacting the many Great Society laws then before it. Because he believed that the North Vietnamese and Vietcong would not fight for long, he tried to mute U.S. involvement in the war as late as 1967.

It was a terrible political error. His failure to encourage a debate about his policy produced a "credibility gap." One comedian joked, "How do you know when Lyndon Johnson is telling the truth? When he pulls his earlobe, strokes his chin, he's telling the truth. When he begins to move his lips, you know he's lying."

The war now became Lyndon Johnson's war instead of America's war. If LBJ had allowed a debate, which had led to national commitments to fight, he would have been in a stronger position to survive the political tumult over what to do about the conflict. But because of his high-handedness, every frustration in the war became his fault and eroded his capacity to govern and lead. The Tet offensive gave the lie to administration predictions that we were winning the war and further undermined Johnson's credibility as a war leader.

As a consequence, Senator Eugene McCarthy of Minnesota led a challenge to Johnson's renomination as the Democratic party's presidential candidate in 1968. After McCarthy came close to upsetting LBJ in the New Hampshire primary, Robert Kennedy, who was now a New York senator, joined the campaign in the middle of March. Fearful that Kennedy would defeat him, exhausted by the tumultuous events of the past four years, and worried that his health would give out in a second term, Johnson announced on March 31 that he would not run again, and he declared a moratorium on the bombing to encourage peace talks.

After the North Vietnamese agreed to begin discussions in Paris in May, Johnson considered reneging on his decision not to run again. Martin Luther King's assassination in April and Robert Kennedy's killing by Sirhan B. Sirhan in June after a primary victory in California had all but assured him of the nomination persuaded Johnson that he should seek a draft at the Democratic Convention in Chicago. Secret inquiries to Johnson's political allies persuaded him that the party would not support the idea, and the nomination went instead to Hubert Humphrey. Demonstrations and riots at the convention, however, blighted Humphrey's candidacy. A campaign largely fought over who could best assure the end of U.S. involvement in Vietnam and who could restore domestic tranquillity gave Richard Nixon a half million plurality out of seventy-two million votes cast.

Although the Paris discussions set the peace process in motion, no agreement had been reached by the end of Johnson's term in January 1969. To prevent an "October surprise," which seemed likely to give Humphrey the election, Nixon secretly discouraged the South Vietnamese government from joining the Paris talks during the U.S. presidential campaign. Nixon, despite a promise of a secret plan to end the war, would take four more years to get America out of the conflict. Two years later, in 1975, South Vietnam fell under Communist control.

The Vietnam disaster was the result of circumstances beyond Johnson's power and the quirks in his personality. Rational calculation suggests that by 1966 and certainly 1967, when it became clear that the war was at a stalemate and that "victory" would require a much larger commitment than Americans were willing to pay, Johnson should have cut his losses and ended U.S. involvement. "What should we do about Vietnam?" a journalist asked Senator George Aiken. "Declare victory and leave," Aiken advised in 1967.

But Johnson couldn't let go of the war. He genuinely believed that American national security was at stake and that a great disservice would result from "defeat" in Vietnam. But as important, he saw failure in the conflict as intolerable: he could not rationalize the loss of some thirty thousand American lives in an unsuccessful war; nor was he prepared to blight his historical reputation by being the first American president to have lost a war.

Johnson lived for four years after leaving the presidency. His concerns about his health were borne out by the onset of angina and the worsening of a heart condition dating back to the 1950s when he had suffered a massive coronary attack. Although he collaborated on his memoirs, set up a presidential library and an LBJ school of public affairs at the University of Texas in Austin during this time, he lived with a sense of frustration at being driven to the political sidelines. His death in January

1973 ended a nearly forty-year career in American politics that stretched from the 1930s to the 1970s.

Lyndon Johnson is an excellent case study in the limits and possibilities of presidential power. As a domestic leader, he seized upon uncommon circumstances to drive a host of major reforms through a receptive Congress. But much of what he did is now in dispute. The war on poverty cut the number of indigent Americans almost in half. However, as critics accurately point out, it did not make most of them productive citizens but dependent members of welfare rolls. During Johnson's presidency, Americans receiving aid to families with dependent children (AFDC) increased between three- and fourfold. The war on poverty was a fine idea, but Johnson's declarations that we could end poverty in America was an exaggeration that undermines his historical reputation. Likewise, Great Society programs like Medicare and Medicaid, environmental protections, and federal aid to education retain substantial support, but other reforms such as public housing, urban renewal, affirmative action, and federal aid to the arts and the humanities are now in bad odor. In general, Johnson's unbounded confidence in social engineering by the federal government has lost its hold on the public's imagination, partly because so much of what he believed in did not work all that well in practice.

Yet whatever one may say thirty years later about the virtues and defects of the War on Poverty and the Great Society, it is clear that Johnson was the greatest presidential legislator in the country's history. It is also clear that public events beyond Johnson's reach also came together with his larger-than-life personality to trap him in a war he would have preferred to avoid.

Whatever future historians may say about his presidency, one thing seems reasonably certain: Lyndon Johnson will be remembered as an exceptional chief, a flawed giant, who will command attention as long as Americans try to understand how presidents succeed and fail in the office.

— ROBERT DALLEK
Boston University

The Master Politician

"What you accomplish in life depends almost completely upon what you make yourself do," Johnson wrote in 1927, as a freshman at Southwest Texas State Teachers College. "Perfect concentration and a great desire will bring a person success in any field of work he chooses." The conviction was prophetic: Johnson was predicting his own, determined elevation from rural isolation and hardscrabble poverty to wealth and international power.

He drove himself hard, but as Johnson plotted his rise through various institutions — from campus government to the U.S. Senate — he also employed a calculated strategy. First, through observation, Johnson studied the levers of power and figured out who really controlled them. Then he ingratiated himself with those in power, impressing them with his determination and ability. As a student, these tactics led Johnson to take a campus job delivering messages for the college president and emerge as his trusted adviser. Once in Washington, Johnson applied himself to learning the intricacies of legislative politics and cultivated mentoring relationships with President Franklin Roosevelt, Speaker of the House Sam Rayburn, and Senator Richard Russell. These men, in turn, aided Johnson's rapid ascension in national politics.

Johnson's accomplishments did not come without a price. His relentless pace contributed to a heart attack at age forty-seven. Because he viewed his relationships as political alliances rather than friendships, he said he felt lonely all his life. Once he achieved the presidency, the thinking that made him a master politician did not always serve him well. For example, he was mistaken when he believed he could negotiate a political compromise with the North Vietnamese government. On the other hand, his astute understanding of Congress lay behind the legislation that brought his dream of a Great Society closer to reality.

Richard Nixon

1969–1974

b. January 9, 1913
d. April 22, 1994

I T IS THE SPRING OF 1940, and a moment of obscure portent for Richard Nixon.

Just twenty-eight years old, chafing at his parochial life and law practice in small-town southern California, he is making a first, lurching run for public office. There are rumors that the Republican state assemblyman from the district may retire. Maneuvering for some time to inherit the seat, the ambitious young Whittier lawyer strikes out that March to win the GOP nomination. In the evenings, he hurries off to speak to civic clubs and small audiences wherever he can arrange an invitation. Through soft, citrus-scented nights, across the bumpy back roads of rural Orange County, ending with long drives home in the darkness alone, Richard Milhous Nixon is beginning to travel what would be over the next thirty-four years literally millions of campaign miles.

The time seems in some ways a last interval of relative innocence for the man and even for the nation. It is, after all, before so much American history in which Nixon would be a major influence, before the cold war, before McCarthyism, before the civil rights revolution and Vietnam, before the historic realignment of both the ideologies and the electoral geographies of the Democratic and Republican parties, before the successive trials of a wealthy but riven America in the last decades of the century, before a rampant distrust of government, before Watergate.

For Nixon, the fledgling assembly candidate who will close his political career more than three decades later as the first American president to resign the office, there would be sharp ironies in that spring. He has no financial backing for this first race. Although he will later become notorious for receiving huge campaign contributions and tainted money, he must now borrow pocket change from his law partners to buy gas for his dusty old Chevrolet coupe. His much rehearsed speech to business-men and women's clubs deplores Franklin Roosevelt's New Deal battle with the Supreme Court, FDR's "unconstitutional acts," as he puts it. The Nixon who will be driven later from the Oval Office for abuses of power begins his political quest with an ardent defense of the Constitution. The eager young Nixon pushes tirelessly, seeing life as a struggle to be won, never to be given up. Fiercely determined, he seems moved by some deeper sense of destiny and righteousness, as he will be so often in the years ahead — although decades later in what he will describe as his last fitful, tormented nights in the White House, he will pray not to wake up, to die in his sleep.

That April of 1940, the popular assembly incumbent decides to run after all. Nixon's prospect vanishes. He dejectedly stops the speeches, but his ambition burns on. He rarely refers to the run, never acknowl-edges it for the fervid, failed effort it was. After the war he will have an-other chance, and not merely for the state legislature but for Congress. Still, seen in a larger history, the image of his Chevy coupe on those de-serted roads evokes a premonition. At the climax of perhaps the most extraordinary rise and fall in modern American politics, Richard Nixon will finish the thirty-seventh presidency much as he ended those long drives home in the fragrant California spring of 1940 — alone, in the darkness.

"Nobody ever reads my biographies," Nixon complained ruefully to his White House aides in 1971, typically urging them to study the first of what would be his own three autobiographies, *Six Crises,* to learn from his shrewd handling of politics and the press. He could scarcely have guessed then how large the literature of "my biographies" would be-come, how often his tale would be retold, if not always so gloriously as in his own version — in dozens of books and films and plays and televi-sion programs, in a bibliography as large as any president's since Abra-ham Lincoln. Still, there remains much that is still hidden, and much nagging irony and paradox in the remarkable career that went before the dark words captured on tape at the end. Richard Nixon may have reached the depths in the difficult history of the modern presidency. But he reached extraordinary heights as well. He was elected twice to Con-gress, once to the Senate, twice to the vice presidency, and twice to the presidency itself — altogether the winner of a national popular vote as many times as any figure in the nation's history. If he was the most disgraced president of the century, he was also one of the most victorious, and nothing less than a political prodigy in his time.

Our most listened-in-on president is also a reminder of how much we do not hear, of the chasm between the public and private presidency regardless of occupant, between the small portion of government and politics carefully performed to be heard aloud and the larger essence uttered in secret. How differently might we see our recent past, our political system, ourselves, if we could eavesdrop on his predecessors and successors equally uncensored?

The earnest young would-be legislator of 1940 was not yet the dark, tortured figure who occupied the White House. But he was, nevertheless, a product of a hard and painful, even if later much romanticized, childhood. Born in 1913 to Quaker parents of modest means — Frank Nixon and Hannah Milhous — he grew up in the craftsman bungalow his father built on a hillside in Yorba Linda, California, and later, in the small frame house behind Frank's small grocery store in Whittier. They were loving parents, but undemonstrative and often severe. And Nixon's youth was shaped to a large degree by their continuous struggle against poverty, by their desperate effort to cling to the lower rungs of the middle class in the raw, new, unforgiving communities of the Los Angeles basin, and by his father's rage and bitterness at his failure to prosper. It was also a youth shaped by grief. At the age of twelve, Richard lost his favorite and much-doted-on seven-year-old baby brother, Arthur, who died of meningitis. Eight years later, Harold, the oldest of the four boys in the family and the parents' and Milhous clan's favorite, succumbed slowly to tuberculosis.

Nixon responded to the trials of his childhood by becoming a serious, pious boy who went to church several times a week and once rose to be "saved" at a revival. Later, he wrote a thoughtful college essay on Christian ethics, and later still taught Sunday school himself as a young bachelor. He strove to become everything his family, church, and community valued, and to a remarkable degree he succeeded. He appeared unperturbed by the pervasive class discrimination and racial bigotry that marked his bucolic hometown, by the exclusion of his own family from the ranks of the privileged and successful in his town — although in later years his continuing resentment of people he considered born to privilege suggests that the impact of his own modest beginnings was not slight.

Nixon was a diligent student, did well in his local high school, became a campus leader at Whittier College, and was a hardworking and successful student at Duke Law School — president of the Student Bar Association and, in the hostile southern climate of the 1930s, a frequent critic of racism. Unable to find work in the great New York law firms to which he aspired, he returned home to Whittier to establish a small local practice in which he almost immediately became restless. His 1940 marriage to Pat Ryan, a popular and attractive local schoolteacher whom Nixon pursued relentlessly despite long periods of rejection and humiliation, seemed only to increase the striving that had already been apparent in his failed run for the assembly earlier that spring.

His search for a foothold in the tough, hard political world of southern California was incessant and, at first, constantly frustrated. Both Nixon and his equally industrious and intent young wife drew from their early setbacks an abiding social resentment and slinking envy of what Pat sometimes called "the moneyed class" and a cynicism about the uses of wealth and power in the political world which never left them. In 1941, with the federal government mobilizing for war, the Nixons moved east to Washington, where Dick worked for a time in the Office of Price Administration, which handled wartime price controls and rationing. But in the spring of 1942, a few months after Pearl Harbor, he left the OPA to join the navy — spurning the pacifism of his family's Quakerism to take part in a war that he knew would define the future of his generation. He served as a supply officer in the Pacific, became uncharacteristically friendly with his fellow officers, and spent many hours playing poker, at which he discovered he was very good.

Back in Whittier in 1946, Nixon embarked on an unpromising race for Congress against the well-established incumbent Jerry Voorhis. Against all odds, he won — by exploiting his status as a veteran, by attacking his opponent for serving the communists, and by taking advantage of the wellsprings of both covert and overt funding from the economic powers of southern California. His backers were the men from oil, agribusiness, insurance, banking, real estate, motion pictures, construction, and other industries — even illegal gambling, drug trafficking, and the other rampant vices of the moment; they and their heirs supported him until the end of his career. Critical to this early political success was a short, pudgy man of thinning hair and trademark clock-face cufflinks: Murray Chotiner, who was hired to write press releases for the thirty-three-year-old congressional candidate in the spring of 1946. Chotiner was a ferocious and reactionary partisan, a disappointed politician consigned to the back room, a crude genius of a new politics of imagery and attack, with an abiding contempt for the public he manipulated so well. Above all, Chotiner epitomized the world of shadowy operators and fixers — a world that flourished amid the regional ambition and animus of the postwar west and south — that Richard Nixon entered and employed from the outset of his career more unabashedly than any other president of his era. Chotiner may have been the single most significant influence in the rise — and perhaps also, even if indirectly, in the fall — of Richard Nixon.

As a new member of Congress, Nixon accepted an assignment to one of the most reviled committees on Capitol Hill: the House Un-American Activities Committee, known above all for the crudeness and recklessness of most of its members. Nixon was a sharp contrast — careful, exacting, unsensational, but alert to the political possibilities of his new position. When Whittaker Chambers accused the former diplomat Alger Hiss of espionage in 1948, Nixon examined the evidence carefully before joining the hunt. But once he did, he pursued Hiss doggedly and successfully and, in the process, made himself one of the most prominent young Republicans in the nation.

Having won an implausible victory over Voorhis as an unknown, Nixon set out on a less quixotic crusade in 1950, when he became the Republican candidate for the U.S. Senate, running against the beautiful former actress and ardent liberal Helen Gahagan Douglas. It was a bitter and dirty campaign on both sides, but most of all on Nixon's. He branded Douglas the "Pink Lady," for what Nixon claimed were her ties to the communists. And yet, in what by later standards would be considered a horrifyingly sexist campaign, Nixon attracted the active support of hordes of southern California women, energized by anticommunism and enlisting in politics as never before. They would remain the shock troops of his political base in California, just as they became the front ranks of other conservative and far-right candidates and causes in years to come.

Without those first victories in 1946 and 1950, there would have been no Vice President or President Nixon. But as ever with Nixon, winning came at a price. These launching races left a legacy of bitterness and worse in California, where Nixon's venom against Douglas, in the reactionary moment of 1950 and the Korean War, had been wantonly needless. The smears inevitably oozed abroad in the nation to raise a singular doubt and resentment about this prodigal young politician, emerging so rapidly yet hooded by questions about technique, integrity, and character. Wunderkind of the GOP, he was already widely known by a name Helen Douglas had given him, "Tricky Dick."

Nixon's own struggle for survival — a recurrent public spectacle actually played out and determined, like most of the essential events of his life, largely in secret — began its life early in 1952, as he was beginning a harsh campaign of denunciation of Adlai Stevenson as the Republican vice-presidential nominee under Eisenhower. The disclosure of a quiet if not clandestine Senate campaign fund maintained by wealthy California donors (the result of a leak not from the Democrats, but from Nixon's even more passionate enemies in the Earl Warren wing of the California GOP) brought the first effort to drive him from the Eisenhower ticket. As much of the country saw at the time in grainy gray-and-white on small glowing screens set proudly in their living rooms, he survived as a result of an artfully disingenuous, maudlin, but brilliantly presented speech. Ever after, devout believers and disgusted critics alike knew it as the "Checkers" speech, by the name of the Nixon daughters' cocker span-iel, the one gift the beleaguered vice-presidential nominee usefully ad-mitted in obscuring and preserving a career already so financed by the favor of strangers. What the public did not see was the untelevised, unreported, equally ominous accompaniment: There were the tactics behind the scenes that made the speech possible, fortifying Nixon's faith in furtive manipulation. But there were also corrosive doubts among Eisenhower and other GOP figures about Nixon's performance, even though it touched them at some level and left them no choice but to keep Nixon. It was not the last time the shrewd and ceaselessly political Ike, who prided himself on maneuver, would find himself outmaneuvered by his running mate. Most telling, the episode fostered a profound, lasting virulence within both Nixons, a bleak conviction at once that although he had saved a career of vast potential, he was now continually vulnera-ble, with only perishable support or loyalties. In the end, they believed, he was always alone.

About the vice presidency, in critical respects the crucible of the Nixon White House, Nixon later wrote and revealed little, as if to hide from the world the many slights and humiliations he absorbed during his eight years in what was still in the 1950s one of the nation's most vacuous jobs. Nixon's intermittent role in the foreign policy of the Eisenhower era in-cluded urging the use of nuclear weapons in French Indochina in 1954, supporting the commitment to the Diem regime in Vietnam, and en-couraging CIA covert interventions in the Middle East, Latin America, Indonesia, and elsewhere. Seldom, however, was he truly at the center of power. When Ike suffered his heart attack in 1955, there was a concerted effort in the GOP, with the president's nod, to remove Nixon from the ticket. (This despite his acquitting admirably some dirty work, in the regime's furtive destruction of a used-up Joseph McCarthy and in inveterate smears in the 1954 congressional races, which the fastidious Ike stood above; having dispatched his vice president on these ugly errands, Eisenhower then claimed that Nixon's rabid partisanship would lose votes in 1956.) "The Dump," as it came to be known, eventually seemed to founder on the clumsiness of the front men, Eisenhower's habitual indirection if not ambivalence, and what appeared to be a popular outcry, including an impressive write-in for Nixon in the New Hampshire primary and a strong vote later in Oregon. As opponents brandished an alarming poll predicting that Nixon would cost the ticket six points, there promptly followed a show of Capitol Hill and grassroots GOP support for the vice president which raised the stakes too high for Ike, and the revolt was over. It would be an oft-told tale of Eisenhower artifice and party cabals, another jot in Nixon's annals of grievance. Uncounted in most versions was the covert money poured into the Nixon cause by Howard Hughes, tens of thousands that quietly bankrolled write-ins, crucial espionage and subversion in the other camp, and conclusive pro-Nixon polling. When Hughes later sought and got Nixon's intervention in an IRS ruling and other government decisions, he was not only the creditor of a notorious $206,000 loan in 1957 to the hamburger drive-in venture of Nixon's brother Donald, but a patron who had helped save Nixon's political life in 1956.

The presidential election of 1960 is one of the most storied in recent American history — the "experienced" Nixon against the vigorous, handsome Kennedy; the dutiful son and loyal vice president against a privileged, charming libertine; a call for stability pitted against a call for action, for "getting the country moving again." Saddled with the record of an administration that had long since run out of steam; with only the tepid support of the beloved Ike, who remained the only truly popular element of the Republican regime and on occasion seemed to go out of his way to belittle his vice president; outmatched in the unprecedented debates by his telegenic rival — Nixon nevertheless came agonizingly close to victory. For the rest of his life, he believed — not without evidence — that the election had been stolen from him, that vote fraud in Illinois and Texas and perhaps elsewhere had robbed him of his margin of victory. He carried a well of bitterness with him into the next stages of his career.

In 1962, a forlorn Nixon — unable to accept his exile from political stardom — launched a doomed bid for the California governorship. Old enemies from the Voorhis and Douglas races lay in wait with his own tactics, and Nixon — having lost to Pat Brown — seemed to ensure his route to oblivion with an exhausted, self-pitying "last press conference" in which he lambasted the press. From defeat in California, he moved to a lucrative, restless exile with a firm of Wall Street bond lawyers. Yet the shattering failures of 1960 and 1962 laid the groundwork for Nixon's extraordinary comeback. They allowed him to reinvent himself in the public eye, a wiser, chastened man, a "new Nixon," a statesman of foreign policy, a stable and reliable leader for a turbulent time. They also helped him nurse lethal old wounds.

The turbulent 1968 presidential race, in which the Democrats seemed to self-destruct in a hail of violence and dissent and bitterness, was Nixon's to lose. He almost did, despite the quiet machinations of Lyndon Johnson in aiding Nixon against LBJ's own vice president, Hubert Humphrey. Nixon campaigned almost as Dewey had done in 1948, exuding statesmanship, avoiding controversy, promising little, explaining nothing, reaching for the passions of the voters only in his strident calls for "law and order." He barely hung on against a late Humphrey surge and found himself the victor by almost as narrow a margin as the one by which he had lost in 1960.

In comparison to the incessant intrigue and burgeoning scandal that marked and ultimately ended Richard Nixon's presidency, the significance of his administration in policy and politics has sometimes seemed almost incidental. But Nixon's was a significant presidency quite apart from Watergate. One of the ironies of his almost obsessive effort to defeat and discredit his enemies is that it destroyed much of the very claim to greatness that it was designed to protect. Like Disraeli, the nineteenth-century British prime minister he admired and at times cited, Nixon hoped to be one of the "Tory men with liberal principles" who had "enlarged democracy in this world." And given his roots in the Republican right and his 1968 campaign stance as a scourge of the left, Nixon was indeed surprisingly progressive in much of what he did as president, even if without deep commitment to his own departures from expectations.

Early in his first term, Nixon introduced a plan to reform the welfare system which may have been the most progressive, even radical, such proposal since the New Deal. It was a plan for creating a guaranteed annual income for every American. Nixon called it the Family Assistance Plan. The plan called for the federal government to assume most of the welfare costs that were then borne by the states and to introduce a system of automatic payments to families whose incomes fell below a certain floor. Welfare payments would not be cut off when a recipient got a job; they would be reduced gradually, according to the amount of the recipient's outside income, so that it would never be more lucrative for anyone to remain on welfare than to go to work. Publicly, Nixon stressed the conservative aspects of the plan: the incentives it provided to get people to work (he was not the first or the last to describe his plan for welfare as "workfare"), the increased efficiency, the reductions of bureaucracy, the ostensibly lower costs (although it seemed likely that the plan would, in fact, eventually have cost more than the existing programs). But in retrospect, at least, the progressive features of the plan are much more striking than the conservative ones. The Family Assistance Plan — awkwardly presented and indifferently supported by the administration — failed in Congress, when a coalition of suspicious liberals and angry conservatives killed it in the Senate.

Another example of Nixon's willingness to move beyond conventional conservatism — and of his relative shallow commitment to his own innovations — came in economic policy. The first years of his presidency coincided with the beginnings of the corrosive economic changes that would plague the American economy for more than fifteen years. The most serious problem at first was inflation; and the federal government under both Johnson and Nixon had made a series of efforts — tax increases, spending cuts, higher interest rates, and others — to bring the inflation back under control in the late 1960s and early 1970s. Nothing worked. By the summer of 1971, inflation had risen to its highest point since the 1940s, 6 percent; and at that point — fearful that if untreated inflation would spiral out of control and jeopardize the president's reelection the following year — the Nixon administration imposed a ninety-day freeze on wages and prices, as an "emergency measure." Then, after the ninety days were up, Nixon announced what he called "Phase II" of the "New Economic Program": selective wage-price controls in major industries.

The controls were controversial from the start. Inflation did subside, although it did not disappear, in 1971 and 1972; but with the drop in inflation came a recession and a significant increase in unemployment. Perhaps as a result, the "New Economic Program" did not survive very long. Early in 1973, after the election, the Phase II controls were replaced by a group of voluntary "guidelines" (Phase III), which proved (like most voluntary guidelines) completely ineffective. By the fourth quarter of 1973, inflation was 12.3 percent (and it remained over 12 percent for a full year — although in large part because in late 1973 the first great American energy crisis had begun, driving up the price of oil). Like the Family Assistance Plan, the New Economic Program, for all its boldness, had no lasting impact on the structure of domestic policy (except, perhaps, to discourage similar efforts by later presidents). But it was a significant program for what it suggests about Nixon — his willingness to abandon past convictions, to innovate, to capture liberal positions while presenting them as tough, conservative measures, and to retreat from his own advances when they seemed no longer to suit his political needs.

The Nixon presidency was in many respects marked by a continuation, even an expansion, of the ambitious liberal reform crusade of the 1960s. Nixon did not initiate but also did not oppose the creation of the Environmental Protection Agency in 1970; supported legislation authorizing the Occupational Safety and Health Administration; permitted his Labor Department to introduce regulations that created the first "affirmative action" programs; and supported a significant expansion of the food stamp program. But, characteristically, Nixon simultaneously moved to establish his credentials with the right by engaging in highly public if at times purely symbolic battles against the most unpopular liberal initiatives of the recent past. The champion of "law and order" in the 1968 campaign, Nixon introduced a number of ineffectual anti-crime measures. Having signaled the white south that he, not George Wallace, would be its most reliable defender against the civil rights revolution, he moved cautiously but in some areas effectively to slow down the progress of the Voting Rights Act and to discourage busing to achieve school integration. A critic of Lyndon Johnson's War on Poverty, he eliminated the Office of Economic Opportunity (its coordinating agency), weakened the Community Action program, and — in a measure he described as the "New Federalism," which he claimed was a bold effort to shift authority away from Washington — initiated a "revenue-sharing" plan, by which the federal government offered "block grants" to the states that in theory they could spend with fewer restrictions than older federal grants. Nixon also continued an attack on the Supreme Court that had begun with the most enthusiastically applauded passage in his acceptance speech at the 1968 Republican Convention. As the liberal justices from the Warren court retired, Nixon replaced them with more conservative men — and twice, with his appointments of Clement Haynesworth and G. Harrold Carswell, so affronted Democrats in Congress that the nominations were defeated in the Senate.

In the end, it was these gestures to the right — more than whatever progressive efforts he made — that solidified Nixon's imprint on national politics, an imprint perhaps larger than any president's since Franklin Roosevelt's. It was he who built the remarkable blue-collar, hard-hat constituency for the Republicans well before Ronald Reagan claimed it in front of the cameras. It was Nixon, the man of the similarly restive west, who bequeathed to the GOP a new Republican south, seized from the Democrats and George Wallace's bullet-quashed insurrection by cunning, ruthless political infighting. The onetime California Assembly candidate, who had once had only a single speech to make and no gas money, played skillfully from the Oval Office to the festering social worries and resentments of both blue- and white-collar middle America, capturing for the GOP the political center stage and the discourse of a generation. It was Richard Nixon, new and old, Eisenhower moderate and secret sharer with the right, who propagated the conservative power of the 1980s and 1990s.

For Nixon himself, however, the real arena — the venue in which he hoped to prove his greatness and win himself not just reelection but a place in history — was foreign policy, the realm in which he considered himself an expert. Aides later reported that he was often bored and inattentive when they discussed domestic policies. But international issues had his full attention, particularly after he forged a complicated relationship (that at times bordered on rivalry) with his brilliant and ambitious national security adviser Henry Kissinger. Together, they sought to move American foreign policy from its rigid bipolar fixation on the Soviet Union toward something like the old balance-of-power system of which Kissinger had long ago written approvingly in one of his first books. Instead of a struggle between the United States and the Soviet Union, Nixon and Kissinger envisioned a world in which the two superpowers would coexist not just with each other, but with other strong power centers: a united Europe, a China drawn out of its self-imposed isolation, and a Japan newly emerged as one of the world's great economic powers.

The most dramatic result of this new policy was Nixon's voyage to China in 1972, a visit preceded by a secret trip to Beijing by Henry Kissinger in 1971 to prepare the way, and by the removal of the American objection to seating communist China in the United Nations. The trip itself became a media triumph. The astonishing meetings between Nixon, the Red-baiter of old, and Mao Tse-tung and Chou En-lai, the aging revolutionaries, produced few specific agreements. But the trip as a whole significantly altered the American relationship with China, beginning the long, rocky movement toward the deep economic and political engagement that continues still.

The China initiative was driven by two additional hopes — that it would give the United States more leverage in its relationship with the Soviet Union, and that it would somehow contribute to a solution to the conflict in Vietnam. Whether or not because of China, Nixon did make considerable progress toward a new relationship with the U.S.S.R. — in part, it appears, because Leonid Brezhnev found in Nixon a fellow believer in hard, pragmatic power politics; because neither of them was as interested in ideology as in their own political fortunes and in the economic benefits that might flow from new agreements. Nixon and Brezhnev developed the most cordial relationship between an American and a Soviet leader since Roosevelt's wartime alliance of convenience with Stalin. Out of prolonged negotiations, and two highly publicized summit meetings, emerged an arms limitation treaty (known as SALT I), later attacked as too favorable to the Soviets, and an infamous grain deal that produced enormous windfall profits for grain dealers at the expense of farmers and that gave the Soviets badly needed grain well below market prices.

But détente did little to help the Nixon administration in other parts of the world. The Soviet Union helped supply the armed forces of Egypt and Syria during the 1973 Yom Kippur War in the Middle East, complicating America's brokerage of a deal favorable to Israel (but not too unfavorable to America's Arab oil suppliers). Soviet assistance to "wars of national liberation" did not flag. Human rights violations within the U.S.S.R., piously decried by Nixon and Kissinger but largely ignored in their negotiations with Brezhnev, continued unabated. Most frustrating of all to them, détente was of little help in solving their most corrosive problem of all: the war in Vietnam.

Nixon inherited what most Americans believed was Johnson's war, a belief Nixon himself did much to encourage. But he inherited, too, a set of assumptions about "credibility," "peace with honor," and "dominoes" that made the war almost as insoluble a problem to him as it had been to his predecessor. In his 1968 campaign, he claimed to have a plan for ending the war, but refused to disclose its details. Once in office, it soon became clear that whatever plan he had was not for ending the war as much as it was for undermining domestic opposition to it. In an early television address, he appealed to the "Silent Majority" that he always believed was the heart of his constituency to make itself heard in defense of the

American effort in Indochina, and he warned that the nation risked being seen as a "pitiful helpless giant" if it could not muster the strength to stand firm in the war. At the same time, he sought to placate some of the war's opponents by beginning a process of "Vietnamization." He gradually reduced American troop strength in Vietnam while working (largely in vain) to strengthen the unreliable army of the South Vietnamese themselves. He also instituted a draft lottery, to reduce the number of young men faced with the draft and hence, he believed, likely to protest the war.

In the spring of 1970, Nixon announced a new and controversial offensive. American and South Vietnamese troops would cross the border into Cambodia to "clean out" communist sanctuaries that the U.S. military believed were hidden there. To many critics of our involvement — unaware that American planes had been secretly bombing Laos and Cambodia for more than a year already — the invasion appeared to be a substantial widening of a war they had begun to believe was moving toward a close. The first days of May 1970 produced the largest and angriest antiwar demonstrations in American history — rallies, marches, and occasionally violence on campuses all over the country, including clashes with the National Guard at Kent State University in Ohio and Jackson State University in Mississippi which left a total of six students dead. Nixon, hunkered in the White House, brooding and bitter, responded to the shootings with what seemed calculated callousness. "When dissent turns to violence," he said coldly, "it invites tragedy." Later, as the crowds in Washington swelled and the air of crisis intensified, he made a late-night visit to the Lincoln Memorial to speak with some of the students. Their stilted conversation seemed to confirm the vastness of the gulf of incomprehension and suspicion between the president and his young critics.

The Cambodian invasion ended inconclusively after a few weeks, but the war continued for more than two years. As the 1972 election approached, Nixon instructed his negotiators to drop their long-standing demand that the removal of North Vietnamese troops from the South be a part of any peace settlement. Kissinger began meeting secretly in Paris with the North Vietnamese foreign secretary, Le Duc Tho, and together they forged the outlines of an agreement, which Kissinger prematurely announced in late October. A few weeks later, the agreement collapsed in the face of adamant resistance from the South Vietnamese government. When the regime in Hanoi refused to reopen negotiations, Nixon ordered an extraordinary new bombing campaign — the largest of a war that had already seen more bombs dropped than had been dropped in all theaters of World War II — focused on the North Vietnamese capital and its principal port, Haiphong. It came to be known as the "Christmas bombing," in the grim parlance of that always grim war.

In January, Kissinger and Le Duc Tho resumed negotiations, while the United States put extraordinary pressure on the forlorn leaders in Saigon to accept an agreement that they correctly considered a death sentence for their regime. The "Paris Accords," finally signed on January 27, 1973, a few days after Nixon's second Inauguration, provided an immediate cease-fire, an exchange of prisoners of war, and a series of murky procedures for resolving differences between the two sides at some unspecified future date. Nixon and Kissinger always insisted that the agreement, had it been enforced, would have assured the survival of a noncommunist South Vietnam. Others believed that it was a formula to provide the United States with a "decent interval" between its own withdrawal from Vietnam and the collapse of the doomed regime in Saigon. The interval, in any case, was relatively brief. In the spring of 1975, the North Vietnamese launched a massive invasion of the south and marched almost unopposed into Saigon — sparking a chaotic, panicked evacuation of the several thousand Americans who were still there and creating a scene that many came to consider a symbol of America's diminished role in the world. Saigon, for twenty years an outpost of America's cold war ambitions, now stood undefended against its bleak future as Ho Chi Minh City in what was soon to be the poorest nation in the world.

The Vietnam War brought down the Nixon presidency as surely as it did Lyndon Johnson's, and in both cases it did so by playing to the worst instincts of the men in power. In Nixon's case, it strengthened what had always been impulses near the center of the man: the fevered resentments, the brooding paranoia, the instinct for furtiveness and intrigue.

The dark side of the Nixon presidency — the secret efforts spawned in rambling late-night tirades in the president's many offices — began almost as soon as he entered the White House and accelerated toward its fervid, pathetic denouement as the 1972 election approached. Out of a continuing attempt to harass and discredit those Nixon considered his enemies came, first, the "Huston Plan," named for the reckless young White House aide who proposed it. It called for a massive, coordinated, illegal effort to use the FBI, the CIA, the IRS, and other agencies to attack the opposition — a plan so drastic and so dangerous that J. Edgar Hoover (not a man usually scrupulous about attacking the left) refused to cooperate with it. One of the proposals in the plan was to bomb the ostensibly liberal Brookings Institution. In the face of resistance, the White House dropped the Huston Plan, but not its efforts to discredit its enemies. Spurned by the FBI, it created its own phalanx of covert operatives, known as the "plumbers" (because they were supposed to plug leaks), consisting of ex-CIA agents and others, among them such figures as G. Gordon Liddy. Their first major assignment, in 1971, was to break into the psychiatrist's office of Daniel Ellsberg. The former Defense Department official had leaked a secret study of the war — the "Pentagon Papers" — to the press, and the plumbers were sent on what turned out to be a futile effort to find information damaging to him. Although formally disbanded after the failed Ellsberg break-in, some of the plumbers continued to operate informally into 1972 and in June helped plan the break-in at the headquarters of the Democratic National Committee at the Watergate hotel-apartment-office complex, which led inexorably to Nixon's resignation a little more than two years later.

Nixon fell because his first instinct, when informed of a burglary in which, as far as we know, he had played no part, was to hide the embarrassing but probably not lethal truth (that high-ranking aides in his campaign and in the White House had been involved) and to make illegal use of government agencies to conceal evidence; and he fell because the extent of his dishonesty ultimately became irrefutably clear when a congressional investigation uncovered the existence of a secret taping system in the White House that had recorded almost all the president's conversations. Little by little, out of bitter clashes with special prosecutors and congressional committees and federal courts, evidence of the president's involvement in the "cover-up" mounted until, in the summer of 1974, impeachment and conviction seemed inevitable. A quiet, desperate effort at persuasion finally penetrated the wall of despair and self-pity with which the president had surrounded himself. On August 8, 1974, he became the first president in history to resign from office, ending his presidential life as he had begun it in his 1968 acceptance speech at the Republican National Convention, by talking about his childhood in California, about his parents, about his own early exclusion from the world of wealth and privilege and his lonely, bitter struggle to belong.

So in disgraced retirement, pardoned for his crimes by his own appointed successor, Gerald Ford, Nixon would be inexhaustible in trying to rescue for himself an honored place in history. Self-aggrandizing, now petty, now magnanimous, always shrewd, ever suspicious and vindictive, relentlessly judgmental about other politicians, he sought the reinstatement of the legacy he was sure he deserved in copious books, as in off-the-record wanderings reported after his death. Nowhere in his post–White House writings and statements is there any suggestion that Nixon doubted that he ranked among the great presidents. And yet over twenty-two years, he and his heirs would fight to keep secret most of the White House tapes, forced only by a lawsuit to give up 201 hours of damning dialogue that was finally published in 1997 and that struck a devastating new blow to his once slowly reviving reputation. The proverbial stonewall would last long enough, however, to assure him a striking funeral service in 1994 by the old house in Yorba Linda, where the Nixon Presidential Library — itself censored and fragmented like the real life of the man — consumes the hillside where his father, Frank Nixon, worked their barren lemon grove. National and world leaders, including President Clinton, eulogized the fallen president very much as he himself would have dictated in one of his ubiquitous Oval Office memos prescribing his prerogatives — as a figure, Republican Senate Leader Robert Dole put it tearfully, who belonged after all, who was "American through and through."

He was, as almost everyone said in 1994, a remarkable man — intelligent, tenacious, bold, a man of some real greatness and a president of lasting, if clouded, importance. But there was also something dark and ugly about his long and never peaceful public career, in which muck and money were inescapable, inseparable from politics or policy.

"My point," the president tells Chief of Staff Bob Haldeman in a typical taped conversation in the White House in 1971, "is that anybody that wants to be an ambassador wants to pay at least $250,000 [to the campaign]. I think any contributor under $100,000 we shouldn't consider for any kind of thing." Reminded of his big donors, Nixon adds that three giant dairy farm groups should be pressed anew for huge contributions in return for the White House arranging an increase in federal milk price supports. They all need prodding, he tells Haldeman. Take Charles Bluhdorn, former Chairman of Gulf+Western Industries: "I want him to be bled for a quarter of a million, too," says the chief executive.

If the Nixon presidency leaves its legacy in bold policy initiatives and in a new national electoral landscape, Nixon the ever anxious man and ever cynical politician is also an authentic godfather of the shadowy intrigue and the bipartisan tyranny of corporate money and private fortunes that have tainted American politics in our own time even more than in most others. Nixon's character and the specters that haunted him were in some measure his own. But like Ulysses, he was also part of all that he met in twentieth-century American life and politics. In a Washington unlikely to erect a Nixon memorial any time soon, his spirit reigns supreme in the end-of-century scandals that have bedeviled his successors and that will likely bedevil still others. *Si monumentum requiris circumspice:* If you would see the man's monument, look around.

— ROGER MORRIS

The Unmaking of the President

The path from the bungled break-in at Democrat campaign headquarters in the Watergate complex to the resignation of a president took surprising twists. It seemed at first as though the story would die until Bob Woodward and Carl Bernstein, two young reporters at the *Washington Post,* uncovered connections between the break-in, other "dirty tricks," and top officials in the Nixon reelection campaign and the White House. Next, a special Senate Committee headed by Sam Ervin, a Democrat from North Carolina, found still more disturbing evidence. And finally, Attorney General Elliot Richardson appointed a special prosecutor, Archibald Cox, to conduct an independent inquiry.

The revelation that Nixon had secretly and routinely taped conversations of meetings in the White House proved devastating to the president. The Ervin Committee asked for the tapes, but Nixon refused. Eventually the committee subpoenaed the tapes — the first time Congress had subpoenaed a president since the treason trial of Aaron Burr. Cox also subpoenaed the tapes and obtained a court order directing the president to comply.

Nixon tried to resist the subpoenas on the grounds of "executive privilege" — a concept not established in the Constitution, but one that had become conventional doctrine. When Cox persisted, Nixon fired him, in the process triggering the resignations of Richardson and his top deputy. The Saturday Night Massacre, as this episode became known, intensified criticism of the president. A second special prosecutor, Leon Jaworski, picked up where Cox left off. Meanwhile, in Congress the House Judiciary Committee began formal impeachment proceedings.

Efforts to pry the White House tapes from the president culminated in *United States v. Nixon,* a unanimous Supreme Court ruling that upheld the principle of executive privilege but rejected its application to protect a president in a criminal case. The ruling was devastating. Nixon was compelled to turn over key evidence that clearly implicated him in the Watergate cover-up. A few days later, he resigned. The crude, cynical inner workings of his administration as revealed by the tapes left a permanent stain on his legacy.

Gerald Ford

1974–1977

b. July 14, 1913

IN 1974 GERALD RUDOLPH FORD, JR., became the only president in American history never elected to national office. He rose to power under the terms of the Twenty-fifth Amendment, which had been ratified in 1967. Ford assumed office under unusual and exceptionally trying circumstances. Grim economic problems, rising energy costs, and a diminished global standing put the country on the cusp of inauspicious change. The national spirit was shattered after the twin blows of the Vietnam War and Watergate, and trust in the presidency had eroded. The office of the presidency itself was weakened, and Congress seemed resolved to recapture political powers it had surrendered to the executive branch. These developments signaled an end to the buoyant optimism associated with America's first decades after World War II.

Ford was born Leslie Lynch King, Jr., on July 14, 1913, in Omaha, Nebraska, to Dorothy Gardner and Leslie Lynch King, Sr., a businessman. Gardner soon divorced her husband, settled in Grand Rapids, Michigan, and married Gerald R. Ford, a paint and varnish salesman. Ford legally adopted and renamed Gardner's young son.

At South High School in Grand Rapids, Ford enjoyed a stellar career as a center for the football team and won a football scholarship to the University of Michigan, which he otherwise would not have been able to afford. As a senior he was named the football team's most valuable player. In 1935 Ford graduated with a degree in economics.

He worked as an assistant football and boxing coach at Yale University and began taking courses part-time at Yale Law School. He eventually enrolled full-time, received his law degree in 1941, and then returned to Grand Rapids to practice law. After serving in the navy during World War II, he returned to his work as an attorney in Grand Rapids. There he whetted his appetite for politics by immersing himself in civic affairs. Ford was one of many young, idealistic Republicans in Michigan who bristled under the iron grip of Frank McKay, an arrogant and dictatorial Republican machine boss. In 1948 Ford decided to challenge Bartel Jonkman, a McKay disciple and five-term congressman, for his seat in Michigan's Fifth District. Jonkman's support of American isolationism gave Ford more reason to mount a challenge, as the war had convinced him that the country needed to play an active role in international affairs. After scoring an upset victory over Jonkman in the Republican primary, Ford cruised to an easy victory in the general election. In 1948 he also married Betty Warren of Grand Rapids.

He entered the Eighty-first Congress in 1949. The following year he won a seat on the prestigious House Appropriations Committee, an assignment that allowed him to gain intimate experience with the federal budget process. Early in his congressional career, Ford also set an ambitious goal for himself: becoming Speaker of the House. In January 1965, he took a decisive step toward realizing this goal when he became the House minority leader.

But with the Republicans continually in the minority, Ford found his reach for the speakership frustrated. In the early 1970s, he decided to run for one more term in 1974 and retire in 1977. His plans changed drastically in October 1973 after Vice President Spiro Agnew resigned and President Richard Nixon chose Ford to replace him. Ford was not Nixon's first choice, but Democratic and Republican leaders of Congress had indicated to Nixon that the popular and affable Ford was probably the only nominee that the Democratic Congress would approve. He won easy confirmation in both the House and Senate to become the nation's first nonelected vice president in December 1973. Ford's inaugural remarks included the memorable line, "I am a Ford, not a Lincoln," which captured his self-effacing and unpretentious character.

Ford served only eight months as vice president. On August 9, 1974, Nixon resigned, and Ford took the presidential oath of office. After the Watergate scandal, Americans found comfort in the palpable decency of their new president, who declared in his abbreviated inaugural address that "our long national nightmare is over."

During his first weeks in office, Ford enjoyed a warm welcome from Congress and the American people and received overwhelmingly favorable press coverage. He moved quickly to reestablish public confidence in the executive branch and to set an unpretentious, amiable tone that marked a refreshing contrast from Nixon's besieged presidency. Striving to maintain an "open" White House, he invited former Nixon enemies to official functions. Whereas Nixon had despised the press, Ford tried to be an accessible president and even introduced cosmetic changes such as a new, more congenial seating arrangement for White House East Room press conferences (an arrangement that all of Ford's successors in office have followed). In reporting on Ford during his first few weeks as president, the media indulged in praise, exulting that the new chief executive was human enough to engage in everyday tasks like fixing his own breakfast of English muffins.

Ford's honeymoon lasted one month. On September 8, 1974, he took the most controversial action of his presidency by granting Richard Nixon "a full, free and absolute pardon" for any offenses he might have committed against the United States while president. Ford was exasperated that he had to devote about 25 percent of his time to the problems of the former president, such as the proper legal disposal of his tapes and papers. He felt that the pardon was the only decisive way to dispel the lingering problems of Nixon and concentrate instead on the domestic and international concerns facing America.

But the nation reacted angrily to the pardon. Overnight, Ford's Gallup approval rating sank from 71 to 49 percent, the largest single drop in presidential polling history. Critics charged (without real evidence) that Ford had arranged a secret "deal" with Nixon to trade the presidency for a pardon. These charges reflected the mood of America after Watergate, a time when the business of politics seemed incorrigibly corrupt. In order to allay suspicions surrounding his decision to pardon Nixon, Ford took the unusual step of appearing before a House subcommittee, one of the only times in history that a president has testified before Congress. Appearing alone at the witness table, he vigorously denied charges that the pardon was the result of a deal, at one point angrily stating, "There was no deal, period, under no circumstances."

The pardon irreparably damaged Ford's presidency. Although he had brought a new atmosphere to the White House, the pardon sullied his credibility and stripped him of the political strength and goodwill he had initially enjoyed. His public approval ratings never recovered, he never again received widespread media adulation, and he was weakened in his political dealings with Capitol Hill.

Despite the storm of protest that the pardon unleashed, Ford still considered himself a "healing" president. His leadership style was steadfast if unspectacular and helped to project stability after the mercurial style of Richard Nixon and the tumult of the 1960s and early 1970s. In an effort to reach out to one disaffected group of Americans, Ford offered a clemency program to draft resisters and deserters from the Vietnam War. He believed one of his most vital functions to be repairing relations between the White House and Congress. Nixon had treated the Democratic Congress contemptuously, leaving legislators in an antagonistic mood. Soon after becoming president, Ford delivered an address to a joint session of Congress in which he pledged "communication, conciliation, compromise, and cooperation" with Capitol Hill. As president, he continued to show the same amiability toward congressmen that he had demonstrated while in the House.

But Ford's good personal relations with Congress were put to a stringent test throughout his presidency. The 1974 midterm elections were a disaster for him and for the Republican party, as voters vented their anger over Watergate and the Nixon pardon. In both houses, the Democrats gained greatly increased majorities. That the new Ninety-fourth Congress was so emphatically Democratic was only one of Ford's political handicaps. As an unelected president, Ford's national support was shallow, and some members of Congress saw his unusual ascension to the White House as a mark of political illegitimacy and a reason to challenge his authority.

By the mid-1970s Congress was clearly a resurgent political force. Through much of the twentieth century, the presidency had steadily accumulated power at Congress's expense. But after having successfully deposed a corrupt president, Congress emerged from the constitutional crisis revitalized and eager to redress what it now considered an imbalance of power. With Ford's claim to the presidency tenuous, many Democrats in Congress saw an opportunity to capitalize on presidential weakness. As a result, Ford found Congress much more assertive than his recent predecessors in office had. His political clashes with Congress became a theme of his presidency and an obstacle to implementing his policies. One early battle was over Ford's vice-presidential nominee, Nelson Rockefeller. Congressional opposition to Rockefeller transcended traditional political lines. Congressmen from both parties resented his enormous personal fortune, and used the nomination as an opportunity to investigate Rockefeller's wealth and his use of it. Conservative Republicans took umbrage at Rockefeller's liberal record as New York governor and still simmered over his divisive challenge to conservative Arizona Senator Barry Goldwater for the 1964 Republican presidential nomination. After three months of debate, Congress finally approved Rockefeller in December 1974. Ford was incensed at the delay, which had deprived him of a vice president during the first critical months of his administration. During Ford's presidency, Congress also investigated illegal Central Intelligence Agency activities and established permanent intelligence committees to oversee covert CIA operations, thereby reining in what had once been a powerful presidential organ in foreign policy.

During Ford's years as president, the office was under severe scrutiny not only from Congress but from the media. The Vietnam War and Watergate convinced many Americans that the nation's presidents were distressingly flawed human beings, and respect for the presidency fell to a low ebb. In the mid-1970s journalists began to practice a new brand of aggressive reporting that aimed to expose the sins and scandals of presidents and other public officials. Moreover, press reporting assumed a more caustic tone in which high officials were no longer immune from savage satire; gone was the deference accorded them in the past. This new style of reporting was painfully evident in stories underscoring Ford's alleged physical clumsiness. In May 1975, Ford tripped and fell down the steps of Air Force One in Austria, a mishap that irrevocably altered press portrayals of him. Thereafter, the media magnified and replayed any incident of Ford falling or stumbling. The press seemed to delight in covering such incidents, creating the unseemly impression that the athletic Ford was a bumbler. He became one of the first presidents to feel the sting of a new press ethic, and his public image frequently took a beating.

The more assertive and abrasive media was only one of the nettlesome domestic problems that Ford faced. When he took office, the American economy was in the midst of a painful transition. After spectacular progress in the post–World War II era, America's economy in the mid-1970s showed signs of entering middle age. Dramatic increases in energy costs crippled basic industries, which faltered against stiff new foreign competition. Especially hard hit were cities in the Northeast and upper Midwest, which imported most of their energy and which struggled with outmoded industries and declining revenues while burdened with increasing populations and social services. The most pernicious indication of America's economic decay was high inflation, which by the year Ford took office was running at 12 percent. This inflation, the product of rising oil prices, food shortages, and expansive fiscal and monetary policies, posed Ford's greatest domestic challenge. As he later reflected, "The fight against inflation provided the basic theme of my administration."

Ford genuinely enjoyed grappling with economic issues and understood the American economy perhaps better than any president in history. His years on the House Appropriations Committee had immersed him in the details of government spending, and as president he devoted considerable time and attention to overseeing the formulation of the federal budget. Rather than introduce new, ambitious legislative programs, Ford's concept of "visionary" leadership centered on making the government operate on a sound fiscal basis. Ford believed that excessive government spending and the deficits it created fueled inflation. He wanted to restore price stability by cutting federal spending, eliminating deficits, and relying on a tight monetary policy at the Federal Reserve. By paring federal spending, Ford signaled that he would tolerate higher unemployment in the short term in exchange for lower inflation over the long term; by contrast, the Democratic Congress was more disposed to lower unemployment through increased federal spending. This philosophical chasm and the daunting Democratic strength in the Ninety-fourth Congress threatened Ford's attempts to pursue fiscal stringency, leading him frequently to use the presidential veto. Ford vetoed sixty-six bills during his presidency, giving him the fourth highest yearly average among presidents; Congress sustained fifty-four of his vetoes. Ford used the veto to reject bills that he considered costly, forcing Congress either to abandon the bill or to remold it into a version that he considered more economical. In this way, he hoped to shape national policy and restrain congressional spending. His emphasis on balanced budgets was plainly evident during the New York City fiscal crisis of 1975, when the city faced bankruptcy. Ford refused to support a federal bailout of New York and instead urged it to curtail its spending and reduce its deficit. New Yorkers were outraged. The *New York Daily News* captured their sentiments with the arresting headline "Ford to City: Drop Dead." In late 1975, after the city had agreed to implement cost-cutting measures, Ford finally agreed to federal aid for the city.

In his fight for price stability, one of his most visible early actions was a series of anti-inflation summit meetings during September 1974. He hosted two summit conferences in Washington, and administration members held meetings in various cities. In October 1974, after evaluating the various proposals that emerged from these summits, Ford unveiled an economic program that rejected wage and price controls and instead called for an anti-inflation surtax, a proposal that immediately met with rigid congressional and public opposition. He included in his economic program a call for a voluntary public anti-inflation campaign, which he called WIN, for "Whip Inflation Now." Ford intended WIN to act as a supplement to his more substantive economic proposals, a bipartisan way to rally public enthusiasm in the battle against inflation. Although initial public and corporate response to WIN was enthusiastic, the program never enjoyed full support even within the administration and soon became a target for ridicule by comedians and political opponents. By the spring of 1975, when recession was a greater economic concern than inflation, the administration quietly abandoned WIN.

The recession that began to grip the country in the fall of 1974 eventually became the worst economic downturn that the country had experienced since the Great Depression. It was also the first recession in the postwar era in which unemployment and inflation — instead of being mutual trade-offs — simultaneously ran high, a new and bewildering economic phenomenon dubbed "stagflation." By late 1974, Ford was forced to abandon his anti-inflation surtax proposal and reverse his economic strategy. In his January 1975 State of the Union address, Ford announced a new antirecession economic program in which he proposed a stimulative tax cut. Within two months, Congress passed a tax cut and Ford signed it.

Ford's State of the Union address also announced his new energy program. The nation's energy crisis was one of the top priorities of Ford's presidency. In October 1973, the Arab nations of OPEC, the Organization of Petroleum Exporting Countries, had instituted a six-month oil embargo against the U.S. in retaliation for American aid to Israel during the Yom Kippur War. America experienced its first modern energy crisis, and the country grew acutely aware of its dependence on foreign oil and its vulnerability to future embargoes. But at the time Ford took office, the United States had no coherent energy policy. He was adamantly opposed to the system of oil price controls on domestically produced oil, which helped artificially to keep prices down but which encouraged wasteful consumption and created shortages, while also acting as a disincentive to oil producers. Instead, his energy program centered on the elimination of oil price controls. Congress opposed that and instead advocated continued controls to shield consumers from high prices. Through most of 1975, the president and Congress were locked in a stalemate over energy policy. Finally, in late 1975, Congress produced a comprehensive energy bill, the Energy Policy and Conservation Act, which represented the first attempt in American history to legislate a national energy policy. But instead of incorporating Ford's plan for quick decontrol, the bill called only for gradual, phased decontrol. Convinced that the bill was the best he could get from Congress, Ford approved it.

Ford struggled to define a coherent foreign policy against the countervailing forces of congressional ascendancy, the apparent limits of American power, and a perception that he was inexperienced in diplomacy and let his inherited secretary of state, Henry Kissinger, make the calls. By the 1970s, America's place in the world was much different from its status during the first postwar decades. For years the dominant military and economic power on earth, America faced critical challenges to its strength not only from the Soviet Union, but also from developing nations (such as those in OPEC) and from allies such as Japan and Germany. Moreover, the Vietnam War had starkly demonstrated the limits of American power. Ford also had to conduct foreign policy in the face of an increased congressional determination to assert its own voice and limit presidential power in American diplomacy. This congressional behavior frustrated Ford, who viewed it as intrusive and even unconstitutional, complaining in his memoirs that "Congress was determined to get its oar deeply into the conduct of foreign affairs. This not only undermined the Chief Executive's ability to act, but also eroded the separation of powers concept in the Constitution."

Congressional ascendancy in foreign policy was apparent from the first days of Ford's presidency. In the fall of 1974, responding to the Turkish invasion of Cyprus (in which the Turks used American weapons), Congress initiated an arms embargo against Turkey. Ford feared that the congressional action would prompt Turkey to retaliate by closing American military bases on Turkish soil, but he was powerless to stop the embargo, anticipating an easy and politically damaging override if he vetoed the bill. The embargo testified to the political will of Congress and also represented a significant foreign policy setback for the new administration.

Congress also resisted Ford's attempts to support South Vietnam. In March 1975 North Vietnam began its final offensive against the South. In April, as the country speedily succumbed to the North Vietnamese advance, Ford requested military and economic aid for South Vietnam, which Congress refused. Later that month Ford ordered the evacuation of the last Americans from Saigon, as South Vietnam fell to the communist advance. Americans, already weary of the long entanglement in Vietnam, witnessed a humiliating spectacle on television as helicopters, overloaded with refugees, tried desperately to take off from the American embassy in Saigon. The scenes seared into the national consciousness and further reinforced the tragedy of the war, erasing traditional beliefs about American power and invincibility.

With the Soviet Union, Ford pursued the policy of détente which Nixon had begun. In November 1974 he went to the Soviet Union to meet with Premier Leonid Brezhnev, and the two leaders agreed upon a tentative framework for a SALT II treaty. The two leaders met again in July 1975 at the European Security Conference in Helsinki, but were unable to make further progress on arms control. But the conference's Helsinki Accords represented an affirmation of détente as the thirty-five signatory nations pledged to increase economic and cultural cooperation and to respect basic human rights.

The most dramatic single foreign crisis that Ford faced came in May 1975, when communist Cambodia seized the American merchant ship *Mayaguez* and its crew of forty men. Ford deployed American marines to rescue the ship's crew and assault Cambodia. He wanted to punish the country for what he considered an act of piracy and demonstrate to the world that the U.S. was still willing to defend its interests militarily despite its recent humiliation in Vietnam. Cambodia released the *Mayaguez* crew shortly after the American military reprisals began, but the operations — which continued even after the crew's release — resulted in the deaths of forty-one American military personnel. Still, the Ford White House touted the military action as a bold display of presidential leadership and American force. The American public rallied behind its president, and Ford enjoyed a brief surge in popularity.

In July 1975, he had announced that he would be candidate for the presidency in 1976. But he recognized that his vice president represented a political liability. The Republican right, which by the mid-1970s was a restive, vociferous force in the party, had a deep animus toward the liberal Rockefeller and had considered his selection an insult. In November 1975, under pressure from Ford, Rockefeller withdrew from the 1976 ticket, a move that was a clear concession to the Republican right.

Despite Rockefeller's withdrawal, conservatives were still unhappy with Ford as their party's standard-bearer. That dissatisfaction became clear later in November 1975 when the former actor and California governor Ronald Reagan, a hero of the right, announced that he would challenge the president for the Republican nomination. After scoring a razor-thin victory in the critical first primary in New Hampshire, Ford rolled to successive victories over Reagan in four states. But in the North Carolina primary, the Californian won an upset victory and broke the president's momentum. Reagan had gone on the offensive by criticizing the president's foreign policy, charging that by pursuing détente Ford had dangerously conceded military strength to the Soviet Union and allowed the U.S. to slip to the status of a "second-rate power." Reagan also attacked Ford for supporting a plan to give up American sovereignty over the Panama Canal. The issue evoked an emotional response from conservatives, who saw relinquishing the canal as another symptom of America's decline in the world. By focusing on foreign policy, Reagan forced Ford on the defensive, and the Republican race tightened. The tough Reagan challenge demonstrated the attraction voters had for a Washington outsider unblemished by the Watergate scandal; it also showed the strength of the conservative wing of the Republican party and the fragility of Ford's support there. When the Republican Convention began in August in Kansas City, neither Ford nor Reagan had enough delegates to clinch the nomination outright, but Ford eventually won on the first ballot. He selected as his running mate Bob Dole, a senator from Kansas whose conservative credentials met Reagan's approval and who Ford thought would help to win midwestern farm states.

The Democratic presidential nominee was the one-term Georgia governor, Jimmy Carter, and in August 1976 he led Ford in public opinion surveys by more than twenty points. Ford realized that he needed to do something dramatic to close the enormous lead Carter enjoyed in the polls. He therefore challenged Carter to a series of televised debates, the first since 1960. Ford's strong performance at the first debate, which concerned domestic issues, enabled him to cut Carter's lead in the polls. But during the second debate, on foreign policy, Ford committed a blunder that haunted him for the rest of the campaign. In response to a question about the political strength of the Soviet Union in Europe, Ford meant to distinguish between the de facto Soviet domination of Eastern Europe and American policy, which affirmed the independence of the countries there and did not officially recognize Soviet domination. But instead he curiously asserted, "There is no Soviet domination of Eastern Europe and there never will be under a Ford administration." The infelicitous wording marred his campaign. Both Carter and the press ridiculed him for the remark, and the public perceived that he had lost the debate by a large margin.

Ford lost the popular vote in the election by only 2 percentage points after a tremendous political comeback in the campaign's last weeks. It was one of the closest presidential contests in history. A number of factors contributed to his defeat. The debate gaffe over Eastern Europe hurt, as did Ford's chronic image problems. The primary fight with Ronald Reagan had damaged Republican unity and drained the Ford campaign's energy and resources. Although Ford had presided over a substantial reduction in the inflation rate, unemployment still ran high after the recession, leaving jobless Americans unhappy with his economic stewardship. Moreover, his vision for America was not one to excite the imagination. Ford's workaday approach to government eschewed new, costly programs and lacked creative flair, thus inadvertently contributing to the perception that his was just a "caretaker" presidency without a blueprint for the country's future.

But one of the greatest burdens that Ford had to bear was the legacy of Watergate. Americans still had not forgiven him for the Nixon pardon, and there was a strong sentiment in the country to clean house and remove Nixon's tainted followers from the executive branch of government. Jimmy Carter was a refreshingly new political face, and he parlayed his status as a Washington outsider into a formidable political asset. Carter won much popular support by using the simple two-word phrase "Trust me" and by promising, "I will never lie to you." To Americans still stunned after the Watergate scandal, these words had a powerful resonance. Moreover, Carter swept the southeast and won strong support from the traditional New Deal coalition of blacks, labor, and ethnic groups.

The 1976 election was the first that Ford had ever lost. When he left office, he became a private citizen for the first time since 1949. He and his wife, Betty, established two homes, a winter residence in Rancho Mirage, California, near Palm Springs, and a summer home in Beaver Creek, Colorado, near Vail. In 1980 Ford declined to enter the presidential race and also turned down an offer from the Republican nominee, Ronald Reagan, to be his vice-presidential running mate. But Ford remained active in Republican politics during retirement, campaigning for congressional and presidential candidates and helping to raise money for the party. He stayed prominent on the lecture circuit, speaking out on issues such as the federal deficit and congressional encroachment on presidential powers. He also sat on the board of directors of more than a half dozen corporations. True to his athletic past, he remained physically active during retirement, swimming nearly every day and golfing three times a week.

Ford served 895 days as chief executive, making his the fifth shortest presidency in history. His greatest achievement, one for which he liked to be recognized, was in helping to bring "healing" to the nation after the tumultuous 1960s and the wrenching experiences of the Vietnam War and Watergate. The relative comity between the executive branch and Congress and the conspicuous absence of presidential abuses during

Ford's term helped to restore probity to the presidency. As he observed, "In the relatively short period of time that I served, the major problem was to restore integrity and confidence to the White House, which we did." Ford's leadership helped to foster the quiet calm that settled over America during the mid-1970s.

— YANEK MIECZKOWSKI
Dowling College

Jimmy Carter

1977–1981

b. October 19, 1924

AMERICANS CELEBRATED the nation's two hundredth birthday on July 4, 1976, with mixed feelings of pride and anxiety. Millions gathered to reaffirm their faith in American ideals with the usual displays of parades and pageantry, but many observers sensed an undercurrent of uncertainty. The *New York Times* reported that Americans were suffering from "self-doubts uncharacteristic of the nation." Urban riots, Vietnam, and Watergate had shaken America's confidence in its government and its leaders, but not in the soundness of the American system or the relevance of American values. What Americans were saying, declared two social scientists, was that "the system" works, but "it is not performing well because the people in charge are inept and untrustworthy."

One week after the bicentennial celebration, Democrats met in New York to nominate Jimmy Carter for president. Carter seemed an odd choice to unite the party and revitalize the nation's spirit. A successful peanut farmer from Plains, Georgia, Carter announced his candidacy in 1972 after serving one term as governor of his home state. Carter's admirers trumpeted him as a representative of a new breed of southern politician known for racial moderation and social progressivism. But even they were stunned by his decision. "President of what?" his mother responded when told of his intentions. Beginning his campaign with a name-recognition of only 2 percent, Carter took advantage of Democratic party reforms initiated after 1968, which increased the role of grassroots activists in the selection process, to roll over a number of better-known candidates on the way to a first-ballot victory at the convention. In an effort to reach out to the party's traditional power brokers, whom he had bypassed in the primaries, Carter selected Minnesota Senator Walter F. Mondale, a protégé of Hubert Humphrey, as his running mate.

Throughout the primary season, Carter played skillfully to the public's conflicting moods. He combined biting attacks on the Washington establishment with uplifting sermons about spiritual renewal. To a public still smarting from Watergate and disenchanted with government, Carter emphasized his rural roots and his lifelong distance from Washington. A Democrat who thought in many ways like a Republican, Carter rejected social experimentation and, instead, emphasized the importance of social efficiency and prudent management of the nation's affairs. A deeply religious man and self-described "born-again" Christian, he seemed to offer a religious salve for the nation's wounds, reassuring audiences that they deserved a government as "decent, honest, truthful, fair, compassionate, and as filled with love as our people are." He possessed a visceral appreciation of the sense of betrayal many people felt toward government, and he conveyed a serene confidence in the relevance of old verities. There was, he seemed to say, nothing ailing America that could not be solved with a little more democracy. "It's time for the people to run the government," he declared.

For the first time since 1964, the Democrats seemed assured of victory in November. In August, polls showed Carter leading his beleaguered Republican opponent, President Gerald Ford, by more than 30 percentage points. Not only was Ford burdened by his party's ties to Watergate and by his own controversial pardon of Nixon, but he had to endure a bruising primary fight against a conservative challenger, Ronald Reagan, which left his image muddied and his party deeply split. By early October, however, a lackluster Democratic campaign and an effective Republican strategy had eliminated Carter's once-formidable lead. On election night, the Carter-Mondale ticket won one of the closest elections in history. In the popular vote, a margin of less than 2 percent separated the candidates — Carter won 40.8 million votes (50.1 percent) to Ford's 39.1 million (48 percent). In the electoral college, too, the outcome was sobering: 297 for Carter, 240 for Ford. It was the narrowest electoral victory since 1916, when Woodrow Wilson had defeated Charles Evans Hughes by 23 electoral votes. The Democrats, however, maintained large majorities in Congress: a Senate margin of 62 to 38 and a House lead of 291 to 142.

The pollster Louis Harris concluded that Carter had won because of "the revival of the old coalition that first sent Franklin D. Roosevelt to the White House in 1932." But Harris failed to appreciate the important changes that were transforming Roosevelt's coalition. The New Deal had built its electoral support on the white south, Catholics, blacks, Jews, and other urban ethnics, low-income voters, and union members. In 1976, the Carter-Mondale ticket scored well among three traditional groups — blacks, Jews, and union members — but it failed to win a majority of either white southerners or Catholics. It was nontraditional groups that made the margin of difference — white Protestants, educated white-collar workers, and rural voters. In a perceptive postelection analysis, Carter's pollster Patrick Caddell argued that the Democratic party "can no longer depend on a coalition of economic division" to guarantee victory. Forty years of sustained economic growth, he argued, had created more "haves" than "have-nots." And he concluded, "In short, the old language of American politics really doesn't affect these voters." The challenge confronting the Carter administration was to create a new philosophy and updated agenda for a Democratic party in transition.

In clear contrast to his two Democratic predecessors, John F. Kennedy and Lyndon Johnson, who used their inaugural addresses as clarion calls for American greatness at home and abroad, Carter struck a tone of limits. "We have learned that 'more' is not necessarily 'better', that even our great nation has its recognized limits, and that we can neither answer all questions nor solve all problems." To demonstrate his new style of openness, Carter, his wife, Rosalynn, and their nine-year-old daughter, Amy, stepped out of their bulletproof limousine to march the mile and a half from the Capitol to the White House. The gesture raised expectations that Carter would bring a new sense of national unity to a nation torn apart by social unrest. The *New York Times* said the Inauguration offered the country "a therapeutic moment of tranquility."

Once in office, Carter continued to make effective use of symbolic politics to demystify the operations of government and create an impression of rule by the common people. He reduced the size of the White House staff by one-third, ordered cabinet officers to drive their own cars, and ended the playing of "Hail to the Chief" at his public appearances. He wore a cardigan sweater for a televised "fireside" chat, and he made a habit of holding "town meetings" in small American communities. The new president showed his disdain for the Washington establishment by appointing outsiders from Georgia — Hamilton Jordan, Jody Powell, and Bert Lance — to key White House positions. He nominated a record number of women and blacks to administration posts. After one hundred days in office, Carter enjoyed rising popularity, with a 75 percent approval rating in the Harris poll. There was, observed the veteran presidential adviser Clark Clifford, "a return of the confidence of the people in our government."

Carter's symbolic politics helped briefly to soothe the nation's troubled conscience, but effective leadership over the longer term required the president to reconcile the conflicting and contradictory claims on government that characterized American politics in the 1970s. The post–World War II generation had absorbed unprecedented expectations about the capacity of the United States to create a better world both at home and abroad. By the mid-1970s, many of the conditions that had created and nurtured those expectations had come to an end. Productivity declined to an average annual rate of 1 percent, well below the 3.2 percent of the 1960s. The drop undermined the social conditions that had allowed the Democratic party to serve as the champion of both corporate America and social generosity. A growing federal budget deficit, which swelled to $66 billion in 1977, and nagging inflation, limited the fiscal maneuverability that had permitted past Democratic presidents to satisfy the demands of the party's constituency groups. In foreign policy, Vietnam had drained the cold war consensus of much of its ideological force. Haunted and constrained by the legacy of Vietnam, Carter had to sustain American internationalism in the face of widespread domestic fears about the costs of military intervention. He confronted a public increasingly preoccupied with domestic economic interests but simultaneously demanding that the United States remain a global leader.

Institutionally, Carter inherited a humbled presidency, an assertive Congress, and an unruly party. In response to the abuses of the Johnson and Nixon years, Congress had moved to limit presidential power while expanding its own capacity for independent analysis. As the historian James MacGregor Burns observed, Congress sought to assert itself in the post-Watergate era, but "its essential power [was] that of obstruction and negation." At the same time, the decline of partisanship and the end of the seniority system destroyed the chain of command a president used to whip his party into shape. Liberalism, the driving force in Democratic party politics since the 1930s, had fallen into disrepair, fractured by debates over race and Vietnam. "If this were France," grumbled House Speaker Tip O'Neill, "the Democratic Party would be five parties." Compounding the problem were the unrealistic and conflicting demands of the party's many constituencies, some of whom viewed the election of a Democratic president as an opportunity to increase spending on social programs starved during eight years of Republican rule.

Instead of adjusting to the new political climate in Washington,

Carter came to power determined to impose his vision on the nation. Because he viewed himself as a steward of the nation's well-being and saw Congress as a captive to narrow special interests, the president felt little need to consult with congressional leaders or to seek their approval. Within the first hundred days of his term, Carter submitted to Congress legislation dealing with energy conservation, government reorganization, changes in immigration policy, social welfare reform, food stamp revision, and an overhaul of election procedures. The centerpiece of his legislative effort was a package of economic reforms designed to stimulate the sluggish economy. His first budget also called for the elimination of eighteen western water projects, which provided patronage and revenue for home districts of Democratic congressmen.

Overwhelmed by the long list of legislation, and feeling slighted by the president and his staff, disgruntled congressional leaders refused to enact most of Carter's agenda, although they did approve a scaled-down version of the economic stimulus bill. "I don't see this Congress rolling over and playing dead," warned Illinois Congressman Dan Rostenkowski. "Carter is going to set up his priorities and we are going to set up ours. We'll see where we go from there." Gleeful Republicans watched from the sidelines as angry Democrats complained publicly about the president's lack of leadership. An exasperated Tip O'Neill lectured Carter about the inner workings of the Washington power structure. But Carter, who came to office campaigning as an outsider, failed to adjust to life on the inside. "Carter didn't seem to understand," O'Neill said.

In 1978, the dark cloud of inflation gathered over the administration, complicating an already precarious political situation. The issue had its roots in Lyndon Johnson's decision to fight the Vietnam War without raising taxes. An overcharged economy saw increases in consumer prices jump from 2 percent in 1965 to 6 percent in 1969. A 1972 election-year loosening of fiscal and monetary policy combined with a worldwide increase in the price of oil, a poor harvest, and a falling dollar had created another surge in inflation in 1973 and 1974. For the next few years, inflation held steady at 6 percent a year, but in 1978, as oil prices surged upward again, it soared to 9.6 percent. By April 1978, a New York Times/CBS poll found that 63 percent of the public viewed inflation as their greatest concern, and only 32 percent approved of Carter's record on the economy.

Inflation forced the administration to make tough, perhaps impossible, choices. Initially, Carter called for voluntary wage and price controls, just as Ford had done; but as inflation surged forward, unions fought for double-digit wage increases to keep ahead of the rising cost of living. In the fall of 1978, following a steep fall in stock prices and a weakening of the dollar on foreign exchange markets, Carter tried to alleviate inflation pressures by narrowing the budget deficit. "I believe that we must firmly limit what the government taxes and spends," he told the nation. "We must face a time of national austerity." In keeping with his theme, Carter submitted to Congress a "lean and austere" budget that called for small cuts in many sensitive domestic programs.

The decision to trim the budget deficit was a reflection both of Carter's fiscal conservatism and his political philosophy. He wanted to reposition the Democratic party, in the words of the journalist James Reston, "in the decisive middle ground of American politics," by proposing policies that appealed to the economic conservatism of independent, middle-class voters. The results of the 1978 midterm elections, where Republicans gained three Senate and twelve House seats, and the success of California's Proposition 13, which slashed local taxes and precipitated a national tax revolt, reinforced Carter's determination to make budget cutting a top priority. That decision, however, placed him on a collision course with his party's vocal liberal wing. "I'm not going to allow people to go to bed hungry for an austerity program," thundered Tip O'Neill. In December 1978, liberals turned the party's midterm convention in Memphis into a grievance session against the administration. In a ringing, podium-pounding speech, Senator Edward Kennedy, a powerful symbol of the party's glorious past, told the faithful to "sail against the wind" of public opinion and reject "drastic slashes" in domestic spending.

Along with trimming the deficit, the administration realized that tackling inflation required progress on another contentious issue — energy. Since the 1973 Yom Kippur War, the price of foreign oil had more than doubled, from about $6 to over $12 a barrel. At the same time, U.S. dependence on foreign oil had risen from about 35 percent to over 50 percent of its total supply. In April 1977, Carter, in a phrase coined by William James, called the issue the "moral equivalent of war" and proposed a comprehensive program to cut energy consumption. Congress, however, refused to act on most of his recommendations. In 1979, the energy outlook turned ominous. A revolution in Iran replaced the pro-American regime of the shah of Iran with a government controlled by Moslem fundamentalists loyal to their religious leader, the Ayatollah Ruhollah Khomeini. The turmoil that resulted created a shortfall in worldwide oil production, driving supply down and prices up. In May, gasoline lines in California ran as long as five hundred cars, and prices at the pump climbed above a dollar a gallon for the first time. Over the Fourth of July weekend, 90 percent of all gas stations in the New York City area were closed, 80 percent in Pennsylvania, 50 percent in Rhode Island.

With public anger reaching a boiling point, the president cut short a vacation and rushed back to Washington to address the nation. Prodded by Patrick Caddell's suggestion that the nation's inability to address the energy problem was symptomatic of a deep crisis of the American spirit, the president abruptly canceled his televised address and retreated to Camp David for eight days of meetings with religious leaders, politicians, poets, and psychiatrists. On Sunday evening, July 15, the president descended from the mountain to give his long-awaited speech to a curious and concerned nation. After describing the "crisis of confidence" that "strikes at the very heart and soul and spirit of our national will," the president proposed a bold new energy plan that included development of alternative energy sources, higher oil and natural gas taxes, and tougher automobile fuel-efficiency standards. The public reacted warmly to the president's speech and to his proposals, but the mood soured a few days later when Carter dismissed half his cabinet. The public viewed the firings as the desperate act of a president unable to control his own administration. Within a week of the cabinet purge, Carter's popularity plummeted, dropping to 74 percent negative, the lowest for any president in modern times.

The administration found no refuge in international affairs. The first president elected in the aftermath of America's defeat in Southeast Asia, Carter found himself trapped by the public's ambivalent attitude toward the world. Americans were determined to avoid another Vietnam, but they refused to question their belief in the universal relevance of American values, and they clung tenaciously to the symbols of world power. During the 1976 campaign, Carter played brilliantly to the ambivalence. Speaking in general terms of reformation and repentance, he emphasized the primacy of moral principles over military might. While renouncing the realpolitik of the Nixon-Kissinger years, and promising to de-emphasize relations with the Soviets, Carter underscored his commitment to an activist foreign policy. Human rights, not anticommunism, would be the organizing principle of Carter's approach to the world in the post-Vietnam era. "We can never be indifferent to the fate of freedom elsewhere," he declared in his inaugural address. "Our commitment to human rights must be absolute."

Despite his stated desire to de-emphasize U.S.-Soviet relations, Carter found himself trapped in the same cold war struggles that had confounded his predecessors. Having declared in his inaugural address that his ultimate goal was the elimination of nuclear weapons from the earth, he ordered the immediate withdrawal of American nuclear weapons from South Korea, delayed development of a new manned bomber, the B-1, and the production of the neutron bomb. In June 1979, he traveled to Vienna to sign a SALT II treaty, which, for the first time, established strategic parity between the United States and the Soviet Union. An ailing and erratic Soviet leadership refused to cooperate with Carter's attempt to reorder American international priorities. Instead of responding to the president's conciliatory gestures, they increased their arms buildup, and used Cuban troops to extend their influence in both the Horn of Africa and in southern Africa.

The effort to decide how to respond to the growing Soviet activism exposed a deep fault line within the administration. The soft-spoken secretary of state, Cyrus Vance, believed that Vietnam had exposed the fallacy of using a superpower context to shape policy toward regional problems. Kremlin actions did not threaten American interests, he argued, and they should not jeopardize the administration's top priority of passing SALT II. His views were at odds with the outspoken and often abrasive national security adviser, Zbigniew Brzezinski, who saw the Russians exploiting postwar American malaise by testing its will around the globe, probing for weak points and moving to take strategic advantage. Brzezinski wanted to use the promise of SALT II as a constraint on Soviet actions. The debate between Vance and Brzezinski often spilled over into public view, creating the impression of an administration incapable of establishing clear foreign-policy objectives. "Incoherence," observed the political scientist Stanley Hoffmann, has "its roots deep in the Carter administration's style of policymaking."

By the end of 1978, however, concern about Soviet aggression had produced a major reordering of administration priorities and a reaffirmation of the cold war policy of containment. Kremlin actions stirred nationalist sentiment in the U.S. and eroded support for détente. Carter, moving closer to the Brzezinski line, called for a significant increase in military spending, the strengthening of NATO forces, development of the neutron bomb, construction of the intermediate-range cruise and Pershing missiles, and the building of a vast new missile — the MX — to be housed in a complex system of underground tunnels. Replacing his idealistic rhetoric about human rights with the lexicon of power politics, Carter reassured America's dictator-allies that U.S. national security commitments would outweigh concern about their internal conflicts.

The confused attitude toward the Soviets distracted attention from Carter's foreign policy accomplishments. In the Middle East, the president's dogged persistence of peace produced the most monumental achievement of his presidency — the Camp David Accords. In July 1978, Carter recorded in his diary that "it would be best, win or lose, to go all out" to bring the feuding Middle East leaders to the peace table. In September, he invited Egyptian President Anwar Sadat and Israeli Prime Minister Menachem Begin to the presidential retreat at Camp David. For thirteen days the president devoted all of his attention to the complicated negotiations of land for peace, shuttling back and forth between cabins, bargaining over the nuance of every word. His efforts paid off. On March 26, 1979, Egypt and Israel signed a treaty that ended thirty years of hostilities. Two months later, Carter completed the process initiated by Richard Nixon by formally recognizing the People's Republic of China.

But even administration successes were sometimes turned into political defeats. In 1978, Carter courageously chose to sign treaties returning to Panama full sovereignty over the Panama Canal. "Fairness, not force," said the president, "should lie at the heart of our dealings with the nations of the world." But at a time of deep uncertainty about American power, conservatives transformed the canal into a reassuring symbol of America's glorious past. Carter prevailed in the Senate, but conservatives won the debate by painting the president as weak and ineffective, unable to defend the national interest. One month after passage of the Panama Canal Treaties, thirty-eight Republican senators boldly accused the administration of conducting a foreign policy "of incoherence, inconsistency and ineptitude."

By the summer of 1979 Carter seemed destined to be a one-term president. Observers complained that he was aloof and arrogant, incapable of seizing control of the levers of power in Washington. A poor public speaker, he failed to arouse voter passion. "He is a soothing flatterer and a sensible president," noted the *New York Times*, "but not yet a leader, or teacher, even for a quiet time." His own party was in open revolt. With polls showing him leading the president by a three-to-one margin, Senator Kennedy, the keeper of the flickering liberal flame, announced that he would challenge Carter for the nomination. Portraying Carter as a weak and ineffective leader who had abandoned the party's liberal tradition, he confidently declared: "The only thing that paralyzes us today is the myth that we cannot move."

Ironically, crises abroad saved Carter from the Kennedy challenge at home. In October 1979, in response to the White House decision to let the deposed shah come to the United States for medical treatment, Iranian nationalists decided to punish the "great Satan" by storming the U.S. embassy, seizing fifty-three American soldiers and diplomats, and holding them hostage. The United States was unable to effect a release by threats or diplomacy, and the hostages languished. As the American hostages entered their eighth week of captivity, another crisis developed. On Christmas Day, Soviet troops invaded neighboring Afghanistan, toppling that nation's bumbling puppet regime. Claiming that the Soviet invasion posed the most serious threat to peace since World War II, the president proclaimed the Carter Doctrine: "An attempt by any outside force to gain control of the Persian Gulf region will be regarded as an assault on the vital interests of the United States of America, and such an assault will be repelled by use of any means necessary, including military force."

Politically, the twin crises in Iran and Afghanistan were a double-edged sword for the administration. As the American people instinctively rallied around the president during a time of crisis, Carter watched his job approval rating double to 61 percent — the sharpest one-month leap in forty-one years of polling. Capitalizing on his sudden surge of popularity, Carter decided to play the role of national leader by standing above the partisan fray and refusing to campaign. In the short run, the strategy worked brilliantly: Kennedy's popularity wilted in the patriotic afterglow and Carter secured his party's nomination on the first ballot.

In the long run, however, the crises in Iran and Afghanistan raised troubling questions about the administration's erratic foreign policy. The Soviet invasion dealt a fatal blow to SALT II, a top administration priority. In April, in a desperate effort to break the deadlock in Iran, Carter ordered an abortive rescue mission that resulted in the death of eight soldiers when two helicopters collided during a sandstorm in the desert. The image of military helicopters burning in the Iranian desert touched a sensitive nerve with Americans worried about the nation's declining position in the world. A tidal wave of nationalist sentiment swamped early expectations that Carter would redefine America's international role in the post-Vietnam era. "For the first time in its history," *Business Week* lamented, "the United States is no longer growing in power and influence among the nations of the world. . . . [The] Pax Americana that shaped modern history since World War II is fast disintegrating."

A failing economy at home presented the most serious obstacle on

Carter's road to reelection. During the summer, with inflation soaring into double digits, newly appointed Federal Reserve Chairman Paul Volcker applied the monetary brakes. The nation's major banks responded by raising their prime rate, which stood as low as 5 percent when Carter took office, to 13 percent and then to 14.5 percent. With the prime rate reaching all-time highs, the economy began its inevitable slowdown. In October, the Dow Jones index of industrial stocks lost nearly one hundred points, auto sales dropped 23 percent compared to the previous year, and rising mortgage rates strangled the housing industry.

Carter struggled to convince a skeptical public that the nation's problems were the inevitable by-products of a new age of limits, both at home and abroad. But Americans nurtured on rhetoric about boundless growth and influence found his Republican opponent, Ronald Reagan, more appealing. Reagan responded to Carter's muddled message of restraint by offering the nation still one more attempt to regain past glory by returning to simple verities. Preaching what the economist Herbert Stein called the "economics of joy," Reagan repudiated traditional Republican economic doctrine of a tight fiscal policy and balanced budgets, and, instead, proposed massive tax cuts and huge increases in military spending. "Are you better off than you were four years ago?" he asked during the lone presidential debate, focusing public attention on the economic issues on which Carter was most vulnerable.

On November 4, a majority of voters expressed their displeasure by rejecting the president's bid for reelection. Reagan received 43.9 million popular votes (51 percent) to 35.5 million (41 percent) for Carter. The Independent candidate John Anderson scored 5.7 million votes, about 7 percent of the total. In terms of electoral votes, Reagan's victory was even more clear-cut: he carried a total of 489 votes to Carter's 49. Not only did the Republicans take possession of the White House, but they gained control of the Senate for the first time in twenty-eight years, and picked up thirty-three seats in the House.

In 1981, Carter returned to Plains, Georgia, an apparently broken man. His public ambitions had been crushed in a massive landslide; the peanut farm he had left behind in a blind trust was $1 million in debt. Facing a "potentially empty life," he decided to devote his energy to building his presidential library. Located on thirty acres of woods and rolling hill in the Virginia Highlands section of Atlanta, near Emory University, the Carter Center became the base of operations for Carter's postpresidential ambitions.

His activities at home and abroad boosted Carter's spirits and his popularity, making him one of the most successful ex-presidents in history. The first president since Herbert Hoover to play an active role in world affairs after leaving the White House, Carter used the center as a shadow state department. His Global 2000 Inc. worked to eradicate diseases and improve agricultural productivity in Third World countries. The ex-president became a roving ambassador, mediating internal conflicts in Ethiopia, Nicaragua, Somalia, Panama, Korea, and Haiti. Along with his official duties, Carter and his wife remained active in church affairs. They spent time each year wielding hammer and saw on behalf of the Habitat for Humanity, a nonprofit Christian enterprise based in Georgia which constructs housing for the poor and homeless in the U.S. and abroad. The former president also served as the finance officer of the 130-member Maranatha Baptist Church. He could occasionally be seen mowing the church lawn while Rosalynn vacuumed the carpets and the upholstered pews. "The church is the center of our social life," he said. "It's like breathing for us."

— STEVEN M. GILLON
University of Oklahoma

Ronald Reagan

1981–1989

b. February 6, 1911

L ONG AFTER RONALD REAGAN rode off into the sunset in January 1989, his administration remains a great American political mystery. How did a washed-up B movie actor end up as president of the United States? Was Reagan a fool or a genius, a success or a failure, an ideologue or a panderer, a budget cutter or a budget buster, a midwestern moralist or a Hollywood libertine, a populist demagogue or the rich man's best friend? And how did Reagan "restore confidence" in the nation and the presidency while intensifying cynicism about politics?

Many of Ronald Reagan's stands were so definitive that partisans easily sanctify and demonize him. Conservatives who forgive his apostasy worship him for saving America. Liberals who do not dismiss this "amiable dunce" damn him for bankrupting the country and abandoning America's disadvantaged.

Surprisingly, Reagan's elusive presidency remained rooted in the New Deal big government consensus. For all his radical conservative rhetoric, his policies were conventional. In his January 1989 farewell address, the "great communicator" called his epoch not the "Reagan revolution," as some others had dubbed it, but "The Great Rediscovery, a rediscovery of our values and our common sense." This pastier label, combined with the pileup of American victories and fiascoes in the 1980s, suggests that Reagan's spasmodic two terms were not one coherent movie with a simple dramatic title. The Reagan administration was more like the future president's Western anthology "Death Valley Days," a series of uneven episodes, loosely connected thematically, which glorified old-fashioned values while actually undermining them.

Ronald Reagan's is a twentieth-century success story, the triumph of a glittering personality more than of a great character. The tall, square-shouldered, happy-go-lucky cowboy wanna-be with a ruddy complexion was in his youth the most popular, not the most likely to succeed. "I'm a plain guy with a set of homespun features and no frills," Reagan told movie fans in 1942. "Mr. Norm is my alias, or shouldn't I admit it?"

Born in Tampico, Illinois, in 1911, Reagan always wanted to live "An American Life," as he titled his post-presidential memoirs. As a mid-westerner, as a Hollywood star, even as a Sunbelt conservative, Reagan remained conventional. His good-natured optimism sprang from the American heartland and found reinforcement as he mastered two brutal but glamorous industries: the movies and politics. The 1930 Hollywood production code captured Reagan's approach to life: "Correct standards of life, subject only to the requirements of drama and entertainment, shall be presented."

Reagan called his childhood "as sweet and idyllic as it could be." In fact, his home was a battlefield between a refined Protestant mother and an alcoholic Irish Catholic father. One "cold, blustery winter's night" when he was eleven, "Dutch" found his father passed out in the snow outside. The angry boy contemplated "going to bed, as if he weren't there." But he feared that "the whole neighborhood" could hear his father's snoring. The boy dragged his father to bed and, following family protocol, "never mentioned the incident to my mother."

Reagan's behavior comported with his contemporaries' code of conduct. Most children of early twentieth-century America followed detailed road maps that emphasized devotion to family, church, community, nation, and God, along with values such as fidelity, honor, grace, patriotism, discipline, and discretion. Even if young people did not always do the right thing, they entertained few doubts about what the right thing was.

Ronald Reagan and his peers learned about politics from *the* president of their generation, Franklin D. Roosevelt. Roosevelt served from 1933 to 1945, when Reagan grew from a twenty-two-year-old Eureka College graduate and rookie radio announcer to a thirty-five-year-old Hollywood star. Reagan "idolized" the president: "During his Fireside chats, his strong, gentle, confident voice resonated across the nation . . . and reassured us that we could lick any problem." Reagan learned what a president who mastered modern media and championed American values could accomplish.

In fifty-three films, from 1937 until 1964, Reagan steeped himself in Hollywood's imaginary universe of quaint ideals and happy endings in which the good guys always win. Yet as a contract player in Warner Brothers' dream-making factory, he learned that stars, like dreams, were man-made. Captain Reagan's wartime service kept him stateside, shooting movies to bolster morale. His "fairy tale" marriage to Jane Wyman was tumultuous. But Reagan distanced himself from ugly or embarrassing realities. He referred often in later years to having "gone off to war." And after his marriage failed in 1949, he would say: "I *was* divorced in the sense that the decision was made by someone else."

While leading the Screen Actors Guild from 1947 to 1952, Reagan mastered the rhetoric of the aggrieved affluent, overlooking his constituents' luxuries when agitating for their "rights." The SAG's role in the Hollywood power struggles with Communists, both real and imagined, crystallized Reagan's hatred for the "Reds," just as the 90 percent tax rate imposed on him and his colleagues shaped his politics. "I was a Democrat when the Democratic party stood for state rights, local autonomy, economy in government, and individual freedom," he later said. "Today it is the party that has changed, openly declaring for centralized federal power and government-sponsored redistribution of the individual's earnings."

Ronald Reagan's marriage to a thirty-one-year-old starlet on March 4, 1952, gave him a second shot at the conventional life he craved. Nancy Davis was devoted to her "Ronnie" and to their domestic illusions. The bond the Reagans developed was particularly striking considering that even Nancy Reagan would admit that "There's a wall around him."

Reagan's movie career sputtered during the early years of his second marriage, even forcing him, at one point, to emcee a Las Vegas revue. Prosperity returned in 1954, when he began hosting CBS's weekly "General Electric Theater." Nancy and Ronnie Reagan became icons of America's postwar prosperity. Yet the consumer-obsessed, leisure-oriented society the Reagans and their peers helped promote fostered the materialism and selfishness that doomed their beloved "Main Street USA."

Reagan emerged into the political limelight with a nationwide sermon urging Barry Goldwater's election in 1964. "The Speech" echoed the hymns to small government which Reagan sang during hundreds of General Electric factory tours. Shortly thereafter, forty-one fat cats

THE AMERICAN PRESIDENCY ★ 470

searching for a new California governor established "The Friends of Ronald Reagan." When asked what kind of a governor he would be, Reagan joked, "I don't know, I've never played a governor." Nevertheless he agreed to run and easily defeated the two-term incumbent, Pat Brown, in 1966.

Governor Reagan's two terms from 1967 to 1975 exemplified the backlash the 1960s rebellion triggered. Reagan's role as what James Cagney would call "the zealous defender of the American dream" linked his Hollywood past with his political future. Detractors dismissed Reagan as an amateur while caricaturing him as a lunatic. The attack on his politics gave him credibility, just as his good looks and sunny approach softened his edge. Reagan was the conservative Kennedy, charismatic enough for the masses, ideological enough for the partisans.

Fed up with moderate Republicans' surrender to the welfare state and Henry Kissinger's compromising foreign policy, Ronald Reagan challenged President Gerald Ford for the 1976 nomination. Ford's narrow nomination victory and eventual loss made Reagan the Republican front-runner in 1980. Reagan's controversial statements and slips of the tongue had 44 percent of those surveyed doubting he understood "the complicated problems a president has to deal with." Jimmy Carter's aide Hamilton Jordan sneered, "The American people are not going to elect a seventy-year-old, right-wing ex-movie actor to be president."

Reagan's brilliant mix of modern image-making and old-fashioned values triumphed. For the first time since 1954, Republicans secured a majority in at least one House of Congress, the Senate. Reagan's Goldwater-conservatism-with-a-smile united evangelical Protestants; southerners; neo-conservative northeastern intellectuals alienated by the sixties; corporate leaders; America's petite bourgeoisie of white-gloved, churchgoing ladies with their country-club-joining chamber of commerce husbands; and "Reagan Democrats," Catholic and working-class New Dealers choked by the high taxes of the Great Society.

Ronald Reagan's paradox was the paradox of the modern presidency. He was the most ideological and the most evanescent candidate of the modern era. Without the conservative grounding, he would have floated away. Without the image-making, he would have sunk under the weight of his rhetoric.

President Ronald Reagan believed that America was ripe for revolution. Vietnam, Watergate, Carter's flaccid presidency, and the Iranian hostage standoff had demoralized Americans politically. The sexual revolution, the youth culture, and the rise of the misanthropic media had unnerved them culturally. And 13.5 percent inflation combined with interest rates soaring to 21.5 percent had devastated them economically.

The January 1981 $16.5 million inaugural gala repudiated the iconoclasm of the 1960s and celebrated all-American ideals of conformity, power, success, and money. Democrats decried the Republican march of minks amid pledges of budget cutting. Iran's release of its fifty-two hostages thirty-five minutes into the new administration silenced the critics and transformed the Carter malaise into the Reagan resurgence.

"In this present crisis, government is not the solution to our problem," Reagan declared in his inaugural address. The new president believed "that if we cut tax rates and reduced the proportion of our national wealth that was taken by Washington, the economy would receive a stimulus that would bring down inflation, unemployment and interest rates, and there would be such an expansion of economic activity that in the end there would be a net increase in the amount of revenue to finance the important functions of government" — such as defense. What he called "common sense," theorists called "supply side economics."

The welfare state had indeed become bloated by the 1980s. Despite billions of dollars spent and thousands of arcane regulations, crime increased, and poverty, though much reduced, persisted. Democratic presidents from Franklin Roosevelt through Lyndon Johnson had been too busy promising to end all misery at no real cost to prepare Americans for the necessary sacrifices — and inevitable disappointments. The backlash against the welfare state spread throughout the Western world, especially in Margaret Thatcher's England and Brian Mulroney's Canada.

Ronald Reagan set out to dismantle Franklin Roosevelt's New Deal by imitating Roosevelt's mix of bold gestures and reassuring compromises. "I place myself in the 'seller' category of leadership," the president confessed. He left the policy nitty-gritty to advisers, especially his first-term "triumvirate," each of whom contributed a key ingredient to his success. Presidential Counselor Edwin Meese transmitted conservatives' ideas, Chief of Staff James Baker moderated them to suit the Republican business establishment, and Deputy Chief of Staff Michael Deaver packaged them to woo the people. Just as Roosevelt told two speechwriters fighting over contradictory policies to "weave the two together," Reagan settled arguments between his aides by smiling and saying, "Okay, you fellows work it out." Reagan's dependence on his staff suited the expanded modern presidency. In Hollywood, Reagan learned that an actor was the public tip of a huge iceberg. One scene in *Knute Rockne, All American* involving only a farmer and a horse, he recalled, required seventy people on location. Yet Reagan had not made it to Washington without some backbone and dedication to principle. "Some in the media delight in trying to portray me as being manipulated and led around by the nose," Reagan would fume. "I'm in charge and my people are helping to carry out the policies I set."

During those first few chaotic weeks in office, Reagan's budget whiz, David Stockman, desperately crunched the numbers so that income taxes could be reduced by 30 percent over three years, defense spending could experience "real growth," entitlement programs like Social Security could be maintained, Reagan's soon-to-be-infamous "social safety net of programs" benefiting "the poverty-stricken, the disabled, the elderly" could be preserved, and the budget could be balanced. On February 18, Reagan presented his "program for economic recovery" to a joint session of Congress. In his nationally televised speech, the president proposed cuts in eighty-three federal programs to trim $41.4 billion from Jimmy Carter's proposed budget.

This attempt at budget cutting ruined the traditional presidential honeymoon. The "permanent government" of bureaucrats, congressional liberals, and media bigwigs remained loyal to the New Deal. And while many Americans believed the welfare state was broken, few agreed just how to fix it. In late March 1981 the president's approval rating of 59 percent was lower than most other modern presidents' "honeymoon" rating. His 24 percent disapproval rating was nearly three times higher than his predecessors' negative rating at the comparable time.

The new president seemed undisturbed, performing affably on cue at stage-managed public appearances. Then on March 30, he heard "what sounded like two or three firecrackers" as he left the Washington Hilton after addressing the Construction Trades Council. In yet another unwelcome tribute to the presidency's unique position in American life, a crazed lone gunman again tried to shoot his way into history.

The oldest man elected president not only defeated death that spring day, but his recovery established a welcome bookend to the murder of the youngest man elected president, John Kennedy. Reagan also emerged as the man most Americans now loved to like as they exorcised the defeatist spirit of the 1960s. "The shooting is probably what most Americans will remember about the one hundred days," NBC's Roger Mudd claimed. "And because he performed under fire that day as if it had been a movie, President Reagan made it difficult for all of us to think of him, ever again, as just another B-grade actor."

On May 8, a congressional coalition of Republicans and mostly southern Democratic "Boll Weevils" passed Reagan's budget. The first round of the Reagan revolution reduced the personal income tax rate by almost one quarter and dropped the capital gains tax from 28 to 20 percent, yielding an unprecedented tax reduction of $162 billion. By the summer, Reagan had eliminated $35 billion in domestic spending from Carter's request, boosted defense spending, and demoralized the labor movement by breaking the air traffic controllers' union, PATCO.

That summer the Reagan revolution peaked. The early success gave Reagan's attack on the New Deal an enduring legitimacy. Yet the assassination attempt helped pass the program without forcing the president to convert the public. The congressional counterattacks, and Reagan's endorsement of "entitlements" and "the social safety net," charted the limits of his revolution. Symbols, not substance, Reagan's personality, not his ideology, triumphed. "It's clear the president's overall strength is greater than the sum of his strengths on individual issues," one 1986 White House memo would acknowledge. "We can never avoid this fact — only turn it to our advantage."

Ronald Reagan proved no more willing than Roosevelt, Kennedy, or Johnson had been to demand sacrifices. The budgetary compromises that Reagan continued to accept reflected the power of the Democratic House and the entrenched power of the welfare state. Some called the impasse between Reagan's budget cutting and the Democrats' "tax and spend" approach "gridlock." Others recognized it as the historic, moderating force of the American consensus, one more proof of the Framers' genius.

Along with Federal Reserve Chairman Paul Volcker's tightening of the money supply, Reagan's measures did stop inflation — by almost stopping the economy. Inflation dropped to 4.6 percent in 1982, and the prime interest rate was halved to 11 percent. Yet despite Reagan's boast that he had cut by half the 1980 federal spending *growth* rate of 17 percent, his tax cut and defense spending increase created a deficit of $100 billion, twice the previous record. The unemployment rate of 7 percent when Reagan took office hit 10.8 percent by December 1982 — the highest since 1941.

Reagan tried to blame this recession on "the failed policies of the past," but his administration was sinking. Barely midway through his first term, Reagan had already fired his secretary of state, his national security adviser, and his arms control chief, reflecting foreign policy disarray. On the domestic side, his secretaries of energy, transportation, and health and human services were gone. In November 1982, with eleven and a half million Americans newly unemployed, the Republicans lost twenty-six seats in the House, while salvaging their slim Senate majority.

Assessing the "totter[ing]" economy, the "ill-planned military build-up," and the "huge deficits" at midterm, the *New York Times* declared, "The stench of failure hangs over Ronald Reagan's White House." The columnist Anthony Lewis said that Republicans had a "government incompetent to govern, a president frozen in an ideological fantasy-land, an administration spotted with rogues and fools." David Broder of the *Washington Post* found himself "nostalgic for the Nixon administration."

The more liberals sniped at "Scrooge" Reagan, the more desperately administration officials repudiated their own revolution. After the midterm pummeling, the White House released a remarkable document, "Ten Myths That Miss the Mark." Myth One was "Reagan has cut social spending to the bone" — and proceeded to boast that "Total nominal spending in 1984 on *15 key social programs* is *45 percent higher*" than the amount spent on the same programs in 1980." Myth Two was "Reagan has slashed taxes, especially for the rich" — and proceeded to boast that taxes would "rise an average of *$345 per year* from 1980 to 1988 for a family with an annual income of *$40,000*. And Myth Three was "Reagan is proposing a massive, unprecedented defense budget" — and proceeded to boast that "By 1988, national defense" would remain "below the pre-Vietnam 1964 level." Debunking Myths Four through Ten transformed Ronald Reagan into Franklin Roosevelt's successor, an antibusiness, pro–civil rights, pro-environment, pro–big government activist.

Such ridiculous postures alienated conservatives without mollifying liberals. Reagan's seesaws between revolutionary rhetoric and consensus pragmatism intensified the civil war raging within his administration. Hundreds of young zealots had overrun Washington to dismantle the bureaucracy that the young New Dealers helped build half a century earlier. Many of these Reaganauts penned lovers' laments to their hero, crying, "What kind of conservative revolution is this?" Increasingly, Ronald Reagan, the bleeding-heart conservative pragmatist, the halfhearted revolutionary, the laissez-faire manager, was looking like a sunnier version of Jimmy Carter, the most recent presidential failure. Presidential trial heats at midterm showed former Vice President Walter Mondale vindicating Carter's honor by beating Reagan 52 percent to 40 percent.

Surprisingly, Reagan's calls to "stay the course" ended in triumph. By 1983, with the inflation rate below 2 percent, with oil prices declining, and with many businesses roused by Reagan's crusade against banking, environmental, and safety regulations, the economy soared. Economists warned that the nation was mortgaging its future with huge budget deficits and easy credit, while Republicans praised "Reaganomics" for saving America.

With the 1984 campaign looming, hawks like Ed Meese wanted a "bold campaign," pushing for a Republican Congress to revitalize the right-wing agenda. Yet Reagan preferred playing to the American mainstream. He listened to the "patriots," who suggested the campaign remain, as his aide Richard Darman put it, "relatively nonspecific programmatically . . . more abstract."

President Reagan loved serving as the high priest of America's civil religion. After Lyndon Johnson's bloviations, Richard Nixon's machinations, Gerald Ford's hesitations, and Jimmy Carter's fulminations, most Americans delighted in Ronald Reagan's old-fashioned, all-American celebrations. In 1984, Reagan would, therefore, ignore his opponent what's-his-name, minimize the right-wing rhetoric, and revel in his America, "where every day is Independence Day." Press stories often were personal or patriotic rather than political or substantive. Thus, Reagan spent the summer of 1984 being photographed at the Los Angeles Olympics, attending his daughter's wedding in Bel-Air, and horseback riding at his Santa Barbara ranch. In an inversion from the nineteenth century, the president's private life was publicized and his public stands were obscured.

The Democratic nominee Walter Mondale was no match for this master of the American popular psyche. Mondale took his issue-oriented campaign based on faith in government to the New Deal coalition in the Rust Belt. Reagan relied on television and carefully crafted appearances to fashion his new coalition anchored in the Sunbelt.

The great "Reagan expansion" justified his "Morning in America" happy-talk campaign. With the stock market skyrocketing, voters found it easy to believe that America was "back." Sixty months of uninterrupted growth from 1982 to 1987 produced 14.8 million new jobs and $20 trillion of new wealth. This greatest spurt of employment growth in American history briefly created the illusion that Reagan's "voodoo economics" (as his one-time opponent George Bush quipped) worked. Billions of dollars streamed into government coffers, even as Reagan spent over $1.5 trillion on the military in seven years and the marginal tax rate plunged from 70 percent to 28 percent, the largest percentage reduction in tax rates in U.S. history.

The 1984 election landslide gave Ronald Reagan an impressive personal victory. He had entered the White House with barely half of the popular vote and with the highest negative ratings pollsters had recorded since they had begun asking such questions three decades before. Four years later, he received a higher percentage of the popular vote than Dwight Eisenhower had, nearly 59 percent, and the most electoral votes ever, 525.

Even though he would lose the Republican Senate in 1986 and failed ever to wrest the House of Representatives from the Democrats, Reagan revitalized the Republican party. By bashing the New Deal while preserving its most popular programs, he wooed southern whites into the GOP and weakened northern workers' ties to the Democrats. Social and fundamentalist conservatism, though not always triumphant, could no longer be dismissed. Republicans outflanked the Democrats for the first time since the New Deal. Even if crime soared, Republicans took a tougher stance on crime than did Democrats. Even if the budget ballooned, Republicans always suggested budget cuts, which Democrats instinctively opposed. Even if the United States suffered humiliations in the Middle East, Republicans always sounded more militant and more patriotic. The rhetoric appealed to millions, even as the gap between the Reagan reality and Republican rhetoric increased popular cynicism.

Ronald Reagan also seemed to have revitalized the presidency. His celebrity presidency made the experts stop talking about Richard Nixon's imperial presidency, and Jimmy Carter's impotent presidency. Reagan showed how, in the age of television, focusing on the well-dramatized big picture and having a strong vision could balance out many smaller failures.

Despite the landslide, Reagan's second term lacked a mandate. The Gramm-Rudman-Hollings Act passed in December 1985 set targets for eliminating the federal budget deficit by October 1, 1990 — but placed many key items "off-budget." Reagan championed the balanced budget as a "goal" — yet never submitted one to Congress. His biggest domestic success, the Tax Reform Act of 1986, closed some tax loopholes and limited the top personal income tax to 33 percent — sweet vindication for the man who still bristled remembering his 90 percent tax rates. Still, as the deficit mushroomed, the president resorted to "revenue enhancements": raising sin taxes on alcohol and tobacco, increasing user fees on government services, and hiking Social Security payroll taxes. Few Americans except for the superrich saw their overall tax bills drop.

Throughout his second term, while searching for a crusade to galvanize the nation, Reagan would often seem befuddled, disengaged, poorly served by his staff, and buffeted by events. His unfortunate trip in May 1985 to the Bitburg cemetery where forty-nine S.S. storm troopers lay buried was followed by terrorist attacks against Americans on a TWA flight and the cruise ship *Achille Lauro,* his bout with cancer, and the Challenger spaceship disaster. Most damaging of all was the harebrained scheme beginning in the summer of 1985 to sell 2,004 TOW antitank missiles, along with crucial spare parts, to Iran and surreptitiously (and illegally) fund the Nicaraguan Contras — a plan exposed in November 1986.

Foreign policy was a centerpiece of the Reagan revolution. Contemptuous of "those liberal elites" who weakened America morally and militarily, Reagan and his followers vowed to save the world by whatever means necessary. The Reagan doctrine called for the United States to check Soviet expansionism throughout the world, be it Angola or Afghanistan. This cold war preoccupation with the "evil empire" blinded Reaganites to the excesses of what they called "authoritarian" regimes in Central America, which became what Cuba was to John Kennedy — America's Maginot Line.

For all his John Wayne–like bluster, Ronald Reagan remained wary of committing American troops to combat. In Central America, Americans did what Richard Nixon tried to do in South Vietnam. Local troops using American weapons and aided by American intelligence did most of the killing and the dying.

In the Middle East, the fear of a quagmire handcuffed Reagan. When a truck bomb killed 241 Marines deployed in Lebanon in 1983, Ronald Reagan, the avenging angel of the United States, whimpered, then retreated. Critics grumbled about a president who spoke "loudly" but carried "a wet noodle."

Reagan's foray into Grenada the same week distracted Americans from their humiliation. The invasion revealed Americans' thirst for any military victory. "We're now one, one, and one," a character says in Clint Eastwood's 1986 movie *Heartbreak Ridge,* balancing the liberation of a Grenada-like island against the loss in Vietnam and the stalemate in Korea. Such movies were loosely based in fact yet perpetuated the nation's greatest fictions and epitomized the values of the Hollywood president.

While Reagan avoided major political damage in Lebanon, the Marine carnage reminded him that Americans in the television age could not abide military casualties. Quick, dramatic, and safe gestures like the April 1986 bombing of Libya were more popular. This fear of casualties, the concern for seven Americans held hostage in Lebanon, and the desire to maintain ties with Iran to trump the Soviets, made Reagan vulnerable to the arms-for-hostages scheme some purportedly moderate Iranians proposed. Meanwhile, frustration with the Boland Amendment prohibiting financial aid to the Nicaraguan Contras justified diverting at least $3.8 million in profits from the arms sales to the "freedom fighters." While they violated Reagan's vow never to negotiate with terrorists, the cowboy stunts of Iran-Contra epitomized Reagan's emphasis on ends, not means, and his contempt for critics as unpatriotic.

A Lebanese magazine, *Al-Shiraa,* first leaked word about the arms-for-hostages deal in November 1986. As White House spokesmen pussy-footed and reporters howled, Colonel Oliver North and fellow National Security Council staff erased computer files and shredded documents. Word of the diversion of profits from the arms sales to the Nicaraguan Contras magnified the scandal into a constitutional crisis. Now the president had to claim that he had been badly served by some subordinates, especially North and two successive National Security advisers, Robert McFarlane and Admiral John Poindexter.

Reagan's defense during Iran-Contra rested on two contradictory and incredible points. He could not boast about efforts to free the hostages in Lebanon and to woo Iranian moderates, then claim that his administration did not negotiate with terrorists. And he could not claim ignorance of the illegal diversion of arms profits to the Contras, while insisting that he, and no one else, was in command of the White House. Two weeks after the president sputtered in a press conference on November 19, pollsters found that his favorability rating had dropped in one month from 67 percent to 46 percent. "For the first time in my life, people didn't believe me," an incredulous Reagan would recall.

With its shady arms dealers and profiteering patriots, Iran-Contra highlighted the odd moral climate in Reagan's America. Reagan was personally honest. Mrs. Reagan's "Just Say No" to drugs campaign and his calls for a moral renaissance revived talk about "family values." Both George Bush and Bill Clinton would strive to emulate Reagan as national preachers. Yet despite all the moralizing there was much immorality. Reagan was the first divorced president, and the father of alienated children who struggled with drugs and bad marriages. Furthermore, corruption flourished amid the "administrative latitude" enjoyed by the president's subordinates, his push for deregulation, and his naive faith in businessmen. Many entrepreneurs who bashed big government, "welfare queens," and "poverty pimps" swilled at the federal trough. The Department of Housing and Urban Development seemed to be one vast welfare scheme for Reaganauts. Hundreds of savings and loan operators stuck the hated government with billions of dollars of bad debts.

Reagan's flamboyant traditionalism failed to reconstruct the community consensus of yesteryear. In fact, his libertarian odes to individualism and untrammeled capitalism intensified the hedonism of America's consumer-oriented society. "Trickle-down economics" diluted the traditional Protestant guilt by suggesting that personal greed served the public good. "At his worst, Reagan made the denial of compassion respectable," New York's Governor Mario Cuomo would mourn. Ultimately, the consumption revolution that Reagan helped foster in Hollywood and celebrated in Washington proved far more damaging to the traditional way of life than a handful of hippie draft dodgers.

Iran-contra revealed a brazen president with noble motives who was also sloppy, aloof, inconsistent, gullible, and delusive. Reagan faced the greatest crisis of his political life as passively as he faced the greatest crisis of his personal life, his divorce. His detachment made it easier for him to believe that his initiative to free the hostages was not related to the arm sales. This time, he blamed hysterical reporters and overzealous aides rather than a moody wife.

Prostate surgery in mid-January 1987 made Reagan the man look as enfeebled as Reagan the president. Some insiders speculated that his infamously bad memory indicated Alzheimer's disease — which Reagan admitted to suffering from years later. The deference accorded the president, and his ability to rise to public occasions, inhibited reporters from speculating openly.

Forced by his defense to declare the president of the United States incompetent or dishonest, Reagan's own internal investigation by the Tower Commission (headed by Texas Senator John Tower), chose to condemn his "management style." When members of the board briefed Reagan in February 1987 the "flustered" president still insisted that he never approved trading arms for hostages. "Mr. President," board member Brent Scowcroft said, "there were occasions when the aircraft loaded with weapons was sitting on the runway, waiting for word that the hostages had been freed."

Dreaming of routing another Republican president, Democrats reassembled the Watergate-era politician's torture chamber of overlapping congressional and criminal investigations. The lengthy special prosecutor's inquiry and the dramatic joint congressional hearings during the summer of 1987 crippled the administration. Yet the public backlash against the Democrats' partisanship also revealed Reagan's success in reinvigorating American patriotism. Oliver North's brilliant performance as the beribboned if overzealous Marine enduring the nitpicking of noncombatants seduced millions. A hero for the 1980s, North parlayed his televised appearance at the hearings into a Reaganesque career straddling the boundaries of popular culture and politics, serving as a best-selling author, highly paid lecturer, Virginia senatorial candidate, and a talk show host.

Eventually, Reagan's popularity recovered, mostly because only 24 percent of those surveyed believed that the president was really in charge. Reagan knew that he was more liked than respected. He hated that "Americans were forgiving me for something I hadn't done." Once again, Reagan's success stemmed from his personality and his vision, not his policies or managerial skills. Still, after November 1986 pollsters usually found more Americans worried that America was on the wrong track rather than convinced it was headed in the right direction.

As 1987 dragged on, other traumas upstaged the Iran-contra scandal. In October 1987 alone, the stock market crashed, the Senate blocked the president's nomination of Judge Robert Bork to the Supreme Court, an Iranian missile struck an American tanker, and the First Lady's mother died ten days after Nancy Reagan's mastectomy.

The five-hundred-point stock market plunge on October 19, 1987, illustrated the economy's structural weaknesses. Exports had flooded the nation, pushing the trade deficit from $24 billion in 1980 to $152 billion in 1987. Many Americans had purchased new Japanese cars on credit, creating individual debts that paralleled the monstrous national debt. Overall, the wealthiest Americans had prospered while the middle class felt squeezed and the poor languished.

The crash alerted Americans to look behind the rhetoric of Reaganomics. In fact, Reagan's government spent wildly. Total federal receipts as a share of the GNP remained almost what they had been under Jimmy Carter, just below 20 percent. "Entitlement programs" like Medicare and Medicaid more than doubled. Farm price supports also soared.

Americans went from owing 16 cents for every dollar in national income in 1981 to owing 44 cents for every dollar in national income in 1988.

The Senate's rejection of Bork proved that even on social issues Reagan's record did not match his rhetoric. The "Religious Right" would never abandon Reagan for the Democratic party; therefore he rarely wasted political capital fighting for their "abc" agenda, their battles against abortion, busing, and crime. Reagan's administration also boasted about its civil rights reforms and its token minority and women appointees — especially Sandra Day O'Connor to the Supreme Court. However, Reagan did derail the progressive agenda by forcing activists to defend programs such as affirmative action rather than expanding them. And, in the long run, Reagan's three Supreme Court appointees, 78 appeals court judges, and 290 district court judges curbed the expansion of civil rights and minority privileges generated by the Warren Court.

The impression of Reagan as "henpecked, manipulated, and oblivious" deepened as publishers paid insiders huge advances to expose the inner workings of the White House. The blurring of politics and entertainment, which had benefited the Reagans so often, backfired as Reagan endured ten kiss-and-tell books while in office. The former budget director David Stockman characterized Reaganomics as a sham, former Press Secretary Larry Speakes admitted that he manufactured quotations from the president, and Michael Deaver portrayed a shrewd First Lady protecting her weak husband. The most damaging revelations came from the embittered second-term Chief of Staff Donald Regan, who claimed that an astrologer shaped much of the advice the meddlesome Mrs. Reagan imposed on the hapless president.

With all the fiascoes, Ronald Reagan ultimately relied on the enduring formula for presidential success: peace and prosperity. The Reagan recession came early enough in his presidency to avoid ruining his reelection prospects (his successor George Bush was not as lucky). At the same time, through luck and skill, Reagan could take credit for what may have been the greatest American achievement of the postwar period: the collapse of the "Evil Empire."

Ronald Reagan's election in 1980 triggered all kinds of doomsday scenarios. By 1983 one hundred million viewers watched ABC's "The Morning After" dramatize the unthinkable — what would happen after the bomb dropped. Reagan's military expansion, his repudiation of détente, his contempt for arms control, and his "Star Wars" antimissile defense system convinced his critics that he was a trigger-happy Neanderthal.

The hostile posturing combined with instability in the Soviet leadership deprived Reagan of a negotiating partner until 1985. Eventually Mikhail Gorbachev, the technocratic reformer, forged a productive if wary alliance with the old cold warrior. The Soviets needed arms control; keeping up with Reagan's defense buildup had drained their economy.

The two leaders finally met in Geneva in November 1985. After talking to the Soviet leader "man to man" for ninety minutes, the visionary president "couldn't help but think something fundamental had changed in the relationship between our countries." The supposedly incompetent and hawkish president would recall having "the pleasure of going into the room to my team and telling them that it was all settled, that there would be a meeting in eighty-six and eighty-seven, the first meeting in Washington, the second one in Moscow. They couldn't believe it."

Reagan's growing sloppiness and his constancy doomed the next summit. Reagan abruptly ended the October 1986 meeting in Reykjavik, Iceland, muttering that his beloved Strategic Defense Initiative was not a "bargaining chip." The Russians, it seems, were among the few who took the Star Wars system seriously and feared it. Nine months later, in June 1987, Reagan stirred the world and further infuriated the Soviets by declaiming at Berlin's Brandenburg Gate: "Mr. Gorbachev, open this gate! Mr. Gorbachev, tear down this wall."

Still, by exchanging ever more sweeping proposals during and after Reykjavik, Reagan and Gorbachev eventually dropped the long-standing "MAD" reliance on the threat of Mutually Assured Destruction to restrain both superpowers. By fall 1987 the two rivals actually agreed to destroy some nuclear devices. The unprecedented INF treaty banned all intermediate and shorter-range nuclear weapons in Europe. The provision for on-site monitoring fulfilled Reagan's commitment to remember the "old Russian maxim: *doverey, no proverey* — trust but verify." The rapid deceleration of the arms race, combined with the peaceful breakup a few years later of the Soviet Union and the Eastern bloc, gave Reagan, then Bush, bragging rights as the winners of the cold war — a bipartisan, multidecade accomplishment that Reagan's tough stance indeed facilitated.

"Politics is just like show business," Ronald Reagan said when he was California's governor. "You have a hell of an opening, coast for a while, and then have a hell of a close." The Gipper's opening salvo against big government, which "overspent, overestimated, and overregulated" overshadowed his own repeated surrenders to the New Deal consensus. Similarly the capitulation of the Soviet Union during the final days of the presidency gave Reagan his storybook ending. The "coasting" in the middle was surprisingly rocky, although the economic boom and the president's brilliant showmanship often camouflaged his many failings.

The postwar boom that made America the first predominantly middle-class nation on earth produced Ronald Reagan's populist conservatism. Whereas Democrats since Franklin Roosevelt had played to the poor and to workers' empathy for the poor, Reagan played to upper-class needs and middle-class dreams of wealth. Politically, this shift led to an orgy of deregulation and tax breaks for the rich. Culturally, it led to a fascination with the lifestyles of the rich and famous, as tycoons like Donald Trump and Michael Milken became celebrities.

For all his paeans to "life" and "liberty," Ronald Reagan epitomized modern Americans' unrelenting "pursuit of happiness." Although he could not resist distributing government goodies, Reagan repudiated the progressive idea that government could solve problems. His optimistic, organic, yet passive nationalism trusted in America's destiny, not policy micromanagement.

Characteristically, the man who defined America as "a sunrise every day — fresh new opportunities, dreams to build" took credit for the boom, the cold war victory, the renewed optimism, while blaming Congress and "those liberal elites" for his failures. "The president doesn't vote for a budget, and the president can't spend a dime," Reagan muttered. But if the "boom" was Reagan's, he must also share the blame for nearly bankrupting the country with a $2.7 trillion debt, for making the United States a debtor nation, and for the growing gap between rich and poor in the land of the free.

In many ways, Ronald Reagan's greatest successes stemmed from the excesses for which he and the 1980s became famous. He unknowingly killed Keynesian economics with runaway Keynesianism; his deficit spending triggered a boom but also bankrupted the welfare state. Only after the Reagan binge did the deficit become a salient political issue. Similarly, his defense buildup helped bankrupt the Soviet Union as well as threatened the financial stability of the United States.

Reagan was at his strongest as the avatar of American consensus, peddling a sunny normalcy, a happy-go-lucky and ultimately accommodating attitude. Reagan's tenure illustrated how traditional values still endured and just how entrenched big government had become. His popularity stemmed from some of the same qualities that attracted derision, from his "averageness," his Trumanlike ability to connect with millions who read the daily horoscope, watched the nightly news, and still believed in the American dream. Sixty million people tuned in to the Reagans' Barbara Walters interview before the 1986 Academy Awards, while barely a million people received the *New York Times* every day.

Reagan's odd mix of ideology and elasticity renewed confidence while deepening cynicism about American politics. As the first show-business president, Reagan helped usher in the age of the manic-depressive presidency, a presidency wired into the media's bipolar world of hysterical headlines that bashed one day and lionized the next. Reagan — and his next two successors — functioned in an unstable political culture where poll ratings zigzagged and headlines clashed. Reagan's spasmodic presidency would pave the way for George Bush, who won a war, then lost an election, and for Bill Clinton, who won reelection only to be bogged down in scandal.

Reagan's "Teflon coating" was manufactured in Washington and sold on TV and at supermarket checkout stands. Just as journalists always tried to fold the day's story into a larger narrative about American society, the Reagan meta-presidency, celebrating America's return, transcended any particular contretemps. Unlike his four immediate predecessors, Reagan could balance out the media criticism and spectacular failures with stirring symbolism, a visionary creed, and some spectacular successes. Reagan's sustained approval ratings of 60 and 70 percent broke the Johnson-Nixon-Ford-Carter losing streak. The media confirmed his morning-in-America message in the upbeat sets on A.M. television chat and news shows, in glamorous afternoon soap operas, and in prime-time programs celebrating American success like "The Cosby Show," "Dynasty," and "Dallas."

At the same time, Reagan followed the tradition of the great "liberal" presidents, from Thomas Jefferson and Andrew Jackson to Woodrow Wilson and Franklin Roosevelt. All championed the people against special interests and big government, yet ultimately aggrandized the presidency and, by extension, big government. Conservatives' newfound love for executive authority, after years of railing against Democratic usurpations, gave this tactical move ideological legitimacy.

Reagan's legacy would be more radical. His successor, George Bush, would exorcise the spirit of Vietnam more effectively during the Persian Gulf War. Domestically, Reagan's assault on the welfare state made *liberal* an epithet by 1988. Still, just as it took Dwight Eisenhower, the Republican, to legitimize the New Deal in the 1950s, it would take Bill Clinton, the Democrat, to legitimize Reagan's counterrevolution in the

1990s. Only when a New Democrat took over and, facing the first Republican Congress in forty years, declared, "The era of big government is over," could Reaganites finally say that their cautious, detached, pragmatic, muddleheaded yet visionary, principled, patriotic, and eloquent hero had truly changed the course of American history.

— GIL TROY
McGill University

The Great Communicator

President Reagan entered office with a substantial majority of the public behind him. He left office eight years later with his standing in the polls still very high. Reagan's enduring popularity angered and baffled his critics and many media commentators, who disagreed with his policies, considered him a political lightweight, made fun of his leadership style, and pointed to burgeoning scandals during his second term.

Reagan had demonstrated his ability to tap into support that transcended his policies or party during his two terms as governor of California — then a heavily Democratic state in which most mainstream Republicans hewed to a more moderate line. As he moved into the White House, he showed that the same ability could work on a national scale. Over the course of his presidency Reagan crafted and preserved a wide political base that drew substantial defections from such traditional Democratic voting blocs as white southerners, conservative Catholics, and blue-collar workers — the so-called "Reagan Democrats." In the early 1980s, political pundits spoke of the "realignment" of American politics and the disintegration of the Democrats' New Deal coalition. But the qualities of Reagan's presidency that voters seemed to respond to, according to opinion polls, had to do mainly with Reagan himself — with his optimism, his serene aura of conviction, his focus on basic principles, and his disinterest in the details.

In his first inaugural address, he articulated an enduring mythic sense of the nation and its promise that resonated with voters across the political spectrum: "To me our country is a living, breathing presence, unimpressed by what others say is impossible, proud of its own success, generous, yes, and naive, sometimes wrong, never mean and always impatient to provide a better life for its people in a framework of a basic fairness and freedom." What Reagan sensed and managed to embody, as *Newsweek* aptly put it, was "America as it imagined itself to be."

George H. W. Bush

1989–1993

b. June 12, 1924

T HE PRESIDENCY of George Herbert Walker Bush left an indistinct mark on America. Except for the Gulf War, it featured no grand event, great speech, dismaying scandal, ideological crusade, or decisive political turn. The greatest transformation linked to his presidency — the end of the cold war — was under way before he took office. His policies and themes resembled those of his Republican predecessor, Ronald Reagan, and his Democratic successor, Bill Clinton. Like Clinton's presidency, Bush's suggests that in lieu of grave national challenges, presidents resist ideological crusades, grope for direction, and compile modest records.

But Bush, not just the times, shaped his presidency. Born in 1924, raised in Connecticut, he enjoyed a privileged childhood. Although his financier father, Prescott Bush, later become a United States senator, the family rarely discussed politics or the Depression. At a prep school isolated from global and national convulsions, he was a popular kid who "never rocked any boats" (as a teacher put it). Heroic service as a young navy combat pilot during World War II left its mark, but his political views and ambitions derived more from his family's patrician sense of service and entitlement. After a Yale education, Bush moved to Texas and, with family financial support, speculated in oil properties before entering politics.

Bush's cluttered political record — failed senatorial candidate in 1964 and 1970, Texas congressman from 1966 to 1970, ambassador to the United Nations, chairman of the Republican National Committee during the Watergate scandal, envoy to Beijing and director of Central Intelligence under President Gerald Ford — established no clear trajectory except that of ambition. As CIA director, he ably mediated between the beleaguered agency and a critical Congress and heeded conservative pressures to issue a warning of Soviet peril that laid the groundwork for the Carter-Reagan arms buildup. But both in appointive positions and in his vice presidency (1981–1989), he was mostly a loyalist to strong-willed superiors — Richard Nixon and Ronald Reagan — and rarely sought or received opportunities for leadership. Critics labeled his performance a résumé, not a record. Bush himself emphasized character and service rather than policy and vision. Asked, amid his 1979 bid for the GOP nomination, what he would do as president, he responded, "I'll bring the best people into government," and he boasted in 1988, "Ready from day one to be a great president." Insistence on American military strength and world leadership, the most consistent element of his outlook, did not distinguish him from many others.

He entered presidential politics as an unstable amalgam of political styles. Like Franklin Roosevelt in family wealth and sheltered upbringing, he lacked FDR's charm, eloquence, and confidence. Like Lyndon Johnson, a product of rough-and-tumble Texas politics, he lacked LBJ's brute force. Like John Kennedy, a World War II combat hero, he never much capitalized on that status. Like Ronald Reagan he was a conservative, but out of background and habit more than conviction.

Bush also lacked the clear cultural definition that would allow Americans to fix him in their affections, animosities, and memories. Eisenhower had his military stature, Kennedy his patrician Irish American veneer, Johnson his Texan earthiness, Reagan his California media cool, Carter and Clinton their southern Protestant manner. As he hopped among several homes, Bush's cultural definition fragmented: was he a New England patrician, a Texas oilman, a Washington bureaucrat, an ambitious politician? Gifted politicians finesse the conflicting elements of their style. Bush juggled them awkwardly, sometimes angrily: it was clumsy for a Yale man to condemn Michael Dukakis's views on defense as "born in Harvard Yard's boutique" or to insinuate that Bill Clinton's time at Oxford University made him a menace. Attempts to talk folksy fared badly.

With these unstable combinations, he most resembled Richard Nixon, just as the Vietnam War and Nixon's presidency honed his sense of presidential politics. From Nixon he learned to preach unity and practice division, and to praise peace but pursue greatness in war. Like Nixon's, his rhetorical style matched those discordant purposes: both men reached unconvincingly for uplifting visions of peace while bearing down on presumed enemies; both tended to equate toughness with meanness. Bush acknowledged his roots in Nixonian politics when he said of the 1988 presidential contest, "It'll be like the Nixon-McGovern race in seventy-two as far as the breadth of differences on issues," even though the differences and the sense of national turmoil were far more muted in 1988.

Bush did lack his mentor's personal hatreds, gnawing insecurity, and conflation of himself with the nation — he was, in a way, a "kinder, gentler" Nixon — and Bush did not face a disastrous war. But like Nixon he was a war president in assessing his office, talents, and place in history. He inherited the eastern patricians' sense that they were the nation's natural leaders in war and foreign policy. World War II taught him how coalition diplomacy should work. The cold war defined his worldview and his most interesting duties: never before 1989 did he depart from its orthodoxies or outline a national mission outside its framework.

Therein lay the greatest tension in his presidency. A product of America's half century of militarization, he was trained to be a war president, not because he relished war but because it seemed his duty and his arena. But the cold war was ending — indeed, he facilitated its end — and militarization was running its historical course. As president, Bush searched for the grand warlike mission he was primed to lead. But despite momentary success, he showed, as the journalist Sidney Blumenthal observed, that "his natural state without war was political collapse."

Bush owed his election in 1988 partly to his vice presidency — which gave him the visibility, ideological direction, durable service, and political clout absent in earlier short-term positions — as well as Reagan's vital blessing. But the vice presidency does not guarantee success; no two-term vice president since John Adams had moved directly to the White House. Clearly, Bush had advantages and appeals beyond mere possession of an office. Physically and psychologically resilient, attractively self-mocking at times, he also had the ideological flexibility to straddle his party's warring factions. His opponent, Massachusetts Governor Michael Dukakis, gained no leverage by proclaiming, like Bush, his mere competence; and Dukakis's whiff of urban liberalism made him vulnerable to GOP left-baiting. Bush came, as the historian Garry Wills put it, from the "America of advantage," but an elitist edge also burdened his opponents. (Although Greek immigrants, Dukakis's father had graduated from Harvard Medical School, his mother from Bates College, and Dukakis himself from Swarthmore. After Clinton's stint as a Rhodes scholar, he went to Yale Law School.) Bush also benefited from the Western triumph in the cold war which the Soviet leader Mikhail Gorbachev

THE AMERICAN PRESIDENCY ★ 490

seemed to grant, without yet facing the downside of triumph — economic dislocations attendant to modest declines in defense spending — that hurt him in 1992.

Given those advantages, Bush's victory — 54 percent of the popular vote — was comfortable but modest, and as most presidents find, what he did to win the presidency hampered his conduct of it. He scarcely offered an agenda beyond consolidating the Reagan revolution. As in Nixon's campaigns, Bush's sunnier appeals for a kinder, gentler America were undercut by shrill nationalism and attack-dog tactics (attributed to campaign aide Lee Atwater) that exploited Americans' racial and cultural divisions and impugned his opponent's loyalty. Bush's verbal clumsiness — sometimes charming, often grating — was also evident. ("I stand for antibigotry, anti-Semitism, antiracism," he once announced.) Even Bush's touted good judgment came into doubt when he chose Indiana Senator Dan Quayle, derided even in his own party, as his running mate (again Nixon, with his choice of Spiro Agnew, had perhaps set the standard). Low expectations accompanied Bush's Inauguration, leaving few goals he could fail to meet, but also few to marshal Americans behind his presidency.

This unsteady beginning was a result in part of how "the vision thing," as he put it in 1988, puzzled him. To his credit, Bush rarely pretended otherwise, seeing himself as a "guardian" rather than an activist president. For that reason, some analysts see Dwight Eisenhower as Bush's kindred spirit. But though cautious in action, Eisenhower had a vision of the nation's strengths and perils, and for good reason his farewell address of 1961 became a fundamental document of modern history, while Bush's has been forgotten.

Surprisingly, Bush stumbled at the start in his chosen arena of national security. His nomination of former Texas Senator John Tower as Secretary of Defense met humiliating defeat in the Senate once allegations of womanizing and drinking by Tower surfaced. With reason, the White House complained of vengeful Democrats, who controlled Congress, but Bush was the first president to have a cabinet nomination defeated in his first term, and Tower was bitterly denounced by some Republicans. As important was growing mistrust — starting in LBJ's presidency and deepened by Bush's 1991 nomination of Clarence Thomas to the Supreme Court — regarding presidential nominations no matter which party controlled which branch.

Bush's uninspired competence and prudence served him best in the tricky diplomacy of ending the cold war and liquidating its legacy. Initially, prudence hampered him: Bush and National Security Adviser

Brent Scowcroft, spooked by memories of how cold war détente had collapsed in the 1970s, admitted more grudgingly than Reagan that the cold war was concluding, and the opening months of Bush's foreign policy were incoherent. But a solid team — including Defense Secretary Richard Cheney and Secretary of State James Baker — soon offered steady leadership that contrasted with the dizzying inconsistencies and deep frictions of the Reagan administration. They dealt ably with the end of Europe's cold war in part because with Europe the grooves of American diplomatic contact, elite cooperation, and cultural affinity were deeply cut; even Soviet-American diplomacy had long assumed a predictable choreography. This ground was familiar even amid changing circumstance.

The fate of Germany — divided since the end of World War II, and the pivot of European diplomacy for the whole century — best displayed that steady, purposeful approach. The East's German Democratic Republic was rigidly authoritarian — as the Berlin Wall it constructed in 1961 symbolized — and dependent on the Soviet Union. But few predicted its early demise, given its imagined prosperity compared to other communist regimes, the reluctance even of Soviet reformers to countenance much change, the apparent docility of East Germans, and West Germany's long-standing collusion with the East's regime.

The Bush administration did not make the East German house of cards collapse. That resulted from the unraveling of the Soviet economic empire, the examples of change elsewhere in Eastern Europe, the sclerosis of the East regime, and the restlessness of its people. But Bush did hasten and shape the process. While others in Washington and Europe hesitated, Bush's team insisted that Soviet-American amity hinged on German reunification on Western terms — its inclusion in NATO, the North Atlantic Treaty Organization — and on liberation elsewhere in Eastern Europe. It did so at some risk that Gorbachev would balk or face a backlash at home, and despite evidence that most Germans preferred a neutral Germany even if their leaders did not. It succeeded because it had the strategic upper hand, but also because Gorbachev proved flexible, Bush avoided rubbing defeat in his face, and the administration skillfully juggled Europeans' conflicting fears (of a united Germany so strong as to overwhelm them or so weak as to drag them down).

Victory went further. Other communist regimes in Europe fell; the Baltic republics slipped from Soviet control; and Soviet-American agreements reduced nuclear arsenals. After Gorbachev's ouster and replacement by Boris Yeltsin in 1991, the Soviet Union itself collapsed, replaced by a Russian state and a host of new nations with names and politics often incomprehensible to most Americans. The administration drew criticism for its role in some stages of this process. It gave restless Soviet republics conflicting messages about seeking independence, clung long to Gorbachev, glibly asserted the liberating value to Soviets of a market economy, poorly anticipated the resurgent nationalisms unleashed by the Soviet collapse, and hesitated as old Soviet nuclear weapons fell into worrisome new hands. Later, it seemed baffled as Yugoslavia descended into civil war. Most of all, the administration rarely found a voice to capture the import of the grand changes under way. Prudence is rarely inspirational. But it was sufficient. The administration never jammed the gears of change.

Indeed, it experienced no foreign policy disaster — nothing like Johnson and Nixon in the Vietnam War, Carter's Iran hostage imbroglio, or Reagan's Lebanon fiasco and Iran-contra scandal — perhaps because the cold war's end removed the occasion for it. Yet something also owed to Bush's skills as diplomatist and crisis manager.

His biggest problem was what to make of the cold war's demise, not how to preside over it. The Berlin Wall's fall in 1989, for example, found him so cautious and inarticulate that reporters wondered to him, "You don't seem elated." Bush responded, "I am not an emotional kind of guy," stressing caution in the face of change. Repeatedly, administration hand-wringing over the consequences of the cold war's end overshadowed joy in its occurrence.

This problem reflected Bush's weakness on the stage of public opinion and his caution about inflammatory triumphalism, but it went deeper. Triumph, hardly unmentioned anyway, was not the only theme available — relief at the cold war's end and reconciliation with old enemies might have served better (awkwardly he tried out those themes as well). His public confusion probably reflected private bafflement about what lay ahead and about the removal of familiar challenges and rationales. His peevish impulse was to assert that little had changed. He foresaw a post–cold war world like its predecessor, simply minus the cold war: NATO would continue, albeit modified; America's militarization would endure, although attenuated; America's global hegemony would remain, only slightly recast. "Notwithstanding the alteration in the Soviet threat," Bush insisted, "the world remains a dangerous place with serious threats to important U.S. interests." Such statements were valid, but also maddeningly unenlightening — applicable to any point in America's modern history. Later talk of a "new world order" seemed vacuous, burdensome, or menacing, and soon disappeared. Worse for his political fortunes,

Bush's statements offered Americans no reward for victory in the cold war and no assurance that he knew what to do next.

That weakness hurt him more because he neither sought nor developed a strong record in domestic matters on which to fall back. Like Reagan, he left little for the federal government to do: after all, he claimed, America's "free enterprise system lets one person's fortune become everyone's gain" and "we've become the most egalitarian system in history and one of the most harmonious." Oblivious to widening class differences, that outlook made his role an essentially negative one, reflected in his record of forty-four vetoes with only one overturned. Bush's domestic policy, grumbled House Majority Leader Richard Gephardt, is "the veto pen." Late in 1990, Bush's acerbic chief of staff, John Sununu, even announced that Bush "doesn't need another single piece of legislation unless it's absolutely right. . . . In fact, if Congress wants to come together, adjourn, and leave, it's all right with us."

Neither doctrinaire nor foolish enough to leave things at that, Bush did eagerly help enact a few key laws: a landmark Clean Air Act and a far-reaching Americans with Disabilities Act. Less came of his boasts to be an "education" president. His greatest impact at home probably came through his judicial appointments, which strengthened Reagan's conservative "revolution." Bush, like Nixon, had little taste for haggling with Congress over domestic policy, sometimes using subordinates like Sununu who only inflamed conflicts. And he disdained public appeals for support even though (as his 1988 campaign showed) media politics had long replaced party structures in shaping public opinion. In such ways, even a sympathetic student of his term, the historian David Mervin, says, he simply "was not fully qualified to meet the demands of the U.S. presidency in the late twentieth century."

Those weaknesses tripped him up when federal deficits (largely products of Reagan-era policies) swelled, tax revenues lagged, trade imbalances persisted, and economic growth stalled. Bush's famous 1988 campaign pledge — "read my lips: no new taxes" — trapped him. In the autumn of 1990 he grudgingly accepted a budget deal that increased some taxes; but he was inept at enlisting congressional and public support and coarse in ways that cracked the facade of patrician grace. Dogged by reporters about one element of the budget deal, Bush, pointing to his buttocks, said, "Read my hips." Conservatives assailed Bush on taxes, liberals gave him little credit, moralists derided the cavalier way he broke his pledge, critics contrasted his feckless leadership on the budget to his aggressive leadership in the Gulf crisis. That contrast highlighted a stubborn problem of image — his apparent insensitivity to the bread-and-butter problems faced by ordinary Americans amid corporate "downsizing," economic sluggishness, and competition from abroad. His pleas that he did care only seemed to compound his problems.

It was also Bush's misfortune to face issues alien to him, as the cold war's end allowed cultural and social conflicts to flare anew. A brittle cultural conservatism, both emboldened and frustrated by Reagan's presidency, found new enemies at home to replace disappearing ones abroad, pitting it against angry cultural minorities and feminist groups. The antiabortion movement intensified its passion and (on its fringes) its violence; venom spewed against gay people for presumably causing AIDS and using federal grants to pollute the arts; foreshadowing controversies to explode under Clinton, male naval personnel allegedly brutalized women at the Tailhook convention, and news surfaced that the Defense Department's leading spokesman, Pete Williams, was gay. Racial and sexual tensions in the 1991 Clarence Thomas hearings (a former subordinate, Anita Hill, accused him of sexual harassment) and the 1992 Los Angeles riots were especially ugly.

For these conflicts Bush lacked the cultural affinity (despite his oft-cited Hispanic grandchildren) and soothing language employed in different ways by Reagan and Clinton. Most elicited snappish responses that reflected less a partisan stance than his simple wish that the issues would go away. His lack of principled views left him buffeted by many forces, especially by such figures as Senator Jesse Helms in his party's cultural right wing, always eager to brand Bush illegitimate as Reagan's heir. By placating that wing yet doing so with obvious discomfort, he failed either to curry its favor or to reach out to other Americans, and he contradicted his "kinder, gentler" rhetoric.

Political opportunism also dogged him. On racial issues, he had been a Goldwaterite in 1964, a moderate from 1966 until assuming the vice presidency, a cynic who allowed political ads about Willie Horton (a black murderer who committed rape on furlough from Dukakis's prison system) in 1988, and a hairsplitter who in 1991 signed a civil rights bill almost identical to one he vetoed in 1990. He earlier had endorsed family planning and abortion rights, only to condemn abortion after joining Reagan's ticket in 1980. Most politicians have inconsistent records, but Bush's was more transparently so, in part because he publicly fretted over his own inconsistencies.

Bush also aggravated cultural and social conflicts by casting them in the rhetoric of war. Partisans of all political stripes had long analogized their animosities and aspirations to the passions of war, but with the fading of an external enemy that reflex intensified, as the shrill language of "culture war" reflected. Bush too displayed that reflex. His first nationwide address, in September 1989, declared war on drugs, branding as virtual traitors "Everyone who uses drugs. Everyone who sells drugs. And everyone who looks the other way." Already long-running, the drug "war" was expensive, ineffective, and fraught with tensions over race and class to which Bush was deaf. But his announcement signaled that war was the symbolic and literal arena in which he naturally operated. A real though brief war soon followed: in December, Bush authorized an invasion of Panama and the kidnapping of its ruler, Manuel Noriega, for trial in the United States on drug charges.

The Mideast crisis of 1990 brought into focus the impulses of the man and his presidency — his apparent strength in war and world leadership, his administration's anxiety about the challenges of the post–cold war world, and his nation's sense of drift at home. Confused signals by the administration to Saddam Hussein may have encouraged the Iraqi leader, but Hussein's actions — a sudden conquest of Kuwait in August and an apparent threat to Saudi Arabia and Mideast oil — were products of his own temperament and the region's politics.

Responding, Bush quickly found his stride. He first drew a line in the Mideast sand, then vowed to liberate Kuwait. Soon he had lined up allies and the UN behind his stance. (Soviet leaders were no longer either eager or able to resist American initiative in the region.) Economic strangulation of Hussein's regime was the first resort, but a huge allied military buildup made clear that Bush would choose war if Hussein failed to back down. In the coalition diplomacy of war — especially in its behind-the-scenes elements — Bush, and lieutenants like Baker and Scowcroft, were confident and masterly. Bush was condescending to Congress — insisting that he could go to war regardless of its wishes — but nonetheless got a resolution, narrowly passed, endorsing military action, which began on January 16, 1991.

In other ways the path to war was rockier. Military leaders, including Joint Chiefs of Staff Chairman General Colin Powell and Desert Storm commander General Norman Schwarzkopf, preferred more time for economic sanctions (Pentagon reluctance to go to war was a defining feature of the post-Vietnam era). They later second-guessed Bush about the war's quick and, some believed, premature termination. The spectacle of American pleas that Germany, Japan, and oil-rich allies finance the war gave the enterprise a desperate, mercenary quality. Worse, the administration's public rationales for war — oil, jobs, aggression, world order — appeared shifting and flimsy. The one sustained rationale — that Hussein's actions resembled Hitler's and the stakes matched those of World War II — was extravagant. By November, a shrill Bush was claiming that Iraqi forces did things "that even Adolf Hitler didn't do," insisting that "we do not need another Hitler in this time of our century." Such claims had resonance for older Americans in whose lives World War II was the central event, but also set the stakes dangerously high.

The allies easily won the war with air attacks followed by a brief assault over land. In weeks, Iraqi resistance crumbled and Kuwait was liberated, resuscitating American pride and Bush's popularity ("Bush was Caesar," Sidney Blumenthal later noted). But Iraq's swift collapse also raised doubt about whether it was comparable to Hitler's regime. And by accepting a conditional surrender that left Hussein in power and some of his military intact, Bush avoided more bloodshed but undercut his rhetoric of a World War II–like crusade and undermined his boast that "By God, we've kicked the Vietnam syndrome once and for all." Unlike Hitler, Hussein survived, to brutalize Kurds and other subjects. The moral clarity of Bush's crusade soon faded.

A more general euphoria about American virtue and victory lasted longer. Commentators, politicians, and ordinary Americans celebrated America's can-do spirit, its triumphant technology, its spirit of unity, its repudiation of the fears and animosities bequeathed by the Vietnam era. "It is as if all the confusion and pain of recent decades have melted," claimed the *New York Times*, "leaving the nation with its reassuring images from World War II intact."

But euphoria also dissipated. Precisely what Americans celebrated, quick and easy victory, also robbed the war of the gravity of sacrifice and accomplishment needed to give it lasting significance. For most Americans it had been a television war — vivid while the screen flickered, remote once the show was over. Economic stagnation and divisions among Americans soon fostered a sour mood. The war seemed more like a diversion from America's troubles than a solution to them.

Earnestly but desperately, Bush worked to rescue something from the war. "We can bring the same courage and sense of common purpose to the economy that we brought to Desert Storm," his final State of the Union address asserted. But invoking a forgettable war did Bush no good, while his leadership in it highlighted his apparent failure to lead at home, leaving the war "a victory from which he never recovered" (as Sidney Blumenthal put it). Knowing his greatest success and sense of purpose in war, Bush kept returning to it. Facing Clinton and maverick tycoon Ross Perot in the 1992 elections, Bush let his agents again wage the "culture war," impugned Clinton's patriotism, and ran as commander in chief. But without a war to command, that stance seemed hollow and his style coarse and flailing. With Perot taking 19 percent of the popular vote, Bush got just 38 percent, less than any sitting president since Taft, and Clinton got an unimpressive plurality.

Soon, Bush's policies appeared less idiosyncratic and more rooted in late-twentieth-century politics. The geopolitical accommodations he helped foster in Europe and the Mideast mostly held. His dispatch of forces to war-torn Somalia set the stage for more such action under Clinton. Only slowly did Clinton reverse Bush's hands-off approach to Yugoslavia's civil war. After fumbled initial challenges, Clinton proved no more inclined than Bush to speed America's demilitarization. Like Bush, Clinton also deferred to corporate capitalism and embraced free trade; and to those ends, he accepted unpleasant realities like the Chinese government's repression of internal challengers, while trade deficits with Japan, a focus of Bush's diplomacy, continued. Championing voluntarism, educational standards, and welfare reform, Clinton soon also resembled Bush in the general temper of his presidency. Neither found lasting causes to give them traction; both practiced a wobbly centrism compromised by efforts to placate a strident conservatism. Beyond his verbal and political skills, however, Clinton enjoyed a comfort with cultural conflicts that baffled Bush, a party less driven by those conflicts, a wariness of outdated war metaphors, and a rebounding economy, always key to presidential success.

By the same token, some of Bush's limitations seemed more like virtues. Though his honesty about the Iran-contra scandal was doubted, no aura of sleaze hung over him like it did over Reagan and, far more, Clinton. His discomfort with public rhetoric seemed less like elitist condescension and more like refreshing avoidance of the cheap, talk-show-host style in which many politicians now indulged. No colossal blunder of action haunted him, though words — the pledge against new taxes, the comparisons of Hussein to Hitler — surely hurt him.

In retirement, Bush imposed on public consciousness or GOP politics rarely (as with his much-praised resignation in 1995 from the National Rifle Association). Herbert Hoover remained a party leader and occasional public servant for decades. Harry Truman was publicly combative in defense of his record, and Nixon even more so. Eisenhower remained a voice in party and national affairs. Jimmy Carter kept a high profile as a humanitarian and broker of ugly conflicts abroad. Genial and low-key, Bush's postpresidential persona resembled Ford's, just as their presidencies bore similarities. As his upbringing taught him to be, Bush was "a good loser," so his prep school coach said in 1997. That year, long after his combat service, he also parachuted from an airplane. His personal vigor and courage were scarcely diminished. The often frustrating task of capitalizing on those virtues in the political arena was over.

Through his sons, George Bush's family dynasty eclipsed the Kennedys'. With Jeb Bush as Florida governor, Texas Governor George W. Bush became the first son of a president since John Quincy Adams to reach the White House, following the contested 2000 election. Father and son shared much: verbal clumsiness, vulnerability in economic matters, impatience with Congress and opposing voices, fulfillment as a war president, fixation on Saddam Hussein, and a Manichean rhetoric drawing on the mother lode of World War II. But Al Qaeda's September 11, 2001, attacks on New York City and the Pentagon, the son's deeper ties to the GOP's economic and religious right wing, and his self-branding as more Texan shaped a different, steadier, but coarser presidency. Cosmopolitan eastern Republicanism — "internationalist" abroad, "moderate" at home — all but disappeared, as did Soviet/Russian constraints on U.S. power. George W. loudly announced America's right to attack enemies preemptively, resisted coalition-building, preached American virtue and empire, revived deficits with tax-cutting and defense increases, and through Attorney General John Ashcroft made federal police power more expansive and punitive. Unclear in 2003 was whether he would find war more redemptive politically than his father (or most presidents) had. Meanwhile, George H. W. Bush remained largely out of the limelight.

— MICHAEL S. SHERRY
Northwestern University

Bill Clinton

1993–2001

b. August 19, 1946

WILLIAM JEFFERSON CLINTON'S PRESIDENCY will likely be remembered as much for the scandals culminating in Clinton's impeachment as for its significant substantive accomplishments. Clinton recorded some important achievements, including balancing the federal budget, presiding over an extended period of economic prosperity, and making incremental but not insignificant changes in a wide range of social policies. But to many these accomplishments are overshadowed by the scandals rooted in Clinton's failure to curb his own worst impulses and exacerbated by the extreme partisanship that characterized his time in office. Clinton took office in 1993 at a time when the political landscape at home and abroad seemed ripe for reworking. The conclusion of the cold war had brought an end to a bipartisan consensus undergirding American foreign policy for more than forty years, and the United States was poised to undertake a new role in world affairs. At home, the nation was coming out of a brief but deep recession and the electorate was demanding change — a theme not only embraced by Clinton in his successful campaign, but also embodied in the strong independent candidacy of H. Ross Perot and countless ballot referenda across the nation. The Republicans, who had held the presidency for a dozen years, were polarizing into factions, with social and religious conservatives increasingly at odds with party moderates. Clinton took office promising to move the Democratic party to the "vital center" of the electorate by combining the party's traditional concern for disadvantaged people with a new emphasis on smaller government, fiscal conservatism, and reliance on market forces.

At the outset of his first term, Clinton seemed well positioned to craft an enduring political realignment. He was likable, well educated, and articulate. Elected four times as governor of Arkansas, he was an experienced public executive (albeit of a small state). As a politician, he possessed an uncanny knack for overcoming adversity and escaping crises. He was also ambitious, setting his sights on the White House as a young man and becoming an avid student of presidential biography. As the first president born after World War II, Clinton brought the youth, energy, and optimism of a new generation of political leaders to the White House.

But the Clinton presidency also faced considerable hurdles, some rooted in Clinton's own political and personal history, and others in an increasingly partisan political atmosphere. The former included allegations of draft-dodging and protests against American involvement in the Vietnam War, a vague association with the 1960s counterculture, and rumors of extramarital affairs. During the 1980s, meanwhile, both major political parties had been gradually reshaped, with many moderate members replaced by those holding more ideologically extreme views. The more polarized party environment interacted with hypercompetitive news coverage to magnify the impact of a series of missteps and mishaps during Clinton's first year in office.

Clinton's presidency began with a bruising budget battle in which he was forced to raise taxes in order to help cut the federal deficit. There was also an ambitious and ultimately doomed effort to overhaul the nation's health care system. To many Americans these steps signified an unwelcome return to the Democratic party's penchant for "big government" solutions to social problems. Republicans capitalized on these perceptions during the 1994 midterm elections to regain control of both the Senate and House of Representatives for the first time in four decades. Thereafter fiercely partisan contests with Congress punctuated the remainder of Clinton's presidency. Tacking with the change in political winds, Clinton defied liberals in his own party and repositioned himself closer to the ideological center in time to win reelection comfortably in 1996, in part by co-opting key Republican issues while portraying others as extremist.

Poised to capitalize on his reelection momentum, Clinton instead found himself embroiled during his second term in a series of legal and political controversies, the most important of which involved his illicit sexual relationship with a twenty-two-year-old White House intern. His evasive and misleading accounts of the affair opened him to charges of perjury and obstruction of justice. In December 1998, the House of Representatives passed two articles of impeachment against Clinton.

Although a majority of Americans disapproved of Clinton's personal behavior, throughout the impeachment inquiry in the House and subsequent trial in the Senate he enjoyed high public approval ratings for his official conduct, and partly as a result the Senate voted to acquit him in February 1999. Nonetheless, politicians of both parties roundly and harshly condemned the behavior that had triggered the impeachment crisis. Moreover, the scandal affected the remainder of Clinton's time in office, with critics arguing that his policy initiatives at home and abroad were motivated as much by Clinton's political self-interest related to his impeachment as by more broad-based substantive concerns. With his second term already tainted by scandal, Clinton left office under still another cloud of suspicion caused by his issuance of a flurry of last-minute presidential pardons. During his post-presidential years, Clinton remained a formidable and polarizing political figure who seemed reluctant to withdraw from the political stage.

Clinton was born August 19, 1946, in Hope, Arkansas, and many of his most visible personality traits — resilience in the face of adversity, empathy for others, an eagerness to please and to be liked — are rooted in his difficult upbringing. He never knew his biological father, William Jefferson Blythe, a traveling salesman who died in an auto accident three months before Clinton's birth. His mother, Virginia, was remarried in 1950 to Roger Clinton, and her son took his stepfather's name. But the older Clinton, a failed automobile salesman and alcoholic, beat Virginia during his drinking binges. Virginia frequented the racetrack in Hot Springs, Arkansas, a mining and resort town not far from Hope, where the family moved when Clinton was five.

Religion and school proved two stabilizing forces in Clinton's formative years. He was an active participant in the local Baptist church, where he developed a distinctive rhetorical style and demonstrated high potential as a public speaker. He was also an excellent student, possessing a quick mind and an almost photographic memory. His high school principal encouraged him to pursue public service, a choice evidently cemented by Clinton's July 1963 trip as a representative of Boy's Nation (a youth organization sponsored by the American Legion) to the White House, where he was photographed shaking hands with President Kennedy.

The next year he left Arkansas to attend Georgetown University, where he again stood out academically, and where he worked for Arkansas Senator J. William Fulbright, who became an important role model. Clinton graduated from Georgetown in 1968, after winning a Rhodes scholarship that permitted him to spend the next two years at Oxford University. His status in the military draft during this period, his brief experiment with marijuana (during which, he later claimed, he did not inhale), and his participation overseas in protests against America's role in the Vietnam War would come back to haunt him during his 1992 presidential campaign. In 1970, he returned to the United States to attend Yale Law School, where he met fellow law student Hillary Rodham, whom he married soon after graduating in 1973.

At age twenty-eight, Clinton went home to Arkansas to begin a political career. After an unsuccessful race in 1974 to unseat the incumbent Republican congressional representative in a historically Republican district, Clinton was elected state attorney general in 1976. Two years later he won his first Arkansas gubernatorial race, becoming the nation's youngest governor in four decades. He was defeated in his 1980 reelection bid, however, when his opponent attacked him for hiking gasoline taxes and license fees to fund a highway spending program, for pushing too many issues, and for being out of touch with the voters' real concerns. Clinton also felt the backlash of the national recession and was hurt by riots among Cuban refugees placed in an Arkansas military installation on President Carter's orders.

Presaging tactics he would employ as president, Clinton won back the governor's office in 1982 by repackaging himself as a "new" Democrat. He embraced policies — including the death penalty and work-oriented welfare reform — more to the center of the political spectrum. This political makeover resonated with Arkansas voters and was an early indication of his political resiliency, which would later earn him the sobriquet "the Comeback Kid." He was reelected Arkansas governor three more times, providing him with a secure political base from which to seek higher office.

In 1985 he cofounded and chaired the Democratic Leadership Council (DLC), an organization composed of moderate governors, members of Congress, and other party officials. He was also an influential participant in the annual meetings of the National Governors Association, serving as its chairman in 1986. In 1988 he was selected to give the nominating speech for Governor Michael Dukakis at that year's Democratic National Convention. Although his speech was remembered primarily for its great length, the opportunity to give it signified his rising status within the party.

BILL CLINTON ★ 503

As a southern governor with a relatively centrist political record, Clinton possessed valuable credentials for reversing the inroads Republican presidential candidates had made in traditional Democratic constituencies in the south and among more moderate voters elsewhere. But his candidacy was also vulnerable on several counts: the recurring allegations of draft-dodging and antiwar activity and the rumors of marital infidelity that had dogged him throughout his career, most recently during his 1990 gubernatorial campaign. Although the media had traditionally refrained from probing into candidates' private lives, the example of the undoing of Gary Hart's campaign for the Democratic presidential nomination in 1988, when he was driven from the race for womanizing, established a sobering new precedent. Thus it was almost inevitable that Clinton would be questioned about his private life during the 1992 campaign. On October 3, 1991, Clinton formally announced his presidential candidacy from the steps of the Arkansas statehouse, pledging to restore "the hopes of the forgotten middle class." Not long after that, he acknowledged having had "problems in his marriage," which he hoped would insulate him from accusations of sexual promiscuity.

But three weeks before the critical New Hampshire primary in February 1992, a supermarket tabloid reported that Clinton had carried on a twelve-year affair with Gennifer Flowers, a former Arkansas television reporter. At a subsequent news conference Flowers played audiotapes of Clinton suggesting she deny any sexual relationship with him. His candidacy now in jeopardy, Clinton made a dramatic appearance on the popular television newsmagazine "60 Minutes." With his wife beside him, he admitted again to having caused "pain" in his marriage, but vigorously denied Flowers's allegations. The high-risk gambit saved his candidacy. A few weeks later, the *Wall Street Journal* ran an article documenting inconsistencies in Clinton's account of his efforts to avoid the draft; once again the campaign had to deal with a massive and embarrassing distraction from its chosen themes. Even so, Clinton still finished a strong second in the New Hampshire primary with 25 percent of the vote. Under other circumstances, a second-place finish for a candidate who had once been an overwhelming front-runner would have been disastrous. But the scandals had caused many observers to write Clinton off, and so his showing was considerably stronger than expected.

Moreover, he had been beaten by a former Massachusetts senator, Paul Tsongas, who was expected to do well by virtue of his regional ties. The self-styled "Comeback Kid" carried the political momentum of his second-place finish into Super Tuesday, when he established himself as the Democratic front-runner by winning eight primaries (mostly in the south) in a single day. Subsequent victories in Michigan, Illinois, and New York effectively clinched the nomination by early April.

At the Democratic National Convention in New York City, Clinton presented himself as a "New Democrat" and delineated the issues with which he would campaign against President George Bush: reducing middle-class taxes, investing in jobs and education, protecting the environment, attacking the deficit, supporting social policies that emphasized personal responsibility, and, above all else, leading America down the road of change. To bolster support among voters in the increasingly Republican south, particularly in the critical border states, Clinton selected Tennessee Senator Albert Gore as his vice-presidential running mate. In addition to solidifying Clinton's "New Democrat" credentials, Gore gave Clinton an experienced voice on Capitol Hill and more credibility on foreign policy and environmental issues.

As a campaigner, Clinton lacked Ronald Reagan's talent for memorable speechmaking. But he rivaled "the Great Communicator" in his ability to establish rapport with audiences, and unlike Reagan he was also remarkably nimble in extemporaneous settings and in conversation with voters. Two days after the convention, Clinton seized a lead in public opinion polls that he never relinquished. The country had been mired in recession for months, and Clinton hammered at Bush on economic issues — a strategy that Clinton political aide James Carville dubbed "It's the economy, stupid!" Clinton's candidacy may also have benefited from the erratic presence and performance of the independent candidate Ross Perot.

Although hostile to both parties, Perot reinforced the theme of change, which was particularly damaging to the incumbent, President Bush. On Election Day, Clinton won 43.5 percent of the popular vote, a 5-percent margin over Bush. Perot showed surprising and impressive strength, gathering nearly a fifth of the total vote. In the electoral college, Clinton won 370 electoral votes to Bush's 168. Perot won none.

Clinton had run a nearly flawless campaign. But he took office with significant handicaps. First, his failure to win a majority of the popular vote, in part due to Perot's presence in the race, denied him a mandate and lessened his influence in Washington, particularly among members of Congress and the media. Although his party still controlled both houses of Congress, Democrats had grown unaccustomed to taking direction from the White House during twelve years of Republican presidents and showed little deference toward Clinton, particularly after he had received fewer votes in the congressional districts than had every congressional Democrat. At the same time, the Republican party had been reshaped by an influx of conservative members, particularly in the House, which made the Republican congressional bloc both more cohesive and more likely to oppose Clinton effectively. In the Senate, Republicans under Robert Dole, who harbored his own presidential ambitions, were determined to force Clinton to collaborate with Republicans to achieve any significant victories. They controlled enough seats to block almost any initiative with a filibuster, and they made use of that parliamentary device more than any party ever had in the past.

Clinton also suffered from his own and many of his aides' inexperience with the intricacies of governing at the national level. His outsider campaign had done little to help him build bridges to the key members of the Washington community. Many of his most senior staff appointments — Vincent Foster, Mack McClarty, David Watkins, Webster Hubbell, William Kennedy — were "FOBs" (Friends of Bill) from Arkansas. With almost no national experience, they had trouble adjusting to the omnipresent media scrutiny and lacked knowledge of the inner workings of politics in Washington. This naivete was to cost the president dearly in the first crucial months of his administration.

Clinton took office determined to "focus like a laser on the economy" and to submit a comprehensive health care reform proposal to Congress within one hundred days. In putting together an economic package, however, it became clear that Clinton's campaign promises to halve the federal budget deficit in four years, to increase long-term investment in human capital, to enact a middle-class tax cut coupled with higher taxes on the wealthy, and to pass a job-stimulus bill were fiscally incompatible. This problem became worse after the outgoing Bush administration released figures showing a widening federal deficit. At a more fundamental level, Clinton's budget exposed the shaky foundation of his New Democrat political philosophy, which many congressional Democrats did not share. With no clear consensus behind him, he had difficulty in choosing among incompatible alternatives. In the end, with his own economic advisers split over the issues of deficit reduction, spending on human capital, and fulfilling the campaign promise of a middle-class tax cut, Clinton allowed party members on Capitol Hill to define budget priorities for him. The first casualty was a short-term $16 billion job-stimulus bill. It was emasculated in Congress, with all but $4 billion in unemployment compensation spending eliminated, after being derided by Senate Republicans and even some Democrats as wasteful "pork" that sent the wrong message about the government's commitment to deficit reduction. Republicans then mobilized against Clinton's deficit-reduction package, arguing that it increased taxes on the middle class, although the tax hikes were targeted at high-income voters. The criticism resonated with the public in part because Clinton had abandoned his promise of a middle-class tax cut. Facing Republican opposition at every stage of the budget process, and with his presidency hanging in the balance, Clinton was forced to cut deals with members of his own party to salvage his budget. He agreed to reductions in proposed expenditures on job training and other human capital programs and, to appease congressional members from Farm Belt and gas- and oil-producing states, to replace a proposed energy tax with a much smaller and more broad-based tax on gasoline. Even then, the final budget passed by the slimmest of margins, by two votes in the House and with Vice President Gore having to cast the tiebreaking vote in the Senate. Despite his many compromises, Clinton could claim some significant victories. The legislation retained the substantial tax increase on upper-income citizens he had proposed, and it included a provision to help the working poor — the Earned Income Tax Credit, which allowed the government to supplement the incomes of working families earning inadequate wages. The EITC gradually earned broad support as an effective and bureaucratically efficient way of reducing poverty.

But Clinton reaped few immediate benefits from his victory. Although the combination of spending cuts and tax increases reduced the deficit in the next year by more than the projected amount and more than halved the projected deficit across his first term, taxpayers did not immediately feel the economic payoff. Instead, influenced by the constant Republican refrain that Clinton's budget had raised taxes on the middle class (although in fact 80 percent of the new tax revenue came from those making more than $200,000 a year), public support for the budget agreement was at best muted. The media interpreted Clinton's last-minute compromises to attract votes as a sign of political weakness.

That perception had been fueled by Clinton's actions in other areas, which suggested to his critics that he would not stay the course in the face of political opposition. In his first week as president, he repeated a campaign pledge to end discrimination against homosexuals in the military as soon as possible. This well-intentioned statement provoked immediate and impassioned opposition from Republicans, conservative Democrats, and leading figures in the military. After rancorous debate, the president produced a political compromise that pleased almost no one. It directed the military not to ask recruits about their sexual preference, but also permitted the military to cashier personnel who were openly gay. Clinton accepted the compromise in part to defuse opposition to the Family Leave Act, which he signed into law on February 5. This popular statute mandated that any firm with more than fifty employees must provide employees with twelve weeks of unpaid leave for childcare or to tend to family illness. Conservative Democrats in the Senate had threatened to block the bill unless Clinton backed away from his pledge regarding gays in the military.

Controversy surrounding several of Clinton's political nominees in this period added to his growing reputation as a leader who lacked political convictions. Hillary Rodham Clinton had urged her husband to name a woman attorney general, but his first two choices were withdrawn from consideration when questions were raised about whether Zoe Baird had paid Social Security taxes for her nanny and her chauffeur and whether Kimba Wood had used an illegal immigrant for childcare. University of Pennsylvania law professor (and former Yale Law School friend of the Clintons) Lani Guinier, tapped to head the Office of Civil Rights in the Justice Department, proved an even more controversial nominee. Critics dubbed her "the quota queen" for advocating a variety of unconventional ideas designed to increase minority representation in government. Evidently caught off guard by the attacks, Clinton let Guinier fend for herself during the confirmation process and then withdrew her nomination, explaining that he had not fully understood the radical nature of her proposals.

The negative impressions created by these and other appointment controversies were compounded by a string of scandals that began plaguing Clinton's presidency in its first months. In May 1993, prodded by the First Lady and White House Counsel Vince Foster, senior White House aides suspended seven career employees of the White House Travel Office and replaced them with an Arkansas-based travel company headed by Clinton's close friends, including one of his third cousins. The fallout from "Travelgate," as it came to be known, was predictable. Within a month, after a hail of negative publicity and threats of a congressional investigation, the original employees of the travel office were rehired — although not before White House aides had instructed the FBI to investigate the employees for information to justify the firings.

Although a relatively minor matter, Travelgate helped precipitate a series of events that came back to haunt the Clintons. On July 20, Foster, evidently depressed by the Travelgate fiasco and the vindictive nature of Washington politics, committed suicide. Foster had been the Clintons' personal attorney for many years; and in the first hours after his body was found, White House aides entered Foster's office and removed some papers. They insisted that they were simply securing privileged personal documents that might otherwise be drawn into the investigation of Foster's death. But their actions fueled speculation that they had tampered with files pertaining to Travelgate and to the Clintons' financial dealings in Arkansas. Three separate investigations of Foster's death failed to establish anything more than a personal tragedy, but for years conspiracy theorists continued to speculate that he had been murdered to prevent damaging revelations about the Clintons' financial affairs.

In his first State of the Union address, Clinton sketched the broad outlines of a health care reform plan designed to both control spiraling health costs and guarantee health care coverage for everyone. He promised to submit legislation embodying these principles to Congress within one hundred days. In hindsight, it was a rash promise, issued in the euphoria of the election victory. Approximately one-seventh of the gross domestic product was spent on some aspect of health care. Comprehensive reform of the type Clinton envisioned promised to be difficult to enact because it affected the interests of virtually everyone, including powerful insurance, hospital, and doctors' groups.

Clinton's plans for health care reform became more controversial still after he appointed his wife to manage the effort. Although few opponents questioned the First Lady's professional credentials, many criticized the propriety of her assuming a leadership role in a major policy initiative. Meanwhile, she ran into the same obstacles — especially naivete about Washington — that had caused her husband to stumble. To mobilize public support she sought to portray health care reform as a battle between privileged special interests, particularly the pharmaceutical and insurance companies, which profited from the status quo, and the general public, which would benefit from comprehensive reform. Critics, however, argued that her stance needlessly alienated influential stakeholders who might otherwise have been more sympathetic. At the same time, to forestall political attack, she isolated the decision-making process from outside scrutiny, a tactic that also ultimately proved counterproductive. Moreover, the decision to present detailed legislation to Congress, rather than simply a set of principles to guide health care reform, made it difficult for the Clintons to compromise with members of Congress during the subsequent legislative debate.

But comprehensive health care reform faced an uphill battle no matter how the plan was developed and presented. The technical and political complexity of the Clinton plan, which was published in a document of more than a thousand pages, was daunting to almost everyone. The plan's centerpiece was the formation of health alliances that augmented consumers' buying power by grouping them into one or more state-based purchasing collectives. Physicians would be asked to join provider networks, and employers were mandated to provide insurance coverage. The plan would be financed by a combination of employer contributions, cigarette taxes, and individuals' copayments, with caps placed on insurance premiums and subsidies provided for low-income citizens.

The Clinton proposal was briefly popular after the president presented it to Congress in a televised speech that was characteristically effective. But it quickly met with well-funded and highly publicized opposition. Doctors and other medical care practitioners opposed the health purchasing alliances. Insurance companies fought the proposed cap on insurance premiums, in part through a provocative series of television spots. Small business interests, meanwhile, opposed the employer mandates, although big companies initially were supportive. Even liberals opposed the subsidy cap for low-income citizens. Popular support for the plan gradually eroded in the face of its bewildering complexity (and because of the effective television commercials denouncing the plan, funded by insurance companies and other health care interests). Ultimately, in September 1994, after more than a year of debate, and facing almost unanimous Republican opposition and uncertain Democratic support, congressional leaders finally abandoned comprehensive health care reform. In pushing for ambitious reform so early in the new administration, the president and First Lady had hoped to capitalize on what they viewed as a brief window of political opportunity. In retrospect, however, the Clintons' failure to adequately consult with Congress during the legislative planning stage, the plan's sweeping scope and lack of bipartisan backing, and the well-funded and highly organized opposition and considerable public confusion and uncertainty combined to assure the demise of health care reform.

Rumors of financial impropriety had dogged the Clintons during the 1992 presidential campaign. Questions focused on their participation as partners in the Whitewater Development Corporation, formed in 1978, when Clinton was attorney general in Arkansas, to develop 220 acres of riverfront property for recreational purposes. The company's eventual collapse raised many questions about its financing, as well as about Mrs. Clinton's legal work on behalf of the lenders in this and other similar real estate transactions. As Clinton emerged as the Democratic front-runner in 1992, the *New York Times* ran a story suggesting that he and his wife had taken advantage of lax oversight by state regulators and benefited from cozy relationships with local bankers. The Clintons denied any wrongdoing, described themselves as passive investors who received no favors in Whitewater or any other real estate transactions, and insisted they had lost money on the investment.

Convinced that the allegations were fueled by right-wing groups out to ruin Clinton's candidacy — a charge with some plausibility — the Clintons instructed campaign aides to reveal as little as possible about the affair. They persisted in this approach after the election. The tactic perhaps made sense in the heat of the presidential campaign. But by dismissing all Whitewater inquiries as the product of right-wing conspirators, the Clintons struck a Faustian bargain, trading short-term electoral success for long-term damage to their political credibility. In 1993, the Resolution Trust Corporation, the government agency created to handle the fallout from the 1980s savings and loan crisis, asked the Justice Department to open a criminal investigation of an Arkansas bank managed by James McDougal, the Clintons' Whitewater partner. As details of the Whitewater deal came to light, they suggested that the Clintons, particularly the First Lady, were more actively involved than initially acknowledged. At the same time, the Clinton administration was slow to respond to media inquiries regarding the matter. As a result, the Clintons became vulnerable to charges of a cover-up, especially after the suicide of Vince Foster and its aftermath raised new suspicions. By mid-March 1994, the major television networks had spent three times more airtime covering aspects of Whitewater than they did Clinton's health care plan, and major newspapers and magazines had run front-page stories on the issue.

Prompted in part by the intensive media coverage, congressional Republicans and many Democrats began to call for an independent investigation. In response, in January 1994 Clinton directed Attorney General Janet Reno to ask a three-judge appeals court to appoint an independent counsel to investigate Whitewater and related charges. A few months later, Kenneth Starr, a conservative federal judge who had served as U.S. solicitor general in the Bush administration, replaced James Fiske, the original appointee. During the next several years, under Starr's direction, the investigation expanded its narrow focus on Whitewater into a wide-ranging inquiry of issues that were only loosely related to it. These included allegations that the Clintons and their aides had abused their power in Travelgate and illegally gathered FBI files on leading Republicans ("Filegate"). The investigations and congressional inquiries, which would expand still further in Clinton's second term, became embarrassing distractions to the administration and continuing fodder for its opponents.

The end of the cold war opened up the possibility for a new conception of America's role in world affairs, and Clinton moved slowly to develop his own distinctive foreign policy. During the campaign he had focused largely on domestic issues, viewing foreign policy primarily in terms of trade and related economic issues. As president, too, he at first acted most purposefully on foreign issues that had domestic economic ramifications. A case in point was the North American Free Trade Agreement (NAFTA), first negotiated by the Bush administration. The agreement sought to create a free trade zone comprising the United States, Canada, and Mexico. Traditional Democratic party allies, including organized labor and environmentalists, opposed it bitterly. Nonetheless, after Vice President Gore effectively defended NAFTA in a televised debate with Ross Perot, it passed the Senate, in November 1993, with strong Republican backing.

On international issues involving force, however, Clinton proved less sure of himself. No longer competing in a bipolar world, the United States found itself entangled in a series of regional military conflicts in Somalia, Haiti, Bosnia, and the Middle East. In dealing with these, Clinton encountered repeated difficulties born of his inexperience and a lack of consensus in the United States. He found himself caught between those who preferred to flex American military might for humanitarian purposes and those who argued the United States could not be the world's "policeman." His early efforts to end an arms embargo in Bosnia and assist Muslims and Croats against Serb aggressors, for example, were thwarted by the unwillingness of the international community to join in the efforts. Clinton fell back to supporting the concept of "safe havens," enforced by American and NATO military forces. When Serb forces in Bosnia threatened to overrun the safe havens in 1994, Clinton found himself hamstrung by congressional resistance to unilateral American military intervention — despite appalling evidence of genocidal acts by the Serbians and vicious treatment of noncombatants by the Muslims and Croats. Finally, in August 1995, NATO forces began a bombing campaign against the Serbians, opening the door to peace talks. Mediated by Assistant Secretary of State Richard C. Holbrooke at a gathering in Dayton, Ohio, representatives of the three warring factions agreed to a cease-fire and the partition of Bosnia into a Serbian republic and a federation of Bosnian Muslims and Croats. Although

NATO and United Nations forces (including U.S. troops) managed to carry out many of the provisions of the Dayton Accord, the Serbians soon resumed their expansion in other areas, and once more Clinton faced challenges in mobilizing the international (and particularly the European) community to respond.

Like his predecessor, Clinton found himself drawn repeatedly into the tumultuous politics of Haiti, where a military government had driven out the elected president, Jean-Bertrand Aristide. After an attempt to land U.S. military forces in Haiti in October 1993 was rebuffed by Haitian demonstrators backing the military government, Clinton spent the better part of the next year feuding with Congress over his authority to intervene to help restore democracy in Haiti. Finally, in September 1994, after a visit by a delegation including Jimmy Carter, coupled with demonstrations of U.S. military power, the military leadership in Haiti stepped down and Aristide was reinstalled. These cases, and the U.S. pullout from Somalia after a failed effort begun by the Bush administration to support a United Nations peacekeeping operation, illustrate Clinton's difficulties in developing a new and coherent foreign policy for the United States. Although he was able in most instances to achieve temporary peace through the use of U.S. military force, usually as part of a multilateral international action, he did not articulate any longer-term plans for addressing the underlying causes of regional conflicts around the world. The result was a foreign policy that appeared reactive and lacking in broader purpose. Clinton did seem to achieve early foreign policy successes — along with the Dayton Accord, he forced Iraq to back down from a troop buildup along the Kuwaiti border and persuaded the North Koreans to desist from nuclear weapons production. And he successfully brokered a mutual recognition agreement between Israeli Prime Minister Yitzhak Rabin and Yasir Arafat, chairman of the Palestine Liberation Organization, which was signed in a White House ceremony in September 1993. But with hindsight it is clear that Clinton's policies did little to bring long-term stability to the affected regions, and each would remain a foreign policy flashpoint for his presidential successor, George W. Bush.

Heading into the November 1994 midterm congressional elections, despite a growing economy at home and relative peace abroad, Clinton was nonetheless in political trouble. Along with Whitewater and the failed health care reform effort, the Republicans' major campaign issue was Clinton's decision to support a major tax increase instead of fulfilling his 1992 campaign pledge to cut middle-class taxes. Even Clinton's legislative successes came back to hurt him: traditional Democratic support from labor unions weakened as a result of NAFTA, and the National Rifle Association used his signing of the Brady bill, which required a waiting period and background check on prospective handgun purchasers, and his backing of a ban on assault weapons, to mobilize gun owners against Clinton. In a historic turnabout, Republicans captured both congressional houses for the first time since 1954 and also won 24 of 36 governors' races. The media interpreted the results, not without reason, as a referendum on the Clinton presidency. Many congressional Democrats who had supported the Clinton budget package that raised taxes subsequently lost reelection bids, while none of the incumbent Republicans, all of whom had opposed the budget deal, were defeated. Even as pundits began speaking openly about a one-term presidency, however, Clinton used the Democrats' losses to undertake a political makeover for himself and his party by erasing his first-term image as a traditional tax-and-spend liberal Democrat who advocated costly government programs. With the left wing of his party no longer holding the balance of power, Clinton was free to become wholly a New Democrat. In his 1995 State of the Union address, Clinton acknowledged "the era of big government is over." Thereafter, however, responding to confidential political advice from Dick Morris, a consultant better known for advising Republican candidates, Clinton increasingly portrayed himself as a moderate bulwark against right-wing Republican zealotry. And the Republicans played into his hands, with new conservatives in Congress committing some of the same mistakes Clinton had made in 1993 by pushing too hard and too fast to implement a radical new agenda that had little public support. Their strong positions, impatience to act, and dismissive attitude toward the administration lent credibility to Clinton's portrayal of them as extremists. The same qualities also made White House assertions that the burgeoning investigations of the administration were politically motivated seem more plausible.

Following Morris's strategy of "triangulation," Clinton gained more strength in part by co-opting traditional Republican positions on a variety of issues. He supported a crime bill that led to the hiring of one hundred thousand new police. He voiced support for the use of school uniforms as a way to increase school discipline. In the 1996 Telecommunications Act, which was mostly designed to reduce regulations in the fast-growing field, he appeared to champion family values by supporting a minor measure to create technology that would allow parents to screen out inappropriate television programs. But he also sensed where to draw the line — as illustrated in frequent clashes with House Republicans, led by Speaker Newt Gingrich of Georgia, over balancing the federal budget. In a series of highly publicized confrontations, including two partial and very unpopular shutdowns of the federal government in late 1995 and early 1996, when congressional Republicans refused to provide interim funding during negotiations on a new budget, Clinton used his veto threat to curb Republican proposals to reduce spending on education and Medicare and to lower taxes. Although both sides agreed to postpone further debate on a long-term budget until after the 1996 election, the government shutdowns redounded to Clinton's advantage by allowing him to erase his first-term image as a man without deep convictions, and to paint Gingrich and the Republicans as supporting reckless policies that were insensitive to the needy. Sensing waning public support for Republican initiatives, Clinton capitalized on opportunities to use the presidency as a bully pulpit to rise above purely partisan politics. In April 1995, he gave a moving eulogy at a memorial service for the victims of the Oklahoma City bombing of a federal office building, and followed that a few days later with a commencement speech that defended the role of government in protecting individual freedom from extremist movements.

In his 1996 State of the Union address, Clinton burnished his new image by promising to veto a Republican plan to overhaul the nation's sixty-year-old Aid to Families with Dependent Children (AFDC) welfare program. House Republicans sought to replace welfare with "workfare," in which states would set their own eligibility requirements for receiving federal welfare payments, including making recipients work after two years of receiving benefits. Clinton argued that the plan would push more children into poverty and unfairly penalize people for working. Clinton's veto sent the bill back to Congress, where, after several months of debate and revision and a second Clinton veto of a second bill, Republicans (with some Democratic support) passed a third version of the bill. Clinton, mindful of his 1992 campaign promise to "end welfare as we know it," signed the revised legislation in August. Although the new law maintained a two-year time limit for receiving welfare benefits before work provisions kicked in, it also extended welfare eligibility to five years overall. However, the bill also cut off food stamps and supplemental Social Security payments to legal immigrants, provisions strongly opposed by members of Clinton's own party. Clinton signed the bill with apparent reluctance and promised to try and revise its most onerous provisions at a later date.

In part to appease disaffected party members on his left, he also sought successfully to boost the minimum wage and to win passage of a bill ensuring the portability of health care when people changed jobs and barring the use of preexisting medical conditions as a disqualification for insurance. Clinton's move to the center effectively preempted his Republican opponent, Senator Robert Dole, on many issues on which Dole had hoped to campaign for the White House. With few issues to debate, Dole largely reprised Bush's unsuccessful 1992 strategy of attacking Clinton on character issues, and with no more success. Clinton, who faced no opposition within his own party, assumed an early lead in public opinion polls and thereafter was never directly threatened. He campaigned against Republican extremism, promised to defend Medicare, education, and the environment (three issues for which government programs remained popular), and made his opponents seem unnecessarily hard-hearted, while sketching only a vague vision of "building a bridge to the next century" to cover his policy objectives.

In the general election, Perot again drew enough votes to prevent Clinton from winning a majority of the popular vote. His victory was nevertheless decisive. He received 49 percent of the vote to Dole's 41, and won the electoral college by 379 to 159. The Republicans lost eighteen seats in the House but still maintained majorities in both congressional chambers.

Clinton was the first Democrat since Franklin Roosevelt to win two terms as president. His victory was in part a result of the blunders and weaknesses of his opponent, in part a result of his own canny political positioning and his exceptional skills as a campaigner. But it also reflected a larger and perhaps more enduring result of his presidency: a reshaping of the Democratic party as a party not of the left, not of minorities and activists, but of the center. Helped by the growing conservatism of the Republican party, Clinton and the Democrats won the middle ground in the political debate, and in the process won substantial numbers of white middle-class voters who had deserted them in the 1970s, while holding on to their base support among minorities, workers, and urban liberals. It would go too far to say that Clinton produced a fundamental realignment of the sort constructed by Franklin Roosevelt — indeed, with the growing number of independent voters and the declining importance of party as a means of organizing the electorate, it is debatable whether such an enduring partisan realignment was even possible. Moreover, the Republicans retained substantial strength and control of Congress through the rest of Clinton's presidency, albeit by

diminishing margins. But the reorientation of the Democratic party relieved future Democratic candidates from carrying much of the old
liberal baggage, which may well be one of Clinton's most important
legacies.

Every modern American president has seen his influence diminish in
the second term, particularly since passage of the Twenty-second
Amendment, which prohibits two-term incumbents from running for
president again. Clinton proved no exception to this rule. Although he
had again run a brilliant tactical race to remain in office, he had done so
without articulating a clear and forceful agenda for his second term.
With Congress still under Republican control, Clinton's second-term
prospects depended on his ability to work with the opposition. But the
difficulty in doing so was brought home soon after his reelection when
Congress rebuffed his efforts to gain "fast track" authority to negotiate
trade agreements not subject to Senate amendment.

The nation's continuing prosperity, combined with the changes in tax
and spending policy that Clinton and the Congress had produced over
the previous four years, raised the possibility that for the first time in
three decades the government could operate without a deficit. In 1997
the Congressional Budget Office projected the first federal budget surplus since 1969. Unemployment fell by early 1998 to its lowest level in
more than twenty years, even as the stock market reached record highs.
An Asian currency crisis in the spring of 1998 hardly seemed to dampen
good economic news at home. With government revenues exceeding
projections, Clinton came to terms with Republican leaders in Congress
on a balanced budget bill. Signed into law in August 1997, the statute reduced middle-class taxes, provided college tax credits, and cut the rate
of growth in Medicare spending.

The new budget agreement would not have been possible without the
controversial first-term tax increases Clinton had negotiated, and it redounded to his benefit in many ways. It enabled the funding of programs in education and health care that the president had long promoted, and bolstered Clinton's image as an effective national leader.
More important, it helped to erase the tax-and-spend image of the Democrats that Republicans since Reagan had so skillfully exploited. Finally,
the agreement opened the way for the president and Congress to pursue
a fresh agenda, including discussions to address the projected shortfall
in Social Security financing. Clinton took the lead in the debate by declaring that he would oppose cutting taxes until he was assured that So-

cial Security would remain on a sound footing well into the twenty-first century.

Clinton's second-term initiatives in foreign policy enhanced his standing in international affairs and boosted his popularity at home. Employing tactics similar to those that had yielded the Dayton Accord, he supported negotiated peace settlements in other conflict-ridden areas of the world. In the spring of 1998, warring Catholic and Protestant factions in Northern Ireland agreed to a compromise proposal for power sharing. Sustained efforts by U.S. envoy George Mitchell and last-minute intervention by Clinton helped keep the talks moving toward a breakthrough.

Yet Clinton's efforts to enforce international peace agreements and respond to terrorist attacks against American targets abroad also generated controversy and frustration. On June 25, 1996, 19 Americans were killed in terrorist attacks on America military barracks in Saudi Arabia. Two years later American embassies in Kenya and Tanzania were struck, with the combined loss of 263 people, mostly Africans. Although the president appeared decisive in using military force to respond to the embassy bombings by authorizing an aerial campaign against suspected terrorist facilities, terrorists struck again in October 2000, killing 17 American sailors in an attack on the USS *Cole* while anchored at Yemen. The September 11, 2001, attacks during his successor's presidency, in which terrorists flew civilian airliners into the World Trade Center in New York and the Pentagon in Washington, D.C., would lead some critics to argue that Clinton should have acted more forcefully against the terrorist threat. It is questionable, however, whether public opinion would have supported a more aggressive antiterrorist campaign in the pre–September 11 political climate.

At the same time, the United States encountered many obstacles in its peacekeeping roles. Serbian aggression and genocidal acts spread to the formerly autonomous Yugoslavian province of Kosovo, but at first the United States could neither muster a coordinated international response nor generate domestic support for taking unilateral action. In 1998, Iraq's rogue leader Saddam Hussein engaged Clinton in a series of showdowns over United Nations inspections of suspected chemical and biological weapons operations. In December, Clinton made good on repeated threats and ordered a brief bombing campaign against Iraq. In a preview of the political divisions that would confront George W. Bush during the second U.S.-led war against Iraq, the United Kingdom joined in the attack, but other former partners in the Gulf War Coalition, including Arab and European allies, distanced themselves from the action. Despite the bombing, Iraq refused to readmit UN inspectors, and there was no evidence that Saddam's authority had been destabilized.

Clinton had little time to enjoy the good feelings that arose from the strong economy, the balanced budget, and his new foreign policy initiatives. Throughout 1997, the investigations that had intermittently plagued his first term marched forward. In addition to the old charges, there were now new allegations of campaign fundraising violations, including illegal influence-peddling by lobbyists working on behalf of the Chinese government, the selling to big campaign donors of "sleepovers" in the White House, and the illegal solicitation, from federal premises, of campaign funds. Although Attorney General Reno decided that the campaign fundraising charges did not warrant the appointment of another independent counsel, congressional Republicans charged ahead with a separate investigation. Hearings in the Senate eventually ended inconclusively, with Republicans issuing a report critical of Clinton's campaign fundraising practices, and Democrats accusing the Republicans of hypocrisy on the issue.

By far the biggest danger to Clinton's presidency, however, resulted from a sexual harassment suit filed against him in April 1994 by Paula Corbin Jones, a former Arkansas state employee. Jones charged that in 1991 Clinton had made crude sexual advances toward her in an Arkansas hotel room and that, after she rejected him, she lost promotion opportunities and suffered mental distress. Clinton immediately denied the allegations, and at first he and his advisers viewed the suit as a mere annoyance — another recurrence of what an Arkansas aide had once referred to as the "bimbo eruptions" that periodically bedeviled Clinton's career. The president, First Lady, and their allies also saw behind the case the shadowy figures of their right-wing opponents, pointing out (correctly) that Jones had assembled a high-powered legal team with connections to well-known conservative activists and donors.

But as the Jones case worked its way through the legal system, it developed into a grave challenge to the president. In May 1997, the Supreme Court unanimously dismissed Clinton's move to delay the case until after he left office. A few weeks later, his attorneys offered to settle the case for $700,000 in damages, but the deal broke down when Jones recruited new, more conservative, more aggressive lawyers, who insisted on including a formal apology from Clinton, which he refused to supply. In January 1998, Clinton submitted a deposition to Jones's lawyers, unaware that independent counsel Starr had received Attorney General

Reno's permission to investigate the possibility that the president was obstructing justice in the case.

Starr had learned of a sexual relationship between Clinton and a twenty-two-year-old White House intern, Monica Lewinsky, and he became suspicious of efforts by Clinton's advisers and friends to find her a job outside Washington. Lewinsky had already submitted an affidavit in the Jones suit in which she denied having a sexual relationship with Clinton. But former White House employee Linda Tripp, who had secretly taped Lewinsky discussing the affair, contradicted Lewinsky's statement. Bolstered by Tripp's evidence, secretly supplied to them by friends of Tripp, Jones's lawyers cross-examined Clinton in detail regarding his relationship with Lewinsky. Clinton denied having a sexual relationship with Lewinsky in contorted, evasive terms. Starr quickly asked for and received permission to expand his investigation of the president to include possible perjury and obstruction of justice in Clinton's handling of the Lewinsky affair.

News of the Lewinsky investigation exploded in the media within days of his deposition. Clinton again denied that he had had a sexual relationship with Lewinsky, and he was bolstered by his wife's strong public defense, including her claim that the charges stemmed from "a vast right-wing conspiracy." Thereafter, evidently on the advice of his lawyer, Clinton made no more substantive comments on the matter, except to deny new accusations by Kathleen Willey, a longtime Clinton supporter, that he had fondled her in the White House when she went to seek his help in finding a job.

Clinton's "Rose Garden" strategy — and his bold performance in delivering a forceful State of the Union address in the midst of the frenzy over early reports of the scandal — proved effective in portraying him as determined to focus on his official responsibilities while giving little time to the politically motivated charges against him. In reality, however, although Clinton and his advisers outwardly sought to portray a sense of normalcy, the ongoing investigation proved to be a major distraction within the White House. Meanwhile, Clinton's supporters launched a sustained attack on Starr, accusing him of carrying out a partisan vendetta. In April 1998, an Arkansas judge momentarily bolstered Clinton's cause by dismissing the Jones suit. The judge ruled that Jones had failed to prove damages from the episode and that Clinton's alleged behavior, no matter how boorish, did not constitute sexual harassment as legally defined. Jones immediately appealed the decision even as Starr's investigation of the president's statements and actions in the Jones case proceeded. Under a grant of immunity, Lewinsky provided an extensive and graphic account of her affair with Clinton to a federal grand jury and provided physical evidence to corroborate her testimony. Soon afterward, on August 17, Clinton also appeared before the federal grand jury. This time, and in a subsequent televised address to the nation, he finally acknowledged an "inappropriate, intimate relationship" with Lewinsky and apologized for "misleading" the public. But he refused to provide further details, and he vehemently denied any illegal actions resulting from the affair. Starr disagreed, and in September 1998 submitted a lengthy referral on the Lewinsky matter to the House of Representatives. The referral claimed to show "substantial and credible information" that Clinton had committed perjury, obstructed justice, and abused his official powers in the Jones case and the investigation of his relationship with Lewinsky. Starr went on to list eleven counts of potentially impeachable offenses. A few weeks later, the House voted, 258 to 156, to begin formal impeachment proceedings. The majority total included 31 Democrats in what would prove one of the few acts of bipartisanship in the entire matter.

All but a few Republicans in the House followed the lead of Judiciary Committee chairman Henry Hyde, an Illinois Republican, who argued that Clinton had violated his oath of office and betrayed the public trust by lying repeatedly under oath and acting illegally to conceal the true nature of his relationship with Lewinsky. On the other side, all but a few Democrats joined the White House in denouncing Starr's investigation and the resulting impeachment inquiry as politically inspired. While Democrats made no attempt to defend Clinton's conduct, they rallied around the claim that his alleged offenses were essentially private and did not qualify as "high crimes and misdemeanors" under the constitutional standard for impeachment. Most Democrats urged instead that the president be censured. Although more than a hundred newspapers and several dozen members of Congress (all but one a Republican) called for Clinton's resignation, public opinion polls showed substantial popular majorities in favor of allowing him to complete his term, perhaps under some form of censure or reprimand. Indeed, the graver the threat to Clinton in Washington, the higher his popularity seemed to rise, at some points exceeding 70 percent.

During the fall of 1998, the House Judiciary Committee took charge of Starr's referral and found itself embroiled in shrill partisan proceedings, with nearly every vote dividing along strict party lines. An unexpectedly strong performance by Democrats in the midterm elections — gaining five seats in the House and losing none in the Senate — seemed at first to dissipate the momentum for impeachment. In the aftermath of the election, Speaker Gingrich resigned from the House. The Judiciary Committee persisted in the investigation, however, and eventually reported out four articles of impeachment.

In a dramatic session on Saturday, December 19, the House approved two of the articles, impeaching the president for perjury before Starr's federal grand jury and for obstruction of justice in the Jones case. On the perjury count, 5 Democrats joined all but 3 Republicans in a vote of 228 to 206; the obstruction count passed more narrowly, 221 to 212. The day's events also featured the unexpected resignation of Speaker-elect Robert Livingston, a Republican from Louisiana, who just days earlier had admitted to adulterous affairs of his own after reports began to emerge in the press. In stepping down before ever becoming the Speaker, Livingston called on the president also to resign. Representative Dennis Hastert, another Illinois Republican and solid conservative with a more restrained style than Gingrich, eventually succeeded Livingston. In the middle of the impeachment trial, on the day his White House counsel made his opening argument for Clinton's acquittal in the Senate, Clinton delivered his 1999 State of the Union address to a joint meeting of Congress. With his formidable gifts as a communicator and politician fully on display, he never mentioned the issue that had consumed his presidency, and instead announced ambitious plans to apply part of the expected federal surplus to new spending on education, health care, and defense while also revamping the basis of Social Security funding. Congressional Democrats greeted the speech with the traditional enthusiastic ovations, while congressional Republicans suffered the speech in stony silence. Overnight opinion polls revealed the address to be yet another in a string of Clinton's public triumphs. Republican leaders countered that Clinton had exposed his true nature as a typical big-government Democrat and announced their own plans to return the surplus through tax cuts.

Less than a month later, the Senate voted to acquit Clinton of both impeachment charges. The Senate trial proved only slightly less overtly partisan than the House proceedings, although the same arguments that had played out in the House surfaced again and again: on one side, the president violated his oath, committed felonious acts that might earn criminal sentences for private citizens, and deserved to lose his job; on the other, the allegations, even if true, originated in a private matter, did not rise to the constitutional standard for impeachable acts, and did not justify overturning the results of a national election. But it was obvious almost from the beginning that conviction was virtually impossible. A two-thirds majority vote was required to remove the president, and

Democrats — who continued to support Clinton almost unanimously — held 45 of 100 seats. The inevitability of acquittal lent to the Senate trial an air of futility. In the end, only 45 senators voted to convict Clinton on the perjury charge, and the Senate deadlocked 50–50 on the obstruction of justice charge — less than the required two-thirds for conviction on either count.

Polled on the matter every few days throughout the entire process, the public gave consistently high marks to the president for his job performance while expressing low regard for his character and private conduct. The polls also revealed dim views of Republican congressional leaders and the independent counsel. These results surprised many political observers, who sought to explain them in various ways. The nation's robust economic health certainly aided the president, as did mostly positive reviews of his domestic and foreign policies. Many people also subscribed to the view that his problems were rooted in a private matter and should not have triggered public investigation, especially one so extensive. Still another explanation was widespread cynicism about politicians generally — "they all do it, so why the fuss?"

The twin public perceptions of Clinton as an effective president but also as a man of low character appeared to provide some support for these explanations. At the same time, the White House — and many other critics of Starr — proved extraordinarily effective in portraying the independent counsel and congressional investigations as partisan, narrow-minded, invasive, and unfair. The timing of Starr's announcement after the 1998 midterm elections that Clinton was no longer under suspicion of illegal conduct in the Travelgate and Filegate matters, combined with his earlier admission that he had no adequate evidence for charging the president with wrongdoing in the Whitewater affair, fed into growing concerns about the fairness of the investigations, the slow pace at which they moved, and the resources they consumed. (And in fact Starr's successor, Robert Ray, would decree in September 2000 that there was insufficient evidence that either Clinton or the First Lady had engaged in criminal wrongdoing in the Whitewater matter as well.)

As the impeachment crisis came to a close, it appeared that its biggest effect on the presidency would be the delineation of new distinctions among reprehensible conduct, impeachable offenses, and just cause for removal from office. Clinton was clearly guilty of the first, the House found him responsible for the second, but the Senate drew the line at the third. A constitutional issue for future generations, then, would be refinement of these distinctions and the development of appropriate remedies for presidential misconduct not deemed worthy of removal from office. Another outcome of the affair was the termination of the independent counsel statute when it came up for renewal in 1999. Although preventing corruption in the executive branch would remain an important national concern, many Republicans and Democrats, as well as Kenneth Starr, concluded that independent counsel investigations, which had affected every president since Richard Nixon, had become too routine, had seldom been truly independent, and had significantly weakened the presidency.

In 1999, Clinton faced the most severe foreign policy crisis of his presidency. Once again the crisis was precipitated by charges that the Serb-dominated Yugoslavian government was systematically "cleansing" an ethnic group from its territory. This time the targets were ethnic Albanians, who comprised more than 90 percent of the population of the southern Yugoslavian province of Kosovo. Throughout the 1990s, as Yugoslavia fragmented into ethnic factions, Kosovo's Albanian population, spearheaded by the Kosovo Liberation Army (KLA), an ethnic Albanian–based paramilitary organization, had agitated for greater autonomy. The Serb president Slobodan Milosevic, however, had come to power in part on his promise never to relinquish Serb control of Kosovo, the site of an ancient massacre of Serbs by invading Turkish forces and thus of great symbolic importance to the Serbian people. As the clashes between the Serbs and the KLA escalated in 1998, amid numerous reports of atrocities against civilians by both sides, members of the U.S.-dominated NATO alliance grew increasingly alarmed at the growing humanitarian cost of the conflict and its potential impact on regional stability. NATO-brokered negotiations between the Serb government and the KLA at Rambouillet, France, beginning in February 1999, failed when Milosevic rejected an interim proposal to introduce twenty-eight thousand NATO ground forces to guarantee limited autonomy for Kosovo while a more permanent settlement was worked out. As talks broke down, Serb forces began massing on the Kosovo border, and NATO warned Milosevic that he risked attack by its air forces unless he accepted the Rambouillet peace plan. When a last ditch appeal by U.S. envoy Richard Holbrooke failed to sway Milosevic, on March 23, 1999, the American-led air campaign began.

From the first, Clinton ruled out the use of NATO ground troops, hoping instead that daily air strikes against the Serb military and economic infrastructure would force Milosevic back to the bargaining table. Instead, the Serbs reacted by immediately moving forty thousand troops into Kosovo to begin systematically driving Albanians from their homes.

Hundreds of thousands of refugees, many reciting accounts of Serb atrocities, streamed out of the war-torn province, threatening the stability of the region. While NATO struggled to cope with the flow of refugees, and as evidence of Serbian-led "ethnic cleansing" mounted, Clinton held firm to his decision not to use ground troops. As the air campaign entered its second month, Milosevic showed no sign of capitulating. And with almost all of the indigent Albanian population forced out of Kosovo, cracks in the NATO alliance became visible as some member nations pushed for a ground campaign and others dissented. In late May, with domestic support for the U.S.-led air attacks threatening to erode, Clinton relented and agreed to begin planning for a ground war, but he continued to insist that he had no intention as yet of introducing NATO ground forces into the region.

While NATO was debating the use of ground troops, the Russians — historically Serbian allies — pressured Milosevic to negotiate a settlement. With the ongoing air strikes threatening to destroy the Yugoslavian economic infrastructure and cripple its military, he finally agreed on June 4 to a peace plan that called for the Serbs to withdraw completely from Kosovo, the Albanians to dismantle the KLA, and NATO to introduce a peacekeeping force, supplemented by Russian troops.

After seventy-two days, as Serb troops began their withdrawal from Kosovo, the NATO air campaign came to an end and its peacekeeping troops, accompanied by thousands of returning refugees and fear of reprisals against civilian Serbs, moved in. Hoping to sway Yugoslavian public opinion, Clinton promised extensive U.S. aid to rebuild the war-ravaged economy on the condition that Milosevic step down. In 2000 Milosevic was voted out of office by a disillusioned public and was subsequently placed on trial before an international tribunal on charges of war crimes against humanity.

In April 1999, a terrible tragedy occurred at Columbine High School in Littleton, Colorado. Two high school students shot to death twelve of their classmates and a teacher before turning their guns on themselves. Clinton responded to this and other incidents of school shootings by advocating stronger measures to reduce gun violence. These included mandatory trigger locks on new guns, extending the Brady law's five-day waiting period and background checks to cover gun sales at gun shows, and the licensing of all handgun owners. But these efforts were strongly resisted by the NRA and its allies in Congress. Indeed, the residue from the impeachment scandal made it difficult for Clinton to work with Republicans on many key domestic issues, most notably addressing the shaky fiscal underpinning of the Social Security and Medicare programs, the two biggest components of the government's social safety net. Democratic and Republican senators had unveiled a bipartisan plan to reform Medicare in 1998, but Clinton had been reluctant to seize the initiative for fear of losing the support of liberal Democrats in the House of Representatives. Nor was Clinton able to capitalize on the budget surplus to craft a plan to reform Social Security.

Much of Clinton's final two years in office were focused on foreign policy. He became the first president to visit Africa in two decades, and also traveled to Asia and Europe. But in foreign policy, too, Clinton found his ability to shape events constrained by the impeachment crisis and its aftermath, as well as by divisions in his own party. Clinton's support for free trade had never sat well with key Democratic interest groups, including labor unions and environmentalists, and they led sometimes violent protests against U.S. participation in the World Trade Organization at the 1999 Seattle meeting of WTO member nations. In the spring of that year, Clinton initially rejected a plan to normalize trade relations with China, which had been negotiated by his own trade representative, in order not to upset liberal Democrats in the House and their allies, who favored more protectionist trade policies (and who were his strongest supporters in the impeachment and trial). Eventually he reversed his position and successfully lobbied for normalization of trade relations with China in 2000. He also pushed for an Israeli-Palestinian peace accord during his last year as president, only to see the effort fail when Arafat rejected a plan to create a Palestinian state in Gaza and portions of the West Bank in return for relinquishing Palestinian claims to possess the right to return to their former homes in what was now Israel. The breakdown of the Camp David talks was followed by a renewal of the Intifada by Palestinians and an escalation in Israeli military reprisals.

Clinton's final days in office were a microcosm of his presidency as a whole, with his substantive accomplishments once again clouded by scandal and congressional investigation. He issued several last-minute executive orders, including the banning of new roads on sixty million acres of public lands and new ergonomic rules designed to protect workers' health. Republicans complained that by making law "with the stroke of a pen" Clinton was ignoring opposition by bypassing the legislative process. Meanwhile, in an effort to prevent his criminal prosecution after leaving office, Clinton publicly agreed, on the day before his term ended, to a plea bargain with Starr's successor Ray in which Clinton admitted that he gave "false" testimony about his sexual affair with former

White House intern Monica Lewinsky. Clinton's lawyer insisted, however, that this was not an admission that he had obstructed justice. In return for Ray's agreement not to press criminal charges, Clinton accepted a five-year suspension of his law license, agreed to pay a $25,000 fine to cover counsel fees, and acknowledged violating an Arkansas rule of professional conduct when he gave testimony in the Paula Jones case deposition. For his part, Ray announced in his final report, issued just before he resigned as independent counsel, that he believed there was sufficient evidence to have brought charges against the former president.

A day after the plea bargain was announced, in his last night as president, Clinton issued 177 presidential pardons and commutations of sentences, including the pardon of fugitive financier Marc Rich. In 1983 Rich was convicted in federal court of tax evasion, racketeering, and other charges related to his oil deals with Iran during a U.S. embargo. Rich had subsequently fled the country and was living in Switzerland. Outraged critics raised the specter of influence-peddling, pointing out that Rich's ex-wife, Denise, had contributed $450,000 to the Clinton Library, $100,000 to Hillary Clinton's senatorial campaign in New York, and $1 million to the Democratic party, not to mention purchasing several thousand dollars' worth of furniture for the Clintons' new home in Chappaqua, New York. Denise Rich's friend Beth Dozoretz was also a prodigious fundraiser for the Clinton Library, and had been checked through White House security almost one hundred times during Clinton's last two years as president. Clinton admitted only that he had been influenced in Rich's case by pleas from Israeli Prime Minister Ehud Barak, who had received campaign contributions from Rich, but to many people Clinton's last-minute pardons seemed a tawdry way to exit the White House, and it led to still another congressional investigation.

As a private citizen, Clinton continues to command public attention. He actively supported his wife's successful candidacy for a U.S. Senate seat from New York, and continues to provide her with behind-the-scenes advice amid speculation that she will someday run for president. He also is active on the financially lucrative lecture circuit, and remains the Democratic party's most potent fundraising weapon and most popular political figure. Indeed, President Clinton left office in January 2001 more popular than when he entered. His final approval rating of 66 percent was higher than any of his post-FDR predecessors' when they stepped down from office. In part this high approval rating testifies to the substantive accomplishments of his eight years as president. His 1993 tax hike, although unpopular at the time, and the 1997 budget agreement with Republicans contributed, along with rising revenues from sustained economic growth, to ending the budget deficits that had largely defined domestic politics since the Reagan era. And by acknowledging the end of "big government," Clinton removed the Republicans' most potent political bogeyman. Instead of the major legislation of the New Deal or the Great Society, Clinton pushed through a flurry of less grandiose but nonetheless impressive programs — the EITC, welfare reform, funding for crime prevention and education spending — that collectively reminded the public that the government still could play an important role. Through NAFTA, normalization of trade with China, and general support for free-market initiatives, Clinton positioned American workers to benefit from an increasingly interconnected and information-based global economy. And by winning reelection, the first Democrat to do so since FDR, and only the third in the twentieth century, he resurrected the Democratic party as a force in presidential politics.

Yet even Clinton's strongest supporters will likely acknowledge that his accomplishments amount to much less than they envisioned at the start of his presidency. In this respect his high popularity seems irrelevant if not misleading, as much a testament to Clinton's public relations savvy, the overreaching of his enemies, and the good times in which he presided as to his leadership skills. Clinton and his advisers proved remarkably adept at adjusting the presidential bully pulpit to the world of cable television and twenty-four-hour news coverage; they understood the deadline-driven rhythms of this hypercompetitive news era, and worked hard to dominate the daily news with a continual flow of statements and information. But Clinton left no lasting rhetorical legacy; his major speeches are typically remembered for their litany of accomplishments and proposals rather than for any clarion call to act on behalf of an overarching goal.

Indeed, his most memorable public utterances — "I did not have sex with that woman," "I didn't inhale," "It depends on what the meaning of the word 'is' is" — are linked to scandals, not to policy accomplishments. He failed in his attempt to reform health care and missed perhaps the best chance in a generation to craft a lasting resolution to what many believe to be the nation's biggest domestic problem: the fiscal uncertainty of Medicare and Social Security. Except for the air campaign in Kosovo, his efforts to stabilize foreign policy hot spots — the Mideast, Iraq, and North Korea — all seem less effective today than when first implemented. And in light of September 11, Clinton's antiterrorist efforts will invariably be second-guessed. Even his political legacy seems less impressive in light of Vice President Al Gore's loss in the 2000 presidential election — a loss due in part to Gore's ambivalence about running too closely to Clinton and his scandal-plagued record. History will judge whether Clinton's not inconsiderable achievements as president would have been even greater had he only been able to restrain his tendency toward personal indulgence. In the end, his was a presidency of substantial accomplishments — and even more of missed opportunities.

— MATTHEW DICKINSON
Middlebury College

The Politics of Presidential Investigation

As impeachment proceedings against Clinton concluded, the mechanism that had first set them in motion — the independent counsel statute — came itself under sharp scrutiny. Congress had authorized the appointment of independent counsels in the wake of the Watergate scandal, citing lingering concerns about the Justice Department's ability to investigate objectively top officials in the executive branch. Although Congress had renewed the statute several times, in 1999 it was disinclined to extend it again. Even Kenneth Starr, the most notorious independent counsel, testified that the statute was of "dubious constitutionality."

The principal objection to the independent counsels was not constitutional, however, but political: they had become *too* independent, with few checks on their investigative powers and budgetary authority. Between 1978 and 1999, some twenty independent counsels looked into a range of allegations of executive abuse. Although some of these investigations all but paralyzed the presidency while they were under way, the outcomes were mostly inconclusive. Before Starr targeted Clinton, the most significant investigation was that of Lawrence Walsh, who looked into the Iran-contra affair. Walsh produced indictments and convictions of senior administration officials, but the matter ended abruptly when George Bush issued pardons as he left office.

Together, the Starr and Walsh investigations showed that independent counsel procedures presented many of the same problems inherent in regular executive branch prosecutions. Certainly they too could become instruments of partisan politics. Presidents Reagan and Bush inveighed against investigations into their administrations as inspired by the Democrats, while Clinton portrayed the Starr investigation as a conservative Republican plot.

From a constitutional perspective, the problem of dealing with a rogue president without impinging on the legitimate prerogatives of the presidency remains. The Constitution does not explicitly endorse any remedy but impeachment, but leaves room to experiment with less drastic solutions. As the history of the independent counsels demonstrates, the experiments have yet to produce a satisfactory result.

George W. Bush

2001–

b. July 6, 1946

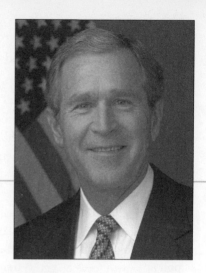

O
N THE MORNING OF SEPTEMBER 11, 2001, President George W. Bush first learned of the terrorist attacks on New York and Washington when his chief of staff whispered in his ear while he was participating in a reading lesson for second-graders at an elementary school in Sarasota, Florida.

A moment later he got up and stepped to a microphone before about two hundred people who had gathered to hear him talk about his education initiative, pending in Congress. Instead he told the group there had been an "apparent terrorist attack on our country."

"This is a difficult moment for America," he went on, his tone somber. He promised "a full-scale investigation to hunt down and find those folks who committed this act." Then his security team whisked him away to spend the rest of the day largely incommunicado, flying around the country on Air Force One until national security officials determined he could safely return to Washington.

While the nation tried to absorb the incomprehensible news of the attacks on the World Trade Center and the Pentagon that terrible day, commentators also shook their heads as they related the president's initial reaction — calling the attackers "those folks" and then disappearing from view for many hours afterward. Few seemed surprised, however. This sort of hapless behavior had come to be what most people expected of their new president, the dauphin of a political family who lost the popular vote by a hair and was ushered into office by a five-to-four decision of the Supreme Court. In some ways he seemed ill suited for the job. The morning of September 11, not quite eight months into his term of office, his national approval rating stood at a strikingly indecisive 51 percent, reflecting the nearly dead-even vote for the presidency the previous year.

But what no one knew that day was that the Bush they had known, and in many cases scorned, would not step off Air Force One when it finally landed at Andrews Air Force Base as dusk fell over the capital. In his place was a determined, clench-jawed president who had found his mission. From that day forward, he spoke with the same single-minded certainty that had long been his hallmark but now, all of a sudden, served him perfectly. As he put it a week after the terrorist attacks, "From this day forward this is the central focus of my administration."

Right away he launched a war, and barely a year later another one — all with strong backing from the American public, as shown by opinion-poll approval ratings that at times touched 90 percent.

A welter of contradictions underlay Bush's presidency at midterm. He had run for office as a self-styled "compassionate conservative," yet his administration aggressively pushed laws and regulations to reduce environmental protection, limit abortion rights, and cut back health and welfare programs for the poor while enacting massive tax cuts that largely benefited the rich. He told fiscally lax government agencies, as no previous president had, to better account for their money — while also proposing budgets that spent all of the surplus left him by his predecessor, recording deficits that set new records. He waged dramatically successful wars in Afghanistan and Iraq, all in the name of defeating terrorism, while seeming to ignore less glamorous antiterrorism needs at home, like securing the nation's ports, which were wide open to terrorists, and equipping and training police around the country to deal with the terrorist threat.

But most striking of all was his repudiation of international treaties, agreements, and alliances, just a decade after his father, George H. W. Bush, the forty-first president, improbably managed to pull together an unarguably wide coalition of nations to fight the first war against Iraq, including even Iraq's Arab neighbors, fellow dictators not unlike Saddam Hussein, the Iraqi leader. At his Inauguration George W. Bush seemed to promise to carry on as his predecessors had: "America remains engaged in the world, by history and by choice, shaping a balance of power that favors freedom," he said. "We will defend our allies and our interests."

Still, none of this seemed to dent his popularity. As he passed midterm, Democratic complaints about the budget, the economy, health care, the environment, and other domestic concerns that normally might rouse the electorate came to be seen as irrelevant bleating, if they were noticed at all. Pausing from his war planning in the autumn of 2002, Bush crisscrossed the country for several weeks, campaigning for Republican congressional candidates in the off year, when administrations can virtually count on losing seats in the Senate and the House. Yet Bush picked up seats in both houses, tipping the Senate back into Republican control — the first time that had happened in a hundred years, since the administration of Theodore Roosevelt, leaving friends and opponents alike shaking their heads in wonder. George W. Bush seemed such an unlikely heir to Teddy Roosevelt's legacy.

Little in Bush's early years suggested he had the capability or desire to lead the nation. He was born July 6, 1946, in New Haven, Connecticut, but at age two moved to Midland, Texas, where his father was seeking his fortune in the oil business. Only twenty-five thousand people lived in Midland then. Segregation was absolute, and the sheriff routinely escorted anyone he regarded as "suspicious" to the town limit. At school the principal paddled students who misbehaved or didn't learn their lessons; Bush received a whacking three times, teachers recalled. Midland then was a community with clear lines of right and wrong, black and white — no shades of gray, an attitude of certainty that seemed in later years to have had a profound effect on young George Bush.

George was the eldest of six children (one sister, Robin, died of leukemia when she was three) and lived a prototypical 1950s suburban life. He collected baseball cards, rode his bike around the neighborhood in jeans and a white T-shirt, and pulled pranks like drawing a beard and sideburns on his face with a pen in the fourth grade and blowing up frogs with firecrackers.

He was the wild son. His father wrote to a friend in 1951, "Georgie has grown to be a near-man, talks dirty once in a while and occasionally swears, aged 4 and one-half. Georgie aggravates the hell out of me at times." At least once, his mother washed his mouth out with soap. His stated ambition through his youth was to be a baseball player, and he behaved in school as if that were to be his future. He was a mediocre student at best and showed no particular interest in any field of study. Reading appeared to be among his lower-priority activities.

For all the outward ordinariness of his childhood, however, even a casual look showed that behind the Bush family's life in Midland was a blueblood clan of wealth and national prominence. Bush's grandfather, Prescott S. Bush, was a U.S. senator from Connecticut who would stir mild interest in Midland when he came for a visit. And in 1964, when

George was eighteen, his father ran for the Senate from Texas. Though he lost, he did run a strong campaign and managed to win a race for the House of Representatives a few years later.

Bush, like his father, attended Phillips Academy, in Andover, Massachusetts, where his father had been a star graduate and was remembered fondly many years later. High schools often give preference to applicants from alumni families, and it seems likely that Bush, whose academic record was less than outstanding, would not have been admitted otherwise. That was even more true when he was accepted at Yale, which normally would not have admitted even a legacy student with a record as poor as Bush's. But by then Bush's father was a successful national politician, just as his grandfather had been.

In an interview with the *New York Times* during his 2000 presidential campaign, Bush bristled at the question of how his life might have proceeded had his last name been Smith, say, instead of Bush. "I think I am asked that all the time," he said. "It's interesting, they always use the word 'Smith,' too." He dropped his voice as if he were mimicking a television interviewer as he said: "'Would you be standing here as a presidential candidate if your name were George Smith?' Well, you know, it's not George Smith. It is George Bush. And how did it influence? I don't have any idea."

Bush did receive significant breaks throughout his life because of his famous father. And during the early months of his 2000 presidential campaign, some voters told reporters that they thought they were voting for the senior George Bush once again.

In any case, Bush did not make much of his opportunities at either Andover or Yale. He was a C student at both schools and demonstrated a proclivity for snobby arrogance — as evidenced by his sarcastic, dismissive smirk and caustic remarks — and for anti-intellectual vapidity. At the same time, however, he did begin to display social leadership skills, best demonstrated by his success at starting a popular stickball league at Andover. "At the time, the whole stickball thing sounded like a grand prank, without political overtones," a classmate, Bryce Muir, observed. "Looking back, it was an inspired scheme with definite political implications." Bush became something of a campus figure as a result of stickball.

At Yale, Bush was often the one who picked up the keg for the fraternity party. Meanwhile, he left little if any impression in the classroom. As one professor told the *Times* years later: "I haven't the foggiest recollection of him."

After graduation, Bush drifted for several years. Yet all the while —

perhaps without being fully aware of it — he was trying to follow the path of the father he idolized and adored. "What makes him tick?" an old friend of both Bushes said in an interview with the *Times*. "It's daddy. Daddy allowed him to do things and achieve ambitions that he would not do on his own. Daddy is his motivation. To please his dad."

And so in 1978 he ran for Congress from a district in West Texas, just as his father had, but unlike his father he ran a wretched campaign. Even though he chewed tobacco and spoke with a thick Texas twang, his Democratic opponent, Kent Hance, managed to portray him as an effete northern interloper from Connecticut. A staple of Hance's stump speech was a story about working in a field as a fancy car drove past.

"It was a Mercedes," he would say with a leering tilt of the eyebrows that suggested he was talking about his opponent. "The guy rolled down the window and wanted to know how to get to a certain ranch." Hance told of giving directions, telling the driver to turn right at the cattle guard, one of those ubiquitous metal grates that keep livestock from straying. "Then," Hance went on, "the driver asked, 'What color uniform will that cattle guard be wearing?'" As the audience erupted in laughter, Hance would add that the Mercedes "had Connecticut plates."

The election was not a rout; Bush got 47 percent of the vote. But Bush had allowed his opponent to set the agenda, a mistake he vowed never to repeat.

The congressional race was a detour from his chosen career — the Texas oil business, the same as his dad's. He started his own company, which merged with another and became the Spectrum 7 Energy Corporation. But he failed to make much of a go at it. Nonetheless, in 1986 — while Bush's father was serving as vice president in the Reagan administration — Harken Energy bought it. Though his track record in the business was uninspiring, Bush became a Harken director and was given Harken stock that was worth more than $500,000. Asked once why he had purchased Spectrum 7 and taken Bush on as a director, Phil Kendrick, Harken's founder, said: "His name was George Bush. That was worth the money they paid for him."

In 1990, while his father was president, he sold Harken stock for $848,000 and used the money from that and future sales to buy into the Texas Rangers baseball team, which was up for sale. A friend had recruited him — and not because of his business acumen, a string of positions in faltering oil companies. His last name "certainly helped," said Robert Brown, then the president of the American League. "I was very friendly with his father." But Bush managed to make something of his new position as an owner. He was credited with doing a good job, and he built a new stadium, an edifice that won accolades. As he walked through the construction site in 1993, he quipped, "When all those people in Austin say 'He ain't never done anything,' well this is it!"

The next year he ran for governor, even as his younger brother, Jeb, ran for governor of Florida. His mother urged him not to do it, as he would be running against Anne Richards, an exceedingly popular governor. Bush's parents did not want to see him fail again. All their hopes fell on Jeb, who had prepared meticulously for his race.

Bush surprised everyone. He ran a skillfully controlled campaign that did not allow Richards to set the agenda. He won — and Jeb lost. As governor, Bush showed the same proclivities he had demonstrated at school. Put simply, he had little interest in policy. In the office, his workday was short: 9 A.M. to 5 P.M., with two hours off at midday for exercise and lunch. Aides said he was impatient with long meetings and thick policy papers. He was remarkably unreflective, about himself or the issues, uninterested in the key political debates of the day. No one who worked with him doubted that he was smart, but he showed little if any intellectual curiosity.

Still, one day in the summer of 1997, two years into his first term, Karen Hughes, one of his senior aides, walked into Bush's office and said, "You're leading in the poll." "What poll?" Bush asked. The first public opinion poll for the 2000 election, she responded. Even then, he was the front-runner among Republicans.

Bush ran for reelection in Texas first, and won handily. For many months he said nothing about running for president, though he formed campaign committees and met with the solons of the Republican party to discuss the possibility. When he finally entered the race, he ran a well-scripted campaign and portrayed himself as a Republican moderate, not a doctrinaire conservative — the image a Republican needed to have to win. He raised prodigious sums of money from Republicans who were eager to usher Bill Clinton and his heir, Vice President Al Gore, from the White House. Senator John McCain, a maverick Republican from Arizona, posed a strong challenge early in the primary season, but Bush overcame that. Still, defeating the vice president at a time of peace and unparalleled economic prosperity was viewed as a difficult challenge. And in fact Bush lost the popular vote, by a mere 539,947 out of 104 million votes cast. But American elections are determined by electoral votes, and on the morning after the election, the electoral vote total was still in question because of voting irregularities in Florida.

The Democrats demanded a partial recount, and so began several weeks of jockeying over misleading ballots and improperly punched voter cards that left "hanging chads." The momentum seesawed between Bush and Gore for five weeks, until December 12, when the Supreme Court, by a vote of five to four, overturned a decision by Florida's supreme court to recount the votes. That handed the victory to Bush — the first time since the election of Rutherford B. Hayes, in 1876, that a president had taken office despite losing the popular vote.

The day after the Supreme Court decision, Gore graciously conceded defeat, and Bush addressed the nation in an attempt to put salve on the wound that the protracted election dispute had opened. "Whether you voted for me or not, I will do my best to serve your interests," he said, "and I will work to earn your respect. I was not elected to serve one party, but to serve the nation." After taking office a month later, some Democrats, at least, came to believe he had betrayed his word.

The first thing many people noticed about the Bush presidency was its singular effort to control the agenda and offer a tightly scripted image to the world. The new president and his staff chose the issue they wanted to discuss on a given day and refused to talk about anything else, almost as if they would not recognize unexpected developments anywhere in the world until they were ready.

Early in April, for example, he allowed reporters into the cabinet room for a few minutes at the beginning of a budget discussion with congressional leaders. As was customary at times like this, the reporters peppered him with questions about the controversies of the moment — in this case the debate over a new campaign finance reform bill.

With a wave of his hand, Bush cut off that question and others, saying, "I'm talking about the budget today." At an event later in the day, as the president posed for pictures with a group of astronauts, a reporter asked him another question about campaign finance. "Zero for two" was all Bush had to say.

This approach fit Bush's nature perfectly, but it quickly became impossible to maintain in the rough and tumble of Washington, particularly after the president made his own agenda known. It was anything but centrist. First he offered an education bill that mandated testing of all students nationwide to measure their achievement. The bill would also give vouchers to parents who wanted to send their children to private schools — a favorite conservative proposal. Opposition to the voucher plan, predictably, was strong and immediate, forcing Bush to drop it. But he breathed life into another long-dormant Republican initiative: opening Alaska's Arctic National Wildlife Refuge to oil drilling.

While Democrats and environmentalists yelped about that, Bush refused to let it go.

He pushed other ideas relentlessly as well, one of which would allow the government to give money to churches that wanted to run community-based social programs, and another for a massive $1.6 trillion tax cut, the fruits of which would fall mostly to the wealthy. That was not surprising, the White House noted, because the wealthy pay the most in taxes. Democrats complained endlessly about that, but to little avail. Congress quickly passed the tax cut after whittling it down a bit, to $1.35 trillion.

All of it caused some Democrats to charge that Bush had pulled a bait and switch. "You certainly don't see the agenda the president promised," complained Senator John F. Kerry, a Massachusetts Democrat. "You run down this list, and you have a right-wing checklist."

While these and other domestic initiatives were classically Republican, Bush's first steps onto the world stage had little parallel in recent American history. Right away he worked to dismantle many of the international agreements that had in some cases served as the bedrock of American foreign policy. Within the first few months in office, he announced that he wanted to build a limited missile defense system, an idea that would obviate the Anti-Ballistic Missile Treaty of 1972. This announcement, accompanied by hostile remarks directed at Russia by the pugnacious new defense secretary, Donald Rumsfeld, caused the Russians to complain that the United States was reopening "the spirit of the cold war."

In short order the administration announced that it would not back the Kyoto Protocol, a worldwide agreement to control the gas emissions that led to global warming. That decision was based on "important new information that warrants a re-evaluation, especially at a time of rising energy prices and a serious energy shortage," Bush said. The president also said he wanted to pull American peacekeeping troops out of Bosnia.

All of that shocked America's European allies, who had been wary of Bush to begin with. But now they cried out, particularly after the announcement that the United States would no longer back the Kyoto Protocol. "To suggest scrapping Kyoto and making a new agreement with more countries involved reflects a lack of understanding of political realities," said Margot Walstrom, Europe's commissioner of environmental affairs. And Dominique Voynet, France's minister for the environment, called the decision "completely provocative and irresponsible."

The criticism left Bush undaunted — until late May, when Senator

James M. Jeffords of Vermont left the Republican party after the White House had struck back at him for failing to support the president's tax cut bill. Jeffords became an independent who said he would vote with the Democrats. That gave the Democratic party control of the Senate.

"Unfortunately," said Scott Reed, a veteran Republican strategist, "the rest of the Bush agenda is in turmoil." And so it was through the summer of 2001. Bush's international initiatives continued apace. But his domestic program stagnated, hamstrung by opposition in the Senate. By late summer the White House had decided that the president, whose job approval ratings were on a steady slide south, should focus on small-bore programs that did not require congressional approval and were more centrist than his earlier ideas.

Among them, he wanted to provide e-mail services so that grandchildren could communicate with their grandparents, add citizenship classes to school curricula, and begin talking about such issues as teen pregnancy, truancy, drugs, school safety, and ministries in prisons. The intention was to allow Bush to regain his footing by making small but unarguably centrist accomplishments. The White House intended to announce the new programs in early September.

Then came September 11, and by the time the president returned to the White House that evening, he had regained his footing. His speeches quickly took on a messianic quality, a far cry from his unscripted comments at the Sarasota elementary school that morning.

"Thousands of lives were suddenly ended by evil, despicable acts of terror," he said in a short televised address from the Oval Office. "I ask for your prayers for all those who grieve, for the children whose worlds have been shattered, for all those whose safety and security has been threatened.

"This is a day when all Americans from every walk of life unite in our resolve for justice and peace," he continued. "America has stood down enemies before, and we will do so this time. None of us will ever forget this day."

The crisis, dreadful as it was, seemed tailored to Bush's preference for single-minded attention to one topic at a time. Even with the nation slipping into recession, the public wanted nothing from the White House but vengeance — and protection from further attack. Bush set out to provide both, with full support from the rest of the world. The outpouring of sympathy and support was nearly universal — even from the European countries that had disparaged Bush just days before. At the same time, nations that the United States had considered warily or with hostility before, such as Pakistan, suddenly became close allies, if they were willing to help with Bush's new war against terror.

American intelligence quickly identified the Islamic terrorist group Al Qaeda, led by Osama bin Laden, as the responsible party. It was head-quartered in Afghanistan, which was under the control of the Taliban, an Islamic fundamentalist regime. Bush ordered the Taliban to turn over bin Laden and Al Qaeda's other leaders or face attack. The Taliban refused. In early October, less than a month after the September 11 at-tacks, the United States began bombing Afghanistan. In short order the Taliban and Al Qaeda's leaders fled; Osama bin Laden disappeared.

In Washington, Bush established an Office of Homeland Security in the White House, headed by Tom Ridge, who had been governor of Pennsylvania. Later it became a cabinet-level department. Its mission was to impose domestic protections against terrorists. Through it all, Bush amassed power that rivaled Franklin Roosevelt's during World War II — which had seemed unthinkable less than a year earlier, when he had lost the popular vote for the presidency. He used the new office to pass a law known as the Patriot Act, which granted law enforcement a wish list of new powers that caused civil libertarians to shudder. Even without that law, the administration asserted that the United States had the right to imprison alien terrorism suspects indefinitely without filing charges or allowing them legal representation, and it did so with hun-dreds of them. Then, even before the war in Afghanistan was won, the White House began talking about invading Iraq and deposing Saddam Hussein, who was left in power after Bush's father had waged the first Gulf War, in 1991, to dislodge Iraq from Kuwait. "The power the presi-dent is wielding today is truly breathtaking," Timothy Lynch, director of the Project on Criminal Justice at the Cato Institute, said in November 2001. "A single individual is going to decide whether the war is going to be expanded to Iraq. A single individual is going to decide how much privacy American citizens are going to retain."

If Bush felt a need, born of political necessity, to return to the political center in the summer of 2001, by early 2002 any hint of that idea had vanished. He announced that the White House would abandon the Comprehensive Nuclear Test Ban Treaty, saying that America had to re-tain the right to test its weapons to make sure they were safe. After a few faltering steps, he dropped any significant efforts to end the ongoing vio-lence between Israel and the Palestinians, even though almost everyone

— the Israelis, the Palestinians, the Europeans — agreed that the problem could not be solved without American involvement. And his administration made clear that the United States would not cooperate with or even recognize the new International Criminal Court, set up to prosecute war criminals. "You might have heard about a treaty that would place American soldiers under the jurisdiction of something called the International Criminal Court," Bush told cheering soldiers at Fort Drum in New York. "The United States cooperates with many other nations to keep the peace, but we will not submit American troops to prosecutors and judges whose jurisdiction we do not accept." This brought the usual round of complaints from Europe, but in the end most Western nations agreed to exempt the United States from the court's jurisdiction.

Meanwhile, at home the administration acknowledged that the federal budget, after several years of surpluses, had fallen into deficit again, the victim of the nation's recession and the reduced tax proceeds resulting from the tax cut of a year earlier. But with the war on terrorism continuing, Bush seemed impervious to harm on domestic issues, at least for the moment.

Rumbling about a war in Iraq began in the spring of 2002, but no one in the White House said anything definitive on the record. Finally, just after Labor Day, when the midterm congressional elections were getting under way in earnest, Bush announced that he intended to take on Iraq. Either Saddam Hussein would disarm or the United States would invade. On September 12 he addressed the United Nations, urging the Security Council to act against Iraq.

"As we meet today," he said, "it's been almost four years since the last UN inspector set foot in Iraq — four years for the Iraqi regime to plan and to build and to test behind the cloak of secrecy. We know that Saddam Hussein pursued weapons of mass murder even when inspectors were in his country. Are we to assume that he stopped when they left? The history, the logic and the facts lead to one conclusion: Saddam Hussein's regime is a grave and gathering danger. To suggest otherwise is to hope against the evidence. To assume this regime's good faith is to bet the lives of millions and the peace of the world in a reckless gamble. And this is a risk we must not take.

"Delegates to the General Assembly, we have been more than patient. We've tried sanctions. We've tried the carrot of oil for food and the stick of coalition military strikes. But Saddam Hussein has defied all these efforts and continues to develop weapons of mass destruction. The first time we may be completely certain he has nuclear weapons is when, God forbid, he uses one. We owe it to all our citizens to do everything in our power to prevent that day from coming.

Politically, the strategy was brilliant. In Congress, the Democrats rushed to pass a resolution authorizing a war so they could get the issue behind them and quickly turn the discussion to the economy, health care, and other issues they favored as the election neared. The resolution passed, but the drumbeat of war grew only louder, drowning out anything the Democrats had to say. That led to their stunning loss at the polls in November.

Meanwhile, the Bush administration had not wholly succumbed to the war frenzy. Through the summer and fall it put forth numerous recommendations and decrees that would have brought howls from the opposition at any other time. The administration, at the request of hospitals, rolled back rules ensuring the privacy of medical records. It proposed deep cuts in Medicaid payments. At the request of lumber companies, it gave federal forest managers the unilateral right to approve logging on public lands without the usual environmental reviews. It gave U.S. housing officials the authority to raise rents dramatically for poor people in federally subsidized housing. And it proposed to reduce or eliminate patients' ability to appeal the denial of benefits under Medicare. But there was so much chaff in the air from the talk of war and terrorism that few people seemed to notice the complaints about these decisions. In fact, on the eve of the midterm elections, the Census Bureau reported that the number of people living in poverty had risen significantly in the past year, the first such increase in eleven years. In normal times that would have been devastating news for the president, but once again it passed by with little comment.

Bush was not having so easy a time with the United Nations. In November the Security Council did pass a resolution mandating new weapons inspections in Iraq and threatening "serious consequences" if the Iraqis did not cooperate. The United States asserted that Iraq maintained vast stocks of chemical, biological, and perhaps nuclear weapons. But after three months of work, the weapons inspectors had found no trace of such weapons. However, they did acknowledge that Iraqi cooperation was grudging at best. Still, the Security Council refused to give authority for war, saying the inspectors should be given more time. France, Germany, Russia, and China banded together to block any resolution authorizing war, even though Bush and his aides bullied and threatened with every tool at their disposal — warning that he intended to go to war no matter what the United Nations did.

But Europe was fed up with the administration's unilateral approach to the world. European leaders had been appalled by the strategic doctrine the administration had published in the fall, saying the United States would take unilateral military action when necessary, adding that "the president has no intention of allowing any foreign power to catch up with the huge lead the United States has opened since the fall of the Soviet Union more than a decade ago."

And here, on the debate over Iraq, the Europeans were going to make their stand. They forced Bush to abandon his quest for UN approval, an embarrassing defeat that opened a dangerous rift with several important European allies. But in late March he went to war anyway, with only Great Britain as an ally on the battlefield. As it turned out, the regime of Saddam Hussein fell like a house of cards — a stunning military victory. The two armies held control of almost the entire country within a month. With that, the United States began the far more difficult task of rebuilding Iraq and fostering democracy in a country and region that had never known it — a task that for many months after the initial victory the administration seemed unable to achieve. Indeed, more American soldiers died in Iraq after the supposed U.S. victory than during the initial combat.

With the military victory Bush's approval rating shot back up to record-high levels, but soon sank back to a level close to Bush's pre–September 11 condition as the problems in Iraq mounted. Bush was also more than aware that his father twelve years earlier had failed to win reelection because he had not paid enough attention to the nation's growing economic problems. The son was determined not to make the same mistake as the father. On the day after his generals declared victory in Iraq, Bush turned to the economy. He tried to use his wartime popularity to push his economic recovery proposal, another massive tax cut — $726 billion — that would primarily favor the rich. Even some Republicans in Congress opposed it, saying that, with growing deficits, the nation could not afford it. "We've spent money like drunken sailors," said Senator George Voinovich of Ohio, one of the Republicans who voted against the tax cut.

Bush disagreed, saying his first tax cut had been highly beneficial, though the evidence to support that was scant. "Make no mistake about it," he said, "the tax relief package that we passed has helped the economy, and the deficit would have been bigger without it."

Democrats pointed out that the Bush administration had argued that a tax cut was the proper economic tool when the country had a surplus, and now it was making the same argument when the country had a deficit. In any case, the nonpartisan Congressional Budget Office issued a report saying that the tax cut would have little effect on the economy even if it passed.

As Bush looked toward reelection in 2004, the White House's chief pollster warned that the president's reelection would be in jeopardy if the economy did not improve. Given his father's history, Bush was certainly aware of the problem, but presidents have little ability to affect the economy. Winning the economic battle and winning reelection stood as challenges at least as difficult as were the wars Bush was fighting in Afghanistan and Iraq.

— JOEL BRINKLEY
New York Times

Acknowledgments

This book derives from and updates an earlier work, *The Reader's Companion to the American Presidency*. Like its forebear, it carries the names of its two editors, but also like its forebear, it is in fact the work of the eminent authors who wrote the forty-two essays that form its heart. And as before, our first and greatest debt is to them.

We are also grateful to the researchers who assisted us, checked facts, and drafted some supporting text: Frederick Dalzell, George C. Eliades, Andrew P. N. Erdmann, Mark H. Haefele, Edward L. Merta, Timothy A. Milford, and Silvana R. Siddali. Deborah Bull expertly carried out the photo research.

Raphael Sagalyn and his colleague Jennifer Graham ably handled relations with Houghton Mifflin, where we thank Brandy Vickers and Susan Canavan for suggesting the revision and change of format and for shepherding the book through production.

For Further Reading

The literature on the American presidency is vast and voluminous. For a larger set of suggested readings, see Alan Brinkley and Davis Dyer, eds., *The Reader's Companion to the American Presidency* (Houghton Mifflin, 2000).

For the institution of the presidency as a whole, over long sweeps of time, several works by historians or political scientists may be of interest:

Sidney M. Milkis and Michael Nelson, *The American Presidency: Origins and Development, 1776–2002* (Washington, D.C., 2003).

Michael Nelson, ed., *The Presidency A to Z*, 3rd ed. (Washington, D.C., 2003).

Richard E. Neustadt, *Presidential Power and the Modern Presidents* (New York, 1991).

Arthur M. Schlesinger, Jr., *The Imperial Presidency* (Boston, 1989).

Stephen Skowronek, *The Politics Presidents Make: Leadership from John Adams to Bill Clinton* (Cambridge, Mass., 1997).

For individual presidents and presidencies, the following may be places to start:

GEORGE WASHINGTON

James Thomas Flexner, *George Washington*, 4 vols. (Boston, 1965–1972).

John H. Rhodehamel, ed., *George Washington: Writings* (New York, 1997).

Barry Schwartz, *George Washington: The Making of an American Symbol* (New York, 1987).

JOHN ADAMS

Ralph Adams Brown, *The Presidency of John Adams* (Lawrence, Kan., 1975).

David McCullough, *John Adams* (New York, 2001).

THOMAS JEFFERSON

Fawn Brodie, *Jefferson: An Intimate History* (1974; reissue, New York, 1998).

James Horn, Jan Lewis, and Peter Onuf, eds., *The Election of 1800* (Charlottesville, Va., 2002).

John Majewski, *The House Dividing* (Baltimore, 2001).
Dumas Malone, *Jefferson and His Times*, 6 vols. (Boston, 1948–1991).
Jack McLaughlin, ed., *To His Excellency Thomas Jefferson: Letters to a President* (New York, 1991).
Merrill D. Peterson, *Thomas Jefferson and the New Nation: A Biography* (Oxford, 1986).
James I. Simon, *What Kind of Nation* (New York, 2002).

JAMES MADISON

Ralph Ketcham, *James Madison: A Biography* (New York, 1971).
Robert A. Rutland, *The Presidency of James Madison* (Lawrence, Kan., 1990).

JAMES MONROE

Harry Ammon, *James Monroe: The Quest for National Identity* (New York, 1971).
Noble E. Cunningham, Jr., *The Presidency of James Monroe* (Lawrence, Kan., 1996).

JOHN QUINCY ADAMS

Allan Nevins, ed., *Diary of John Quincy Adams, 1794–1845* (New York, 1951).
Lynn Hudson Parsons, *John Quincy Adams* (Madison, Wis., 1998).

ANDREW JACKSON

Donald B. Cole, *The Presidency of Andrew Jackson* (Lawrence, Kan., 1993).
John F. Marszalek, *The Petticoat Affair: Manners, Mutiny, and Sex in Andrew Jackson's White House* (New York, 1997).
Robert V. Remini, *Andrew Jackson*, 3 vols. (New York, 1977–1984), condensed into *The Life of Andrew Jackson* (New York, 1988).
Harry L. Watson, *Liberty and Power: The Politics of Jacksonian America* (New York, 1990).

MARTIN VAN BUREN

John Niven, *Martin Van Buren: The Romantic Age of American Politics* (New York, 1983).
Major J. Wilson, *The Presidency of Martin Van Buren* (Lawrence, Kan., 1984).

WILLIAM HENRY HARRISON

Freeman Cleaves, *Old Tippecanoe: William Henry Harrison and His Time* (1939; reissue, Newtown, Conn., 1990).
Richard P. McCormick, *The Presidential Game: The Origins of American Presidential Politics* (New York, 1982).

JOHN TYLER

Oliver Perry Chitwood, *John Tyler: Champion of the Old South* (1939; reissue Newtown, Conn., 1991).

Norma Lois Peterson, *The Presidencies of William Henry Harrison and John Tyler* (Lawrence, Kan., 1989).

JAMES K. POLK

Paul H. Bergeron, *The Presidency of James K. Polk* (Lawrence, Kan., 1987).

Charles Sellers, *James K. Polk*, 2 vols. (Princeton, 1957–1966).

ZACHARY TAYLOR

K. Jack Bauer, *Zachary Taylor: Soldier, Planter, Statesman of the Old Southwest* (Baton Rouge, 1985).

Elbert B. Smith, *The Presidencies of Zachary Taylor and Millard Fillmore* (Lawrence, Kan., 1988).

MILLARD FILLMORE

Benson Lee Grayson, *The Unknown President: The Administration of Millard Fillmore* (Washington, D.C., 1981).

Robert J. Rayback, *Millard Fillmore: Biography of a President* (Buffalo, 1959).

FRANKLIN PIERCE

Larry Gara, *The Presidency of Franklin Pierce* (Lawrence, Kan., 1991).

Roy F. Nichols, *Franklin Pierce: Young Hickory of the Granite State* (Philadelphia, 1958).

David M. Potter, *The Impending Crisis, 1848–1861*, completed and edited by Don E. Fehrenbacher (New York, 1976).

JAMES BUCHANAN

Philip S. Klein, *President James Buchanan* (University Park, Pa., 1962).

Elbert B. Smith, *The Presidency of James Buchanan* (Lawrence, Kan., 1975).

ABRAHAM LINCOLN

David Donald, *Lincoln* (New York, 1995).

Philip Shaw Paludan, *The Presidency of Abraham Lincoln* (Lawrence, Kan., 1994).

James Randall, *Lincoln the President*, 4 vols. (New York, 1945–1955). The last volume was completed by Richard Current.

ANDREW JOHNSON

Eric Foner, *Reconstruction: America's Unfinished Revolution, 1863–1877* (New York, 1988).

Hans L. Trefousse, *Andrew Johnson: A Biography* (New York, 1989).

ULYSSES S. GRANT

William S. McFeely, *Grant: A Biography* (New York, 1982).
Geoffrey Perret, *Ulysses S. Grant: Soldier and President* (New York, 1997).

RUTHERFORD B. HAYES

Ari Hoogenboom, *Rutherford B. Hayes: Warrior and President* (Lawrence, Kan., 1995).
T. Harry Williams, *Hayes: The Diary of a President, 1875–1881* (New York, 1964).

JAMES A. GARFIELD

Harry James Brown and Frederick D. Williams, eds., *The Diary of James A. Garfield*, 4 vols. (East Lansing, Mich., 1967–1981).
Allan Peskin, *Garfield: A Biography* (1978; reissue, Kent, Ohio, 1999).

CHESTER ARTHUR

Justus D. Doenecke, *The Presidencies of James A. Garfield and Chester A. Arthur* (Lawrence, Kan., 1981).
Thomas C. Reeves, *Gentleman Boss: The Life of Chester Alan Arthur* (New York, 1975).

GROVER CLEVELAND

H. Wayne Morgan, *From Hayes to McKinley: National Party Politics* (Syracuse, N.Y., 1969).
Allan Nevins, *Grover Cleveland: A Study in Courage* (New York, 1932).
Richard Welch, *The Presidencies of Grover Cleveland* (Lawrence, Kan., 1988).

BENJAMIN HARRISON

Harry J. Sievers, *Benjamin Harrison*, 3 vols. (Indianapolis, 1952–1968).
Homer E. Socolofsky and Allen B. Spetter, *The Presidency of Benjamin Harrison* (Lawrence, Kan., 1987).

WILLIAM McKINLEY

Lewis L. Gould, *The Presidency of William McKinley* (Lawrence, Kan., 1980).
Margaret Leech, *In the Days of McKinley* (New York, 1959).

THEODORE ROOSEVELT

John Morton Blum, *The Republican Roosevelt* (Cambridge, Mass., 1977).
H. W. Brands, *T.R.: The Last Romantic* (New York, 1997).
H. W. Brands, ed., *The Selected Letters of Theodore Roosevelt* (New York, 2001).
Edmund Morris, *Theodore Rex* (New York, 2001).

WILLIAM HOWARD TAFT

Donald F. Anderson, *William Howard Taft: A Conservative's Conception of the Presidency* (Ithaca, N.Y., 1973).

Paolo E. Coletta, *The Presidency of William Howard Taft* (Lawrence, Kan., 1973).

WOODROW WILSON

John Milton Cooper, Jr., *The Warrior and the Priest: Woodrow Wilson and Theodore Roosevelt* (Cambridge, Mass., 1983).

Arthur S. Link, *Woodrow Wilson*, 5 vols. (Princeton, 1947–1966).

John A. Thompson, *Woodrow Wilson* (London and New York, 2002).

WARREN G. HARDING

William E. Leuchtenberg, *The Perils of Prosperity* (Chicago, 1958).

Robert K. Murray, *The Harding Era: Warren G. Harding and His Administration* (Minneapolis, 1969).

CALVIN COOLIDGE

Calvin Coolidge, *Autobiography of Calvin Coolidge* (New York, 1933).

Donald R. McCoy, *Calvin Coolidge: The Quiet President* (New York, 1967).

HERBERT HOOVER

David Burner, *Herbert Hoover: A Public Life* (New York, 1978).

Herbert Hoover, *The Memoirs of Herbert Hoover*, 3 vols. (New York, 1952).

George H. Nash, *The Life of Herbert Hoover*, 3 vols. to date (New York, 1983–).

FRANKLIN D. ROOSEVELT

Kenneth Sydney Davis, *FDR*, 5 vols. (New York, 1971–2000).

Frank Freidel, *Franklin D. Roosevelt: A Rendezvous with Destiny* (Boston, 1990).

David M. Kennedy, *Freedom from Fear* (New York, 1999).

William E. Leuchtenberg, *Franklin D. Roosevelt and the New Deal, 1932–1940* (New York, 1963).

George McJimsey, *The Presidency of Franklin Delano Roosevelt* (Lawrence, Kan., 2000).

HARRY S. TRUMAN

Alonzo Hamby, *Man of the People: A Life of Harry S. Truman* (New York, 1995).

David McCullough, *Truman* (New York, 1992).

Harry S. Truman, *Memoirs*, 2 vols. (Garden City, N.Y., 1955–1956).

DWIGHT D. EISENHOWER

Stephen E. Ambrose, *Eisenhower*, 2 vols. (New York, 1983–1984).

Dwight D. Eisenhower, *Mandate for Change, 1953–1956* (New York, 1963), and
 Waging Peace, 1956–1961 (New York, 1965).
Chester J. Pach, Jr., and Elmo Richardson, *The Presidency of Dwight D. Eisen-
 hower* (Lawrence, Kan., 1991).

JOHN F. KENNEDY

Hugh Brogan, *Kennedy* (London, 1996).
Robert Dallek, *An Unfinished Life: John F. Kennedy, 1917–1963* (New York,
 2003).
Theodore Sorenson, *Kennedy* (New York, 1965).

LYNDON B. JOHNSON

Vaughn Bornet, *The Presidency of Lyndon Johnson* (Lawrence, Kan., 1983).
Robert A. Caro, *The Years of Lyndon Johnson,* 3 vols. to date (New York,
 1982–).
Robert Dallek, *Flawed Giant: Lyndon Johnson and His Times, 1961–1973* (New
 York, 1998).

RICHARD NIXON

Stephen E. Ambrose, *Nixon,* 3 vols. (New York, 1988–1991).
Roger Morris, *Richard Milhous Nixon: The Rise of an American Politician*
 (New York, 1990), the first volume in a projected three-volume biography.
Richard Nixon, *RN: The Memoirs of Richard Nixon* (New York, 1978), and *In
 the Arena: A Memoir of Victory, Defeat, and Renewal* (New York, 1990).
Melvin Small, *The Presidency of Richard Nixon* (Lawrence, Kan., 1999).

GERALD FORD

Gerald R. Ford, *A Time to Heal: The Autobiography of Gerald R. Ford* (New
 York, 1979).
John Robert Greene, *The Presidency of Gerald R. Ford* (Lawrence, Kan., 1995).

JIMMY CARTER

Jimmy Carter, *Keeping Faith: Memoirs of a President* (New York, 1982).
Burton Kaufman, *The Presidency of James Earl Carter* (Lawrence, Kan., 1993).

RONALD REAGAN

Lou Cannon, *President Reagan: The Role of a Lifetime* (New York, 1991).
Edmund Morris, *Dutch: A Memoir of Ronald Reagan* (New York, 1999).
Ronald Reagan, *An American Life* (New York, 1990).
Ronald Reagan et al., *In His Own Hand: The Writings of Ronald Reagan That
 Reveal His Revolutionary Vision for America* (New York, 2001).

GEORGE H. W. BUSH

John Robert Greene, *The Presidency of George Bush* (Lawrence, Kan., 2000).

David Mervin, *George Bush and the Guardianship Presidency* (New York, 1996).

BILL CLINTON

Elizabeth Drew, *On the Edge: The Clinton Presidency* (New York, 1994), and *Showdown: The Struggle Between the Gingrich Congress and the Clinton White House* (New York, 1996).

Joe Klein, *The Natural: The Misunderstood Presidency of Bill Clinton* (New York, 2002).

Stanley Renshon, *High Hopes: The Clinton Presidency and the Politics of Ambition* (New York, 1996).

GEORGE W. BUSH

David Frum, *The Right Man: The Surprise Presidency of George W. Bush* (New York, 2003).

Molly Ivins and Lou Dubose, *Shrub: The Short but Happy Political Life of George W. Bush* (New York, 2003).

Illustration Credits

Washington: Pennington portrait by Gilbert Stuart, Library of Congress. John Adams: portrait by Mather Brown, 1788, collection of the Boston Athenaeum. Jefferson: *Thomas Jefferson* by Rembrandt Peale, 1805, collection of the New-York Historical Society, accession number 1887.306. Madison: *Portrait of James Madison* by Gilbert Stuart, 1805–1807, Bowdoin College Museum of Art, Brunswick, Maine, bequest of the Honorable James Bowdoin III. Monroe: *Portrait of James Monroe* (detail), attributed to Rembrandt Peale, c. 1824–25, courtesy James Monroe Museum and Memorial Library, Fredericksburg, Virginia. John Quincy Adams: daguerreotype by Philip Haas, 1843, Metropolitan Museum of Art, gift of I. N. Phelps Stokes, Edward S. Hawes, Alice Mary Hawes, Marion Augusta Hawes, 1937 (37.14.34). Jackson: *General Andrew Jackson, President of the United States* by Ralph E. W. Earl, 1833, Memphis Brooks Museum of Art, Memphis Park Commission Purchase 46.2. Van Buren: portrait by G. P. A. Healy, National Archives. W. H. Harrison: portrait, Library of Congress. Tyler: portrait, Library of Congress. Polk: daguerreotype by Mathew Brady, Library of Congress. Taylor: portrait by Southworth and Hawes, c. 1850, Metropolitan Museum of Art, gift of I. N. Phelps Stokes, Edward S. Hawes, Alice Mary Hawes, and Marion Augusta Hawes, 1937 (37.14.32). Fillmore: portrait, Library of Congress. Pierce: portrait, Library of Congress. Buchanan: portrait, Library of Congress. Lincoln: portrait, Library of Congress. Andrew Johnson: portrait, Library of Congress. Grant: portrait, 1876, Library of Congress. Hayes: portrait, Rutherford B. Hayes Presidential Center, Fremont, Ohio. Garfield: portrait, Library of Congress. Arthur: Library of Congress. Cleveland: portrait, 1892, Library of Congress. Benjamin Harrison: portrait, c. 1889–93, Library of Congress. McKinley: Library of Congress. Theodore Roosevelt: portrait, Brown Brothers. Taft: portrait, Library of Congress. Wilson: portrait, Library of Congress. Harding: Library of Congress. Coolidge: Library of Congress. Hoover: Herbert Hoover Presidential Library-Museum, West Branch, Iowa. Franklin Roosevelt: portrait courtesy Franklin D. Roosevelt Library, Hyde Park, New York. Truman: courtesy Harry S. Truman Library, Independence, Missouri. Eisenhower: courtesy Dwight D. Eisenhower Presidential Library, Abilene, Kansas. Kennedy: John F. Kennedy Library collection, photo by Ted Spiegil, 1960 (detail). Lyndon Johnson: portrait in the Lyndon Baines Johnson Library collection, photo by Arnold Newman. Nixon: courtesy Richard Nixon Library and Birthplace Foundation. Ford: courtesy Gerald R. Ford Library. Carter: courtesy Jimmy Carter Library. Reagan: National Archives, ACR: 558523. George H. W. Bush: courtesy George Bush Presidential Library & Museum. Clinton: portrait, the White House. George W. Bush: the White House, photo by Eric Draper.

Index

Realists, 313
Reconstruction, era of presidential,
192–93, 195
Reconstruction Act, 196
Reconstruction Finance Corporation,
339
"Red Fox of Kinderhook." *See* Van
Buren, Martin
Red Scare, 37
Reed, Scott, 538
"Refrigerator." *See* Harrison,
Benjamin
Regan, Donald, 481
Reminiscences (Carlyle), 231
Reno, Janet, 510
Republican landslide of 1946, 374
Republicans: as opponents of Feder-
alists, 13
Resumption Act of 1875, 222
Revenue Act of 1926, 329
Revenue Act of 1932, 339
Revenue Act of 1935, 355
Rhodes, James Ford, 205
Richards, Anne, 535
Richmond Recorder (newspaper), 44
Ridge, Tom, 539
Robards, Lewis, 86
Robards, Rachel, 86
Roberts, Kenneth, 319
Robertson, William H., 237
Rockefeller, Nelson
as vice-presidential nominee, 446
Rockefeller, John D., 287
Rolling Thunder bombing campaign,
420
Roosevelt, Anna Eleanor, 345, 365
Roosevelt, Franklin Delano
childhood, 344–45
death of, 363
as Democratic party candidate for
vice presidency, 347
and fireside chat, 351
and fourth term, 368
greatest disappointment, 345
inauguration, 350
as James M. Cox's running mate,
315
and Lucy Mercer, 347
marriage, 345

and New Deal, 351–63
and polio epidemic, 347
political life, 346
quoted, 335, 341, 351
restoring Jefferson's image, 36
World War II, 359–63
Roosevelt, James, 344
Roosevelt, Sara Delano, 344
Roosevelt, Theodore
allegations made by, 253
campaign to reform Army, 282
childhood, 269
as contributing editor of *Outlook*
(magazine), 283
death of first wife, 270
as Depression president, 338
inauguration, 269
and Nobel Peace Prize, 282
quoted, 266, 277, 278, 281, 283
race relations, 279
response to industrialization, 277
and stuffed toy teddy bear, 273
Root, Elihu, 260
Ross, John, 93
Rostenkowski, Dan, 459
Rouge Assembly Plant, 340
Rough Riders, 284
*Rules of Civility & Decent Behaviour
in Company and Conversation*
(George Washington), 2
Rumsfeld, Donald, 537
Rural Electrification Administration,
353
Rush, Benjamin, 17, 115
Russo-Japanese War, 282
Rutherfraud. *See* Hayes, Rutherford
B.

SALT I, 436
SALT II, 451, 464
Saturday Evening Post, 319
"Savior of capitalism." *See* Roosevelt,
Franklin Delano
Schlesinger, Arthur, Jr., 317, 323
Schurz, Carl, 219
Schwarzkopf, Norman, 495
Scott, Winfield, 111, 136, 140, 141, 150,
182
Scowcroft, Brent, 491